AF396477

STATISTICAL MECHANICS AND STOCHASTIC THERMODYNAMICS

OXFORD GRADUATE TEXTS

Statistical Mechanics and Stochastic Thermodynamics

A textbook on modern approaches in and out of equilibrium

David T. Limmer

University of California, Berkeley

OXFORD
UNIVERSITY PRESS

Great Clarendon Street, Oxford, OX2 6DP,
United Kingdom

Oxford University Press is a department of the University of Oxford.
It furthers the University's objective of excellence in research, scholarship,
and education by publishing worldwide. Oxford is a registered trade mark of
Oxford University Press in the UK and in certain other countries

Published in the United States of America by Oxford University Press
198 Madison Avenue, New York, NY 10016, United States of America

British Library Cataloguing in Publication Data
Data available

Library of Congress Control Number: 2024936393

ISBN 9780198919858

DOI: 10.1093/oso/9780198919858.001.0001

Printed and bound by
CPI Group (UK) Ltd, Croydon, CR0 4YY

Links to third party websites are provided by Oxford in good faith and
for information only. Oxford disclaims any responsibility for the materials
contained in any third party website referenced in this work.

The manufacturer's authorised representative in the EU for
product safety is Oxford University Press España S.A. of el Parque
Empresarial San Fernando de Henares, Avenida de Castilla, 2 –
28830 Madrid (www.oup.es/en).

In memory of Phillip Geissler.

Preface

This book presents the material I have used in my year-long graduate course in statistical mechanics, taught in the Department of Chemistry at the University of California, Berkeley. Students entering this course come to it with a wide array of backgrounds and preparations. Most have seen classical thermodynamics, and I assume have a healthy understanding of multivariable calculus. I expect that everyone starting this course has had some exposure to classical and quantum mechanics at an undergraduate level, and have a reasonable understanding of probability and statistics. However, many have not been exposed to a Boltzmann distribution and many fewer still know the relationship between microscopic fluctuations and macroscopic quantities. For this reason, in this book I try to build basic principles of statistical mechanics up from scratch, and through postulates and definitions make connections to macroscopic theories. Further, as most students taking this course come from a chemistry background, many have not had exposure to the more advanced mathematics used in physics such as Fourier analysis, variational calculus, and complex analysis. To ameliorate this to some extent, I have included judicious asides on these topics where they are first needed. These asides are enough to guide students through the calculations presented, but undoubtedly more study is needed in these topics than presented here for students to master the techniques.

As for the content, the text is loosely broken up into two parts. Chapters 1–5 focus on equilibrium statistical mechanics, while chapters 6–10 focus rather on nonequilibrium statistical mechanics. The progression for both sections is the same. First some basic postulates, definitions, and theorems are presented. In the case of equilibrium statistical mechanics, these are the ergodic hypothesis and the subsequent thermodynamic definitions. For nonequilibrium statistical mechanics, these are Liouville's theorem and Jarzynski's equality. The introductory chapters are then followed by examples of linear phenomena where exact results are possible to obtain. In equilibrium these are the canonical noninteracting systems of traditional presentations in Chapter 2, and then Gaussian theories largely brought into chemistry by David Chandler in Chapter 3. Out of equilibrium, these are the linear response theories and ideas of fluctuation-dissipation from Onsager and Kubo, followed by the theories of irreversibility from Zwanzig in Chapter 7. As so many experimental techniques employ linear response relations to learn about molecular behavior, some emphasis is given to scattering and spectroscopy. The latter chapters in each section are devoted to nonlinear phenomena and the numerical techniques that can be used to study them. In equilibrium, strong nonlinearities owe their origin to phase transitions, and much of Chapter 4 is devoted to Landau's theory, while Chapter 5 discusses Monte Carlo techniques that have been important in the study of phase behavior in molecularly detailed models. For dynamical systems, nonlinear phenomena come in the form of reactive dynamics discussed in Chapter 8 and systems driven far from equilibrium as discussed in Chapter 9. Molecular dynamics simulations are the primary contemporary means of studying such systems and are discussed in Chapter 10. Throughout, examples from both quantum and classical mechanics are discussed where their developments are sufficient. Exercises within the chapters range from simple generalizations of presented results to more complicated proofs of relationships employed. Additional exercises presented at the end of the chapters serve mostly to prove the generality of the arguments by developing additional concrete examples.

Intellectually, this book owes a great deal to David Chandler. His *Introduction to modern statistical mechanics* is the origin of many of the early arguments presented and the philosophy developed. His *little green book* and the never-published *white notes* serve as the template for the equilibrium material presented. Indeed, I hope that this book serves to faithfully distill the perspective of the Berkeley school of statistical mechanics he built. Pedagogically, this book is indebted to Phillip Geissler. I hope that I have successfully approximated his careful, insightful approach to the subject. Much of the presentation throughout the book has been adapted from Phill's own material that he taught in the same sequence of courses at Berkeley. Indeed, his courses are where I learned much of the subject, and his lectures are where I saw firsthand how to guide students through complex arguments. The origin of this book stems from conversations with Phill on the need for an update to the little green book, and in particular to write a text that includes the recent development of stochastic thermodynamics that has taken place in the last 20 years and the incorporation of large deviation theory into the subject. Throughout the text, distinctions between heat and work, as well as notions of trajectory ensembles reflect the incorporation of stochastic thermodynamics into the modern language of statistical mechanics. Similarly, ideas of scaling, the concentration of measures, and generalized theories of ensemble equivalence represent the important contribution of the mathematics of large deviations.

Acknowledgements

A number of people were instrumental in seeing this book come to fruition. First, a set of graduate students from my research group initially helped to check and edit the class notes that served as the basis for this book. In particular, Chloe Gao and Yoonjae Park were responsible for establishing the accuracy and precision of the more advanced material as graduate student instructors at UC Berkeley. Layne Frechette, Julia Rogers, and Avishek Das provided useful feedback and refinement of some of the introductory material also as graduate student instructors. Anthony Poggioli and Seokjin Moon gave extensive technical feedback on the material in the last few chapters. Additionally, Sam Oaks-Leaf, Leonardo Coello, Melanie Huynh, Songela Chen, Rohit Rana, Jorge Rosa, Aditya Singh, and Michelle Anderson all provided useful feedback at various stages of the preparation of the text, as did many students in my Chemistry 220A course in the Fall of 2023. More senior colleagues including Gavin Crooks, Kranthi Mandadapu, Eran Rabani, Tim Berkelbach, and Benjamin Rotenberg provided helpful suggestions concerning the content of book and terminology. Most importantly, Jenna Jeffrey acted as a sounding board, copy editor, muse, and art critic throughout the writing of this book. It would certainly not exist without her.

Contents

1

Fundamental postulates and definitions

The world around us is not infinitely divisible, but rather is made up of discrete units we know as atoms. This fact was intuited by ancient Greek philosophers, hypothesized by early scientists, and in the twentieth century, confirmed through direct observation. More than any other fact—that matter is made of atoms in constant motion, attracting each other when they are apart, but repelling upon being squeezed into one another— informs our understanding of the universe. It was Boltzmann and later Einstein that used the atomic hypothesis to develop a theoretical framework to explain the origins of phenomenological theories like thermodynamics and hydrodynamics that are widely applicable and universally successful in describing the macroscopic observable world. Together with other more abstract work by Maxwell and Gibbs, these scientists laid the foundation for the theory of statistical mechanics. Throughout the remainder of the twentieth century, this theory was refined and extended to include a description of the passive dynamics of atoms and molecules, and in the early part of the twenty-first century the theory captured the consequences of our ability to directly manipulate matter on its smallest scales.

As a theory, statistical mechanics serves as a bridge between an underlying acknowledgement of the atomic composition of matter, and its macroscopic consequences. In this text, we will first build the theory formally, through postulates, definitions and mathematical inference, in an effort to understand properties of systems at rest, and their emergent thermodynamics. Later, we will generalize the theory to confront systems' motion, employing principles of stochastic thermodynamics, and making connections to hydrodynamics. Of primary concern in this theory is the use of the mathematics of probability, as the tremendous numbers of atoms in any macroscopic object lend themselves naturally to a statistical description. As a consequence, the primary objective is to first deduce the likelihood of a specific arrangement of atoms. The organization of likelihoods of molecular arrangements is encoded in a probability distribution. The establishment of this probability distribution is our first goal.

1.1   The molecular hypothesis

Our endeavor to understand macroscopic phenomena from the molecular hypothesis begins in ignorance. Specifically, we will assume that *in an isolated system, each microstate is equally probable.* By an isolated system, we mean one in which the energy, E, and global mechanical quantities, $\mathbf{X}$, are conserved. The conservation of E and

$\mathbf{X}$ can be considered a consequence of Hamiltonian mechanics. By a microstate, denoted ν, we mean a complete description of the pertinent dynamical variables. For a classical, non-relativistic system the microstate would be determined by the full set of positions and momentum of all particles. For a quantum mechanical system, the microstate would be the wavefunction. Finally, by *equally probable*, we mean that the natural dynamics of the system generate a statistical distribution that is independent of the precise initial condition, but dependent on the globally conserved quantities.

We will take this statement, which is not possible to prove generally, as a fundamental postulate of statistical mechanics. It can be mathematically considered as a principle of indifference, that absent any information, we can only assign equal probabilities to each microstate that is physically allowable. This statement of equal or uniform probability is often referred to as the *ergodic hypothesis*, as under assumptions concerning the chaotic nature of systems of many interacting degrees of freedom, a dynamical system will be ergodic in that its evolution over time will explore all accessible microstates uniformly.

> **Aside: Uniform probabilities.** Consider flipping a fair coin six times. Two possible sequences of heads (H) and tails (T) that could occur are
>
> $$HTTHHT \qquad \text{and} \qquad HHHHHH$$
>
> each of which is equally probable, with probability equal to $1/2^6$. This result runs counter to our intuition. This is because there is only one way to create a sequence with all heads, while there are many ways to create a sequence with equal heads and tails. While each specific sequence is equally probable, the aggregate probability of any sequence with equal heads and tails is much greater than the probability of any sequence with all heads.

Mathematically, we can denote our postulate for the probability of a microstate, $P(\nu)$, with associated energy $E(\nu)$ and mechanical variables $\mathbf{X}(\nu)$,

$$P(\nu) = \begin{cases} 1/\Omega(E, \mathbf{X}) & \text{if} \quad E(\nu) = E, \ \mathbf{X}(\nu) = \mathbf{X} \\ 0 & \text{else} \end{cases}$$

where $\Omega(E, \mathbf{X})$ is the total number of microstates satisfying the global constraints of fixed energy and mechanical variables. The form of this expression hides potential complications associated with the countability of microstates. For systems that evolve within a discrete space, this form is strictly interpretable as a probability. For systems that evolve within a continuum, this definition should be considered a statement of the probability density, with some measure that will be made explicit as needed.

An immediate consequence of equal likelihoods of microstates is that the most probable collections of states will be those that have the largest number of ways of occurring. Take, for instance, grouping states with some shared global property we will denote as B. The aggregate probability of observing a state in that group, denoted $P(B)$, is computable by summing the probability of each microstate in that group, or

$$P(B) = \sum_{\nu \in B} P(\nu)$$

$$= \frac{1}{\Omega} \sum_{\nu \in B} 1 = \frac{\Omega(B)}{\Omega}$$

where we have introduced $\Omega(B)$ as the number of microstates consistent with B. If we were to allow B to vary its value, the most probable value would be that which maximizes $P(B)$ or alternatively $\Omega(B)$. Here, we learn a general lesson of probability— the more ways something can happen the more likely it will be. This basic inference lies at the heart of macroscopic thermodynamics as we will soon see.

Exercise 1.1: Consider a volume V that is partitioned into cells of volume ℓ^3, such that there are $M = V/\ell^3$ cells. Count the number of ways, $\Omega(N, V)$, of arranging N particles into each of the M cells acknowledging that particles are not labeled and are thus indistinguishable, and assuming that only one particle can fit in each cell. If the volume is able to freely change at fixed N between $V_{\min} \leq V \leq V_{\max}$, what is the most likely volume of the system provided our fundamental postulate?

Exercise 1.2: Consider a system for which the number of microstates factorizes, $\Omega(E, V) = \Omega_{\mathrm{s}}(V)\Omega_{\mathrm{th}}(E)$, where $\Omega_{\mathrm{s}}(V)$ can be taken as your result in Exercise 1.1. If the energy E depends on V as $E = -pV$ where p is a constant, determine a relationship for the most likely value of V. Hint: the maximum of $\Omega(E, V)$ is also the maximum of $\ln \Omega(E, V)$, and you might use an approximation due to Stirling, $\ln N! \approx N \ln N - N$, for $N \gg 1$ and define $\beta = \partial \ln \Omega_{\mathrm{th}}(E)/\partial E$.

Before we go there, let us reconcile just how big $\Omega(E, \mathbf{X})$ is. Let us imagine that our system is large, so large that we can block it up into M subregions of volume ℓ^3 that are statistically independent. For concreteness, imagine that there are N particles distributed throughout a volume V such that $M = V/\ell^3 = N/\rho\ell^3$ where $\rho = N/V$ is the density particles in each subregion. If there are $\tilde{\omega}$ states in each of the M subregions, then we would expect that the total number of microstates is

$$\Omega = \tilde{\omega}^M$$

$$= \left[\tilde{\omega}^{1/\rho\ell^3}\right]^N = \omega^N$$

where we have introduced $\omega = \tilde{\omega}^{1/\rho\ell^3}$. The product structure results from the independence of each region. The number of microstates, ω, depends on the density, the size of the region ℓ^3, and other quantities like the energy density $\epsilon = E/N$. In other words, all quantities that do not depend on the global scale of the system, or are *intensive*. The total number of microstates, $\Omega = \omega^N(\epsilon)$, depends exponentially on the size or extent of the system. This specific dependence is known as a large deviation form, and implies that probability rapidly concentrates in the vicinity of the typical behavior as the scale of the system is increased. When N is the size of Avogadro's number,

$N_A \approx 10^{24}$, only the mean is observable. This is why that while atoms and molecules are in constant motion around us, the world looks relatively boring, as deviations from the average are overwhelmingly improbable.

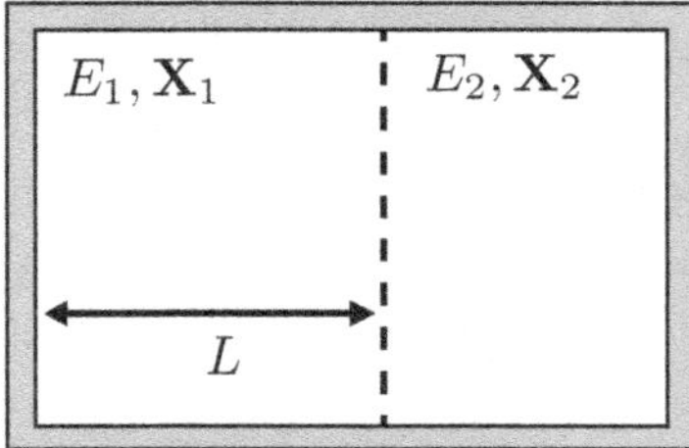

Fig. 1.1 Illustration of an isolated system partitioned into two regions, 1 and 2.

1.2 Partitioning, probability, and spontaneity

In order to begin making inferences about collections of states, we will need to motivate a few definitions. Let us consider for example, an isolated system, closed off from its environment such that its energy and mechanical variables, things like the volume and number of particles, are fixed. Now introduce a partition into this system, dividing it unequally into two regions like that pictured in Figure 1.1. Because the total E and $\mathbf{X}$ are fixed, the amounts in each region obey a sum rule,

$$E = E_1 + E_2 \qquad \mathbf{X} = \mathbf{X}_1 + \mathbf{X}_2$$

where E_i and $\mathbf{X}_i$ are the energy and mechanical variables partitioned into region $i = 1, 2$. Let us label the location of the partition by L.

If the partition were able to move freely, the likelihood of finding it at any particular location could be evaluated by

$$P(L) = \sum_{\nu \in L} P(\nu) \propto \Omega(L)$$

which as before is proportional to the number of microstates consistent with the partition at L, or $\Omega(L)$. The most likely value of L can be denoted L^* and determined by maximizing $\Omega(L)$, or alternatively $\ln \Omega(L)$. For a large system, we can assume that the fluctuations about the average are small, so that we are overwhelmingly likely to find the system with $L = L^*$. We can refer to this value, the one which maximizes $\ln \Omega(L)$, as its equilibrium position. An example of $\ln \Omega(L)$ is shown in Figure 1.2.

Now imagine making a virtual displacement, $L^* \rightarrow L^* + \Delta L$ keeping the total energy and mechanical variables fixed. It must be that

$$\ln \Omega(E, \mathbf{X}|L^* + \Delta L) \leq \ln \Omega(E, \mathbf{X}|L^*)$$

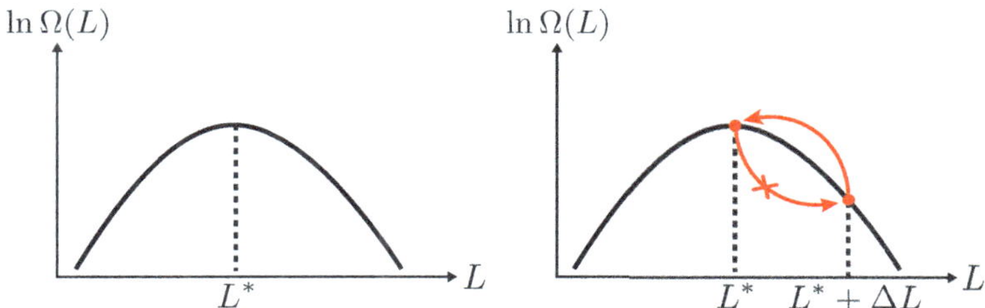

Fig. 1.2 Expectation for how the number of microstates changes with the length of two otherwise equivalent regions with L^* the most likely state.

or in other words, there are fewer states away from the most probable partitioning of the system. In fact more generally, if we were to introduce a constraint that repartitions the system in any way

$$\ln \Omega(E, \mathbf{X}|\text{constraint}) \leq \ln \Omega(E, \mathbf{X}|\text{no constraint})$$

there would be a non-positive change in the number of consistent microstates of the system. That is, $\ln \Omega$ must decrease with a displacement away from equilibrium. Equivalently, imagine beginning with the system at $L^* + \Delta L$, again with the same E and $\mathbf{X}$. Such a situation is unlikely and subsequently if the partition is able to move, it would with high probability go to L^*. One would say that such a change is *spontaneous*. If the system is large such that $\ln \Omega$ is sharply peaked, the reverse does not happen, or is very improbable. Mathematically, this is a statement of the regression toward the mean. Said in another way, for a spontaneous process near equilibrium with fixed E and $\mathbf{X}$, $\ln \Omega$ must increase.

It was Boltzmann who took the bold step to define the non-negative quantity that strictly increases in a spontaneous process as S, for entropy,

$$S = k_{\mathrm{B}} \ln \Omega$$

the natural logarithm of the number of accessible microstates. The constant k_{B} that gives entropy units now bares his name and is equal to R/N_{A} the ideal gas constant divided by Avogadro's number, or $k_{\mathrm{B}} = 1.38 \times 10^{-23}$ J/K. From our large deviation scaling demonstration earlier, $\Omega = \omega^N$, the entropy is easily seen as an *extensive* quantity, $S = k_{\mathrm{B}} N \ln \omega$, one that increases in proportion to the size of the system, a unique property of its definition as a logarithm. With this definition we can recognize that the probabilistic statement of regression toward the mean is identical to the *second law of thermodynamics*, that entropy increases in a spontaneous process.

1.3 Exchanging energy with a bath

Most systems are not isolated, but rather can exchange energy with its surroundings. In order to understand what our fundamental postulate has to say about such cases, we will consider a system in contact with a surrounding environment we will refer to

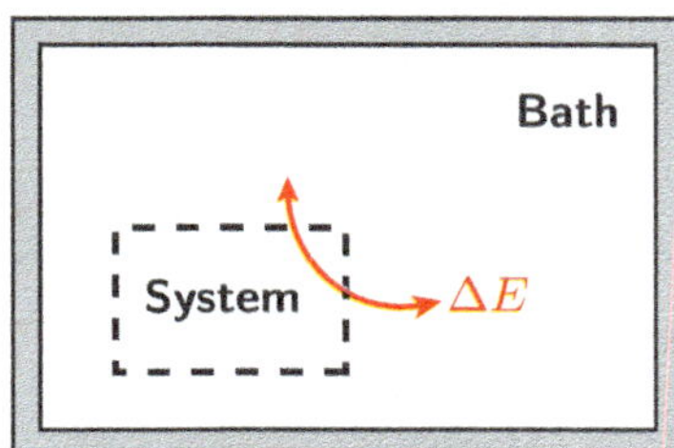

Fig. 1.3 Isolated region partitioned into a system and a bath that can exchange energy, but not mechanical variables.

as a *bath*. The composite system and bath is isolated, such that the total energy, E_T, and mechanical variables are constant and shown schematically in Figure 1.3.

Imagine that the system and bath can only exchange energy, such that the total energy can be partitioned,

$$E_T = E(\nu) + E_B(\nu_B)$$

where $E(\nu)$ is the energy in the system in microstate ν, while $E_B(\nu_B)$ is the energy of the bath in microstate ν_B. Implicit in this description is an assumption of negligible interaction between the system and the bath and a bipartite structure to the combined microstate, ν_T, such that it decomposes into two composite parts, $\nu_T = \{\nu, \nu_B\}$. Because the composite system and bath are isolated, each ν_T is equally likely with $P(\nu_T) = 1/\Omega(E_T)$. The marginal probability associated with a system microstate can be computed by summing over all of the allowable bath states, those with the right amount of energy,

$$P(\nu) = \sum_{\nu_B} P(\nu, \nu_B) = \frac{\Omega_B[E_T - E(\nu)]}{\Omega(E_T)}$$

resulting in a likelihood that is proportional to the number of bath microstates with energy $E_B(\nu_B) = E_T - E(\nu)$, which we denote as Ω_B.

While this relation is correct for the composite system plus bath, it is not useful as it requires detailed knowledge of the number of bath microstates with $E_B = E_T - E$. Let's assume that the bath is large compared to the system, so that E/E_B is infinitesimally small, since typically energies scale with system size. In this limit, we need only understand the behavior of Ω_B near E_B. As the number of microstates varies exponentially with system size, it is natural to expand $\ln \Omega_B$,

$$\ln \Omega_B[E_T - E(\nu)] \approx \ln \Omega_B[E_T] - E(\nu) \left. \frac{\partial \ln \Omega_B}{\partial E_B} \right|_{E(\nu)=0} + \ldots$$

where we have truncated at first order in E/E_B. The zeroth-order term is a constant for all system microstates, while the first-order term depends only on the energy of the microstate times a number that we can define as

$$\beta_B = \frac{\partial \ln \Omega_B}{\partial E_B}$$

which encodes how the number of bath microstates varies with the energy in the bath, and is thus a property of the bath. Putting these together we find,

$$P(\nu) \propto e^{-\beta_\mathrm{B} E(\nu)}$$

the distribution of microstates in the system is no longer uniform, but is biased by the energy associated with each microstate. For $\beta_\mathrm{B} \geq 0$, higher energy states are exponentially less likely. This probability distribution for a system that can exchange energy with its surroundings is known as the *Boltzmann distribution*.

> **Exercise 1.3:** Assuming that the number of microstates obeys a large deviation form, $\Omega_\mathrm{B}(E_\mathrm{B}) = [\omega_\mathrm{B}(E_\mathrm{B}/N_\mathrm{B})]^{N_\mathrm{B}}$, show that directly expanding Ω_B for small changes to its energy yields the same Boltzmann distribution as we have arrived through expanding $\ln \Omega_\mathrm{B}$, in the limit that $N_\mathrm{B} \to \infty$.

We can gain some insight into the parameter β_B by asking for the most likely partitioning of energy between the system and the bath. The probability of any specific amount of energy in the system, $P(E)$, is

$$P(E) = \sum_{\nu \in E} P(\nu) \propto \Omega(E) e^{-\beta_\mathrm{B} E}$$

$$\equiv \exp[-N\,(\beta_\mathrm{B}\epsilon - \ln\omega)]$$

which is naturally a product of the degeneracy of microstates with energy E in the system and an exponential weight factor. Employing the large deviation scaling of $\Omega(E)$, we find that for an open system, the distribution of energies obeys a large deviation scaling form, meaning that as the number of particles N becomes large, it will become increasingly sharply peaked at its mean value. The rate at which the probability condenses around its mean is given by the rate function, $\phi(\epsilon) = \beta_\mathrm{B}\epsilon - \ln\omega(\epsilon)$, illustrated in Figure 1.4.

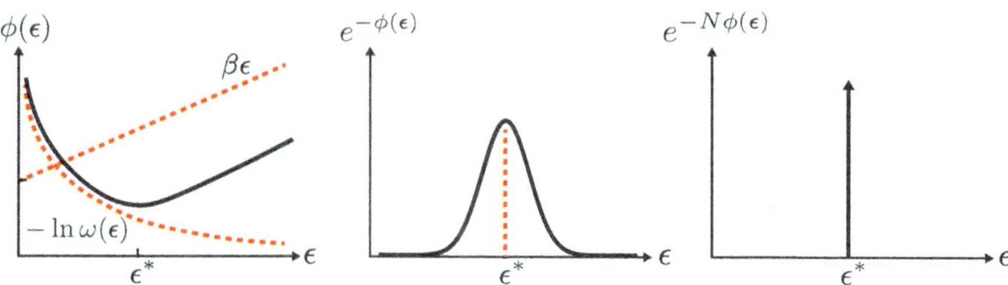

Fig. 1.4 Illustration of the large deviation scaling of the distribution of energy, $\epsilon = E/N$, in a system that can exchange heat with its surroundings.

The most likely value, or correspondingly the equilibrium energy, can be found by minimizing $\phi(\epsilon)$,

$$\frac{\partial}{\partial \epsilon}\left[\beta_{\mathrm{B}}\epsilon - \ln \omega(\epsilon)\right] = 0$$

leading to the relationship

$$\beta_{\mathrm{B}} = \frac{\partial \ln \omega(\epsilon)}{\partial \epsilon}$$

$$= \left(\frac{\partial \ln \Omega(E)}{\partial E}\right)_N = \beta$$

where in the second line we have multiplied by N/N and taken the derivative at fixed N, to observe that the equilibrium energy of the system is obtained when β corresponding to the system is equal to β_{B}, a property of the bath. In other words, the parameter β controls the partitioning of energy between the system and the bath, or the flow of *heat*. Employing Boltzmann's definition of the entropy, we find

$$\beta = \left(\frac{\partial S/k_{\mathrm{B}}}{\partial E}\right)_{\mathbf{X}} = 1/k_{\mathrm{B}}T$$

a natural, microscopic, definition of *temperature T*.

> **Exercise 1.4:** Consider a system that can exist in two states, a ground state "g" and an excited state "e", with energies $E(g) = 0$ and $E(e) = \varepsilon$. Using the Boltzmann distribution, $P(\nu) \propto \exp[-\beta E(\nu)]$, determine the probability of being in the excited state and plot its temperature dependence. What is the probability to be in the excited state in the limit that $T \to \infty$?

1.4   Potentials for spontaneous change

We have motivated a few definitions for well-known thermodynamic quantities. From Boltzmann, the entropy is given for a closed system by the number of accessible microstates times Boltzmann's constant, $S = k_{\mathrm{B}} \ln \Omega$, while the temperature is how the number of states change with energy, $1/k_{\mathrm{B}}T = \partial \ln \Omega/\partial E$. The definition of entropy was motivated by a probabilistic interpretation of a spontaneous process, namely that of the regression toward the mean— initially improbable states evolve to more probable states. Taken with our fundamental postulate that in an isolated system all microstates are equally likely, higher likelihood collections of microstates must result simply from having more ways of happening. From this we deduced what would happen to a system that could exchange energy with an infinite surroundings, resulting in a bias for observing particular microstates based on their energies, encoded in the Boltzmann distribution. The strength of this bias we identified as the temperature, and decreases the likelihood of fluctuations to microstates with energy larger than $k_{\mathrm{B}}T$. We will explore this definition of the temperature further, to make sure it makes sense.

Let's begin delving deeper into our definition of T by reconsidering an isolated system, in which the total energy E and mechanical variables $\mathbf{X}$ are fixed. In such a system, both E and $\mathbf{X}$ can be partitioned in many ways. Consider a partitioning in which $E = E_1 + E_2$ and $\mathbf{X} = \mathbf{X}_1 + \mathbf{X}_2$, as illustrated in Figure 1.5. If such a partitioning is specified, the number of microstates available to each partition can be counted, and as a consequence the entropy $S = S_1 + S_2$ determined. Initially, imagine the partition is such that no E or $\mathbf{X}$ can exchange. If after preparing the system in such a state, we allow heat to flow, in the form of energy exchanging between the two partitions with fixed $\mathbf{X}_1$ and $\mathbf{X}_2$, such a spontaneous process must reflect the system evolving to a more probable state. Consequently, it must be that

$$(\Delta S)_{E,\mathbf{X}} \geq 0$$

the entropy change is non-negative. This again is a statement of the second law of thermodynamics. If no spontaneous process occurs, the system must have already been at its most likely or equilibrium position, and $\Delta S = 0$. Otherwise, the entropy must increase. Let's work out what this implies about our definition of the temperature.

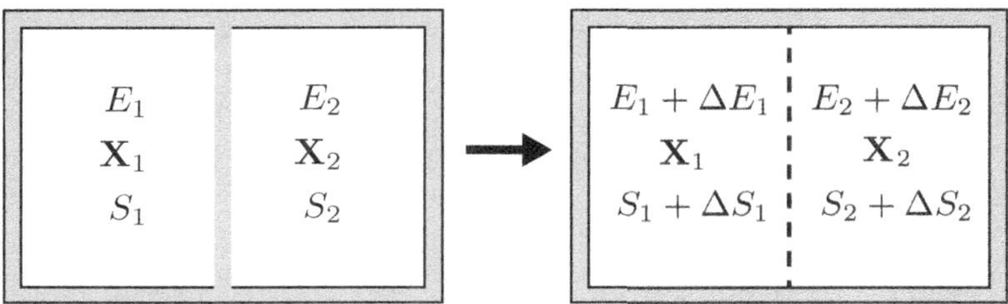

Fig. 1.5 Illustration of an isolated system in which energy is spontaneously exchanged upon the removal of a partition.

Since the whole system is isolated, it must be that there is no net change in energy, and thus $\Delta E_1 = -\Delta E_2 = \Delta E$. Entropy is not conserved, but it can be decomposed into contributions from changes to both partitions,

$$\Delta S = \Delta S_1 + \Delta S_2 \geq 0$$

which if the change in energy is small can be approximated for each partition as

$$\Delta S_1 = \Delta E_1 \frac{\partial S_1}{\partial E_1} = \frac{\Delta E}{T_1}$$

where we have employed our definition of the temperature. Analogously, $\Delta S_2 = -\Delta E/T_2$. Putting these two together, we find a rephrasing of the second law,

$$\Delta S = \Delta E \left(1/T_1 - 1/T_2\right) \geq 0$$

which implies that for $\Delta E > 0$, then $T_2 > T_1$ or alternatively if $\Delta E < 0$, then $T_2 < T_1$. In other words, heat (energy) flows spontaneously from high temperature to low temperature. The temperature, as we have defined it, is a measure therefore of the potential for heat to flow. Heat flow is said to be *irreversible* in the sense that the reverse does not happen. A reversible process therefore has $\Delta S = 0$. These are reasonable statements to make of macroscopic systems where the typical behavior is dominant, but more nuance is required if we are to apply such reasoning microscopically where deviations from the mean behavior can be significant.

Exercise 1.5: Consider a collection of N noninteracting particles that can each exist in two states, where the energy for the i'th particle is denoted $\epsilon_i = \{0, \varepsilon\}$. If the total energy of the system is given by $E = \sum_{i=1}^{N} \epsilon_i$, compute the number of microstates, $\Omega(E, N)$, consistent with a fixed energy as a function of the number in the excited states N_e. Is $\Omega(E, N)$ a monotonic function of E? What does this imply about the sign of $1/T = \partial S/\partial E$?

Let us now turn to the question of spontaneous changes of mechanical variables. We will proceed much like we did with our previous heat flow calculation by first imagining we can prepare a system with E and $\mathbf{X}$ fixed, but partitioned in a particular way. As before, imagine that $E = E_1 + E_2$ and $\mathbf{X} = \mathbf{X}_1 + \mathbf{X}_2$. However, let's pick out a special mechanical variable, X, so that $\mathbf{X}$ decomposes as $\mathbf{X} = \{X, \mathbf{X}'\}$ where $\mathbf{X}'$ denotes the remaining mechanical variables. Taking our initially partitioned system as shown in Figure 1.6, what happens if the partition is released in such a way that both energy and X are allowed to exchange between the two regions?

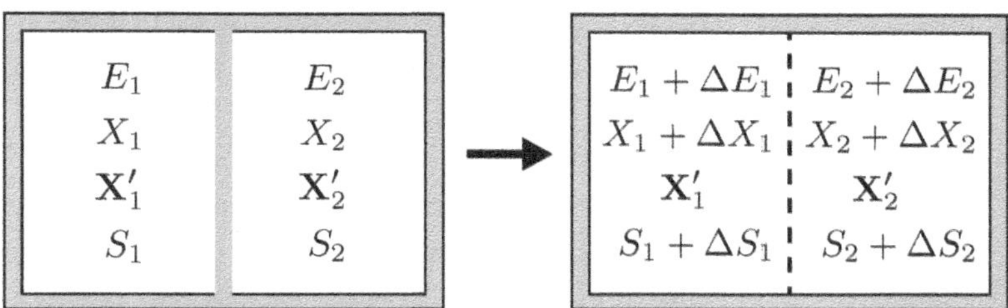

Fig. 1.6 Illustration of an isolated system in which energy and a specific mechanical variable are spontaneously exchanged upon the removal of a partition.

In general, the second law demands that $\Delta S \geq 0$. Since the total energy E and X are conserved, changes to the two regions are related, $\Delta E_1 = -\Delta E_2 = \Delta E$ and $\Delta X_1 = -\Delta X_2 = \Delta X$. If changes to both are small, then

$$\Delta S_1 = \Delta E \left(\frac{\partial S_1}{\partial E_1} \right)_{\mathbf{X}} + \Delta X \left(\frac{\partial S_1}{\partial X_1} \right)_{E, \mathbf{X}'}$$

$$= \Delta E \frac{1}{T_1} + \Delta X \frac{-f_1}{T_1}$$

where we have introduced a definition $\partial S_1/\partial X_1 = -f_1/T_1$. For simplicity, let's assume that the temperatures of the two sides are the same, $T_1 = T_2$. Then the total change in entropy $\Delta S = \Delta S_1 + \Delta S_2$ is

$$\Delta S = -\frac{\Delta X}{T}\left(f_1 - f_2\right) \geq 0$$

which for $\Delta X > 0$ requires $f_1 < f_2$, while for $\Delta X < 0$ requires $f_1 > f_2$. Thus, we can infer that f controls the spontaneous change of X, in such a way that X flows from large f to small f. Concretely, if X is the volume V, we will associate f with the pressure, $f \to -p$. If X is a number of particles, we will associate f with a chemical potential $f \to \mu$. These thermodynamic definitions for pressure and chemical potential thus encode the potential for volume and mass to flow, in analogy with temperature.

1.5 Changing between equilibrium states

Thus far we have considered autonomous systems, whose changes were spontaneous. These changes were a consequence of the random motion that particles undergo. However, we can also affect a change to a system directly by imposing it. We will consider now what happens to a system if we impose a change in some controllable mechanical parameter, such as $X \to X + \Delta X$. We will restrict our attention to cases where ΔX is small, and ask how does changing X affect the energy and entropy of the system.

Since the probability of microstates depends on mechanical variables, affecting a change in X, results in $P \to P + dP$, and correspondingly, since probabilities change, so does the entropy $S \to S + dS$, and average energy $E \to E + dE$. Let us begin by considering how the entropy changes. To do so, we need to generalize our definition of the entropy for cases where the system can exchange energy with its surroundings. This definition is due to Gibbs,

$$S = -k_{\mathrm{B}} \sum_{\nu} P(\nu) \ln P(\nu)$$

which is consistent with Boltzmann's definition when microstates are distributed with equal probability, as they are in an isolated system.

Exercise 1.6: Show explicitly that if $P(\nu) = 1/\Omega$, the Boltzmann and Gibbs definitions of the entropy are consistent.

Using Gibbs' definition of the entropy, the first order change to the entropy from an imposed change to a mechanical variable can be evaluated from the total derivative of S,

$$\begin{aligned}
dS &= d\left(-k_{\mathrm{B}} \sum_{\nu} P(\nu) \ln P(\nu)\right) \\
&= -k_{\mathrm{B}} \sum_{\nu} dP(\nu) \ln P(\nu) - k_{\mathrm{B}} \sum_{v} dP(\nu) \\
&= -k_{\mathrm{B}} \sum_{\nu} dP(\nu) \ln P(\nu)
\end{aligned}$$

where in the third line we have made use of the conservation of probability that requires $\sum_\nu dP = 0$. If the system begins in equilibrium with a bath at constant temperature, such that microstates are Boltzmann distributed, $P(\nu) = \exp[-\beta E(\nu)]/Q$ where Q is a normalization constant, then the change in entropy is

$$dS = k_{\mathrm{B}} \sum_\nu dP(\nu) \left[\beta E(\nu) + \ln Q\right]$$

$$= \frac{1}{T} \sum_\nu dP(\nu)\, E(\nu)$$

where we have again used the conservation of probability to eliminate the second term. From this we find that changes in entropy in equilibrium result from how probabilities of microstates change due to the mechanical manipulation.

To compute the change in energy, we need to acknowledge that macroscopically the average energy is what is observed. The average energy is

$$E = \sum_\nu P(\nu) E(\nu)$$

like any other expectation value. The total change to the observable energy results from two contributions,

$$dE = \sum_\nu P(\nu) dE(\nu) + \sum_\nu dP(\nu) E(\nu)$$

the first term is how the energy changes directly through manipulation of X, while the second term is how the distribution of microstates change. The first term can be written more clearly as,

$$dE = \frac{\partial E}{\partial X} dX \equiv f_{\mathrm{ext}} dX$$

where we have introduced an externally applied force f_{ext} acting on variable X. In the case that the system proceeds through a set of equilibrium states, the second contribution to the change in E can be identified with our previous expression for the change in entropy. Putting these together in the general case of multiple mechanical variables, we find

$$dE = T dS + \mathbf{f}_{\mathrm{ext}} \cdot d\mathbf{X}$$

which is a statement of the *first law of thermodynamics*. The requirement that the distribution of microstates maintain a Boltzmann form, restricts this particular statement to reversible changes of the system.

There are two different statements embedded in this form of the first law of thermodynamics. The first is a statement of book-keeping— energy is conserved. Once we understand the world is made of atoms and molecules, and they evolve with Hamiltonian mechanics, this conservation is a consequence of Noether's theorem, which states that accompanying the time reversal symmetry of the dynamics is a conserved quantity. The other statement is more clarifying— there are two fundamentally distinct ways in which the energy of a system can change. The first, $\mathbf{f}_{\mathrm{ext}} \cdot d\mathbf{X}$, are changes to the internal energy of a system resulting from processes we *control*. Through the application of an applied force, we can manipulate the system. These controlled processes are what we mean by performing work on the system, $\mathcal{W}$, in its differential form, $đ\mathcal{W}$. The second, TdS, are changes to the internal energy of a system resulting from processes we *do not directly control*. These are processes that take energy in and out of the system in ways we do not prescribe. These uncontrolled processes are what we mean by heat flow into the system, $\mathcal{Q}$, in its differential form, $đ\mathcal{Q}$. Together, the first law states that while energy is conserved,

$$dE = đ\mathcal{Q} + đ\mathcal{W}$$

the way it is conserved is generally complicated by processes both in and out of our direct ability to control.

Note the notation we have adopted for a small increment in the heat or work, $đ\mathcal{Q}$ and $đ\mathcal{W}$. These refer to each as inexact differentials. This is to distinguish them from the increment in the energy, dE, which is an exact differential. For an exact differential, the fundamental theorem of calculus allows us to compute the change in energy, ΔE, between and initial and final state as

$$\Delta E = \int_{E_i}^{E_f} dE = E_f - E_i$$

where E_f and E_i are the final and initial energy, respectively. This is distinct from the heat and work,

$$\mathcal{Q} = \int đ\mathcal{Q} \neq \mathcal{Q}_f - \mathcal{Q}_i \qquad \mathcal{W} = \int đ\mathcal{W} \neq \mathcal{W}_f - \mathcal{W}_i$$

where as inexact differentials, their integral depends explicitly on the path taken. One says that the energy is a *state function*, as one needs only to specify the conditions in which microstates are sampled in order to be able to evaluate the energy of that state. Heat and work are not state functions as they depend on the way the system is manipulated between two states. For example, one typically expects that

$$\int_{\text{fast}} đ\mathcal{W} \;\geq\; \int_{\text{slow}} đ\mathcal{W}$$

the amount of work required to affect a transition quickly is larger than the amount of work required to perform the same transition slowly. Because the total energy change is the same in both processes, provided the start and end points are the same, the first law implies,

$$\int_{\text{fast}} đ\mathcal{Q} \;\leq\; \int_{\text{slow}} đ\mathcal{Q}$$

or that the amount of energy released as heat, energy lost, or dissipated from the system, is larger in a fast process than a slow process.

> **Exercise 1.7:** Determine which of the following differentials are exact, and which are inexact:
>
> $$df = \sin(x)\cos(y)dx + \frac{1}{2}\sin^2(x)dy \qquad df = x^3ydx + \frac{1}{4}x^4dy$$
>
> using the property of exact differentials that $\partial f^2/\partial x\partial y = \partial f^2/\partial y\partial x$.

1.6 Adiabatic processes and statistical forces

If the system is manipulated in the absence of an environment, then while the energy may change due to the work being done on it, there will be no heat flow, as there is nowhere for the energy to go. Such a process is termed *adiabatic*, and characterized by zero change in the entropy of the bath, $(\Delta S_{\mathrm{bath}})_{\mathrm{Ad}} = 0$. Consequently, for a spontaneous, adiabatic process,

$$\Delta S = \Delta S_{\mathrm{sys}} + \Delta S_{\mathrm{bath}} \geq 0$$
$$= (\Delta S_{\mathrm{sys}})_{\mathrm{Ad}} \geq 0$$

the entropy of the system must increase. This generalizes slightly our understanding of the second law of thermodynamics to a specific case where we can confine our attention to what is happening to a system, absent of what is happening in the surrounding environment.

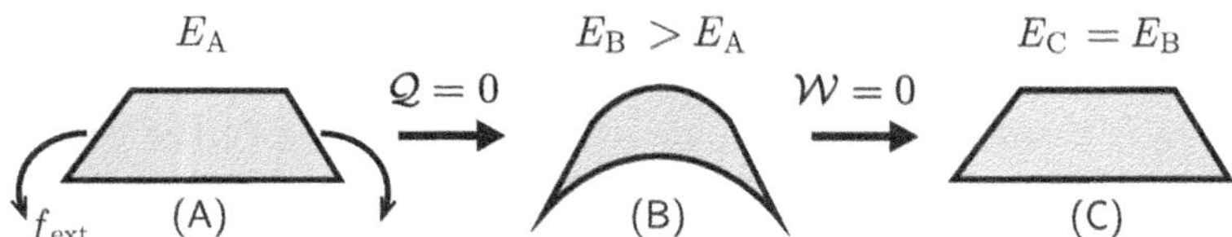

Fig. 1.7 Bending an elastic sheet in air is an example adiabatic process.

For an example of an adiabatic process, consider compressing a gas in a piston. If the gas is in a volume, wrapped in insulation, to a good approximation compression could be done without heat flow to the surroundings. If after the gas is compressed, the pressure is reduced below that which is needed to hold it at that volume, then the gas will spontaneously expand. That expansion would occur adiabatically, and thus is driven by the system adopting a more probable state, or one with more entropy. Analogously, one could envision bending an elastic sheet, as illustrated in Figure 1.7. If done in air, then to a good approximation the heat flow to the environment might be neglected. If the bending is done carefully, such that the process could be considered as proceeding through a succession of equilibrium states, then the process is described as both adiabatic and reversible. If the force applied to bend the sheet is released, then the sheet will snap back. Such a process is spontaneous and thus irreversible, as well as still adiabatic.

Exercise 1.8: Consider the elastic sheet described earlier. Bending the sheet required an applied force, and thus work, $\mathcal{W} > 0$, is done on the sheet. If it is done adiabatically, then there is no heat flow so $\Delta E = \mathcal{W} > 0$, and if it is done reversibly then $\Delta S = 0$. Once the force is released, a spontaneous process occurs that restores the sheet to its initial flat structure. There is no applied force, so $\mathcal{W} = 0$, and it is adiabatic so $\mathcal{Q} = 0$, and thus $\Delta E = 0$, however it is spontaneous so $\Delta S > 0$. Provide a microscopic explanation for the deduced increase in entropy.

In considering spontaneous processes where mechanical variables can be repartitioned, we have introduced a thermodynamic definition of a force conjugate to a mechanical variable X_i, $\partial S/\partial X_i = -f_i/T$. This implies that a mechanical force, in an equilibrium system, is equivalent to how the number of microstates changes with a conjugate mechanical variable. In order to confirm that this makes sense, let us consider the total change in the entropy. Since the entropy is a function of E and $\mathbf{X}$, its total differential is given by

$$dS = \left(\frac{\partial S}{\partial E}\right)_{\mathbf{X}} dE + \sum_i \left(\frac{\partial S}{\partial X_i}\right)_{E,X_{j\neq i}} dX_i$$

where the coefficient in front of changes of the energy, dE, is the temperature $T^{-1} = (\partial S/\partial E)_{\mathbf{X}}$. We would like to check if the statistical definition of the force,

$$-\frac{f_i}{T} \overset{?}{=} \left(\frac{\partial S}{\partial X_i}\right)_{E,X_{j\neq i}}$$

makes sense. Note that the differential form of the entropy written above is valid only for reversible changes between equilibrium states. For an infinitesimal, reversible, and adiabatic change, $dS = 0$, and rearranging the total derivative for the entropy, we find a relation for the change in the energy

$$dE = -T \sum_i \left(\frac{\partial S}{\partial X_i}\right)_{E,X_{j\neq i}} dX_i$$
$$= d\mathcal{W}_{\mathrm{rev}}$$

which for an adiabatic change is equal to the work. In the reversible limit, the work is

$$d\mathcal{W}_{\mathrm{rev}} = \mathbf{f}_{\mathrm{ext}} \cdot d\mathbf{X} = -T \sum_i \left(\frac{\partial S}{\partial X_i}\right)_{E,X_{j\neq i}} dX_i$$

from which follows for each component i,

$$\frac{f_i}{T} = -\left(\frac{\partial S}{\partial X_i}\right)_{E,X_{j\neq i}}$$

a relationship between the force and the change of the entropy. At equilibrium, f_i is a real force, and equal to the external force. The equilibrium relationships between the force, mechanical variables, and temperature is known as an *equation of state*.

1.7   Reversible work theorem

What determines a spontaneous change when heat can flow? To analyze this, we can consider the composite system plus bath in isolation, such that the condition for a spontaneous change is that the overall entropy increases. Decomposing the entropy change as a part from the system and a part from the bath,

$$\Delta S_{\text{sys}} + \Delta S_{\text{bath}} \geq 0$$

their sum must be nonnegative for a spontaneous process. If we assume the bath is not directly manipulated, its mechanical variables are fixed, $d\mathbf{X}_{\text{bath}} = 0$, then any change to its energy is just heat

$$dE_{\text{bath}} = d\!Q_{\text{bath}} = TdS_{\text{bath}}$$

and assuming the bath much larger than the system, any change to it due to the system will be reversible. The reversible heat flow is just the change in the entropy. Substituting the heat flow into the system for its change in entropy,

$$\Delta S_{\text{sys}} + \frac{d\!Q_{\text{bath}}}{T} \geq 0$$

and recognizing that any heat flow into the bath is equal to and opposite of the heat flow from the system, $d\!Q_{\text{sys}} = -d\!Q_{\text{bath}}$, we find

$$\Delta S_{\text{sys}} \geq \frac{d\!Q_{\text{sys}}}{T}$$

that the change in the entropy of the system is always greater than or equal to the heat flow into the bath at fixed T. This inequality is Clausius' statement of the second law, the first to be written down mathematically in 1862. Clausius uncovered this bound by considering an ideal heat engine. An engine operates in a cycle, so that $\Delta S = 0$, and in the reversible limit there is no loss of energy due to heat, but for an irreversible cycle, there is loss since $Q < 0$.

An implication of Clausius' inequality can be discerned by considering two processes involving the same heat bath at a temperature T, with the same initial and final states. In one process, the transformation is done irreversibly while the other is done reversibly. An example of such a comparison for pulling on a chain molecule is shown in Figure 1.8. Since the initial and final states are conserved, the change in the energy is the same,

$$dE = d\!Q + d\!W = d\!Q_{\text{rev}} + d\!W_{\text{rev}}$$

so we can equate the sum of the heat and work in both. Since the process is done at fixed temperature, $d\!Q_{\text{rev}} = TdS$. Rearranging the second equality,

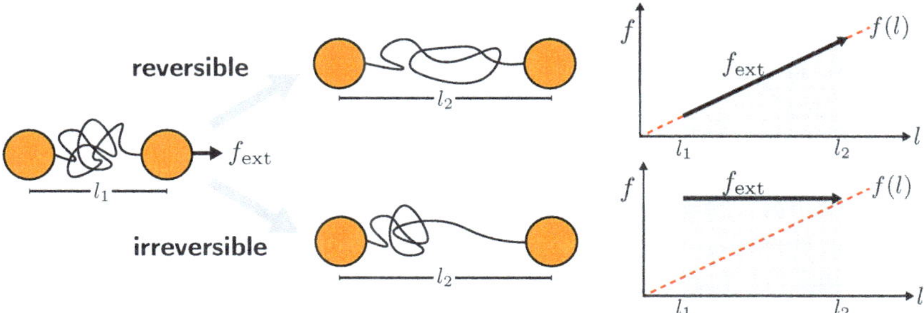

Fig. 1.8 Comparison of a reversible and an irreversible pulling process of a chain molecule with a linear force-extension equation of state.

$$\bar{d}\mathcal{W} = \bar{d}\mathcal{W}_\text{rev} + T dS - \bar{d}Q$$

noting the Clausius' inequality $T dS - \bar{d}Q \geq 0$, if we drop the last two terms, we change the equality into an inequality.

$$\bar{d}\mathcal{W} \geq \bar{d}\mathcal{W}_\text{rev}$$

and we find the so-called *reversible work theorem*. The reversible work theorem clarifies that for a process at constant temperature, a so–called *isothermal* process, the minimal amount of work required for a transformation is the reversible work. For our chain molecule, this can be understood graphically. If the molecule has an equilibrium force–extension relationship, or equation of state, that is linear like shown in Figure 1.8, the work $\mathcal{W} = \int dl f_\text{ext}$, or the area under the force-extension curve is clearly the minimum amount required, under the requirement that the force is large enough to actually affect the transition.

Exercise 1.9: The work for extending the chain molecule shown previously is given by $\bar{d}\mathcal{W} = f_\text{ext} dl$ where l is the extension and f_ext is the applied external force. If the equation of state for that molecule is $f(l) = \alpha T l$ where α is a positive constant, compute the work for extending the chain molecule from l_1 to $l_2 > l_1$ in the reversible limit, where $f_\text{ext} = f(l)$ and irreversibly with constant force $f_\text{ext} = \alpha T l_2$.

Another implication of Clausius' statement of the second law can be found from considering a system at constant T by rewriting the original statement using the first law,

$$\Delta S \geq \frac{1}{T}\left(dE - \bar{d}\mathcal{W}\right)$$

where we have use $\bar{d}Q = dE - \bar{d}\mathcal{W}$. Rearranging this expression,

$$\bar{d}\mathcal{W} \geq dE - T dS$$
$$\geq d\left(E - TS\right) \equiv dA$$

we find that the work is greater than the change in the quantity $E - TS$, which we introduced as $A = E - TS$, and refer to as the *Helmholtz free energy*. If no work is done on the system, so that $\mathbf{X}$ is held constant,

$$(\Delta A)_{T,\mathbf{X}} \leq 0$$

we find a compact statement for a spontaneous process in cases where energy can be exchanged with the surroundings at fixed T. In other words, the equilibrium partitioning of energy and mechanical variables for a system in contact with a heat bath minimizes A. It is referred to as a free energy because at fixed T,

$$\mathcal{W}_{\mathrm{rev}} = (\Delta A)_T$$

the Helmholtz free energy is the energy available to do work.

1.8   Auxiliary thermodynamic functions

In the previous section, we were motivated to introduce the Helmholtz free energy, A, as it is a natural quantity that dictates the direction of spontaneous change for a system that can exchange energy with its surroundings. We defined A as a particular transformation of the energy E,

$$A = E - S\left(\frac{\partial E}{\partial S}\right)_{\mathbf{X}}$$
$$= E - ST$$

which results in differential changes in A being equal to

$$dA = -SdT - pdV + \mu dN$$

given a particular set of mechanical variables $\mathbf{X} = \{V, N\}$. This differential relationship provides alternative definitions of thermodynamic quantities in the case where the temperature of the system is controlled rather than the total energy. For example,

$$\left(\frac{\partial A}{\partial T}\right)_{V,N} = -S \qquad \left(\frac{\partial A}{\partial V}\right)_{T,N} = -p \qquad \left(\frac{\partial A}{\partial N}\right)_{T,V} = \mu$$

first derivatives of A with respect to its natural variables T, V and N, generate either thermodynamic potentials like μ, or extensive variables like S.

The particular transformation of the energy that yielded the Helmholtz free energy is known as a *Legendre transform*. Microscopically we saw this was a consequence of moving between ensembles in the thermodynamic limit. In this case, the relationship between the energy and the entropy was recast to one that is convenient when the temperature is controlled. Specifically, the energy, which naturally depends on the entropy, volume, and number of particles, $E(S, V, N)$, is exchanged for the Helmholtz free energy, $A(T, V, N)$. In this sense, A is an auxiliary function in that is contains the same information as E.

> **Aside: Legendre transforms.** Legendre transforms are linear transformations of convex functions. Consider a function $f(x)$ whose derivative with respect to x is $y = df/dx$. Let us define a new function g, as a Legendre transform of f, such that
>
> $$g = f - x\frac{df}{dx} = f - xy$$
>
> whose total derivative is
>
> $$dg = \frac{df}{dx}dx - ydx - xdy = -xdy$$
>
> so that g depends *naturally* on y. The derivative of g with respect to y is $dg/dy = -x$, so
>
> $$g - y\frac{dg}{dy} = g + xy = f$$
>
> the Legendre transform of g returns f. Legendre transforms are thus invertible. Geometrically they recast the relationship between a function and its argument by a family of tangent lines.

The specific relationship between T and S that admits a Legendre transform is known as a *conjugate* pair. These are pairs of variables that include one extensive variable and one thermodynamic potential that controls its flow. Writing the energy as a function of its natural variables, S, V, and N,

$$dE = \left(\frac{\partial E}{\partial S}\right)_{V,N} dS + \left(\frac{\partial E}{\partial V}\right)_{S,N} dV + \left(\frac{\partial E}{\partial N}\right)_{S,V} dN$$
$$= TdS - pdV + \mu dN$$

we can establish other conjugate variable pairs, $p - V$, and $\mu - N$.

Other conjugate variable pairs can serve as a basis of other useful auxiliary thermodynamic functions. For example, we can construct a function, H, the *enthalpy*, defined as

$$H(p, S, N) \equiv E + pV$$

which is a Legendre transform of the energy with respect to the volume. As a consequence of the Legendre transform, the enthalpy is a natural function of S, p, and N,

$$dH = TdS + Vdp + \mu dN$$

and supplies useful definitions of T, V, and μ when the pressure of the system is controlled. For example, consider changes to the enthalpy when p and N are fixed,

$$(dH)_{p,N} = TdS = đQ_{\text{rev}}$$

we find that the enthalpy reports directly on the heat. In fact at fixed pressure and number of particles, it is equal to the heat even in an irreversible transformation.

To round out the traditional auxiliary thermodynamic functions, we can construct a function, G, the *Gibbs free energy*, defined as

$$G(p, T, N) \equiv E + pV - TS$$

which is a Legendre transform of the energy with respect to the entropy and with respect to the volume. As a consequence of the Legendre transform, the Gibbs free energy is a natural function of T, p, and N,

$$dG = -SdT + VdP + \mu dN$$

and thus is useful when both the temperature and the pressure are controlled. When both T and p are fixed, the Gibbs free energy reports on the amount of chemical work. Generalizing to the case of multiple particle species,

$$(dG)_{p,T} = \sum_{i=1} \mu_i dN_i$$

or holding all of the particle numbers fixed

$$(dG)_{p,T,\mathbf{N}} \leq 0$$

we have a statement of a spontaneous process in cases where the temperature and the pressure are fixed. The Gibbs free energy has a simple interpretation. Changes in G are equal to

$$\Delta G = \Delta H - T\Delta S$$
$$= -T\Delta S_{\text{bath}} - T\Delta S \leq 0$$

which is nothing more than a compact way of stating the total change in the entropy of a system and the bath it is in contact with, using the fact that the enthalpy is the heat released which is minus the entropy change in the idealized bath within which all changes are reversible.

> **Exercise 1.10:** By taking the Legendre transform of the energy with respect to the entropy and number of particles, construct an auxiliary function $Y(T, V, \mu)$ that is a natural function of T, V, and μ.

1.9 Thermodynamic stability

We have seen how first derivatives of the energy with respect to its natural variables define thermodynamic potentials for their spontaneous change. These potentials, $T, p,$ and μ, however, are not signed, they can in principle be positive or negative. We expect, however, to be able to say something about how each potential changes with its conjugate quantity. For example, if energy is spontaneously transferred from a hot region to a cold region of an isolated system, it should be the case that the hot region cools and the cold region warms. If this were not the case, if when energy is added to the cold region it cools further, then the system would not get closer to an equilibrium state, or rather that the equilibrium state would not be stable.

To understand this sharply, consider taking an isolated system initially at equilibrium, in its most probable state, and perturbing it by redistributing some energy, δE. For such a perturbation from equilibrium,

$$S(E, \mathbf{X}|\delta E) \leq S(E, \mathbf{X}|\delta E = 0)$$

the entropy must decrease as we have displaced the system from its most probable, highest entropy state. Let's unpack the consequences of this decrease in the entropy by moving the energy into some region 1, from region 2. If δE is small, then we can expand the entropy,

$$S(E, \mathbf{X}|\delta E) - S(E, \mathbf{X}|0) = \delta E \left[\left(\frac{\partial S_1}{\partial E_1} \right)_{\mathbf{X}_1} - \left(\frac{\partial S_2}{\partial E_2} \right)_{\mathbf{X}_2} \right]$$
$$+ \frac{1}{2} \delta E^2 \left[\left(\frac{\partial^2 S_1}{\partial E_1^2} \right)_{\mathbf{X}_1} + \left(\frac{\partial^2 S_2}{\partial E_2^2} \right)_{\mathbf{X}_2} \right] \leq 0$$

where we have included terms up to second order. If the system is initially at equilibrium, then $T_1 = T_2$ and the first term on the right-hand side vanishes, so we are left with just the second-order term. Since $\delta E^2/2$ is a nonnegative number, the inequality is determined by the term in the square brackets. The second derivative of the entropy with respect to the energy can be written as

$$\left(\frac{\partial^2 S}{\partial E^2} \right)_{V,N} = \left(\frac{\partial 1/T}{\partial E} \right)_{V,N}$$
$$= -\frac{1}{T^2} \left[\left(\frac{\partial E}{\partial T} \right)_{\mathbf{X}} \right]^{-1} = -\frac{1}{T^2} C_V^{-1}$$

where we have introduced the constant volume heat capacity C_V as how the energy changes with temperature. From the inequality above, we find that $C_V \geq 0$, or that $(\partial E/\partial T)_{\mathbf{X}} \geq 0$, also that the entropy is a convex function $\partial^2 S/\partial E^2 \leq 0$. This is a condition of stability of equilibrium. Analogous statements could be derived employing other auxiliary functions. For example, using the Helmholtz free energy we could determine,

$$\left(\frac{\partial^2 A}{\partial V^2} \right)_{T,N} \geq 0 \quad \rightarrow \quad -\left(\frac{\partial p}{\partial V} \right)_{T,N} \geq 0$$

or that A too is a convex function, and there is a definite relationship between changes in p and V. Indeed, generically because of the stability requirements of equilibrium, conjugate variable pairs, $E - T$, $p - V$, or $\mu - N$ have definite signed derivative relationships.

In addition to convexity, we typically assume that thermodynamic functions are smooth. If a function $f(x, y)$ that depends on multiple variables x and y is smooth, then

$$\left(\frac{\partial}{\partial x} \left(\frac{\partial f(x, y)}{\partial y} \right)_x \right)_y = \left(\frac{\partial}{\partial y} \left(\frac{\partial f(x, y)}{\partial x} \right)_y \right)_x$$

the order of its derivatives are interchangeable. While this is a simple mathematical statement, it has profound implications thermodynamically. Consider constructing an analogous expression employing the Gibbs free energy, taking derivatives with respect to two of its natural variables, p and T, in different orders,

$$\left(\frac{\partial}{\partial T} \left(\frac{\partial G}{\partial p} \right)_{T,N} \right)_{p,N} = \left(\frac{\partial}{\partial p} \left(\frac{\partial G}{\partial T} \right)_{p,N} \right)_{T,N}$$

$$\left(\frac{\partial V}{\partial T} \right)_{p,N} = - \left(\frac{\partial S}{\partial p} \right)_{T,N}$$

while the first line is just arrived at by exchanging the order of the derivatives, the second line inserts in definitions of the first-order derivatives. The statement of the second line is non-trivial, as it relates two seemingly distinct experiments. These relationships are known as *Maxwell relations*, and are of ubiquitous utility.

Similarly for a smooth function of multiple variables, there is a mathematical identity known as the cyclic chain rule. For a function $f(x, y)$, this can be written as

$$\left(\frac{\partial f}{\partial x} \right)_y = - \left(\frac{\partial f}{\partial y} \right)_x \left(\frac{\partial y}{\partial x} \right)_f$$

for which a thermodynamic example might be

$$\left(\frac{\partial p}{\partial T} \right)_{V,N} = - \left(\frac{\partial p}{\partial V} \right)_{T,N} \left(\frac{\partial V}{\partial T} \right)_{p,N}$$

a complex relation between thermodynamic variables. Together, Maxwell relations and the cyclic chain rule can be used to deduce complex thermodynamic statements, such as bounds on observables held under different constraints.

> **Exercise 1.11:** By writing S as a function of V, T, and N, and invoking the thermodynamic stability criteria, determine which is larger, the constant pressure heat capacity, $C_p = T(\partial S/\partial T)_{p,N}$ or the constant volume heat capacity, $C_V = T(\partial S/\partial T)_{V,N}$.

A final important aspect of the mathematics of thermodynamics concerns the expected extensivity of particular thermodynamic variables, like E, S, V, and N. Because each of these is expected to grow in proportion to the scale of the system, mathematically this implies that the energy is a so-called homogeneous function of order one.

This means that as E is a function of S, V, and N, if λ copies of a system with energy E are placed together it is equal to a system with energy λE, or

$$E(\lambda S, \lambda V, \lambda N) = \lambda E(S, V, N)$$

for any $\lambda > 0$. Euler's theorem of homogeneous functions implies an integrated form of the energy,

$$E = TS - pV + \mu N$$

which holds provided only the extensiveness of the energy. A consequence of this integrated form is the so-called *Gibbs–Duhem relationship*

$$-SdT + VdP - Nd\mu = 0$$

which relates changes of the intensive properties of a system to each other.

1.10   Partition functions

For a system that can exchange energy with its surrounding, microstates are distributed according to the Boltzmann distribution

$$P(\nu) = \frac{e^{-\beta E(\nu)}}{Q} \qquad Q = \sum_\nu e^{-\beta E(\nu)}$$

where Q is a normalization constant. This constant is known as a *partition function* and encodes a significant amount of thermodynamic and probabilistic information. For example, its name derives from its ability to relate likelihoods of groups, or partitions, of microstates. Consider a protein in solution that can be in either a folded or unfolded configuration, like those illustrated in Figure 1.9. We can group microstates that are characterized by being folded into group B, and those that are unfolded into group C. The total probability of being in group B is computable as

$$P(B) = \sum_{\nu \in B} P(\nu) = \sum_{\nu \in B} \frac{e^{-\beta E(\nu)}}{Q}$$
$$= \frac{1}{Q} \sum_{\nu \in B} e^{-\beta E(\nu)} = \frac{Q(B)}{Q}$$

a sum of microstates in group B weighted by their Boltzmann weight. Manipulating that sum by pulling Q out, and identifying the partition sum over microstates $Q(B)$, we find that the likelihood of observing a microstate in group B is proportional to $Q(B)$. The same is true for group C, resulting in the ratio of likelihoods of being in one of those groups as

$$\frac{P(B)}{P(C)} = \frac{Q(B)}{Q(C)}$$

or a ratio of partition functions. This should remind you of similar inferences made of Ω for our isolated system. Indeed, Ω plays the same role as a partition function for an

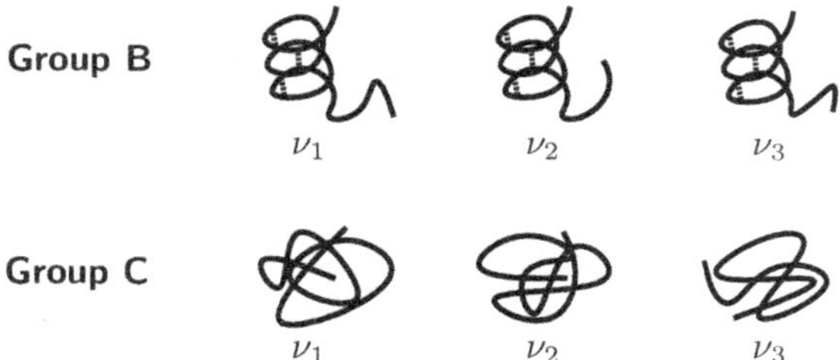

Group B ν_1 ν_2 ν_3

Group C ν_1 ν_2 ν_3

Fig. 1.9 Examples of grouping microstates of a single polymer based on their fold.

isolated system as Q does for our open system. We refer to the case of constant energy as the *microcanonical* partition function and that when energy can be exchanged as the *canonical* partition function.

Similarly, just as Ω is related to a thermodynamic quantity, the entropy, so too is Q. This can be seen by rewriting Q as a sum over energies,

$$Q = \sum_{\nu} e^{-\beta E(\nu)} = \sum_{E}\sum_{\nu \in E} e^{-\beta E(\nu)} = \sum_{E} \Omega(E) e^{-\beta E}$$

$$= \sum_{E} e^{-\beta E + \ln \Omega(E)} = \sum_{E} e^{-\beta[E - TS(E)]}$$

where in the third equality we have identified the number of microstates consistent with a fixed energy as Ω, and incorporated it into the argument of the exponential using the definition of the entropy. As we have already shown, the argument of the exponential obeys a large deviation form, so that in the large N limit the sum is dominated by its most likely value, E^*, or

$$Q \sim e^{-\beta[E^* - TS(E^*)]}$$

which is a relationship previously defined in thermodynamics. Specifically, we identify $A = E - TS$ as the Helmholtz free energy, and so

$$A = -k_{\mathrm{B}}T \ln Q \qquad Q = e^{-\beta A}$$

we find that the partition function Q is the microscopic definition of A. In fact, the relationship between A and E can be considered an implication of a transformation in the limit that E is continuous,

$$Q = \int dE \, \Omega(E) e^{-\beta E}$$

which is a Laplace transform of $\Omega(E)$ with Laplace variable β. Evaluating the Laplace transform with Laplace's saddle point method recovers a Legendre transform relationship between A and E, $A = E - S(dE/dS)$, discussed previously.

To demonstrate that $A = -k_{\mathrm{B}}T \ln Q$ is a sensible definition even outside of the thermodynamic limit, let us check for consistency. Taking the derivative of $\ln Q$ with respect to β, employing this definition for A, we have

$$-\left(\frac{\partial \ln Q}{\partial \beta}\right)_{\mathbf{X}} = \frac{\partial}{\partial \beta}(\beta A) = A + \beta \left(\frac{\partial A}{\partial \beta}\right)_{\mathbf{X}}$$

$$= A - \beta k_{\mathrm{B}} T^2 \left(\frac{\partial A}{\partial T}\right)_{\mathbf{X}} = A + TS$$

which implies that for consistency that derivative should be equal to E. Performing that derivative explicitly,

$$-\left(\frac{\partial \ln Q}{\partial \beta}\right)_{\mathbf{X}} = -\frac{1}{Q}\frac{\partial}{\partial \beta}\sum_\nu e^{-\beta E(\nu)}$$

$$= \frac{1}{Q}\sum_\nu e^{-\beta E(\nu)} E(\nu) = \sum_\nu P(\nu) E(\nu) = \langle E \rangle$$

we find that, as expected, the derivative is the energy, on average.

That the first derivative of $\ln Q$ returns the average E could be anticipated, as $\ln Q$ is the Laplace transform of a probability distribution, it functions as a *cumulant generating function*. A cumulant generating function is a statistical object whose derivatives return the cumulants of a random variable. In this case, the random variable is the energy. Testing this explicitly, we can take the second derivative,

$$\left(\frac{\partial^2 \ln Q}{\partial \beta^2}\right)_{\mathbf{X}} = -\frac{\partial}{\partial \beta}\left(\frac{1}{Q(\beta)}\sum_\nu e^{-\beta E(\nu)} E(\nu)\right)$$

$$= \frac{1}{Q(\beta)}\sum_\nu e^{-\beta E(\nu)} E^2(\nu) + \frac{1}{Q^2}\frac{\partial Q}{\partial \beta}\sum_\nu e^{-\beta E(\nu)} E(\nu)$$

$$= \langle E^2 \rangle - \langle E \rangle^2$$

and find we obtain the variance of the energy,

$$\left(\frac{\partial^2 \ln Q}{\partial \beta^2}\right)_{\mathbf{X}} = \langle \delta E^2 \rangle \qquad \delta E = E - \langle E \rangle$$

which is the second cumulant. Interestingly, we can observe that there is an alternative way to evaluate the second derivative of $\ln Q$, which is by acknowledging that it is equivalent to the first derivative of the first derivative,

$$\left(\frac{\partial^2 \ln Q}{\partial \beta^2}\right)_{\mathbf{X}} = -\frac{\partial}{\partial \beta}\langle E \rangle$$

$$= k_{\mathrm{B}} T^2 \frac{\partial \langle E \rangle}{\partial T} = k_{\mathrm{B}} T^2 C_V$$

where we have identified the heat capacity $C_V = (\partial E/\partial T)_{N,V}$. Equating the two,

$$C_V = \frac{1}{k_\mathrm{B} T^2} \langle \delta E^2 \rangle \geq 0$$

we find our first example of a so-called *fluctuation–dissipation relationship*. The heat capacity is a quantity that encodes the response of a system to a perturbation. In this case C_V relates how the energy of a system responds to a change in temperature. The right-hand side of the equality however, encodes the scale of fluctuations of an unperturbed system. Those fluctuations therefore determine the size of the response, and equivalently, the measurement of a response portrays the scale of a system's fluctuations. Put another way, the spontaneous fluctuations of a system measure its potential to respond.

1.11 Exchanging mass with a bath

We have shown that the likelihood of a microstate is different if the energy is fixed, as compared with the case when the energy can be exchanged with a surrounding bath. A natural question is what happens in other cases, where mechanical variables can be exchanged as well. Let us consider the case where particles as well as energy can be exchanged.

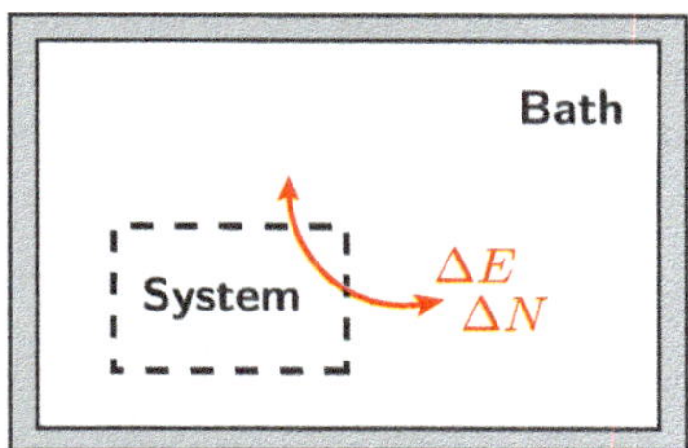

Fig. 1.10 Composite system and bath where both energy and particle number can be exchanged.

We can proceed as before, by constructing a composite system and bath that together conserve the total energy and particle number pictured in Figure 1.10. This results in expressions for the total energy, E_T and particle number N_T,

$$E_T = E(\nu) + E_B(\nu_B) \qquad N_T = N(\nu) + N_B(\nu_B)$$

where the subscripts B denote quantities of the bath. The total microstate is determined by a composition of the state of the system ν and the state of the bath ν_B, which jointly have a probability $P(\nu, \nu_B)$ that is uniform under our fundamental postulate. Therefore, just as before, the marginal probability of a system microstate $P(\nu)$ is proportional to the number of bath microstates satisfying the joint conservation laws of constant energy and particle number $\Omega_B(E_T - E(\nu), N_T - N(\nu))$. Again, assuming that the bath is large so that E and N are small compared to E_T and N_T, we can expand $\ln \Omega_B$ as

$$\ln \Omega_B(E_B, N_B) \approx \ln \Omega_B(E_T, N_T) - E(\nu)\frac{\partial \ln \Omega_B}{\partial E_B} - N(\nu)\frac{\partial \ln \Omega_B}{\partial N_B}$$

$$= \ln \Omega_B(E_T, N_T) - \beta E(\nu) + \beta\mu N(\nu)$$

where in the second line we have inserted our previous definition for $\beta = (\partial S/\partial E)_{N,V}/k_B$ and introduced the chemical potential, $\beta\mu = -(\partial S/\partial N)_{E,V}/k_B$. Exponentiating that result provides us with the probability of a microstate when both E and N fluctuate,

$$P(\nu) = \frac{e^{-\beta E(\nu) + \beta\mu N(\nu)}}{\Xi(T, \mu, V)} \qquad \Xi(T, \mu, V) = \sum_\nu e^{-\beta E(\nu) + \beta\mu N(\nu)}$$

where we have explicitly normalized the probability with a new partition function Ξ, known as the *grand canonical* partition function.

> **Exercise 1.12:** In the earlier expression for $P(\nu)$ in the grand canonical setting, we have dropped the subscript of μ on the bath. By solving for the most likely number of particles in the system, demonstrate that the condition of equilibrium requires $\beta\mu = \beta\mu_B$ where $\beta\mu = -(\partial \ln \Omega/\partial N)_{E,V}$ and $\beta\mu_B = -(\partial \ln \Omega_B/\partial N_B)_{E,V}$.

The grand canonical partition function can be expressed as

$$\Xi(T, \mu, V) = \sum_N \sum_{\nu \in N_\nu = N} e^{-\beta E(\nu) + \beta\mu N(\nu)}$$

$$= \sum_N Q(N, V, T)e^{\beta\mu N(\nu)}$$

or a Laplace transform of Q. Thus, provided Q is a smooth function, we should expect that there is no loss of information in moving between the two sets of constraints, or *ensembles*, and that the way we decide to count microstates is a matter of convenience. Evaluating the Laplace transform for large N using a saddle point approximation and remembering that $\ln Q = -\beta(E - TS)$, we can equate Ξ to a thermodynamic property,

$$\Xi(T, \mu, V) \sim e^{-\beta(E - TS - \mu N)}$$

$$= e^{\beta pV}$$

which in this case is pV the pressure times the volume of the system, which we see follows from $E = TS - pV + \mu N$. Taking derivatives of Ξ with respect to $\beta\mu$,

$$\left(\frac{\partial \ln \Xi}{\partial \beta\mu}\right)_{T,V} = \frac{1}{\Xi}\sum_\nu N_\nu e^{-\beta E(\nu) + \beta\mu N(\nu)} = \langle N \rangle$$

we observe that $\ln \Xi$ is a cumulant generating function for N. Indeed, while the first derivative is equal to the mean particle number, the second derivative

$$\left(\frac{\partial^2 \ln \Xi}{\partial \beta \mu^2}\right)_{T,V} = \langle \delta N^2 \rangle$$

$$= \left(\frac{\partial \langle N \rangle}{\partial \beta \mu}\right)_{V,T}$$

yields its variance, which is also equal to the response of $\langle N \rangle$ to a change in chemical potential. This is another fluctuation–dissipation relationship relating the fluctuations in particle number in a system to its compressibility. With this example of the generality of the formalism, we are now in a position to begin using equilibrium statistical mechanics to understand some physical phenomena.

Exercise 1.13: Using the Gibbs–Duhem relationship, $N d\mu - V dp + S dT = 0$, show that

$$\left(\frac{\partial N}{\partial \beta \mu}\right)_{V,T} = k_{\mathrm{B}} T \frac{N^2}{V} \kappa_T$$

where $\kappa_T = -(\partial \ln V/\partial p)_{T,N}$ is the isothermal compressibility.

Further reading

Much of the material in this chapter is available in canonical statistical mechanics textbooks. Of particular usefulness for more details are David Chandler's *Introduction to modern statistical mechanics*, Richard Tolman's *The principles of statistical mechanics*, and Terrell Hill's *An introduction to statistical thermodynamics*. Ideas of heat and work are best read from the *Collected works of J. Willard Gibbs*.

Additional exercises

Exercise 1.14: We will make judicious use of the factorial function,

$$N! = N \times (N-1) \times (N-2) \ldots \times 2 \times 1.$$

in these exercises and the next chapter. In order to manipulate this function, it is very convenient to consider an approximation due to Stirling,

$$\ln N! \approx N \ln N - N, \quad N \gg 1.$$

Here, you will derive this result and the leading order correction by estimating the integral

$$\Gamma(n+1) \equiv \int_0^\infty dt \, t^n e^{-t}$$

by the method of steepest descent, or Laplace's saddle point approximation.

1. Show that $\Gamma(n+1) = n!$ for positive integer values of n.
2. Rewrite the integrand in the equation for $\Gamma(n+1)$ as $\exp\{[f(t)]\}$, and locate the maximum value of $f = f(t_0)$.

3. Approximate $f(t)$ using a second-order Taylor expansion about t_0. Then evaluate the integral, justifying any necessary modification of integration limits. You might find it helpful to recall that

$$\int_{-\infty}^{\infty} dx\, e^{-ax^2} = \sqrt{\frac{\pi}{a}}$$

Why should your approximations be reasonable for large n?

4. Identify the leading order correction to Stirling's approximation.

5. Make a plot of $\ln n!$ as a function of n, together with Stirling's approximation both with and without the correction term. Over what range of n would you consider the approximation to be accurate?

Exercise 1.15: Consider a mixture with two molecular species, 1 and 2, in a container with volume V that is connected to a bath at temperature T. For counting purposes, imagine dividing this volume into M cells of a cubic lattice, each with microscopic volume ℓ^3. In a given microstate, each lattice cell may be empty, may be occupied by n molecules of type 1, or may be occupied by n molecules of type 2. Let N_1 be the total number of type 1 cells and N_2 be the number of type 2 cells.

1. Calculate the number Ω_{spatial} of spatial arrangements of this system as a function of N_1, N_2, and M. You may neglect any correlations between the states of different cells, but do not assume that N_1 and N_2 are much smaller than M.

2. Using Boltzmann's formula, calculate the entropy per cell S_{spatial}/M associated with this variety of spatial arrangements. Using Stirling's approximation, $\ln N! \approx N \ln N - N$, write your result solely in terms of the fractions $\phi_1 = N_1/M$ and $\phi_2 = N_2/M$, and fundamental constants.

3. Associated with a single lattice cell containing molecules of type 1 is an energy ϵ_1 and number of molecular configurations w_1. Similarly, a type 2 cell has energy ϵ_2 and degeneracy w_2. Calculate the Helmholtz free energy A of this system, accounting both for the multiplicity of spatial arrangements and internal fluctuations within each cell.

4. By differentiating A appropriately, compute the chemical potentials μ_1 and μ_2 of species 1 and 2, respectively.

5. Now imagine that each molecule can interconvert between identities 1 and 2, leading to fluctuations in N_1 and N_2. Let $nN_i^{(\text{equ})}$ be the average number of type i molecules at equilibrium. Calculate the change in Helmholtz free energy δA that would result from a small virtual displacement about the equilibrium partitioning, so that $N_1 = N_1^{(\text{equ})} + \delta N$ and $N_2 = N_2^{(\text{equ})} - \delta N$. Write your answer in terms of δN and the chemical potentials μ_1 and μ_2.

6. Using all of your results, apply the second law of thermodynamics to determine the relative proportions of the two molecular species at equilibrium. Specifically, calculate the ratio $N_1^{(\text{equ})}/N_2^{(\text{equ})}$ in terms of the quantities ϵ_1, ϵ_2, w_1, and w_2.

7. What is the probability P_1 for observing a particular lattice cell occupied by type 1 molecules, relative to the probability P_2 for observing the cell occupied by type 2 molecules? Write your answer for P_1/P_2 in terms of the quantities ϵ_1, ϵ_2, w_1, and w_2, and explain your reasoning.

Exercise 1.16: Work can be done on a membrane by bending it. The restoring force F exerted by the membrane depends on the curvature κ, how far it is bent, and on temperature. Let's take the equation of state for a membrane to be

$$F = \alpha \kappa e^{-T/T_0}.$$

where α and T_0 are positive constants.

1. A small amount of bending requires an amount of work $d\!W = F_{\text{ext}} d\kappa$, where F_{ext} is the force imposed on the membrane. Compute the total work required to bend an initially straight membrane ($\kappa = 0$) *reversibly* and *isothermally* to a final curvature κ. By isothermal we mean a process that maintains the system's temperature at a fixed value throughout.

2. The membrane's energy is

$$E = E_0(T) + \frac{1}{2}\alpha\kappa^2 \left(1 + \frac{T}{T_0}\right) e^{-T/T_0},$$

   where $E_0(T)$ depends *only* on temperature not on κ. Use this equation, and your result from part 1, to compute the total heat flow into the membrane during the reversible, isothermal bending.

3. Compute the change in entropy ΔS resulting from the reversible, isothermal bending.

Exercise 1.17: There are many ways to perform work on a system. For an electrochemical cell consisting of two electrodes, the work to charge the electrode is $\Psi_i dQ_i$ where Q_i is the charge on electrode i and Ψ_i its applied voltage. For a system of two electrodes, this results in a first law of the form

$$dE = TdS - pdV + \mu dN + \Psi_1 dQ_1 + \Psi_2 dQ_2$$

where the other variables take on their usual meaning. Below we will work out the consequences of this additional form of work.

1. Global electroneutrality requires that $Q_1 + Q_2 = 0$. Using this constraint, show that the first law can be written as

$$dE = TdS - pdV + \mu dN + \Delta\Psi dQ$$

   where $\Delta\Psi = \Psi_1 - \Psi_2$ and $Q = Q_1 = -Q_2$.

2. Using the above form of the first law, show that if the electrodes are able to exchange charge with an ideal bath with a constant applied voltage that the probability of a microstate is given by

$$P(\nu) = \frac{1}{\Gamma}e^{-\beta E(\nu)+\beta\Delta\Psi Q(\nu)} \qquad \Gamma = \sum_{\nu} e^{-\beta E(\nu)+\beta\Delta\Psi Q(\nu)}$$

   where Γ is the associated partition function at fixed T, V, N, and $\Delta\Psi$.

3. From the definition of the partition function, evaluate the first and second derivatives of $\ln\Gamma$ with respect to $\beta\Delta\Psi$ and determine a fluctuation-dissipation relation by equating the differential capacitance $C = (\partial \langle Q \rangle /\partial\Delta\Psi)_{T,N,V}$ to the charge fluctuations.

2

Ideal systems, chemical and mass equilibrium

The consideration of all possible microstates of a system is complicated by their tremendous number for many particle systems. Nevertheless, strategies exist to systematically count states and assign their probabilities. In this chapter, we will consider the application of statistical mechanics to the simplest of systems, those consisting of components that do not interact. Such ideal systems, like gases, dilute solutions, or collections of quasiparticles, provided an initial concrete testing ground for many of the earliest ideas in statistical mechanics. Many concepts born from ideal systems are widely useful, like the laws of mass action and chemical equilibrium. Moreover, ideal systems often serve as a useful reference state from which to develop more sophisticated descriptions. The strategy for the evaluation of partition functions of ideal systems is based on factorization, the idea that for systems that can be approximated as not interacting, the enumeration of the states is equivalent to summing the states of a single particle up to factors that account for particle indistinguishability. We will explore this extensively in the following. Throughout we will consider systems described by both quantum and classical mechanics. While in general such descriptions will yield observable differences, we will discuss considerably in what limits they are the same.

2.1 Ensemble equivalence

To apply the formalism of statistical mechanics to concrete systems, it is useful to convince ourselves that the ensemble we use to count microstates does not matter, and can be chosen out of convenience. As a concrete example, consider a chain of N spins evolving under the action of an applied magnet field, separated sufficiently far from each other that they do not interact. Each spin can either be aligned or anti-aligned with the applied field, $s_i = \pm 1$. The microstate of the system is determined by the N values of the spins, $\nu = \{s_1, s_2, \ldots s_N\}$. Let us associate an energy scale ϵ that favors aligning the spin in the field, such that the energy of the system shown in Figure 2.1 is $E = -\epsilon/2 \sum_{i=1}^{N} s_i$.

For a system that can exchange energy with its surroundings, the likelihood of observing the microstate of the system is given by the Boltzmann distribution. Noting that each spin is statistically identical,

$$P(\nu) \propto \exp\left[\beta\epsilon/2 \sum_i s_i\right] = \prod_i \exp\left[\beta\epsilon s_i/2\right] = P(s_1)P(s_2)\cdots P(s_N)$$

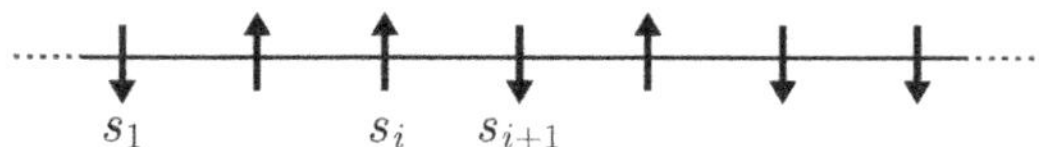

Fig. 2.1 System of N noninteracting spins held in an applied magnetic field.

we find that the probability of observing the full microstate of the system factorizes into a product of individual spin probabilities. This is a generic consequence of the assumption that the spins do not interact and the resulting additive energy function. Analogously the partition function associated with the Boltzmann distribution,

$$Q = \sum_{s_1=\pm 1}\sum_{s_2=\pm 1}\cdots\sum_{s_N=\pm 1} e^{\beta\frac{\epsilon}{2}\sum_i s_i} = \sum_{s_1=\pm 1} e^{\beta\frac{\epsilon}{2}s_1}\sum_{s_2=\pm 1} e^{\beta\frac{\epsilon}{2}s_2}\cdots\sum_{s_N=\pm 1} e^{\beta\frac{\epsilon}{2}s_N}$$

$$= \prod_{i=1}^{N}\left(\sum_{s_i=\pm 1} e^{\beta\epsilon s_i/2}\right) = q^N$$

also factorizes into N factors of a single spin partition function, q. Normalizing the full probability distribution is accomplished by carrying out a single sum over the microstates of an individual spin, yielding

$$q = e^{\beta\epsilon/2} + e^{-\beta\epsilon/2} = 2\cosh(\beta\epsilon/2)$$

where in the second equality we have utilized a hyperbolic trigonometry identity. The likelihood of an individual spin aligning with the field is

$$P(s) = \frac{e^{\beta\epsilon s/2}}{2\cosh(\beta\epsilon/2)} \qquad \langle s\rangle = \sum_{s=\pm 1} sP(s) = \tanh(\beta\epsilon/2)$$

from which we can determine its average value as a function of the temperature relative to the energy scale ϵ.

Exercise 2.1: Demonstrate that if a joint probability of two random variables, x and y, factorizes $P(x,y) = P(x)P(y)$, then $\langle xy\rangle = \langle x\rangle\langle y\rangle$.

Exercise 2.2: If two random variables are statistically independent, $P(x,y) = P(x)P(y)$, demonstrate that the marginal distribution, $P(x)$, is normalized independently of $P(y)$.

An alternative way to evaluate the temperature dependence of the average spin is to perform the counting for a system isolated from its surroundings and employ consistent thermodynamic definitions. To proceed in this way, we can rewrite the energy, fixed for an isolated system, in terms of the number of up spins $N_\uparrow$,

$$E = -\frac{\epsilon}{2} \sum_i s_i = -\frac{\epsilon}{2} \left[N_\uparrow - (N - N_\uparrow) \right]$$

$$= \frac{\epsilon}{2} N \left(1 - 2\frac{N_\uparrow}{N} \right) = \frac{\epsilon}{2} N \left(1 - 2f \right)$$

where we have introduced in the last equality, the fraction of spins pointing up, $f = N_\uparrow/N$. If the number of up spins is allowed to change, we can find the most likely arrangement by maximizing the entropy. The number of ways of arranging $N_\uparrow$ spins in N spins is given by the binomial coefficient

$$\Omega = \frac{N!}{(N - N_\uparrow)!\, N_\uparrow!}$$

and assuming that both N and $N_\uparrow$ are large, we can employ Stirling's approximation, $\ln N! \approx N \ln N - N$, to simplify the entropy. The resultant entropy is

$$S = k_{\mathrm{B}} \ln \Omega = -k_{\mathrm{B}} N \left[f \ln f + (1 - f) \ln(1 - f) \right]$$

which is a relation we will often see concerning the entropy of mixing two components, in this case up and down spins. The thermodynamic definition of the temperature is

$$T^{-1} = \left(\frac{\partial S}{\partial E} \right)_N = \left(\frac{\partial f}{\partial E} \right)_N \left(\frac{\partial S}{\partial f} \right)_N$$

$$= \frac{1}{\epsilon N} k_{\mathrm{B}} N \left[\ln f - \ln(1 - f) \right]$$

where in second line we have employed the chain rule to decompose the change in both S and E through their dependence on f. Solving this expression for f,

$$f = \frac{e^{\beta\epsilon}}{1 + e^{\beta\epsilon}} \quad \rightarrow \quad s = \frac{e^{\beta\epsilon/2} - e^{-\beta\epsilon/2}}{e^{\beta\epsilon/2} + e^{-\beta\epsilon/2}} = \tanh(\beta\epsilon/2)$$

which can be further rearranged noting that $s = 2f - 1$ to yield the most likely value of s as a function of the temperature. This is precisely what we determined for our open system, demonstrating an equivalence of the two procedures in the limit of the mean behavior. One tends to refer to the collections of global constraints imposed on a system as determining the ensemble, for example, fixed NVE or NVT, and thus we have found in the thermodynamic limit that ensembles are equivalent. We can choose the ensemble to do the calculation out of convenience.

2.2 Counting on a lattice

As another example of ensemble equivalence, consider a system of independent particles in a box, where both mass and energy are allowed to fluctuate. If we assume that the particles are sufficiently dilute that they do not interact with each other, then we

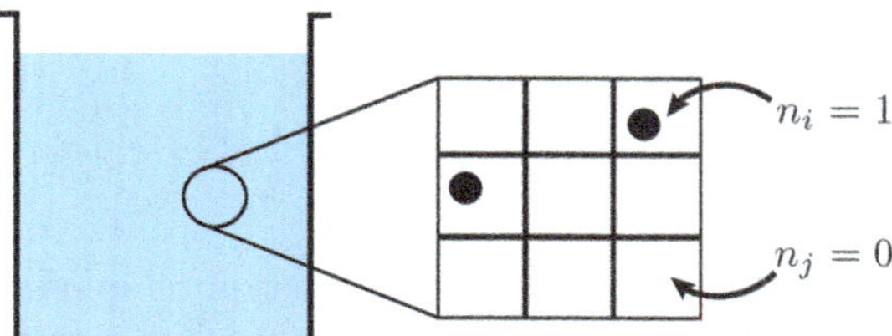

Fig. 2.2 System of dilute solutes in a fixed observation volume V held at constant T and with a fluctuating number of solutes N kept at constant $\beta\mu$.

can set their energy to a constant, taken here to be 0. We need only concern ourselves with how N fluctuates,

$$E(\nu) = 0 \qquad N(\nu) = \sum_{i=1}^{M} n_i$$

which we can define as a sum over M cells of an indicator function for cell i, $n_i = 0, 1$ depending on whether the cell is occupied or not. Let's set the volume of each cell as ℓ^3 such that the total volume is $V = \ell^3 M$. This system is illustrated in Figure 2.2. Starting with fixed energy, volume, and number of particles, the number of accessible microstates is

$$\Omega = \frac{M!}{(M - N)!\, N!}$$

from which, in the dilute limit $M \gg N \gg 1$, results in an entropy that is

$$S = k_{\mathrm{B}} N \left(\ln M - \ln N + 1 \right)$$

extensive in N. We can evaluate the pressure the solutes exert on the walls of the container by asking how the entropy varies with volume,

$$p = -T \left(\frac{\partial S}{\partial V} \right)_{N,E} = -T \left(\frac{\partial S}{\partial M} \right)_{E,N} \left(\frac{\partial M}{\partial V} \right)_{E,N}$$
$$= k_{\mathrm{B}} T \frac{N}{M} \frac{1}{\ell^3} = \frac{k_{\mathrm{B}} T N}{V}$$

where we find the *ideal gas law, $pV = N k_{\mathrm{B}} T$*, as just a consequence of counting particle configurations on a lattice.

Alternatively, we could evaluate the grand canonical partition function in the case that μ and T are fixed. The partition function requires that we sum over microstates defined as the string of M binary variables, $\nu = \{n_1, n_2, \ldots, n_M\}$,

$$\Xi(T, \mu, V) = \sum_{\nu} e^{-\beta E(\nu) + \beta \mu N(\nu)}$$

$$= \sum_{n_1=0,1} \sum_{n_2=0,1} \cdots \sum_{n_M=0,1} e^{\beta \mu \sum_1^M n_i}$$

$$= \left(\sum_{n_1=0,1} e^{\beta \mu n_1} \right)^M = (1 + z)^M$$

where in the third line we have employed the factorization of the exponential to recognize that the partition function can be written as M factors of the single cell partition function. In carrying out that sum, we have introduced the fugacity, $z = \exp[\beta \mu]$.

We can inquire about specific average properties of the system by evaluating derivatives of the partition function. For example, the average number of particles in the system is computable from,

$$\left(\frac{\partial \ln \Xi}{\partial \beta \mu} \right)_{T,V} = \langle N \rangle = M \frac{z}{1 + z}$$

which grows in proportion to the number of cells. Defining an intensive quantity, $f = \langle N \rangle / M$, the fraction of occupied cells, we can rearrange and find $f = z/(1 + z)$, or $z = f/(1 - f)$. Inserting what we mean by z, we find a relation for the chemical potential of the solutes,

$$\beta \mu = \ln[f/(1 - f)]$$

which for small f is just the logarithm of the fraction of occupied cells. The logarithm of the partition function is equal to the pV work available to the system, so

$$\beta p V = \ln \Xi$$
$$= -\ln(1 - f)^M = -M \ln(1 - f)$$

where in the second line we have substituted our relationship between z and f to find an expression for the pressure in terms for the fraction of occupied cells. In the dilute limit, $f \ll 1$ and $\ln(1 - f) \approx -f$, resulting in

$$\beta p V = M f \qquad pV = \langle N \rangle k_B T$$

nothing but the ideal gas law. In the thermodynamic limit of large systems, where fluctuations can be ignored, we have found another example of ensemble equivalence.

2.3 Counting in the continuum

Thus far we have considered partition functions that reduce to simple discrete counting problems. As a first step into more realistic systems, we can consider the thermal fluctuations associated with a collection of structureless gas particles held at fixed temperature T, volume V, and number of particles N. To tackle such a system, let's build up complexity slowly, and start simply with a single particle in a one-dimensional box. Already however, if we are to attempt a calculation of its partition function, we have a problem. How do we describe the microstate of the system, and associate with that microstate an energy? This ambiguity arises because we can describe the motion of the particle classically, associating a position and velocity to it, and thus $\nu = \{x, v_x\}$. Alternatively, we could describe its motion quantum mechanically, for which we have particle in a box states indexed by some quantum number, in which $\nu = n$.

Aside: Summing continuous variables. The midpoint rule for evaluating an integral states that for smooth functions $f(x)$,

$$\int_{x_1}^{x_2} dx\, f(x) = \Delta x \sum_i f(x_i)$$

which is valid in the limit that $\Delta x \to 0$, and the discrete sum indexes positions on the x line Δx apart between limits of integration. Rearranging this equality

$$\sum_i f(x_i) = \frac{1}{\Delta x} \int_{x_1}^{x_2} dx\, f(x)$$

we have a means of converting a discrete sum to an integral by introducing a discrete scale Δx, which is accurate if $f(x_i)$ varies slowly over Δx.

Let us start by proceeding classically. Classically, the energy function for a free particle in a box with mass m is

$$E = \frac{1}{2} m v_x^2 \qquad 0 \le x \le L$$

where we impose the constraint that the particle stays on the interval of the line in the x direction of length L. A picture of this setup is shown below in Figure 2.3. To

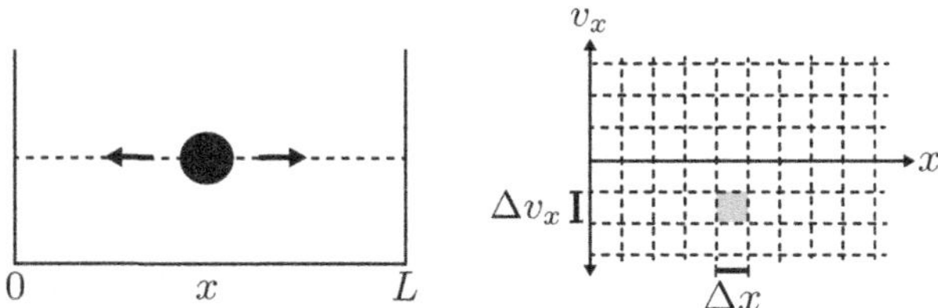

Fig. 2.3 Illustration of one-dimensional translational motion in a box of length L and the introduction of a scale in the phase space of x and v_x.

evaluate the particle's partition function we need to sum over all possible microstates, which in this case means all possible positions and all possible velocities. Here we have another problem, because there is an uncountably large number of both, as both are continuous quantities. For the time being, we will abstract that problem away by explicitly introducing a scale Δx to count unique positions and an analogous scale Δv

to count unique velocities. The joint x and v_x space we will refer to as phase space, and this discretization is illustrated in Figure 2.3.

Converting the sum over states into an integral we have

$$q_{\mathrm{CM}} = \sum_x \sum_{v_x} e^{-\beta E(v_x)}$$

$$= \frac{1}{\Delta x \Delta v} \int_0^L dx \int_{-\infty}^{\infty} dv_x e^{-\beta m v_x^2/2} = \frac{1}{\Delta x \Delta v} L \sqrt{\frac{2\pi}{\beta m}}$$

where in the second line we introduced an identity associated with so-called Gaussian integrals, allowing us to evaluate the integral over v_x. Note the domain of integration for x consisted of the length of the line L while the domain of the velocity integral contained all possible real numbers. While a particle's velocity cannot exceed the speed of light, for particles of finite mass and temperature the likelihood of observing a velocity fluctuation of relativistic speeds is so small we can neglect those contributions. These integrals provide a functional dependence of the partition function on β and L of $q_{\mathrm{CM}} \propto L\beta^{-1/2}$. This β dependence implies immediately a temperature-dependent average energy,

$$\langle E \rangle = -\left(\frac{\partial \ln q_{\mathrm{CM}}}{\partial \beta}\right)_V = -\frac{\partial}{\partial \beta}\left(\ln \beta^{-1/2} + \dots\right)$$

$$= \frac{1}{2} k_{\mathrm{B}} T$$

which is linear in T and noticeably independent of the arbitrary scale we used for counting. It is also independent of the mass of the particle, so this is equivalent for any structureless particle, regardless of identity. In this way, a classical particle's kinetic energy is a reporter on the temperature of the environment.

Aside: Gaussian integrals. Gaussian integrals show up often in physical science, so it is useful to know a few of them. Three particularly important ones are

$$I_1 = \int_{-\infty}^{\infty} dy\, e^{-(y-y_0)^2/2\sigma^2} = \sqrt{2\pi\sigma^2}$$

$$I_2 = \int_{-\infty}^{\infty} dy\, y\, e^{-(y-y_0)^2/2\sigma^2} = I_1 y_0$$

$$I_3 = \int_{-\infty}^{\infty} dy\, y^2\, e^{-(y-y_0)^2/2\sigma^2} = I_1\left(\sigma^2 + y_0^2\right)$$

which provide the normalization of a Gaussian distribution, or evaluate its mean or variance.

Now let us perform the calculation quantum mechanically. The microstate of the system is now the index $\nu = \{n\}$, which determines which energy eigenstate the particle resides in, with corresponding energy for a particle in a box,

$$E_n = \frac{h^2}{8mL^2} n^2$$

where h is Planck's constant. In this case there is not an explicit problem with writing down the partition function, as n is a countable infinity,

$$q_{\mathrm{QM}} = \sum_{n=1}^{\infty} e^{-\beta h^2 n^2 / 8mL^2}$$

however, evaluating this sum is not easy. Assume that the particle is confined to a macroscopic length, so that at room temperature $h^2/8mL^2 \ll k_{\mathrm{B}}T$. In this case the variation of the argument of the sum is slowly varying in n and the sum can be reasonably approximated as an integral, yielding,

$$q_{\mathrm{QM}} = \int_0^{\infty} dn \, e^{-\beta n^2 h^2 / 8mL^2} = \frac{1}{2}\sqrt{\frac{8mL^2}{\beta h^2}}$$

which in the second equality is evaluated using the same Gaussian integral identity as employed in the classical calculation. We find that the functional dependence of q_{QM} on β and L is $q_{\mathrm{QM}} \propto L\beta^{-1/2}$, identical to that found classically. It follows directly that $\langle E \rangle = k_{\mathrm{B}}T/2$.

It is peculiar that the results of the partition function we have obtained quantum mechanically and classically are so similar. In general, they are not the same, however if we consider the specific limit we are working in, this can be clarified. A fundamental distinction between quantum and classical mechanics is the discrete nature of variables in the former. The energy for the particle in the box obtains only a particular set of finite values, while classically since the particle's velocity is a continuous number, its kinetic energy is also a continuously varying quantity. However, when the thermal energy $k_{\mathrm{B}}T$ is much larger than the energy spacing, $h^2/8mL^2 \ll k_{\mathrm{B}}T$, the spectrum approaches an effective continuum, and this distinction vanishes. Indeed, in this limit, if we choose our phase space resolution in a particular way

$$\Delta x \Delta v = h/m \quad \rightarrow \quad q_{\mathrm{QM}} = q_{\mathrm{CM}}$$

the partition functions are equal. Choosing $\Delta x \Delta v = h/m$ of course carries a special meaning. This is nothing but *Heisenberg's uncertainty principle*, which suggests that at a fundamental level one cannot distinguish states whose phase space density is resolved beyond units of h. Note that if T or L were very small, we would expect to see deviations from the classical result, manifesting the discreteness of the energy levels.

Exercise 2.3: Evaluate the claim explicitly that $h^2/8mL^2 \ll k_{\mathrm{B}}T$ using a particle of mass 1 amu, held at room temperature, and a box of length $L = 1\mathrm{m}$. Specifically, calculate the quantum number n^* for which $\beta E_{n^*} = 1$, and estimate the number of states, M_x that are accessible in the course of thermal fluctuations.

Given that for a macroscopic system at room temperature, the classical approximation to the partition function is accurate, let us proceed with generalizing it. First let's scale up to three spatial dimensions. For a three-dimensional system, the microstate is described by $\nu = \{x, y, z, v_x, v_y, v_z\}$, a vector of six numbers. The energy associated with each state is

$$E = \frac{1}{2}m\left(v_x^2 + v_y^2 + v_z^2\right)$$

provided the position of the particle in within the volume. If we choose the same phase space resolution for counting states in each direction, the partition function becomes

$$q = \frac{1}{\Delta x^3 \Delta v^3} \int_0^L dx \int_0^L dy \int_0^L dz \int_{-\infty}^{\infty} dv_x \int_{-\infty}^{\infty} dv_y \int_{-\infty}^{\infty} dv_z e^{-\beta m v^2/2}$$

$$= q_x q_y q_z q_{v_x} q_{v_y} q_{v_z} = \frac{V}{\Delta x^3 \Delta v^3}\left(\frac{2\pi}{\beta m}\right)^{3/2}$$

where in the second line we note that q factorizes into individual contributions from each dynamical degree of freedom. This is due to the fact that the energy is additive for each of the components of the velocity. This results in a product of three one-dimensional partition functions. Written compactly, and using the uncertainty principle to define the resolution, $\Delta x \Delta v = h/m$, we find

$$q = \frac{V}{\lambda(T)^3} \qquad \lambda(T) = h/\sqrt{2\pi m k_B T}$$

where a natural lengthscale λ results. The length λ has the interpretation as the new effective resolution for counting positional states and is known as the *thermal wavelength*. It quantifies the uniqueness of a positional state, having accounted for quantum uncertainty and that due to velocity fluctuations.

For a system of N structureless particles, the kinetic energy is given by a sum of each particle's kinetic energy,

$$E = \frac{1}{2}m \sum_{i=1}^{N} |\mathbf{v}_i|^2 = K_1 + K_2 + \ldots K_N$$

meaning that velocity fluctuations of each particle are independent, or that the partition function again factorizes. Denoting q as the partition function for an individual particle, Q becomes

$$Q = \frac{q^N}{N!} \qquad q = \frac{V}{\lambda^3}$$

where we have accounted for the indistinguishability of the identical particles by dividing by $N!$, something we will consider further later. Very generally, a classical system of interacting particles has an energy function that is separable into a part that depends on the momentum and a potential that depends on position,

$$E = K(\mathbf{v}_1, \mathbf{v}_2, \ldots, \mathbf{v}_N) + U(\mathbf{r}_1, \mathbf{r}_2, \ldots, \mathbf{r}_N)$$

which means that even for an interacting system, velocities are uncorrelated with positions. While dynamically Newton's equations require that positions and velocities depend intimately on each other, for a classical thermal system, chaos results in them decoupling. Because of this, the integrals over the velocities can always be done independently of those over the positions,

$$
\begin{aligned}
Q &= \frac{1}{(h/m)^{3N}} \left(\int d\mathbf{v}_1 \int d\mathbf{v}_2 \ldots e^{-\beta K} \right) \left(\int d\mathbf{r}_1 \int d\mathbf{r}_2 \ldots e^{-\beta U} \right) \\
&= \frac{1}{N!} \frac{Q_{\text{config}}}{\lambda^{3N}} \qquad Q_{\text{config}} = \int d\mathbf{r}_1 \int d\mathbf{r}_2 \ldots e^{-\beta U}
\end{aligned}
$$

and all of the nontrivial information concerning the interparticle correlations is embedded into the configurational part of the partition function, Q_{config}. Consequently, the energy of the system is additive,

$$
\begin{aligned}
\langle E \rangle &= - \left(\frac{\partial \ln Q}{\partial \beta} \right)_{N,V} = \left(\frac{\partial \ln \lambda^{3N}}{\partial \beta} \right)_{N,V} - \left(\frac{\partial \ln Q_{\text{config}}}{\partial \beta} \right)_{N,V} \\
&= \frac{3}{2} N k_{\text{B}} T + \langle U \rangle
\end{aligned}
$$

and each velocity degree of freedom contributes $k_{\text{B}}T/2$ to the average kinetic energy providing a robust microscopic relation to the temperature.

2.4 Maxwell–Boltzmann distribution

The fact that the energy function of a classical system is a sum of a kinetic energy function that depends only on the particle's velocities and a potential energy that depends only on the particle's position,

$$
E = K(\mathbf{v}_1, \mathbf{v}_2, \ldots, \mathbf{v}_N) + U(\mathbf{r}_1, \mathbf{r}_2, \ldots, \mathbf{r}_N)
$$

means that the velocities are uncorrelated with the positions. Further, that the kinetic energy is a sum of individual particle components,

$$
K(\mathbf{v}_1, \mathbf{v}_2, \ldots, \mathbf{v}_N) = \sum_{i=1}^{N} \frac{1}{2} m \mathbf{v}_i^2
$$

implies that the probability of the velocities factorizes

$$
P(\mathbf{v}_1, \mathbf{v}_2, \ldots, \mathbf{v}_N) = P(\mathbf{v}_1) P(\mathbf{v}_2) \cdots P(\mathbf{v}_N)
$$

and so each particle's velocity is independent of the others. The distribution of velocities are also independent for each direction,

$$
\begin{aligned}
P(\mathbf{v}) &= P(v_x) P(v_y) P(v_z) \\
&= \frac{1}{(2\pi/\beta m)^{3/2}} e^{-\beta m v_x^2/2} e^{-\beta m v_y^2/2} e^{-\beta m v_z^2/2}
\end{aligned}
$$

and given by a Gaussian distribution. This means that the average velocity in any direction is 0, while the variance of each component is $\langle v_x^2 \rangle = k_{\text{B}}T/m$. The distribution of particle speeds, the magnitude of the velocity vector, $v = |\mathbf{v}|$, is given by a product

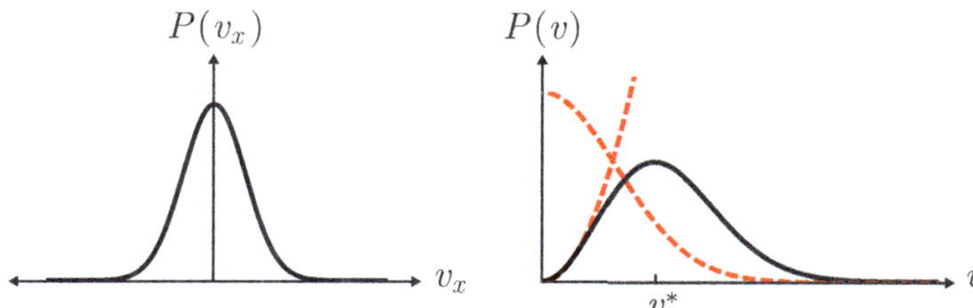

Fig. 2.4 Distribution of molecular velocities, v_x, and speeds, $v = |\mathbf{v}|$.

of the Boltzmann factor and a factor that accounts for the number of ways to point that velocity vector, which in three dimensions grows as v^2, yielding

$$P(v) \propto v^2 e^{-\beta m v^2/2} \qquad v = \sqrt{v_x^2 + v_y^2 + v_z^2}$$

the so-called *Maxwell–Boltzmann distribution*. Maximizing that distribution, the most probable speed is found to be $v^* = \sqrt{2k_{\mathrm{B}}T/m}$. Both distributions of speeds and velocities are shown Figure 2.4.

Exercise 2.4: Evaluate the distribution of speeds, $P(v)$, for a particle confined to 2d, where $v = \sqrt{v_x^2 + v_y^2}$, in order to generalize the Maxwell–Boltzmann distribution.

2.5 Harmonic oscillator

Another model system we can consider explicitly is a collection of harmonic oscillators. As before we have a choice for how to proceed—treating the dynamics classically or quantum mechanically. From our translational motion calculation, we have an expectation that at temperatures higher than a characteristic energy spacing that the quantum and classical results should agree. As we did previously, we will first compute the partition function for a single harmonic oscillator classically in one dimension and then proceed quantum mechanically.

For a classical system, the microstate is defined by a position and a conjugate momentum $\nu = \{x, p_x\}$. The corresponding energy function is a sum of kinetic and potential energies

$$E = \frac{1}{2m}p_x^2 + \frac{1}{2}m\omega^2 x^2$$

where m is the mass of the particle, ω the characteristic frequency of the harmonic motion. Again, to count microstates, we have to introduce a scale. We will count positions with a resolution Δx and momentum with a resolution Δp. In this resolution the classical partition function is

$$q_{\mathrm{CM}} = \frac{1}{\Delta x \Delta p} \int_{-\infty}^{\infty} dx \int_{-\infty}^{\infty} dp_x \, e^{-\beta m \omega^2 x^2/2 - \beta p_x^2/2m}$$

$$= \frac{1}{\Delta x \Delta p} \sqrt{\frac{2\pi}{\beta m \omega^2}} \sqrt{\frac{2\pi m}{\beta}}$$

where we have evaluated the two Gaussian integrals yielding a functional dependence $q_{\mathrm{CM}} \propto \beta^{-1}$. From this dependence, the average energy follows as

$$\langle E \rangle = k_{\mathrm{B}} T$$

which, as we saw with translational motion, is an integer multiple of $k_{\mathrm{B}}T/2$. This general result is called the *equipartition theorem*, which states that any independent, classical, Gaussian degree of freedom contributes $k_{\mathrm{B}}T/2$ to the average energy and $k_{\mathrm{B}}/2$ to the heat capacity. For the harmonic oscillator, we have a position and momentum, both of which are Gaussian and are independent of each other yielding an average energy of $2 \times k_{\mathrm{B}}T/2$.

> **Exercise 2.5:** Compute the average energy and heat capacity for a system with energy
> $$E = ay^2 \qquad -\infty \leq y \leq \infty$$
> where $a > 0$ is a positive constant an y the dynamical variable and show it obeys the equipartition theorem.

To proceed quantum mechanically, we note that the harmonic oscillator states are indexed by an integer, $\nu = \{n\}$, between 0 and ∞, with corresponding energy levels,

$$E_n = \hbar \omega (n + 1/2)$$

where $\hbar = h/2\pi$. The partition function follows as a discrete sum,

$$q_{\mathrm{QM}} = \sum_{n=0}^{\infty} e^{-\beta \hbar \omega (n+1/2)}$$

$$= e^{-\beta \hbar \omega/2} \sum_{n=0}^{\infty} e^{-\beta \hbar \omega n} = e^{-\beta \hbar \omega/2} \frac{1}{1 - e^{-\beta \hbar \omega}}$$

where in the second line we observe that the sum is a geometric series, which can be evaluated exactly. Generally, the quantum mechanical partition function is not equal to its classical counterpart. However, in high-temperature limit, we find

$$\lim_{\beta \to 0} q_{\mathrm{QM}} = \frac{1}{\beta \hbar \omega} \qquad q_{\mathrm{CM}} = \frac{2\pi}{\beta \omega \Delta x \Delta p}$$

where once again choosing a phase space resolution of $\Delta x \Delta p = h$ we arrive at an equivalence between q_{QM} and q_{CM}. In the opposite limit of low temperature, the partition function can be expanded as,

$$\lim_{\beta \to \infty} q_{\mathrm{QM}} = e^{-\beta\hbar\omega/2} + e^{-3\beta\hbar\omega/2} + \ldots$$

where only the first few terms would contribute significantly as each subsequent term is exponentially smaller. The temperature dependence of the average energy and heat capacity are shown in Figure 2.5. These demonstrate that corrections to the classical limit are important for $k_{\mathrm{B}}T \lesssim \hbar\omega$.

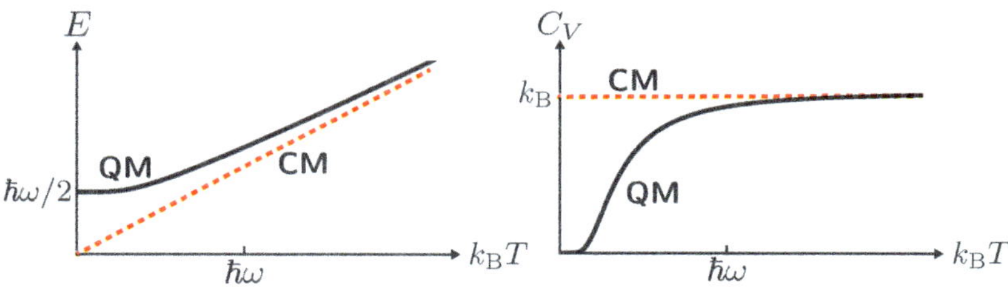

Fig. 2.5 Energy and heat capacity of a single harmonic oscillator as a function of temperature as it transitions between quantum and classical behavior.

At low temperatures, only the ground state and the first excited state have appreciable probability. In this limit the harmonic oscillator is equivalent to a two-level system. Consider, for example, a two-level system with a ground state energy set to 0 and a first excited state at $\Delta\epsilon$. The partition function would be

$$q = 1 + e^{-\beta\Delta\epsilon}$$

and its corresponding average energy is

$$\langle E \rangle = \Delta\epsilon \left(1 + e^{\beta\Delta\epsilon}\right)^{-1}$$

while the fluctuations of the energy are

$$\langle \delta E^2 \rangle = \Delta\epsilon^2 \frac{e^{\beta\Delta\epsilon}}{\left(1 + e^{\beta\Delta\epsilon}\right)^2} \approx \Delta\epsilon^2 e^{-\beta\Delta\epsilon}$$

where in the second equality we have taken the low temperature limit.

Harmonic oscillators are effective approximations for physical systems fluctuating with small amplitudes around their ground states. For example, consider a diatomic, with intramolecular potential $U(r)$ that depends on the separation distance r. While in general $U(r)$ has a complex functional form, for small displacements around its minima,

$$U(r) \approx U_m + \frac{1}{2}m\omega^2(r - r_m)^2$$

where U_m is the minima of the potential, r_m the location of the minima, and $m\omega^2$ related to the curvature around the minima as pictured in Figure 2.6. Typically, intramolecular vibrations are very stiff, so harmonic approximations are accurate. Further, typical frequencies are 1000s of cm^{-1}, while $k_{\mathrm{B}}T = 210$ cm^{-1}, which means that most intramolecular vibrations are in their ground states with excited states rarely populated, since $\exp[-\beta\hbar\omega] \ll 1$.

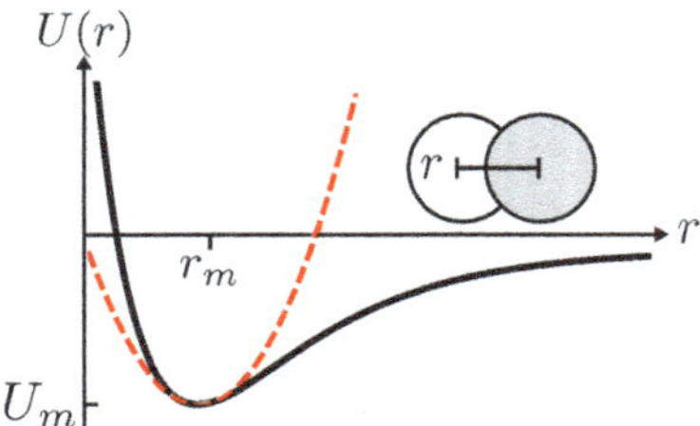

Fig. 2.6 Harmonic approximation to an intramolecular bonding potential.

Another system for which a harmonic approximation is valid is a low-temperature solid. Whether a crystal with regular structure, or an amorphous solid which is aperiodic, at low temperature the size of displacements is expected to be small. Therefore, if $\mathbf{r}_i$ is the position of the ith particle in the solid, we need to only consider fluctuations, $\delta\mathbf{r}_i = \mathbf{r}_i - \mathbf{r}_i^\mathrm{o}$, around a local minimum energy position, $\mathbf{r}_i^\mathrm{o}$. In that limit, the potential energy function for a collection of particles can be approximated as

$$U(\{\mathbf{r}_i\}) \approx U_0(\{\mathbf{r}_i^\mathrm{o}\}) + \frac{1}{2}\sum_{i,j}\delta\mathbf{r}_i\mathbf{K}_{ij}\delta\mathbf{r}_j$$

where U_0 is local minimum potential energy, and $\mathbf{K}_{ij}$ is a stiffness matrix of second derivatives. While in general the stiffness matrix is complicated, it is symmetric, and so can be diagonalized as explored next chapter. The linear combination of displacements that diagonalize $\mathbf{K}$ are known as normal modes, which for a solid are called *phonons*. Let ξ_α be the α'th normal mode, then the energy of the system can be written as

$$E = \frac{1}{2}\sum_{\alpha}m_\alpha\dot{\xi}_\alpha^2 + m_\alpha\omega_\alpha^2\xi_\alpha^2$$

which is a sum of independent harmonic oscillators with mass m_α and frequency ω_α. It follows from the equipartition theorem that the energy of a collection of N uncoupled harmonic oscillators in three dimensions is

$$\langle E \rangle = 3Nk_\mathrm{B}T \qquad C_V = \left(\frac{\partial\langle E\rangle}{\partial T}\right)_{V,N} = 3Nk_\mathrm{B}$$

where the corresponding molar heat capacity is $3R$. This limiting behavior is routinely observed and is known as the *Dulong Petit law.*

While at relatively high temperatures, the Dulong Petit law states that the heat capacity is a constant as a function of temperature, at low temperatures, quantum mechanical effects could render the heat capacity temperature dependent. In fact, measurements of many solids, particularly insulators, exhibit a strong temperature dependent C_V at low temperatures, with a characteristic T^3 scaling. This is shown in Figure 2.7.

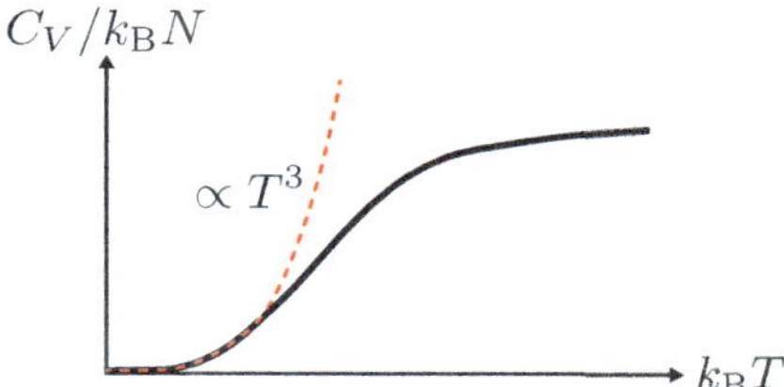

Fig. 2.7 Schematic temperature dependence of the heat capacity of a semiconducting solid.

In order to rationalize the observed T^3 scaling at low temperature, we can assume that at low temperatures the solid is effectively a collection of two-level systems. The heat capacity of a single two-level system with energy gap $\Delta\epsilon$ is

$$C_V/k_\mathrm{B} = (\beta\Delta\epsilon)^2 e^{-\beta\Delta\epsilon}$$

If we view a solid as a collection of independent two levels systems, then the heat capacity is just a sum over each contribution with gap $\hbar\omega$

$$C_V/k_\mathrm{B} = \int d\omega\, g(\omega)(\beta\hbar\omega)^2 e^{-\beta\hbar\omega}$$

where $g(\omega)d\omega$ is the density of states, or the number of modes with a frequency between ω and $\omega + d\omega$. In this light, understanding the low-temperature behavior is equivalent to understanding the density of states.

Shortly after the experimental observation of the strong decrease in heat capacity at low temperatures, prominent scientists proposed models for $g(\omega)$ and therefore C_V. Einstein proposed a model, now known as the Einstein crystal, in which each atom vibrated independently around its average position. As each atom is statistically indistinguishable, for a three-dimensional system of N particles, the low-frequency modes are all the same. Taking the characteristic frequency as ω_0, and the density of states to be

$$g(\omega) \approx 3N\delta(\omega - \omega_0)$$

where $\delta(\omega - \omega_0)$ is the delta function, the heat capacity is

$$C_V/k_\mathrm{B} = 3N(\beta\hbar\omega_0)^2 e^{-\beta\hbar\omega_0}$$

and goes to 0, as in the experiment. However, it decreases faster than T^3.

An alternative model, proposed by Debye, envisioned the low-frequency modes as collective, in which long-wavelength oscillations propagated throughout the whole solid. Such *phonons* were envisioned as standing waves, known now as acoustic modes, resulting in an increased density of low-frequency modes due to the spreading out of the oscillation over the full crystal. An illustration of the Einstein and Debye models is shown in Figure 2.8. Accounting for the number of standing waves in a

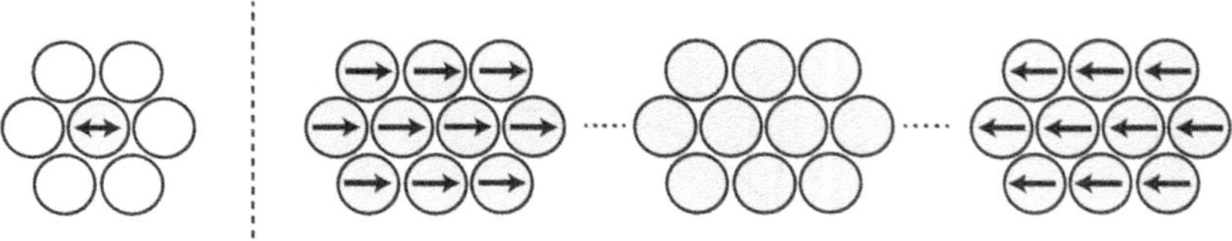

Fig. 2.8 Einstein (left) and Debye (right) models of solid vibrations.

three-dimensional volume and assuming a linear dispersion relating the frequency and wavelength, Debye arrived at a model for the density of states as,

$$g(\omega) \approx \begin{cases} 3N(\omega/\omega_0)^2 & \omega \le \omega_0 \\ 0 & \omega > \omega_0 \end{cases}$$

where ω_0 is a cutoff, or Debye frequency, which when inserted into the heat capacity expression yields,

$$C_V/k_B = 3N \int_0^{\omega_0} d\omega \left(\frac{\omega}{\omega_0}\right)^2 (\beta\hbar\omega)^2 e^{-\beta\hbar\omega}$$

$$= 3N(\beta\hbar)^{-3} \int du \frac{u^4}{\omega_0^2} e^{-u} \propto T^3$$

where in the second line we have employed a substitution $u = \beta\hbar\omega$, to find an agreement with the observed T^3 scaling up to a factor weakly dependent on T.

2.6 Molecular gases

Realistic models of gases must account for the complex internal structure associated with both the nuclei and the bound electrons. Overwhelmingly for most atoms and molecules, electronic excitation energies are large compared to $k_B T$ at room temperature. As a consequence, the partition function is dominated by their ground state contributions and electronic fluctuations can be ignored. The nuclei in principle can access different spin configurations, but likewise these excitation energies are large and so do not contribute meaningfully to the partition function. Within the Born–Oppenheimer approximation, the ground electronic state depends parametrically on the nuclear coordinates, forming an effective potential for the nuclei, which with the Hellman–Feynman theorem provides a means of solving for the motion of the nuclei. Fluctuations on this potential surface are in general expected to contribute at room temperature.

The effective energy function for a single molecule consisting of n atoms, for which the Born-Oppenheimer potential surface is $U(\mathbf{r}_1, \ldots, \mathbf{r}_n)$, is a sum of it and a contribution from the kinetic energy,

$$E = K(\mathbf{p}_1, \ldots, \mathbf{p}_n) + U(\mathbf{r}_1, \ldots, \mathbf{r}_n)$$

$$= \frac{1}{2M}\mathbf{P}^2 + K(\mathbf{p}_1 - \mathbf{P}, \ldots) + U(\mathbf{r}_1 - \mathbf{R}, \ldots)$$

where in the second line we have moved to a center of mass coordinate system. Introducing the total mass, $M = \sum_{i=1}^{N} m_i$, these coordinates are given by

$$\mathbf{R} = \frac{1}{M} \sum_{i=1}^{n} m_i \mathbf{r}_i \qquad \mathbf{P} = M\dot{\mathbf{R}}$$

where $\mathbf{R}$ is the center of mass position and $\mathbf{P}$ its corresponding momentum. The center of mass is free to translate, and thus behaves just like our previously considered structureless particle. Therefore the partition function for the molecular gas factorizes into two terms

$$q = q_{\text{trans}} q_{\text{int}}$$

where using the resolution for counting states in the continuum informed by quantum mechanics, $q_{\text{trans}} = V/\lambda^3$ and $\lambda = h/\sqrt{2\pi M k_{\text{B}} T}$. The second term, q_{int} counts the number of thermally accessible microstates from the internal motion of the molecule. For typical small molecules, this includes rotations and vibrations. As each typically evolves on a distinct timescale, to a good approximation each is uncoupled from each other, and thus $q_{\text{int}} = q_{\text{rot}} q_{\text{vib}}$ the internal partition function factorizes into a rotational part, q_{rot}, and a vibrational part, q_{vib}. If there are n atoms in the molecule, there are $3n$ degrees of freedom, 3 of which will be associated with the center of mass translations. Each degree of freedom has an associated position and a conjugate momentum. For a diatomic, $n = 2$, the remaining 3 will be partitioned into 2 rotational modes and 1 vibration. More generally, for $n > 2$, there will be 3 rotations about the center of mass, and the remaining $3n - 6$ degrees of freedom will be vibrations.

Since the partition function factorizes, the average energy of a molecular gas is a sum of the ground state energy and independent averages of the translational kinetic energy, rotational kinetic energy, and vibrational energies. As discussed previously, most molecular vibrations are stiff, and excitation energies are large compared to $k_{\text{B}} T$ at room temperature. As a consequence, vibrational energies do not fluctuate. Rotations and translations are well approximated for most heavy atoms as classical, and as a consequence both fluctuate appreciably and thus contribute to the heat capacity of the gas. From the equipartition theorem, the classical independence of each angular momentum term results in

$$C_V = C_{\text{trans}} + C_{\text{rot}}$$
$$= \frac{3}{2} k_{\text{B}} T + \frac{1}{2} n_d k_{\text{B}} T$$

where $n_d = 2$ or $n_d = 3$ depending on whether the molecule is a linear diatomic ($n_d = 2$) or has three distinct rotational axes ($n_d = 3$). This is the contribution for one structured gas molecule. As in a dilute gas the molecules are independent, the total heat capacity of N identical molecules is just N factors of this single molecule heat capacity. Indeed, the constant volume heat capacities of most small molecule gases, like Ar, Ne, N_2, CO, CO_2, H_2O, and others, are well-approximated by these simplifying forms.

Exercise 2.6: A diatomic molecule with a fixed bond length R and point masses m_1 and m_2 is mechanically equivalent to a three-dimensional rigid rotor, with moment of inertia $I = m_1 m_2 R^2 / (m_1 + m_2)$. Quantum mechanically, the energy levels for such a rigid rotor are denoted by the quantum number J and equal to

$$E = J(J+1)\hbar^2 / 2I \qquad J = \{0, ..., \infty\}$$

where the J'th level is $2J + 1$-fold degenerate. By approximating the sum as an integral, compute the quantum mechanical partition function, q_{rot}.

Exercise 2.7: Evaluate the classical rotational partition function for the rigid rotor whose energy function depends on free angles $0 \leq \phi \leq \pi$ and $0 \leq \theta \leq 2\pi$ and their conjugate angular momentum L_ϕ and L_θ

$$E = \frac{1}{2I} \left(L_\phi^2 + L_\theta^2 \right)$$

where I is the moment of inertia, and compare your result to the quantum mechanical partition function.

For a collection of N indistinguishable, classical particles, with a factorized single particle partition function of the form $q = q_{\mathrm{trans}} q_{\mathrm{int}}$, the full partition function is

$$Q = \frac{q^N}{N!} = \left(\frac{V}{\lambda^3} \right)^N \frac{q_{\mathrm{int}}^N}{N!}$$

or N factors of q divided by the number of ways to permute the N particles. As a consequence the Helmholtz free energy is

$$\beta A = -\ln Q$$
$$= -N \ln\left(V/\lambda^3 \right) + N \ln N - N - N \ln q_{\mathrm{int}}$$

where we have used Stirling's approximation assuming N is large. This particular form of the free energy isolates the N, V, and T dependences, lending to particularly simple, universal thermodynamic properties. For example, the average energy

$$\langle E \rangle = -N \frac{\partial}{\partial \beta} \ln\left(q_{\mathrm{int}}(T)/\lambda^3 \right)$$
$$= \frac{3}{2} k_{\mathrm{B}} T N - \frac{\partial}{\partial \beta} \ln q_{\mathrm{int}}(T)$$

will yield the results of the equipartition theorem plus its corrections, a function dependent on the temperature but not the volume. The pressure follows the V dependence,

$$\beta p = \left(\frac{\partial \ln Q}{\partial V} \right)_{T,N} = \frac{N}{V}$$

and is equal to the ideal gas law. We will explore the N dependence shortly.

2.7 Quantum indistinguishability

In considering the evaluation of partition functions quantum mechanically, we have restricted our calculations to single particles. This is because of a very important subtlety associated with distributing multiple quantum particles among discrete energy levels. To explore this, let us consider a collection of non-interacting identical particles. The Hamiltonian, $\mathcal{H}$, for a set of N independent particles is separable

$$\hat{\mathcal{H}} = \sum_{i=1}^{N} \hat{h}_i(\mathbf{r}_i)$$

where $\hat{h}_i$ is a single particle Hamiltonian. For a single particle Hamiltonian, the solution to the time-independent Schrödinger equation,

$$\hat{h}_i(\mathbf{r})\phi_j(\mathbf{r}) = \epsilon_j \phi_j(\mathbf{r})$$

defines a set of energy levels ϵ_j and eigenstates $\phi_j(\mathbf{r})$. These energy levels are common for each particle. The microstate of the system therefore is a set of numbers that convey how many particles are in each state. In order to evaluate the energy of the system, we need only sum the number of particles in each level, n_i, times the energy of that level, or

$$E = \sum_i n_i \epsilon_i \qquad N = \sum_i n_i$$

where we note the sum over all n_i should equal the total number of particles N. This description of the system presents two challenges for evaluating the canonical partition function. First, the constraint $N = \sum_i n_i$, will couple all of the sums, meaning that we will not be able to factorize the calculation, complicating its computation. This difficulty can be overcome by evaluating the grand canonical partition function instead. Second, and more fundamental, we must know what, if any, restrictions are on the occupation numbers.

To understand the restrictions on the occupation numbers, we need to think more carefully about what microstates are available to the set of quantum particles. As the particles are identical, they are indistinguishable, and so the state of the system should be invariant to swapping the particle labels. Let $\hat{\mathcal{P}}_{ij}$ be the permutation operator that takes a wavefunction and swaps the labels associated with the i'th and j'th particles. Since the energy is invariant to permutations of the labels these operators commute, $[\mathcal{H}, \hat{\mathcal{P}}_{ij}] = 0$ and both operators share a set of eigenstates. Acting $\hat{\mathcal{P}}_{ij}$ on an eigenstate $\Psi(\mathbf{r}_1, \ldots, \mathbf{r}_i, \ldots, \mathbf{r}_j, \ldots)$, returns

$$\hat{\mathcal{P}}_{ij}\Psi(\mathbf{r}_1, \ldots, \mathbf{r}_i, \ldots, \mathbf{r}_j, \ldots) = p\Psi(\mathbf{r}_1, \ldots, \mathbf{r}_j, \ldots, \mathbf{r}_i, \ldots)$$

where p is some eigenvalue associated with the operation of $\hat{\mathcal{P}}_{ij}$. Acting $\hat{\mathcal{P}}_{ij}$ on the state twice,

$$\hat{\mathcal{P}}_{ij}^2 \Psi(\mathbf{r}_1, \ldots, \mathbf{r}_i, \ldots, \mathbf{r}_j, \ldots) = p^2 \Psi(\mathbf{r}_1, \ldots, \mathbf{r}_i, \ldots, \mathbf{r}_j, \ldots)$$

returns the original state, so that we can conclude that $\hat{\mathcal{P}}_{ij}$ is unitary and $p^2 = 1$ or $p = \pm 1$.

There are thus two fundamentally different kinds of solutions to the many-particle Schrödinger equation, those that are even under particle permutations and those that are odd. Together with the linearity of the Schrödinger equation, this means that multiparticle eigenstate wavefunctions are permutation symmetrized superpositions of single particle wavefunctions, like $\phi_j(\mathbf{r})$. Take for example a $N = 2$ particle wavefunction,

$$\Psi(\mathbf{r}_1, \mathbf{r}_2) = \frac{1}{\sqrt{2}} \left[\phi_i(\mathbf{r}_1)\phi_j(\mathbf{r}_2) + p\phi_i(\mathbf{r}_2)\phi_j(\mathbf{r}_1) \right]$$

which generalized to N particles can be written as either determinates ($p = -1$), or permanates ($p = 1$), of single particle functions. In the simple two-particle case is it clear that for the permutationally odd wavefunction, $i \neq j$, otherwise the state would be 0. This implies that for $p = -1$ symmetric wavefunctions, particles cannot occupy the same state. There is no such restriction for $p = 1$ wavefunctions. These two cases refer to distinct kinds of particles—Fermions which obey the Pauli exclusion principle, $n = \{0, 1\}$, are odd under permutation, and Bosons that have no restriction on their occupations, $n = \{0, 1, \ldots, \infty\}$ and are even under permutation. For a given particle, determining whether it is a Fermion or Boson is the domain of the so-called spin statistics theorem of relativistic quantum mechanics, which is beyond the scope of this text. Nevertheless, we can explore the consequences of these occupation restrictions.

2.8 Fermi–Dirac statistics

Let us first consider the case of Fermions, particles like electrons, protons, and neutrons—indeed all half integer spin particles. The statistics associated with Fermions are called Fermi–Dirac statistics. The grand canonical partition function can be written as a sum over all possible strings of occupation numbers $n_i = 0, 1$ for all M energy levels, indexed by i, as

$$\Xi = \sum_{n_0=0,1} \sum_{n_1=0,1} \cdots \sum_{n_M=0,1} e^{-\beta \sum_i n_i \epsilon_i + \beta \mu \sum_i n_i}$$

$$= \prod_{i=0}^{M} \sum_{n_i=0,1} e^{-\beta(\epsilon_i - \mu)n_i} = \prod_{i=0}^{M} z_i$$

where in the second line we have used our same factorization trick to decouple the sums over each energy level and introduce a single level partition function, z_i. It can be easily evaluated

$$z_i = 1 + e^{-\beta(\epsilon_i - \mu)}$$

a function of the temperature, and energy of the i'th level relative to the chemical potential. Because the partition function factorizes in the grand canonical ensemble, each level is statistically independent, and properties of each level can be computed without regard to the other levels. For example, the average occupation of the i'th level is

$$\langle n_i \rangle = \left(\frac{\partial \ln z_i}{\partial \beta \mu} \right)_T$$
$$= \frac{1}{1 + e^{\beta(\epsilon_i - \mu)}} \equiv F(\epsilon_i)$$

which is commonly referred to as a *Fermi function*. Notice that for energies large compared to μ, $F(\epsilon_i) \approx \exp[-\beta(\epsilon_i - \mu)]$, or the occupation is a usual Boltzmann factor. Due to the binary nature of occupation numbers of Fermions, a reflection of the Pauli exclusion principle, the fluctuations about the average occupation are simple,

$$\langle \delta n_i^2 \rangle = \langle n_i^2 \rangle - \langle n_i \rangle^2$$
$$= \langle n_i \rangle \left(1 - \langle n_i \rangle \right)$$

because $n_i^2 = n_i$ for $n_i = 0, 1$.

Exercise 2.8: Using the property of the grand canonical partition function that $\beta p V = \ln \Xi$, show that in the high temperature limit where $\epsilon \gg \mu$, the ideal gas law holds for a box of independent electrons

$$\beta p V = \langle N \rangle$$

while in the low temperature limit where $\mu > \epsilon$, $\beta p V \to \infty$ even though the particles are not interacting. Explain the origin of incompressibility.

An important application of Fermi–Dirac statistics is the Fermi gas theory of electrons in a metal. The likelihood that a state with a given energy ϵ is occupied is given by the Fermi function $F(\epsilon)$, which depends on the temperature and chemical potential. A macroscopic system will have a typical number of electrons, $\langle N \rangle_\mu$ that depend on the temperature and chemical potential. Let's define this number at zero temperature as $\langle N \rangle_{\mu_o}$. For Fermions at zero temperature, this means the lowest $\langle N \rangle_{\mu_o}$ energy levels are filled, which is similar at finite but low temperatures. The *Fermi Energy* is the chemical potential μ_o, that returns the lowest $\langle N \rangle_{\mu_o}$ levels. At electron high density, when $\langle N \rangle$ is large, the corresponding Fermi energy will be large. Indeed, for most metals $\mu_o / k_B \approx 10,000K$. Since, $\langle \delta n^2 \rangle = \langle n \rangle (1 - \langle n \rangle)$ fluctuations are confined to $\epsilon = \mu \pm k_B T$, and if μ is large, then the density of the electrons that can participate in thermal fluctuations is small. This implies that electrons can be treated to a good approximation as non-interacting.

Using this insight that only electrons in a metal near the Fermi energy fluctuate, we can understand the heat capacity of a metal at low temperatures. To proceed, we can decompose the Fermi function into

$$F(\epsilon) = F_o(\epsilon) + \Delta f(\epsilon)$$

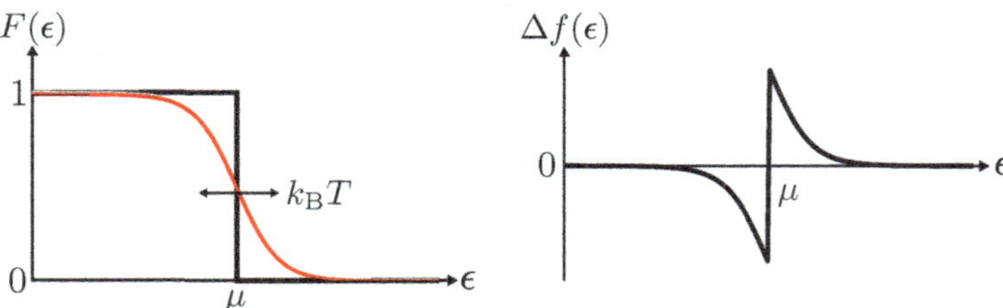

Fig. 2.9 Fermi function and its decomposition into a zero temperature and temperature-dependent part, $\Delta f(\epsilon)$.

where $F_\mathrm{o}(\epsilon)$ is the Fermi function at $T = 0$ and $\Delta f(\epsilon)$ the correction due to finite temperature. Note that $\Delta f(\epsilon)$ is odd about μ as shown in Figure 2.9. Given that we can assume that the electrons in the metal behave ideally, we can associate with them only kinetic energy. Their eigenstates thus correspond a particle in a three-dimensional box and can be indexed by three integers

$$\epsilon_\mathbf{k} = \frac{h^2}{8mL^2}\left(n_x^2 + n_y^2 + n_z^2\right) \qquad n_i = 1, 2, \ldots, \infty$$

$$= \frac{\hbar^2}{2m}\mathbf{k}^2 \qquad \mathbf{k} = \frac{\pi}{L}\left(n_x\hat{k}_x + n_y\hat{k}_y + n_z\hat{k}_z\right)$$

or as used in the second line, a wavevector. These eigenstates are standing waves that extend over the length of the metal. The system's energy is computable by summing over all of the eigenstate energies times their occupation numbers

$$\langle E \rangle = \sum_\mathbf{k} \epsilon_\mathbf{k}\langle n_\mathbf{k} \rangle = \sum_k G_k \epsilon_k \langle n_k \rangle$$

where the first sum over all wavevectors is replaced in the second sum over the magnitude of the wavevector noting that all wavevectors of the same magnitude have the same energy and thus the same average occupation number. We have also introduced G_k as the number of wavevectors with a given magnitude, k. In the large L limit, this sum can be replaced by an integral,

$$\langle E \rangle = \int dk\, g(k)\epsilon_k \langle n_k \rangle$$

where

$$g(k)dk = \left(\frac{L}{\pi}\right)^3 k^2 \frac{4\pi}{8} dk = \frac{V}{2\pi^2} k^2 dk$$

is the density of wavevectors. As the wavevectors are indexed over all possible combinations of positive integers in three dimensions, their number grows like k^2. An illustration of the density of states is shown in Figure 2.10.

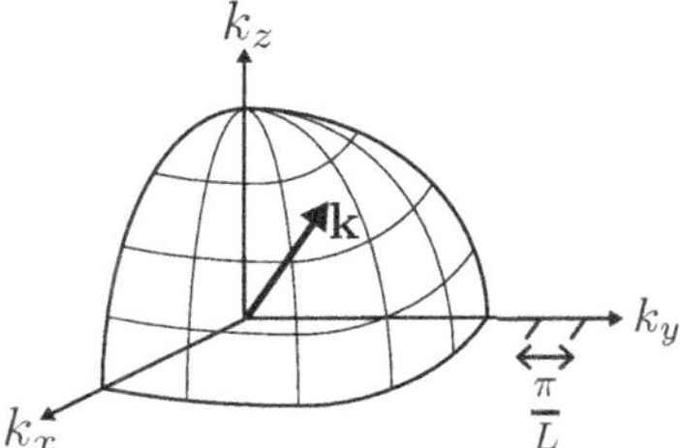

Fig. 2.10 Illustration of the degeneracy of wavevectors for standing wave modes.

Changing integration variables from k to ϵ, changes the density of states from $g(k)dk$ to $g(\epsilon)d\epsilon$. Introducing the Fermi function, the average energy is

$$\langle E \rangle = \int d\epsilon\, g(\epsilon)\epsilon F(\epsilon)$$

which is what we will need to take the temperature derivative of to evaluate the heat capacity. If we focus on the low-temperature behavior, only $\Delta f(\epsilon)$ for $\epsilon \approx \mu_{\mathrm{o}} \pm k_{\mathrm{B}}T$ is finite, so we can Taylor expand $g(\epsilon)$ in the vicinity of μ_{o},

$$\langle E \rangle = \int d\epsilon\, g(\epsilon)\epsilon \left[F_{\mathrm{o}}(\epsilon) + \Delta f(\epsilon)\right]$$

$$= E_{\mathrm{g}} + \int d\epsilon\, \left[g(\mu_{\mathrm{o}}) + g'(\mu_{\mathrm{o}})(\epsilon - \mu_{\mathrm{o}}) + \ldots\right]\epsilon \Delta f(\epsilon)$$

where we have identified E_{g}, the ground state energy, as the integral over $F_{\mathrm{o}}(\epsilon)$. In order to work out the low-temperature behavior of the heat capacity, we have only included the lower order terms in T. Let us make a substitution to factor out the temperature dependence completely, $x = \beta(\epsilon - \mu_{\mathrm{o}})$, or equivalently $\epsilon = \mu_{\mathrm{o}} + k_{\mathrm{B}}Tx$, yielding

$$\langle E \rangle = E_{\mathrm{g}} + k_{\mathrm{B}}T \int dx\, \left[g(\mu_{\mathrm{o}}) + g'(\mu_{\mathrm{o}})k_{\mathrm{B}}Tx + \ldots\right](\mu_{\mathrm{o}} + k_{\mathrm{B}}Tx)\Delta f(x)$$

$$= E_{\mathrm{g}} + (k_{\mathrm{B}}T)^2 \int dx\, x\, \left[g(\mu_{\mathrm{o}}) + \mu_{\mathrm{o}}g'(\mu_{\mathrm{o}})\right]\Delta f(x) + \mathcal{O}(T^4)$$

where in the last line we have noted that because $\Delta f(x)$ is an odd function, only when it is multiplied by another odd function will the integral provide something nonzero. This means that

$$C_V = \left(\frac{\partial \langle E \rangle}{\partial T}\right)_{V,N} \propto T$$

or that for a metal at low temperature, the heat capacity scales linearly in T. This scaling from the electronic degrees of freedom dominates the contributions from the phonons we worked out previously that scaled like T^3. Deviations from this linear behavior in a metal signal a breakdown in the Fermi gas theory, and correspondingly is an indication of correlated electron physics.

2.9 Bose–Einstein statistics

Now let us consider the case of Bosons, quasiparticles like photons, phonons, and excitons—indeed all integer spin particles. The statistics associated with Bosons are called Bose–Einstein statistics. The grand canonical partition function can be written as a sum over all possible strings of occupation numbers $n_i = 0, .., \infty$ for all energy levels, indexed by i, as

$$\Xi = \prod_{i=0}^{M} z_i \qquad z_i = \sum_{n_i=0}^{\infty} e^{-\beta(\epsilon_i - \mu)n_i}$$
$$= \frac{1}{1 - e^{-\beta(\epsilon_i - \mu)}}$$

where the single-level partition function is the form of a geometric series, which provided $\epsilon_i > \mu$, converges. As in the case of Fermions, the factorization of the total partition function implies that the occupation statistics of each level are independent of the others. The average occupation of the i'th level is

$$\langle n_i \rangle = \left(\frac{\partial \ln z_i}{\partial \beta \mu} \right)_T$$
$$= \frac{1}{e^{\beta(\epsilon_i - \mu)} - 1}$$

which is of a peculiar form. At low temperature, and as $\epsilon_i \to \mu$ from above, $\langle n_i \rangle \to \infty$! That is under these conditions, particles pile into the i'th level, and provided an infinite particle reservoir, an infinite number of particles can *condense* into this single level. This phenomena is known as *Bose–Einstein condensation*, and is the reason underlying such exotic behaviors as superconductivity and superfluidity. Both the Fermi–Dirac and Bose–Einstein distributions are shown in Figure 2.11.

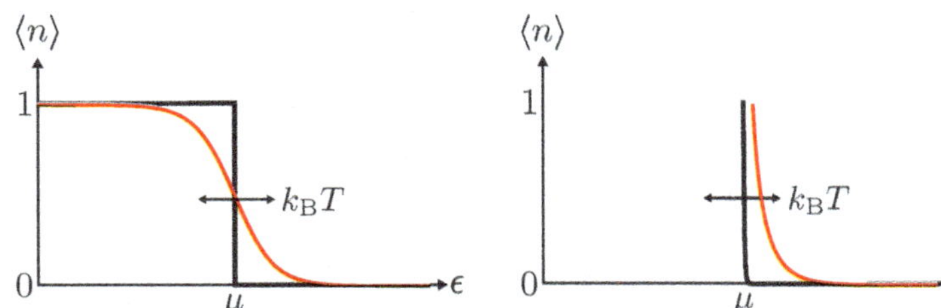

Fig. 2.11 Fermi–Dirac (left) and Bose–Einstein (right) distribution for a high (red) and low (black) temperature. Both decay as $\exp[-\beta(\epsilon - \mu)]$, for $\epsilon \gg \mu$.

In the limit that the occupancy of the i'th level is very small, which is achievable if $\epsilon_i \gg \mu$, then the average occupancy is equal to

$$\langle n_i \rangle \approx e^{-\beta(\epsilon_i - \mu)}$$

just as we found for Fermions. Employing this result for the average number of particles in the system, which is computable from summing over the individual average level occupancy,

$$\langle N \rangle = \sum_i \langle n_i \rangle$$

$$\approx e^{\beta \mu} \sum_i e^{-\beta \epsilon_i}$$

we can take a ratio of $\langle n_i \rangle$ to $\langle N \rangle$ to arrive at an estimate for the probability that the i'th level is occupied

$$p(\epsilon_i) = \frac{\langle n_i \rangle}{\langle N \rangle} \approx \frac{e^{-\beta \epsilon_i}}{\sum_i e^{-\beta \epsilon_i}}$$

which is nothing but the result afforded by classical statistical mechanics ignoring the potential restrictions from the Pauli exclusion principle. This result follows from a dilute limit in energy space, a limit in which the number available eigenstates is large compared to the number of particles so the likelihood that two particles occupy the same level is small.

Another consequence of the low-density limit can be understood by relating the canonical and grand canonical ensembles,

$$\ln \Xi = \ln Q + \beta \mu \langle N \rangle$$

which follows from the Laplace transform structure of changing ensembles in the thermodynamic limit where fluctuations can be ignored. Rewriting Ξ in the limit that the average occupancy is small, for either Bosons or Fermions,

$$\ln \Xi = \pm \sum_{i=0}^{M} \ln[1 \pm e^{-\beta(\epsilon_i - \mu)}]$$

$$\approx \sum_{i=0}^{M} e^{-\beta(\epsilon_i - \mu)} = \sum_{i=0}^{M} \langle n_i \rangle$$

and inserting this into the expression for Q,

$$\ln Q = -\beta \mu \langle N \rangle + \langle N \rangle$$

we have a relation dependent only on $\langle N \rangle$ and μ. However, these two values are themselves related, rearranging our previous expression for $\langle N \rangle$,

$$\beta \mu = \ln \langle N \rangle - \ln \sum_i e^{-\beta \epsilon_i}$$

and plugging that into our equation for Q,

$$\ln Q = - \left(\ln \langle N \rangle - \ln \sum_i e^{-\beta \epsilon_i} \right) \langle N \rangle + \langle N \rangle$$

$$= -(\ln \langle N \rangle - \ln q) \langle N \rangle + \langle N \rangle = \langle N \rangle - \langle N \rangle \ln \langle N \rangle + \langle N \rangle \ln q$$

we can recognize the single-particle partition function q as a sum of Boltzmann factors over single particle microstates. Further in the large system limit we are working in, fluctuations are negligible $\langle N \rangle \to N$, and reminding ourselves of Stirling's approximation $\ln N! = N \ln N - N$, we can rewrite the expression for Q as precisely the classical partition function of N identical, non-interacting particles

$$Q = \frac{q^N}{N!}$$

where in this case the $N!$ comes from particle indistinguishability. Thus we find that in the limit that the accessible set of microstates is dense compared to the available number of particles, so that the likelihood that any individual state is occupied is low, we recover classical statistics.

The classical limit for free particles can be considered concretely. As employed in the Fermi gas theory, the energy of a free quantum particle can be written as

$$E = \frac{\hbar^2}{2m}\mathbf{k}^2 \qquad g(k) = \frac{V}{2\pi^2}k^2 \qquad \mathbf{k} = \frac{\pi}{L}\mathbf{n}$$

where we have fixed the volume at L^3. We can compute the average number of particles in the system in the high-temperature dilute limit

$$
\begin{aligned}
\langle N \rangle &= \int_0^\infty dk\, g(k) e^{-\beta(E_k - \mu)} \\
&= \frac{V}{2\pi^2} \int_0^\infty dk\, k^2 e^{-\beta \hbar^2 k^2 / 2m} e^{\beta \mu} \\
&= V e^{\beta \mu} \left(\frac{2\pi m}{\beta h^2} \right)^{3/2}
\end{aligned}
$$

where we have introduced the density of states and performed the Gaussian integral. As we have seen with the classical gas, the thermal wavelength naturally non-dimensionalizes the volume, $\lambda = h/\sqrt{2\pi m k_B T}$. Defining the density as $\rho = \langle N \rangle / V$, we find that the average density is

$$\rho \lambda^3 = e^{\beta \mu}$$

and the condition for the high-temperature limit implies a condition on the density, $\rho \ll 1/\lambda^3$. For a small molecule gas, $\lambda = \mathcal{O}(\text{Å}/10)$ at room temperature and atmospheric pressure, and so for typical gas densities, $\rho \approx 1/1000\text{Å}^3$, this condition is easily met. Even for most liquid densities, $\rho \approx 0.1/\text{Å}^3$ this is satisfied. For particularly light elements like He at low temperature, this condition can break down, signaling the importance of quantum statistics.

2.10 Mass equilibrium

For a collection of independent particles with internal structure, we have found that the single particle partition function factorizes

$$q = q_{\text{trans}} q_{\text{int}}$$

into a translational part dependent on the volume

$$q_{\text{trans}} = \frac{V}{\lambda^3} \qquad \lambda = h/\sqrt{2\pi m k_{\text{B}} T}$$

and an internal part that is independent of the volume, though dependent on the temperature. The total partition function for N classical particles treated as indistinguishable, is

$$Q = \left(\frac{V}{\lambda^3}\right)^N \frac{q_{\text{int}}^N}{N!}$$

where q_{int} is the partition function for the internal set of states of a single particle. Note that while the quantum mechanics of fundamental particles informs the factor of $N!$, it should be more generally considered a consequence of how we measure properties of particles. In the following we will assume that we choose not to distinguish particles even if in principle we could, and unpack its consequences. The resulting N dependence of Q provides an ideal form for the chemical potential

$$\beta\mu = -\left(\frac{\partial \ln Q}{\partial N}\right)_{T,N}$$
$$= -\ln(V/N) - \ln\left(q_{\text{int}}/\lambda^3\right)$$

where μ separates into a part dependent on the density, and a part dependent on the details of the particle's internal structure. Introducing a concentration, $1/v_{\text{o}}$,

$$\mu = k_{\text{B}} T \ln(\rho v_{\text{o}}) + k_{\text{B}} T \ln\left(\lambda^3/q_{\text{int}} v_{\text{o}}\right)$$
$$= k_{\text{B}} T \ln(\rho v_{\text{o}}) + \mu^{(\text{o})}$$

we define the so-called standard state chemical potential $\mu^{(\text{o})}$ that contains all of the complexities associated with the specific particle considered.

This ideal form for the chemical potential is universal for any set of independent particles. Consider an ideal solution, in which N particles can occupy any one of M cells of volume ℓ^3. Let the partition function for a cell with a solute in it be q_l while the partition function of a cell without a solute be q_0, which acknowledges that the presence of the solvent contributes a finite number of thermally accessible microstates. The N particle partition function is then,

$$Q = \frac{M^N}{N!} q_l^N q_0^{M-N} = \frac{M^N}{N!} \left(\frac{q_l}{q_0}\right)^N q_0^M$$

in the limit that $M \gg N$. Evaluating the chemical potential of the solutes yields

$$\beta\mu = -\ln(M/N) - \ln(q_l/q_0)$$

which can be put into the same ideal form by introducing a concentration, $1/\ell^3$,

$$\mu = k_{\text{B}} T \ln\left(\rho\ell^3\right) + k_{\text{B}} T \ln(q_0/q_l)$$
$$= k_{\text{B}} T \ln\left(\rho\ell^3\right) + \mu^{(\text{o})}$$

which defines a different standard state chemical potential. While the computation of $\mu^{(\text{o})}$ may be cumbersome, in this dilute limit one could imagine tabularizing it by measuring μ relative to $1/\ell^3$.

An important consequence of the ideal form of the chemical potential is the ease with which it can be used to establish the conditions of mass equilibrium, or how mass is spontaneously distributed through space, phases, and chemical components. As a concrete example, consider the partitioning of a solute between a liquid and vapor. If the solute is volatile, what determines the relative concentrations in the liquid and vapor phases? A schematic of such a system is shown below in Figure 2.12.

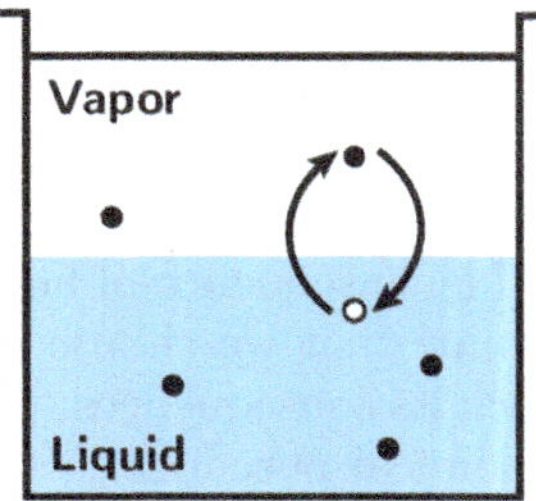

Fig. 2.12 Mass equilibration of a volatile solute between a liquid and a vapor.

Let us proceed by fixing N_v solutes in the vapor and N_l solutes in the liquid, such that the total number $N = N_v + N_l$ is fixed. If the vapor has a total volume V, and the liquid is gridded up into M cells, then the total partition function for the two-phase system is

$$Q = Q_v Q_l$$
$$= \left(\frac{V}{\lambda^3}\right)^{N_v} \frac{q_{\text{int}}^{N_v}}{N_v!} \frac{M^{N_l}}{N_l!} \left(\frac{q_l}{q_0}\right)^{N_l} q_0^M$$

where we have used our previous results for the partition functions of an ideal gas with internal partition function q_{int} and the partition functions of a filled and empty cell in solution, q_l and q_0. If the number of solutes in the liquid and vapor are both large, the Helmholtz free energy of the system is

$$\beta A = -N_v \ln\left(q_{\text{int}} V/\lambda^3\right) + N_v \ln N_v - N_v$$
$$-(N - N_v)\ln(M q_l/q_0) + (N - N_v)\ln(N - N_v) - (N - N_v) - M \ln q_0$$

where we have used Stirling's approximation and eliminated N_l using $N_l = N - N_v$. The Helmholtz free energy is parametrically dependent on N_v. If N_v is allowed to change, we can find the most probable value of N_v by minimizing A with respect to N_v,

$$\frac{\partial \beta A}{\partial N_v} = 0$$
$$= \ln N_v - \ln(N - N_v) + \ln(M q_l/q_0) - \ln\left(q_{\text{int}} V/\lambda^3\right)$$

rearranging, we recognize

$$\ln\left(N_v\lambda^3/q_{\text{int}}V\right) = \ln(N_l q_0/Mq_l) \quad \rightarrow \quad \mu_v = \mu_l$$

using the previously determined forms of the chemical potential. Thus, equilibrium is achieved when the chemical potential of the solutes in solution is equal to the chemical potential of the solutes in the vapor. Defining the concentrations in the vapor and solution as $\rho_v = N_v/V$ and $\rho_l = (N - N_v)/M\ell^3$, we find

$$\frac{\rho_v}{\rho_l} = \frac{\ell^3}{\lambda^3}\frac{q_0 q_{\text{int}}}{q_l} = K_{\text{H}}(T)$$

which is *Henry's law*, where $K_{\text{H}}(T)$ is the Henry's law constant that determines the solubility of solutes in solution provided a fixed concentration in the vapor. Defining a common standard state for the liquid and vapor, $q_v = \ell^3 q_{\text{int}}/\lambda^3$, the condition of equilibrium becomes

$$\frac{\rho_v}{\rho_l} = \frac{q_0 q_v}{q_l} = e^{-\beta \mathcal{W}_{\text{rev}}}$$

a ratio of partition functions for starting with the solute in the vapor plus the solution empty, $q_0 q_v$, and then moving the solute into the liquid, q_l. A ratio of partition functions is equivalent to a free energy change, or equally a reversible work, so the solubility of solutes in solution is given by the reversible work $\mathcal{W}_{\text{rev}}$ to drag a solute from a fixed point in the vapor to a fixed point in the solution, often referred to as the *solvation free energy*.

2.11   Chemical equilibrium

The insight gleaned from understanding how mass distributes itself spatially can be abstracted into the thermodynamics associated with repartitioning of mass into distinct chemical species. Specifically, consider a reaction in which b B molecules are converted into c C molecules,

$$bB \rightleftharpoons cC$$

where the double arrows signify the ability of the reaction to convert reactants to products or products back to reactants. If the reaction were initially at equilibrium at fixed T and p, the Gibbs free energy would be minimized. Starting from that equilibrium position, imagine repartitioning a small number of molecules, ΔN, between B and C, such that

$$N_B \rightarrow N_B - b\Delta N \qquad N_C \rightarrow N_C + c\Delta N$$

where the stoichiometry results in a loss of $b\Delta N$ B's and a gain of $c\Delta N$ C's. The corresponding change in the Gibbs free energy is

$$(\Delta G)_{T,P} = \left(\frac{\partial G}{\partial N_B}\right)_{T,p,N_C}(-b\Delta N) + \left(\frac{\partial G}{\partial N_C}\right)_{T,p,N_B}(c\Delta N)$$
$$= \mu_B\left(-b\Delta N\right) + \mu_C\left(c\Delta N\right) = 0$$

where we identify the thermodynamic definition of the chemical potential for B and C. Back at equilibrium, the Gibbs free energy change is 0, which occurs if,

$$b\mu_B = c\mu_c$$

namely, equilibrium is reached when the chemical potentials of reactants and products weighted by their stoichiometric factors are equal.

Exercise 2.9: Consider a slightly more complex reaction, also at fixed T and p,

$$aA + bB \rightleftharpoons cC + dD$$

and following the same analysis as before, demonstrate that

$$a\mu_A + b\mu_B = c\mu_C + d\mu_D$$

is the more general statement of equilibrium.

If both B and C are present under dilute solution conditions, we can employ our result for the ideal form of the chemical potential. Choosing a common standard state implies,

$$b\left[k_{\mathrm{B}}T\ln(\rho_B\ell^3) + \mu_B^{(\mathrm{o})}\right] = c\left[k_{\mathrm{B}}T\ln(\rho_C\ell^3) + \mu_C^{(\mathrm{o})}\right]$$

or rearranging,

$$\frac{\rho_B^b}{\rho_C^c} = (\ell^{-b+c})^3 e^{-\beta\Delta\mu_{\mathrm{rxn}}^{(\mathrm{o})}} = K_{\mathrm{eq}}(T) \qquad \Delta\mu_{\mathrm{rxn}}^{(\mathrm{o})} = b\mu_B^{(\mathrm{o})} - c\mu_C^{(\mathrm{o})}$$

we find the *law of mass action*, that the ratio of the concentrations of reactants and products is given by a composition independent equilibrium constant, $K_{\mathrm{eq}}(T)$. With the ideal form of the chemical potential, the equilibrium constant is a product of factors that define the standard states, like the length ℓ, and the chemical potential change of the reaction at standard state, $\Delta\mu_{\mathrm{rxn}}^{(\mathrm{o})}$. If the reaction were in the gas phase, and still dilute,

$$K_{\mathrm{eq}}(T) = \frac{\left(q_{\mathrm{int}}^{(B)}/\lambda_B^3\right)^b}{\left(q_{\mathrm{int}}^{(C)}/\lambda_C^3\right)^c}$$

the equilibrium constant could be evaluated by dividing the partition functions for the internal degrees of freedom for the reactants and products.

2.12 Potentials of mean force

A chemical reaction can be considered more microscopically by relating the equilibrium constant to a reversible work. For concreteness, consider a reaction in which reactants B and C can bind together and establish an equilibrium

$$B + C \rightleftharpoons BC$$

with a concentration of complexes BC. As before, we will imagine that all species are in dilute solution. Let us define the *complex BC* molecularly as B and C being within a cutoff distance, $r \leq r^*$. A cartoon of this setup is shown in Figure 2.13.

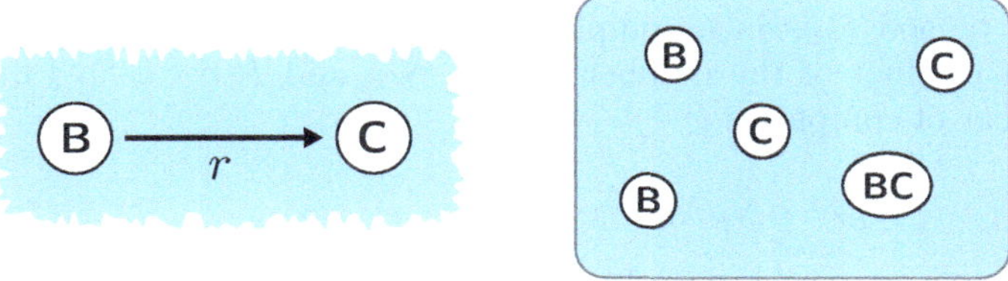

Fig. 2.13 Mass equilibration of a complexation reaction between free particles B and C, and complex BC.

The way in which B and C interact in general will be complicated, however, we can abstract that away by defining a partition function for a pair of B and C reactants at a fixed separation distance r,

$$Q(r) = e^{-\beta w(r)}$$

where the r dependent partition function $Q(r)$ is equivalent to a free energy $w(r)$. This free energy is the reversible work surface for moving a tagged B and C relative to each other, averaged over all of the motion of the surrounding solvent. Such a reversible work surface is also known as a potential of mean force and computable from the logarithm of the probability distribution of r.

Consider first an isolated pair. The probability of finding r less than r^* is given by

$$P(r < r^*) = \frac{\int_{r<r^*} d\mathbf{r}\, Q(r)}{\int_{r \in V} d\mathbf{r}\, Q(r)}$$

$$= 4\pi \frac{\int_0^{r^*} dr\, r^2 e^{-\beta w(r)}}{V e^{-\beta w(\infty)}} = \frac{4\pi}{V} \int_0^{r^*} dr\, r^2 e^{-\beta \Delta w(r)}$$

which is a sum over all of the states $r \leq r^*$, weighted by their Boltzmann factors, normalized by the total partition function. The solution has rotational symmetry and $w(r)$ depends only on the separation distance, so the angular dependence can be integrated out as done in the second line. Also in the second line, we have noted that for a macroscopic volume, and a potential of mean force that has a finite extent, the integral in the denominator will be dominated by the volume, V, times the limiting value of the free energy, $w(\infty)$. Finally, we have introduced $\Delta w(r) = w(r) - w(\infty)$.

Exercise 2.10: Assuming the potential of mean force is finite only over a distance r^*, or $w(r > r^*) = -\epsilon \Theta(r^* - r)$ where $\Theta(r)$ is a Heaviside step function, evaluate the integral

$$4\pi \int_0^R dr\, r^2\, e^{-\beta w(r)}$$

and demonstrate that for $R \gg r^*$ that it is well approximated by $4\pi R^3/3$.

If now there are many B's and C's, the number of complexes N_{BC} is just the number of B's, N_B, times the probability that a tagged B is in a complex. The probability that a tagged B is in a complex is the probability of finding a C with $r \leq r^*$, which is a product of the number of C's, N_C, and $P(r < r^*)$. Putting this together, the number of complexes is

$$N_{BC} = N_B N_C P(r < r^*)$$
$$= N_B N_C \frac{4\pi}{V} \int_0^{r^*} dr\, r^2 e^{-\beta \Delta w(r)}$$

where we have introduced our explicit form for the probability that a tagged B and a tagged C are within $r \leq r^*$. Dividing both sides by the volume,

$$\frac{\rho_{BC}}{\rho_B \rho_C} = 4\pi \int_0^{r^*} dr\, r^2 e^{-\beta \Delta w(r)} \equiv K_{\text{eq}}$$

we find an explicit microscopic expression for the equilibrium constant as an integral over the potential of mean force.

As an example, let's imagine that the potential of mean force is a simple square-well potential,

$$\Delta w(r) = \begin{cases} -\epsilon & r < r^* \\ 0 & r \geq r^* \end{cases}$$

where ϵ is an effective strength of binding between B and C. Inserting that into our expression of the equilibrium constant, we can perform the integration and

$$K_{\text{eq}} = \frac{4\pi}{3}(r^*)^3 e^{\beta \epsilon}$$

providing an interpretation of the standard state chemical potential change of the reaction as the reversible work to bring a tagged pair of reactants initially infinitely separated into contact. Also, we obtain an interpretation of the lengthscale defining the standard state in solution $\ell^3 = 4\pi(r^*)^3/3$.

Further reading

The calculation of partition functions of ideal systems is discussed in David Chandler's *Introduction to modern statistical mechanics*, Donald McQuarrie's *Statistical mechanics* and Ken Dill and Sarina Bromberg's *Molecular driving forces*. The original discussion regarding potentials of mean force is in John Kirkwood's *Statistical mechanics of fluid mixtures* in the Journal of Chemical Physics, 1935.

Additional exercises

Exercise 2.11: Consider a dilute vapor, for which the ideal gas law, $pV = N k_B T$, is a good approximation.

1. Ideality implies that internal energy E depends only on T and N, and not on p. Show that this is true by evaluating the derivative $(\partial E/\partial p)_{T,N}$. Use as a starting point the fundamental relation

$$dE = TdS - pdV + \mu dN$$

and divide by dp at fixed T and N:

$$\left(\frac{\partial E}{\partial p}\right)_{T,N} = T\left(\frac{\partial S}{\partial p}\right)_{T,N} - p\left(\frac{\partial V}{\partial p}\right)_{T,N}$$

From here you will need to employ a Maxwell relation and differentiate the ideal gas law.

2. Argue based on your result from part 1 and the property of extensivity that the energy of an ideal gas can be written in the form

$$E = N\epsilon(T)$$

3. The entropy of an ideal gas can be divided into two parts: that associated with intramolecular fluctuations, $S_{\text{intra}} = Ns(T)$, and that associated with distributing molecules in space, center of mass translation, S_{trans}. In a previous exercise you calculated S_{trans} for a dilute system to be

$$S_{\text{trans}} = -k_B V[\rho \ln(\rho v_0) - \rho],$$

where $\rho = N/V$ and v_0 is the volume of microscopic lattice cells introduced for the purpose of counting. Using that result, write the total entropy $S = S_{\text{intra}} + S_{\text{trans}}$ of our ideal gas as a function of p, N, and T. Your answer will involve the unspecified function $s(T)$.

4. Combining these results, compute the Gibbs free energy, $G = E - TS + pV$, of an ideal gas. Differentiate appropriately to determine the chemical potential $\mu(T, p)$.

5. Show that $\mu(T, p)$ can be written

$$\mu(T, p) = \mu^{(o)}(T) + k_B T \ln\left(\frac{p}{p_0}\right),$$

where p_0 is a reference value of pressure (e.g., 1 atm). Identify the quantity $\mu^{(o)}(T)$ in terms of T, v_0, and p_0.

6. Determine a condition of chemical equilibrium for H atoms and H_2 molecules interconverting through the reaction $2H \rightleftharpoons H_2$. Using that result, together with the ideal gas chemical potential you just computed, write a relationship between the partial pressures p_{H_2} and p_H of hydrogen molecules and atoms, respectively (assuming them to both be very dilute). Your answer should involve only p_{H_2}, p_H, and a quantity $K(T)$ that depends on temperature alone. Identify $K(T)$ in terms of T, $\Delta\epsilon_{\text{rxn}}(T) = \epsilon_{H_2} - 2\epsilon_H$, $\Delta s_{\text{rxn}}(T) = s_{H_2} - 2s_H$, and the parameter v_0.

Exercise 2.12: Consider a classical degree of freedom x, which can take on any value between $x = 0$ and $x = \infty$. (x might describe the position of a particle in one dimension, but could also represent something more complicated.) Variations away from $x = 0$ cost potential energy $u(x) = ax$, where a is a positive constant.

1. Calculate the canonical partition function q at temperature T.
2. By differentiating q appropriately, determine the average value of x, as well as the average potential energy $\langle u \rangle$.
3. By further differentiating, calculate the heat capacity C_V, as well as the variance $\langle \delta u^2 \rangle$ of energy fluctuations.

Exercise 2.13: Consider a one-dimensional model fluid in which N hard rods of width σ are located at lattice positions $x_1, x_2, \ldots, x_N$. The spacing between lattice sites is Δ. The rods are impenetrable, so that the possible values of x_j (with its neighbors fixed) are $x_{j-1} + \sigma, x_{j-1} + \sigma + \Delta, x_{j-1} + \sigma + 2\Delta, \ldots, x_{j+1} - \sigma - \Delta, x_{j+1} - \sigma$. The total length of the system is

$$L = \sum_{j=1}^{N-1} \ell_j,$$

where $\ell_j = x_{j+1} - x_j$ is the distance between centers of adjacent rods j and $j + 1$. There are no interaction energies other than those that prohibit different rods from overlapping.

1. Derive a partition function, $\zeta(\beta p, N)$, for an isothermal, and isobaric ensemble, where the temperature and pressure are fixed, and L fluctuates.
2. Compute the partition function of this fluid $\zeta(\beta p, N)$ for an ensemble in which $N \gg 1$ is constrained, but length is allowed to fluctuate under an applied pressure p. Take the first rod to be fixed at the origin. You may assume that the lattice spacing Δ is sufficiently small for sums to be well approximated by integrals:

$$\sum_{\ell=a,a+\Delta,a+2\Delta\ldots}^{b} f(\ell) \approx \frac{1}{\Delta} \int_a^b dx\, f(x).$$

   Write your answer in terms of βp, N, σ and Δ.
3. Explain why $\zeta(\beta p, N)$ does not depend on temperature.
4. Differentiate $\zeta(\beta p, N)$ to obtain the average density $\rho = N/\langle L \rangle$ as a function of βp and the parameter σ.
5. Compute the mean square fluctuation in length $\langle (\delta L)^2 \rangle$ in terms of ρ and σ. How does your result compare with fluctuations of a one-dimensional ideal gas, in particular when ρ approaches σ^{-1}?
6. Compute the isothermal compressibility $\kappa_T = (\partial \rho / \partial p)_{T,N}/\rho$ as a function of ρ.
7. Now consider attractive interactions between the hard rods. For simplicity, imagine that these attractions are very short in range, acting only when two rods are immediately adjacent to each other:

$$u_{\text{att}}(\ell) = \begin{cases} -\epsilon, & \text{if } \ell = \sigma \\ 0, & \text{otherwise} \end{cases}$$

   Calculate the partition function $\zeta(\beta, \beta p, N)$ in the presence of these attractions. For integrals that do not involve u_{att}, you need not worry about the distinction between σ and $\sigma + \Delta$ in the limits of integration,

$$\int_{a+\Delta}^{b} d\ell \; f(\ell) \approx \int_{a}^{b} d\ell \; f(\ell)$$

8. As before, differentiate $\zeta(\beta, \beta p, N)$ to obtain the average density $\rho = N/\langle L \rangle$ as a function of βp and the parameters σ, ϵ, and Δ.

9. Now differentiate ζ to determine the probability P_{contact} that a rod is contacted by its rightward neighbor, $x_j - x_{j-1} = \sigma$.

10. Rewrite your answers to parts 8 and 9 in terms of dimensionless variables: $p^* = \beta p \sigma$, $\rho^* = \rho \sigma$, and the parameter $a = (\Delta/\sigma)e^{\beta \epsilon}$ characterizing the strength of attractive forces. Make a plot of p^* vs. ρ^* over the density range $0 < \rho^* < 1$, for several values of a between 0 and 100. Plot P_{contact} as a function of p^* over the same range of conditions and attraction strengths.

11. What do your results suggest about the possibility of liquid-vapor phase transitions in one-dimensional fluids? Explain.

Exercise 2.14: Consider the simplest model for conformational fluctuations of a long chain molecule, namely, an ideal gas of n polymer segments connected end to end. Each segment $i = 1, 2, \ldots, n$ has a fixed length ℓ and an orientation $\hat{b}_i$ that is parallel to one of d Cartesian axes ($\hat{x}$, $\hat{y}$, or $\hat{z}$ in 3 dimensions). In other words the molecular configuration traces a random walk on a d-dimensional cubic lattice. Imagine that the orientations of different segments are statistically independent, and that there is no preferred orientation, $\langle \hat{b}_i \rangle = 0$ and $\langle \hat{b}_i \cdot \hat{b}_j \rangle = \delta_{ij}$, where $\delta_{ij} = 1$ if $i = j$ and vanishes otherwise.

1. Show that the entropy of such an ideal chain molecule has the form $S = k_{\text{B}} n \ln a$. Determine the parameter a as a function of dimensionality d.

2. Calculate the mean squared end-to-end distance of the chain molecule, $\langle R^2 \rangle$, where $\mathbf{R} = \ell \sum_{i=1}^{n} \hat{b}_i$ and $R = |\mathbf{R}|$. Your result should indicate that the typical distance between ends of the molecule grows with chain length as $\sqrt{\langle R^2 \rangle} \sim n^{\nu}$. Identify the exponent ν. How does your result depend on dimensionality d?

3. For a polymer in *good* solvent, experiments yield $\nu \approx 3/5$. (A *good* solvent is one that prevents the molecule from collapsing onto itself.) Compare this measured value with the one you calculated, and comment on the discrepancy.

4. Now imagine that there is an energetic bias for segments to point in a particular direction $\hat{a}$, described by a potential energy $u = -\gamma \hat{b}_i \cdot \hat{a}$ for each segment i. Compute the mean segment orientation $\bar{b} = \langle \hat{b}_i \rangle$ in the canonical ensemble. (Assume that fluctuations in orientations of different segments remain uncorrelated, $\langle (\hat{b}_i - \bar{b})(\hat{b}_j - \bar{b}) \rangle = 0$ for $i \neq j$.) Then compute $\langle |\mathbf{R}|^2 \rangle$ and sketch its dependence on the bias strength γ. For simplicity you can take $\hat{a}$ to be aligned with a Cartesian axes.

3

Linear response and Gaussian field theories

Most systems are not ideal and in condensed phases interparticle interactions cannot typically be neglected. This renders the direct evaluation of the partition function using the methods of the last chapter intractable. Nevertheless, statistical mechanical theories can still be brought to bear on these systems provided the aim is slightly reduced. In particular, instead of trying to solve for the full parametric dependence of the free energy on all relevant control parameters, we can try to understand the dependence around some reference using the formal tools of perturbation theory. If instead of describing systems with full molecular detail, we adopt reduced descriptions, these coarser-grained models may be solvable. This is the perspective adopted in this chapter. We will consider in particular cases where a system can be thought of as behaving close to specific reference, and develop the theory of linear response around those references. Linear response theory, a perturbation theory, relies on the evaluation of measures of correlations between particles to employ. Such correlation functions can be evaluated using Gaussian field theories, which can be thought of as simple effective models of complex systems. So the combination of linear response theory and Gaussian models can be used synergistically to deduce the behaviors of such systems. Here we will largely focus on examples drawn from theories of solvation and liquid state structure, but the techniques are general. We will mostly consider classical systems as these techniques are most useful in cases where both energetic and entropic effects are important, though an extension to quantum mechanical phenomena is worked out at the end of the chapter.

3.1 Forced harmonic oscillator

To understand the structure of the perturbation theory we will develop, let us consider a concrete system where everything can be evaluated exactly. Let's ask, how does a harmonic oscillator respond to an external force? We have considered the statistics of a classical harmonic oscillator previously. To add a force to the system, we add a linear potential energy of the form,

$$U = x^2/2\alpha - fx$$

which we recognize as the work done on a particle by a constant force. This is confirmed by looking at the total force on the system $-\partial_x U(x) = -x/\alpha + f$. We can write down the partition function with this additional term,

$$Q(f) = \int_{-\infty}^{\infty} dx\, e^{-\beta x^2/2\alpha + \beta f x} = \frac{Q(0)}{Q(0)} \int_{-\infty}^{\infty} dx\, e^{-\beta x^2/2\alpha + \beta f x} = Q(0) \left\langle e^{\beta f x} \right\rangle_0$$

where the last equality follows from defining $\langle \dots \rangle_0$ as an expectation value taken within an ensemble with $f = 0$. This simple rewriting of the partition function is profound. It implies that the information of the system under an applied force is completely encapsulated within a system fluctuating without the applied force. We have written it in this way in order to identify $Q(f)$ as a *generating function*. Evaluating the expectation value explicitly we get,

$$Q(f) = Q(0)e^{\beta \alpha f^2/2}$$

where we make use of another Gaussian integral identity,

$$I = \int_{-\infty}^{\infty} dy\, e^{-ay - y^2/2\sigma^2} / \sqrt{2\pi\sigma^2} = e^{\sigma^2 a^2/2}$$

which follows from our previous identity by completing the square in the argument of the exponential. This is another very useful relationship to remember.

A generating function generates properties of a statistical distribution by taking derivatives with respect to a counting parameter, in this case f. Specifically, the logarithm of $Q(f)$ is a cumulant generating function and as such the first cumulant, the mean, can be computed as

$$\frac{d \ln Q(f)}{d\beta f} = \langle x \rangle_f$$
$$= \alpha f$$

where $\langle \dots \rangle_f$ is an ensemble average taken with finite f. Taking a second derivative provides the second cumulant, or the variance,

$$\frac{d^2 \ln Q(f)}{d\beta^2 f^2} = \langle (\delta x)^2 \rangle_f$$
$$= \alpha/\beta = \langle (\delta x)^2 \rangle_0$$

where we find a unique property of a Gaussian distribution, or harmonic oscillator. For arbitrary values of the force, f, the variance does not change. An illustration of this invariance is shown in Figure 3.1.

This variance reports on the size of fluctuations around an equilibrium position. Rearranging the previous two equations,

$$\langle x \rangle_f = \beta \langle (\delta x)^2 \rangle_0 f$$

we find that the variance also sets the scale of the response of the mean to the applied force. This is a pretty spectacular result. It says that the way a system responds to a perturbation away from equilibrium, is the same as how its fluctuations relax around its equilibrium position. The differential form of this result is a fluctuation-dissipation

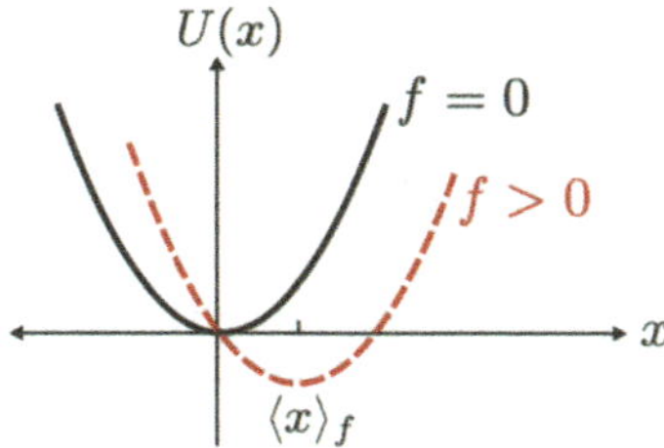

Fig. 3.1 Illustration of a harmonic potential and how it changes with an additional constant force, f.

relation, and is a microscopic analog to those we derived for thermodynamic quantities, like the relationship between the heat capacity and energy fluctuations. The equation is an example of a linear response relation that we will derive in other more complicated situations throughout this chapter.

> **Exercise 3.1:** Using the partition function for the forced harmonic oscillator, compute the change in the average energy of the oscillator $\Delta U = \langle U \rangle_f - \langle U \rangle_0$ with and without the force. Demonstrate that the change in the entropy is $\Delta S = 0$. Provide an interpretation of this latter fact.

3.2 Linear response in nonlinear systems

The relationship between microscopic fluctuations and response is very general. Consider a system with an arbitrary, stable potential $U_0(x)$,

$$U_0(x) = x^2/2\alpha + bx^3 + cx^4 + \ldots$$

which is expanded about $x = 0$ with $\alpha > 0$. For arbitrary coefficients α, b, c, the partition function is not analytically computable. Nevertheless, we can ask how the system responds to a perturbation of the form of a constant applied force f,

$$U = U_0(x) - fx.$$

We begin by writing the partition function of the system in the presence of the force as,

$$Q(f) = \int dx e^{-\beta U_0(x)} e^{\beta f x}$$

$$= Q(0) \frac{\int dx e^{-\beta U_0(x)} e^{\beta f x}}{Q(0)} = Q(0) \left\langle e^{\beta f x} \right\rangle_0$$

where,

$$Q(0) = \int dx e^{-\beta U_0(x)}$$

and $\langle \ldots \rangle_0$ denotes the ensemble average with respect to U_0.

As we found for the harmonic oscillator, $\ln Q(f) = \ln Q(0) + \ln \left\langle e^{\beta f x} \right\rangle_0$ is identified as a cumulant generating function,

$$\frac{d \ln Q(f)}{d\beta f} = \langle x \rangle_f \qquad \frac{d^2 \ln Q(f)}{d(\beta f)^2} = \left\langle (\delta x)^2 \right\rangle_f \quad \cdots \quad \frac{d^n \ln Q(f)}{d(\beta f)^n} = c_n$$

with c_n the n'th cumulant. In order to compute the first-order response to the applied force, f, we Taylor expand the partition function

$$\ln Q(f) \approx \ln Q(0) + \beta f \langle x \rangle_0 + (\beta f)^2 \left\langle (\delta x)^2 \right\rangle_0 /2 + (\beta f)^3 c_3/6 + \mathcal{O}((\beta f)^4)$$

and taking a derivative with respect to f to compute the average position. We arrive at

$$\frac{d \ln Q(f)}{d\beta f} = \langle x \rangle_f$$
$$\approx \langle x \rangle_0 + \beta f \left\langle (\delta x)^2 \right\rangle_0 + \mathcal{O}(f^2)$$

where the first equality follows directly, and after rearranging,

$$\langle x \rangle_f - \langle x \rangle_0 = \beta f \left\langle (\delta x)^2 \right\rangle_0 + \mathcal{O}(f^2)$$

which is the same linear response relationship we found for a harmonic system. However, here we see that for a nonlinear system, this relationship is the result of neglecting terms $\mathcal{O}(f^2)$ and higher. This derivation makes it clear that the higher-order response comes from higher-order cumulants that vanish for a Gaussian distribution. It also serves to delineate the source of the response. Namely, for a nonlinear system, $\left\langle (\delta x)^2 \right\rangle_f \neq \alpha/\beta$ in general, which is to say that the local curvature around $x = 0$, is not equal to $1/\langle (\delta x)^2 \rangle$ unless $b = c = \cdots = 0$. Put another way, to the extent that fluctuations are Gaussian, we can use a linear response theory to understand the system away from its unperturbed state.

Exercise 3.2: Consider a two-dimensional potential under an applied force. Take as a reference potential $U_0(x, y)$ and perturb it linearly in the y direction,

$$U(x, y) = U_0(x, y) - fy.$$

and demonstrate that to first order in the applied force,

$$\langle x \rangle_f - \langle x \rangle_0 = \beta f \langle \delta x \delta y \rangle_0 + \mathcal{O}(f^2)$$

so that there is no response if x and y are uncorrelated.

3.3 Central limit theorem

In general, particles interact and potential energy surfaces are not quadratic. Nevertheless, for small fluctuations around the most typical value of a random variable, quite often the probabilities are Gaussian and thus admit linear statistical restoring forces. This is because of a general result in mathematics known as the *Central Limit Theorem*. In probability theory, the central limit theorem establishes that the distribution of a sum of randomly distributed variables tends toward a Gaussian distribution, independent of how the original variables themselves are distributed, provided they have finite variance. To prove the Central Limit Theorem, consider a set of randomly chosen variables, $\{\tilde{x}_i\}$'s, which are correlated over a finite range R, but have mean $\langle \tilde{x}_i \rangle = 0$. We define a collective variable X such that,

$$X = \sum_{i=1}^{\tilde{N}} \tilde{x}_i = \sum_{i=1}^{\tilde{N}/R} \sum_{j=1}^{R} \tilde{x}_{j+iR}$$

where in the second line we have explicitly factored out the part of the sum that is correlated. We now introduce a variable, x, defined as this partial sum,

$$x_i = \sum_{j=1}^{R} \tilde{x}_{j+iR}$$

so that the x_i's are uncorrelated random variables with mean 0. Now our collective variable X is given by

$$X = \sum_{i=1}^{N} x_i$$

where $N = \tilde{N}/R$, is the size of the sum reduced by the range of correlations assumed $R_i = R$ for all i.

In order to compute the probability distribution associated with this collective variable, we introduce the delta function representation of a probability, $P(X) = \left\langle \delta \left(X - \sum_{i=1}^{N} x_i \right) \right\rangle$ where the expectation value is taken over the distribution of the x_i's. To convert this into a usable expression, we note the Fourier representation of the delta function,

$$P(X) = \left\langle \delta(X - \sum_{i=1}^{N} x_i) \right\rangle$$
$$= \left\langle \int \frac{dk}{2\pi} \exp\left(-ikX\right) \exp\left(ik \sum_{i=1}^{N} x_i\right) \right\rangle$$
$$= \int \frac{dk}{2\pi} \exp\left(-ikX\right) \langle \exp\left(ikx_i\right) \rangle^{N}$$

and in the third line we use the fact that the exponential of the sum factorizes into N identical terms. The expectation value of $\langle \exp\left(ikx_i\right) \rangle$ is known as the *characteristic function*, and can be defined by

$$\hat{p}(k) = \langle e^{ikx} \rangle$$
$$= \int dx\, p(x) e^{ikx}$$

The characteristic function has the property that the n'th derivative with respect to k generates the n'th moment of the distribution, up to factors of i^n,

$$\hat{p}(0) = 1 \qquad \left.\frac{d\hat{p}}{dk}\right|_{k=0} = \langle ix \rangle = 0 \qquad \left.\frac{d^2\hat{p}}{dk^2}\right|_{k=0} = \langle (ix)^2 \rangle = -\langle x^2 \rangle$$

where the first follows from normalization of the probability distribution and the others from the definition of $\hat{p}(k)$. The fact that the first derivative evaluates to the mean, which is 0, and the second is negative, means that $\hat{p}(k)$ is peaked around $k = 0$.

Because $\hat{p}(k)$ is peaked around $k = 0$, $\hat{p}(k)^N$ is strongly peaked around 0 if N is large. In order to compute the collective distribution, we define

$$\begin{aligned} f(k) &= [\hat{p}(k)]^N \\ &= e^{N \ln \hat{p}(k)} = e^{Ng(k)} \end{aligned}$$

where $f(k)$ is the characteristic function for the collective variable X and $g(k) = \ln \hat{p}(k)$ is a scaled cumulant generating function or another large deviation function, which we will Taylor expand in order to approximate $f(k)$.

Expanding $g(k)$ around its maximum at $k = 0$,

$$g(k) \approx g(0) + kg'(0) + k^2 g''(0)/2 + \mathcal{O}(k^3)$$

we can evaluate each expansion coefficient,

$$g(0) = 0 \qquad g'(0) = \left.\frac{d \ln \hat{p}(k)}{dk}\right|_{k=0} = \frac{\langle ix \rangle}{\hat{p}(0)} = 0 \quad g^2(0) = -\langle (\delta x)^2 \rangle \quad \ldots$$

and find that these are the cumulants of the distribution $p(x)$ up to factors of i. By truncating at second order,

$$f(k) \approx \exp\left[N\left(-k^2 \langle (\delta x)^2 \rangle/2 + \mathcal{O}(k^3)\right)\right]$$

we have an approximation to $f(k)$. In order to obtain $P(X)$ we just take an inverse Fourier transform,

$$\begin{aligned} P(X) &= \int_{-\infty}^{\infty} \frac{dk}{2\pi} \exp\left(-ikX\right) f(k) \\ &\propto \exp\left[-X^2/2N\langle (\delta x)^2 \rangle\right] \end{aligned}$$

which yields a Gaussian distribution with mean 0 and variance $N\langle (\delta x)^2 \rangle$, or N times the variance of the uncorrelated variables. Because the standard deviation or the relative size of fluctuations go as $1/\sqrt{N\langle (\delta x)^2 \rangle}$, in the large N limit, the higher k terms in the expansion of $g(k)$ are expected to become insignificant. Starting from an initial distribution $p(x)$, which could be structured, we find that independent of its form,

$P(X)$ is always Gaussian near its typical value. We should expect a Gaussian approximation to be valid for values of the collective variable up to $\mathcal{O}(N)$. In terms of the distribution, fluctuations of

$$P(X) = \mathcal{O}\left(e^{-N}\right)$$

are expected to show deviations from a Gaussian distribution. In other words, only fluctuations near the mean are Gaussian distributed according to the central limit theorem with that nearness depending on the size of N.

> **Exercise 3.3:** The central limit theorem dictates that fluctuations of sums of random variables are Gaussian around their mean. Deviations are expected for sufficiently rare fluctuations. In order to understand the range of validity of the central limit theorem, work out the leading order correction. You should find that for consistency with the truncation to second order,
>
> $$X \ll 2Nc_2^2/c_3 \sim \mathcal{O}(N)$$
>
> where c_2 and c_3 are the second and third cumulants of x.

3.4 Linear theory for density fluctuations

Provided the central limit theorem, we can expect that collective variables of interacting systems can be considered as locally Gaussian. For a system like a dense fluid, consisting of strongly interacting particles whose partition function is not directly computable, we can use its expected Gaussian statistics to construct an approximate description. The goal will be to have a framework with which we can compute properties of liquids like their structure and their ability to solvate other particles. We will start on a lattice of m cells, each of volume v, that grid up space. We will take the bold position that within each of those m cells, the density fluctuations $\{\delta\rho(\mathbf{r}_1), \delta\rho(\mathbf{r}_2), \ldots, \delta\rho(\mathbf{r}_m)\}$ are correlated Gaussian random variables. Certainly, in the limit that v is large, this assertion makes sense. However, in a dense liquid even at the finest scales, this assertion is valid. This setup is illustrated in Figure 3.2.

As particles in adjacent cells interact, fluctuations of the densities in two cells may be correlated. By fluctuation, we mean the deviation of the density $\rho(\mathbf{r}_i)$ from its average, $\bar{\rho}$, $\delta\rho(\mathbf{r}_i) = \rho(\mathbf{r}_i) - \bar{\rho}$. A quantification of correlation between cells is a correlation function, like a covariance, which for densities is

$$\langle \delta\rho(\mathbf{r}_i)\delta\rho(\mathbf{r}_j)\rangle = \chi(\mathbf{r}_i, \mathbf{r}_j)$$
$$= \chi(|\mathbf{r}_i - \mathbf{r}_j|) \equiv \chi_{ij}$$

which is 0 if cells i and j are independent but in general is finite. Assuming that the liquid is isotropic, the correlation function will depend only on the distance between the two cells. Moreover, assuming that correlations persist over a finite scale, if $|\mathbf{r}_i - \mathbf{r}_j|$ is very large, $\chi(\mathbf{r}_i, \mathbf{r}_j)$ will approach zero. The third equality serves to simplify the

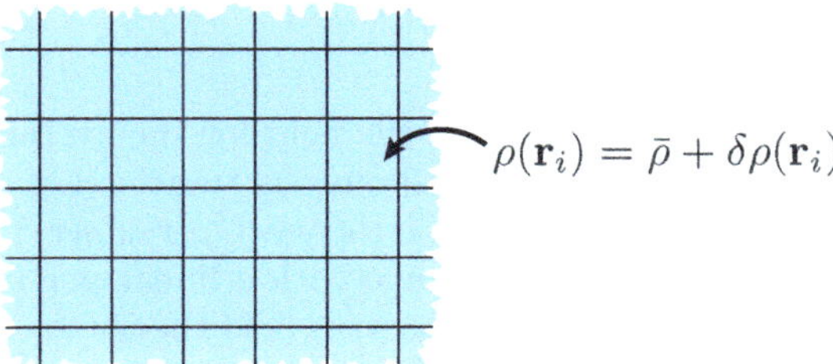

Fig. 3.2 Construction of a model of density fluctuations on a grid, with m cells each of volume v where $\bar{\rho}$ is the bulk density.

notation. Assuming density fluctuations are Gaussian, an effective Hamiltonian for those fluctuations can be written as

$$\beta\mathcal{H}_0[\{\delta\rho(\mathbf{r}_1)\}] = \frac{1}{2}\sum_{i,j}\delta\rho(\mathbf{r}_i)\chi_{ij}^{-1}\delta\rho(\mathbf{r}_j)$$

where the matrix inverse χ_{ij}^{-1} enters through the definition of the correlation function. This effective Hamiltonian is really a free energy function, defined through the probability of a density field that through the Boltzmann distribution will ensure that density fluctuations are Gaussian,

$$P[\{\delta\rho(\mathbf{r})\}] \propto e^{-\beta\mathcal{H}_0[\{\delta\rho(\mathbf{r}_1)\}]} = e^{-\frac{1}{2}\sum_{i,j}\delta\rho(\mathbf{r}_i)\chi_{ij}^{-1}\delta\rho(\mathbf{r}_j)}$$

and in practice would result from counting all microstates consistent with a given density distribution $\{\delta\rho(\mathbf{r})\}$. As a consequence of being a function of an average, $\mathcal{H}_0$ will in general depend on the thermodynamic state of the system.

Exercise 3.4: A special property of Gaussian theories are that they are completely determined by 1 and 2 point functions. Using the decomposition for the inverse of the matrix $\boldsymbol{\chi}^{-1} = \chi_o^{-1}\mathbf{1} - (\boldsymbol{\Delta\chi})^{-1}$ where $\mathbf{1}$ is the identity matrix, prove the following recursion relation

$$\boldsymbol{\chi} = \chi_o\mathbf{1} + \chi_o\mathbf{1}(\boldsymbol{\Delta\chi})^{-1}\boldsymbol{\chi}$$

for $\boldsymbol{\chi}$. This self-consistency is a hallmark of Gaussian theories.

From this model, we can use the form of the proceeding section to see how the density field responds to a perturbation like an applied potential,

$$\beta\mathcal{H}[\{\delta\rho(\mathbf{r}_i)\}] = \beta\mathcal{H}_0 - \beta\sum_{i}\delta\rho(\mathbf{r}_i)\phi(\mathbf{r}_i)v$$

where $\phi(\mathbf{r}_i)$ is an added external field that acts on the density within cell i. The linear response of the density in cell i to this sum of perturbations is,

$$\langle \delta\rho(\mathbf{r}_i)\rangle_\phi = \beta \sum_j \langle \delta\rho(\mathbf{r}_i)\delta\rho(\mathbf{r}_j)\rangle_0 \phi(\mathbf{r}_j)v$$

where the subscript ϕ denotes the value of the field for the average. This clarifies the role of the correlation function in linear response as transferring a perturbation from one point in space to another. For a Gaussian theory, $\langle \delta\rho(\mathbf{r}_i)\delta\rho(\mathbf{r}_j)\rangle_0 = \langle \delta\rho(\mathbf{r}_i)\delta\rho(\mathbf{r}_j)\rangle_\phi$ since the scale of fluctuations does not change with a linear perturbation.

Let us now go off-lattice and build a Gaussian field theory for density fluctuations. While above we had a vector of instantaneous values of the density in each of m cells, in the limit of $v \to 0$,

$$\{\delta\rho(\mathbf{r}_1), \delta\rho(\mathbf{r}_2), \ldots, \delta\rho(\mathbf{r}_m)\} \to \delta\rho(\mathbf{r})$$

where $\delta\rho(\mathbf{r})$ is interpreted as a function of a continuum of points defined on the region of $\mathbf{r}$. Similarly, the effective Hamiltonian that was a function of the m points on the grid now becomes

$$\beta\mathcal{H}_0[\{\delta\rho(\mathbf{r}_1), \ldots, \delta\rho(\mathbf{r}_m)\}] = \frac{1}{2}\sum_{i,j} \delta\rho(\mathbf{r}_i)\chi_{ij}^{-1}\delta\rho(\mathbf{r}_j)$$

$$\to \quad \beta\mathcal{H}_0[\delta\rho(\mathbf{r})] = \frac{1}{2}\int d\mathbf{r}\int d\mathbf{r}'\delta\rho(\mathbf{r})\chi^{-1}(\mathbf{r},\mathbf{r}')\delta\rho(\mathbf{r}')$$

where $\mathcal{H}_0[\delta\rho(\mathbf{r})]$ is a *functional* of the density field and the sums in the discrete case have become integrals. The matrix χ_{ij} and its inverse are defined by

$$\sum_k \chi_{ik}^{-1}\chi_{kj} = \delta_{ij} \quad \to \quad \int d\mathbf{r}''\chi^{-1}(\mathbf{r},\mathbf{r}'')\chi(\mathbf{r}'',\mathbf{r}') = \delta(\mathbf{r} - \mathbf{r}')$$

and now satisfy a functional inverse relationship. Finally, with the addition of an external field acting on the density

$$\beta\mathcal{H}[\delta\rho(\mathbf{r})] = \beta\mathcal{H}_0[\delta\rho(\mathbf{r})] - \beta\int d\mathbf{r}\phi(\mathbf{r})\delta\rho(\mathbf{r})$$

we can intuit the linear response relationship

$$\langle \delta\rho(\mathbf{r})\rangle_\phi = \beta \int d\mathbf{r}'\,\langle \delta\rho(\mathbf{r})\delta\rho(\mathbf{r}')\rangle_0 \phi(\mathbf{r}')$$

$$= \beta \int d\mathbf{r}'\,\chi(\mathbf{r},\mathbf{r}')\phi(\mathbf{r}')$$

where the integral of the field point $\mathbf{r}'$ has taken the place of the sum.

3.5 Fluid structure and thermodynamics

Correlation functions serve two distinct roles. They encode the statistical interdependence of particles separated through space and thus are a measure of structure that is particularly useful in a system like a liquid that has no long-ranged order. However, as we have already seen, they also encode the way a system responds to a perturbation. Indeed, for a system that interacts with pairwise decomposable forces, the pair correlation function uniquely determines the thermodynamics of that system. As such, they deserve some elaboration.

Let us start with some definitions. Specifically, we define the microscopic density *field* from the particle positions as

$$\rho(\mathbf{r}) = \sum_{i=1}^{N} \delta(\mathbf{r} - \mathbf{r}_i)$$

where the sum is over all N particles, and Dirac's delta function adds up the contributions to the density at $\mathbf{r}$ provided there is a particle at that point in space. In an isotropic system, the expectation value of the density can be computed,

$$\langle \rho(\mathbf{r}) \rangle = \left\langle \sum_{i=1}^{N} \delta(\mathbf{r} - \mathbf{r}_i) \right\rangle$$
$$= N \langle \delta(\mathbf{r} - \mathbf{r}_1) \rangle = \bar{\rho}$$

where in the second line we have assumed that all N particles are identical and in the third line, provided the system is translationally invariant, the probability that a particle is at any given point in space is just $1/V$, where V is the volume of the system. Throughout this section we will denote the bulk density by $\bar{\rho}$. This procedure also clarifies that the single particle density contains no structural information about a liquid, it is just a number independent of $\mathbf{r}$. This is a generic consequence of a disordered system.

To characterize the structure of a liquid, a higher-order function of the density must be computed. The density–density correlation function can be defined from our density field,

$$\chi(\mathbf{r}, \mathbf{r}') = \langle \rho(\mathbf{r})\rho(\mathbf{r}') \rangle - \langle \rho(\mathbf{r}) \rangle \langle \rho(\mathbf{r}') \rangle$$
$$= \left\langle \sum_{i,j=1}^{N} \delta(\mathbf{r} - \mathbf{r}_i)\delta(\mathbf{r}' - \mathbf{r}_j) \right\rangle - \bar{\rho}^2$$
$$= \bar{\rho}\delta(\mathbf{r} - \mathbf{r}') + \bar{\rho}\frac{\left\langle \delta(\mathbf{r} - \mathbf{r}_1) \sum_{j=2}^{N} \delta(\mathbf{r}' - \mathbf{r}_j) \right\rangle}{\langle \delta(\mathbf{r} - \mathbf{r}_1) \rangle} - \bar{\rho}^2$$
$$= \bar{\rho}\delta(\mathbf{r} - \mathbf{r}') + \bar{\rho}^2 \left[g(\mathbf{r} - \mathbf{r}') - 1 \right]$$

which decomposes into the $i = j$ and $i \neq j$ parts. The first term on the right-hand side comes from self-correlations. The second term comes from interparticle correlations. The part of that term, $g(r)$, is known as the *radial distribution function*. Its meaning can be understood by looking more closely at its definition. If we set $\mathbf{r}_1$ to the origin, $\mathbf{r}_1 = 0$, and use the fact that the delta function is even, the definition can be rewritten as

$$\bar{\rho} g(\mathbf{r}) = \frac{\left\langle \delta(\mathbf{r}_1) \sum_{j=2}^{N} \delta(\mathbf{r} - \mathbf{r}_j) \right\rangle}{\langle \delta(\mathbf{r}_1) \rangle} = \langle \rho(\mathbf{r}) \rangle_{\mathbf{r}_1 = 0}$$

which has a natural interpretation as a conditional probability, defined in the second line. From *Bayes' theorem*, a joint probability divided by a marginal probability is a conditional probability. A joint probability is the likelihood of two random variables like A and B occurring, $P(A, B)$, while a conditional probability is the probability of a random variable given some condition, $P(A, B)/P(A) = P(B|A)$. In this way, $g(r)$ is the conditional probability that a particle exists a distance r away from the origin, conditioned on another particle being at the origin, relative to the mean density.

An equivalent way to define the $g(r)$ is through particle distances,

$$g(r) = \frac{N-1}{\bar{\rho}} \langle \delta(\mathbf{r} - \mathbf{r}_{12}) \rangle = \frac{1}{\bar{\rho}N} \left\langle \sum_{i \neq j}^{N} \delta(\mathbf{r} - \mathbf{r}_{ij}) \right\rangle$$

which is an equation that implements a conditional probability implicitly. Specifically, the $g(r)$ is the probability of two particles being a distance r apart, or alternatively, the density of particles away from the origin given a particle is fixed at the origin. Two limits of the $g(r)$ are clear. First, due to Pauli's principle a particle excludes volume, it has a hard core, so that as $r \to 0$, $g(r) \to 0$. Another limit which is applicable to disordered phases (e.g., liquids and gasses) is that as $r \to \infty$, $g(r) \to 1$. This is to say that at large distances, particles' positions become uncorrelated. If the system exhibits long-ranged order, like a crystal, then as $r \to \infty$, $g(r)$ will still depend on r.

An external potential that acts on the particles in a system can also be written in terms of the density field. For any field that can be decomposed as a sum of contributions acting on individual particles,

$$\Phi(\mathbf{r}) = -\sum_{i=1}^{N} \phi(\mathbf{r}_i) = -\int d\mathbf{r} \, \phi(\mathbf{r}) \sum_{i=1}^{N} \delta(\mathbf{r} - \mathbf{r}_i) = -\int d\mathbf{r} \, \phi(\mathbf{r})\rho(\mathbf{r})$$

we see that such a field is a conjugate variable to the microscopic density. This linearity should remind you of the force, f, acting on the position of the harmonic oscillator. Provided this definition, and a reference system whose Hamiltonian is known $\mathcal{H}_0[\delta\rho(\mathbf{r})]$, we can define the partition functions for different microscopic density patterns. In the presence of an external field,

$$Q(\phi) = \int \mathcal{D}\left[\rho(\mathbf{r})\right] \exp\left\{ -\beta\mathcal{H}_0[\delta\rho(\mathbf{r})] + \beta \int d\mathbf{r} \, \phi(\mathbf{r})\rho(\mathbf{r}) \right\}$$

$$= Q(0) \left\langle \exp\left\{ \beta \int d\mathbf{r} \, \phi(\mathbf{r})\rho(\mathbf{r}) \right\} \right\rangle_0$$

where $\mathcal{D}\left[\rho(\mathbf{r})\right]$ is functional integral over the density and,

$$Q(0) = \int \mathcal{D}\left[\rho(\mathbf{r})\right] \exp\left\{ -\beta\mathcal{H}_0[\delta\rho(\mathbf{r})] \right\} .$$

Again the partition function is identified as a generating function, in this case, for characterizing density fluctuations. Taking the appropriate functional derivatives,

$$\frac{\delta \ln Q(\phi)}{\delta \beta \phi(\mathbf{r})} = \langle \rho(\mathbf{r}) \rangle_\phi \qquad \frac{\delta^2 \ln Q(\phi)}{\delta \beta \phi(\mathbf{r}) \beta \phi(\mathbf{r}')} = \frac{\delta \langle \rho(\mathbf{r}) \rangle_\phi}{\delta \beta \phi(\mathbf{r}')} = \langle \delta\rho(\mathbf{r})\delta\rho(\mathbf{r}') \rangle_\phi$$

we see the types of response relationships we have uncovered in simpler contexts.

Aside: Variational calculus. In order to derive the earlier relationships, a working knowledge of *functionals* and *calculus of variations* is required. Here we start by making an analogy to vector spaces, as a function can be considered as a vector with continuous index. So by analogy,

$$\mathbf{f} = \{f_1, f_2, \ldots, f_n\} \quad \rightarrow \quad f(\mathbf{x})$$

and similarly, a function of a vector becomes a functional of a function

$$g(f_1, f_2, \ldots, f_n) \quad \rightarrow \quad G[f(\mathbf{x})].$$

Derivatives with respect to a component, becomes a functional derivative

$$\left(\frac{\partial g}{\partial f_i}\right)_{f_j \neq i} \quad \rightarrow \quad \frac{\delta G}{\delta f(\mathbf{x})}.$$

To evaluate a functional derivative consider first variations in a vector space,

$$\mathbf{f} = \{f_1, f_2, \ldots, f_n\} \quad \rightarrow \quad \mathbf{f} = \{f_1 + \delta f_1, f_2 + \delta f_2, \ldots, f_n + \delta f_n\}$$

which result in variations of functions of that vector space

$$g(f_1, f_2, \ldots, f_n) \quad \rightarrow \quad \delta g = \sum_i^n \left(\frac{\partial g}{\partial f_i}\right)_{f_j \neq i} \delta f_i.$$

Analogously, variations in a function space,

$$f(\mathbf{x}) \quad \rightarrow \quad f(\mathbf{x}) + \delta f(\mathbf{x})$$

result in variations in the functional of that space,

$$\delta G = \int d\mathbf{x} \, \frac{\delta G}{\delta f(\mathbf{x})} \delta f(\mathbf{x}).$$

The basic rule for vector spaces, $g = f_i \rightarrow \frac{\partial g}{\partial f_j} = \delta_{ij}$ can be translated into a rule for functionals,

$$G = f(\mathbf{x}) \quad \rightarrow \quad \frac{\delta G}{\delta f(\mathbf{x}')} = \delta(\mathbf{x} - \mathbf{x}')$$

Examples of $g(r)$'s for a simple monatomic liquid, solid, and gas are shown in Figure 3.3. As anticipated, for both a gas and a liquid, the $g(r)$ decays to 1, while it continues to oscillate for a solid. In all cases, however, given a diameter of the particle σ, each phase exhibits a region of exclusion between $r = 0$ and $r \approx \sigma$ where the $g(r)$ is 0. Indeed, excluded volume dominates the density correlations even away from $r = 0$

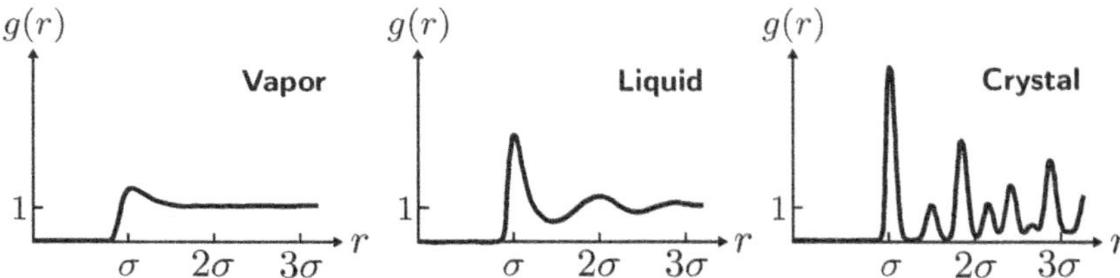

Fig. 3.3 Example radial distribution functions for a structureless particle in different states of matter, with distances measured relative to the particle diameter, σ.

for a liquid and a solid, where the deviations from 1 occur at regular values of σ. Note that, a system of purely hard spheres will have $g(r)$'s nearly identical to those with isotropic attractions.

Exercise 3.5: Evaluate the functional derivative of G with respect to $f(\mathbf{x})$ for the following functionals

$$G = \int d\mathbf{x}\, a(\mathbf{x}) e^{b(\mathbf{x}) f(\mathbf{x})} \qquad G = \int d\mathbf{x}\, a(\mathbf{x}) [f(\mathbf{x} - \mathbf{x}_o)]^2$$

using $\delta G[f(\mathbf{x})]/\delta f(\mathbf{x}') = \delta(\mathbf{x} - \mathbf{x}')$ and other rules of calculus.

Not only does the radial distribution function contain important structural information, but it also encodes thermodynamic information. For example, imagine a system that interacts through pairwise forces, so that the potential energy is $U = 1/2 \sum_{i \neq j} u(|\mathbf{r}_i - \mathbf{r}_j|)$, where $u(r)$ is the potential between two particles separated by a distance r. Such a model of inter-atomic interaction is approximate, as in principle the interaction between atoms can include many-body effects, but is often a very good approximation, as many-body effects in liquids are often negligible. Within the pair-potential approximation, the full partition function is

$$Q[u(r)] = \frac{1}{\lambda^{3N} N!} \int d\mathbf{r}^N \exp\left[-\beta/2 \sum_{i \neq j} u(|\mathbf{r}_i - \mathbf{r}_j|)\right]$$

and its free energy follows as,

$$A[u(r)] = -k_\mathrm{B} T \ln Q[u(r)]$$

which can be identified as a functional of that pair potential. More explicitly, we can relate the free energy of this system, or its configurational part, to changes in its pair potential by taking a functional derivative,

$$\frac{\delta A}{\delta u(r)} = \frac{-k_{\mathrm{B}}T}{Q} \int dr^N \left[-\beta/2 \sum_{i \neq j} \delta(r - |\mathbf{r}_i - \mathbf{r}_j|) \right] e^{-\beta/2 \sum_{i \neq j} u(|\mathbf{r}_i - \mathbf{r}_j|)}$$

$$= \frac{1}{2} \left\langle \sum_{i \neq j} \delta(r - |\mathbf{r}_i - \mathbf{r}_j|) \right\rangle = \frac{1}{2} N \bar{\rho} g(r)$$

where we can identify the expectation value as constants times the $g(r)$. So the $g(r)$ encodes how the interactions amongst particles change their free energy.

We can make this more concrete by defining a pair potential that changes from one form to another smoothly as a function of a control parameter λ, $u_\lambda(r) = u_{\mathrm{o}}(r) + \lambda \Delta u(r)$ where $\Delta u(r) = [u(r) - u_{\mathrm{o}}(r)]$. Formally integrating the functional differential from above with these boundary conditions we find

$$A[u(r)] - A[u_{\mathrm{o}}(r)] = \int_0^1 d\lambda \int dr \left. \frac{\delta A}{\delta u(r)} \right|_\lambda \Delta u(r) = \bar{\rho} N \int_0^1 d\lambda \int d\mathbf{r} \, g_\lambda(r) \Delta u(r)/2$$

where $g_\lambda(r)$ is the radial distribution function with configurations distributed according to a potential energy function $u_\lambda(r)$. This is a direct connection between structure and thermodynamics, and is an example of an adiabatic connection equation. It states that the free energy changes as the interactions amongst particles change, by changing the pair correlations. If we evaluate that function with end points that are $\lambda = 0, u_0(r) = 0$, and $\lambda = 1, u_1(r) = u(r)$,

$$A[u(r)] = A_{\mathrm{id}} + \bar{\rho} N \int d\mathbf{r} \, g(r) u(r)/2$$

we find that the free energy has an ideal piece, $A_{\mathrm{id}} = N k_{\mathrm{B}} T [\ln(\bar{\rho} \lambda^3) - 1]$, dependent on the density and thermal wavelength, but which is independent of particle interactions, and an additional configurational piece due to interactions. This follows from the fact that the A must be a state function, so the integral can be evaluated at its end points. This relation serves as the basis for an approximate description of liquid state thermodynamics, as one can construct approximations to $g(r)$ and then the free energy or its derivatives are evaluated by quadrature. Analogously,

$$\langle U \rangle = \bar{\rho} N \int d\mathbf{r} \, g(r) u(r)/2$$

is the average configurational energy.

The chemical potential can also be computed as an integral over $g(r)$. The chemical potential of a fluid is $\mu = (\partial A / \partial N)_{T,V}$ or the change in A when one particle is added to the system $\mu = A(N+1, T, V) - A(N, T, V)$. A reversible work path that can accomplish a particle addition is to add a particle to the system by gradually turning on its interactions with the other particles. If the new particle is placed at the origin, we can modify our previous result using $u_\lambda(r) = \lambda u(r)$. It follows that the chemical potential for a particle that has no internal structure and interacts with the fluid in a pairwise additive way is

$$\mu = k_{\mathrm{B}}T\ln(\bar{\rho}\lambda^3) + \bar{\rho}\int_0^1 d\lambda \int d\mathbf{r}\, g_\lambda(r)u(r)$$

where the first term is the ideal part of μ, while the second is an excess or, equivalently, a standard state piece. Collectively these relationships codify that the radial distribution function relates structure to thermodynamics.

Exercise 3.6: From the relationship between the free energy and the $g(r)$, which is exact for particles that interact with pairwise potentials, show that the pressure $p = -(dA/dV)_{T,N}$ can be written as

$$p = \bar{\rho}k_{\mathrm{B}}T - \frac{\bar{\rho}^2}{6}\int d\mathbf{r}\, r\, g(r)\frac{du(r)}{dr}$$

where you will find it useful to introduce a scaled coordinate system $r = V^{1/3}x$ to get rid of the volume in the limits of the integral. This relationship is known as the *virial theorem*.

The radial distribution function is experimentally observable from scattering experiments. In order to probe structure which varies on lengthscales of Å's, scatterers with wavelengths of comparable size must be used. These are typically either high-energy neutrons or x-rays. Taking x-rays as an example, imagine an atom at position $\mathbf{r}$ scatters an incoming wave with wavevector $\mathbf{k}'$ into a wave with wavevector $\mathbf{k}''$ and is measured at a detector at position $\mathbf{R}_{\mathrm{d}}$. Assuming that the wave is scattered elastically, and that the sample center $\mathbf{R}_{\mathrm{c}}$ is far from the detector, the amplitude of the wave at the detector is

$$\mathcal{A} = f(k)\frac{1}{|\mathbf{R}_{\mathrm{d}} - \mathbf{R}_{\mathrm{c}}|}e^{i\mathbf{k}\cdot\mathbf{r}}e^{i\mathbf{k}''\cdot\mathbf{R}_{\mathrm{d}}}$$

where $f(k)$ is the atomic form factor, a measure of the scattering amplitude by an isolated atom, and $\mathbf{k} = \mathbf{k}' - \mathbf{k}''$. The two complex exponentials represent the incoming and scattered waves. The geometry of this experiment is shown in Figure 3.4.

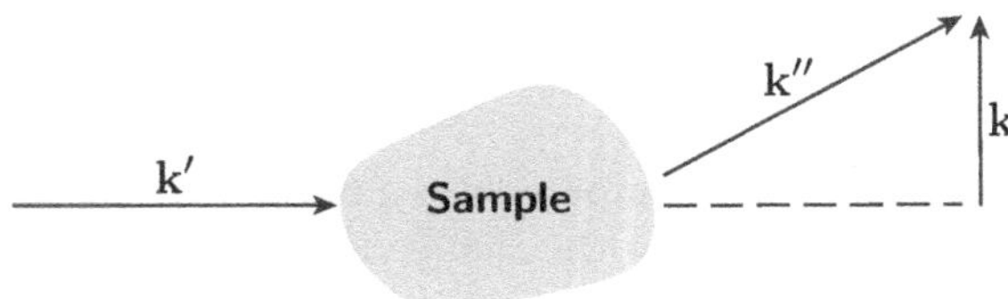

Fig. 3.4 Typical x-ray scattering experiment with incoming wave $\mathbf{k}'$ and outgoing wave $\mathbf{k}''$.

Aside: Fourier Transforms. In order to derive the above relationships, an understanding of how to manipulate *Fourier transforms* is helpful. Let us define a Fourier transform and its inverse on an arbitrary function $f(\mathbf{r})$ in d dimensions as

$$\hat{f}(\mathbf{k}) = \int d\mathbf{r}\, e^{i\mathbf{k}\cdot\mathbf{r}} f(\mathbf{r})\,, \qquad f(\mathbf{r}) = \int \frac{d\mathbf{k}}{(2\pi)^d}\, e^{-i\mathbf{k}\cdot\mathbf{r}} \hat{f}(\mathbf{k})$$

$$\equiv \mathcal{F}[f(\mathbf{r})] \qquad\qquad\qquad \equiv \mathcal{F}^{-1}[\hat{f}(\mathbf{k})]$$

where the operation and its inverse are denoted $\mathcal{F}[..]$ and $\mathcal{F}^{-1}[..]$, respectively, and we denote the Fourier transform of the function $f(\mathbf{r})$ with a hat as, $\hat{f}(\mathbf{k})$ where $\mathbf{k}$ is the Fourier variable conjugate to position, often in physical situations referred to as the wavevector.

The exponential structure of the Fourier transform results in simple relationships between the function and its derivative in Fourier space

$$\nabla f(\mathbf{r}) = \int \frac{d\mathbf{k}}{(2\pi)^d}\, e^{-i\mathbf{k}\cdot\mathbf{r}} \left(-i\mathbf{k}\right) \hat{f}(\mathbf{k})$$

which results in the following relationships

$$\mathcal{F}[\nabla f(\mathbf{r})] = -i\mathbf{k}\hat{f}(\mathbf{k}) \qquad \mathcal{F}[\nabla^2 f(\mathbf{r})] = -\mathbf{k}^2 \hat{f}(\mathbf{k})\,.$$

We will make judicious use of the convolution theorem,

$$\mathcal{F}\left[\int d\mathbf{r} f(\mathbf{r})g(\mathbf{r}-\mathbf{r}')\right] = \hat{f}(\mathbf{k})\hat{g}(\mathbf{k})$$

which can be thought of as a consequence of translational invariance.

The total intensity I at the detector is the absolute squared value of the amplitudes from each scattering event, which could come from any of N particles in the sample. On average, the intensity is

$$I(k) = |f(k)|^2 \frac{1}{|\mathbf{R}_\mathrm{d} - \mathbf{R}_\mathrm{c}|^2} \left\langle \left| \sum_{\alpha=1}^{N} e^{i\mathbf{k}\cdot\mathbf{r}_\alpha} \right|^2 \right\rangle$$

which is accomplished by integrating the scattering over time. While it is clear that $I(k)$ contains information on the positions of the particles, it is in fact the Fourier transform of the density–density correlation function. To see this let us first consider the Fourier transform of the density, $\hat{\rho}(\mathbf{k})$,

$$\hat{\rho}(\mathbf{k}) = \int d\mathbf{r}\rho(\mathbf{r})e^{i\mathbf{k}\mathbf{r}} = \sum_{\alpha=1}^{N} e^{i\mathbf{k}\mathbf{r}_\alpha}$$

where we have used the delta function representation of the density. Squaring $\hat{\rho}(\mathbf{k})$ and taking its thermal average,

$$\left\langle |\hat{\rho}(\mathbf{k})|^2 \right\rangle = \left\langle \sum_{\alpha,\alpha'} e^{i\mathbf{k}(\mathbf{r}_\alpha - \mathbf{r}'_\alpha)} \right\rangle = N S(\mathbf{k})$$

we find that the Fourier transformed density is N times a quantity known as the *structure factor*, $S(\mathbf{k})$. Writing out the transform explicitly, we find

$$S(\mathbf{k}) = \frac{1}{N} \int d\mathbf{r} \int d\mathbf{r}' \, \langle \rho(\mathbf{r})\rho(\mathbf{r}') \rangle \, e^{i\mathbf{k}(\mathbf{r}-\mathbf{r}')}$$
$$= \frac{1}{N} \int d\mathbf{r} \int d\mathbf{r}' \, \left(\bar{\rho}\delta(\mathbf{r}-\mathbf{r}') + \bar{\rho}^2 g(\mathbf{r}-\mathbf{r}') \right) e^{i\mathbf{k}(\mathbf{r}-\mathbf{r}')}$$
$$= 1 + \bar{\rho} \int d\mathbf{R} \, g(\mathbf{R}) e^{i\mathbf{k}\mathbf{R}}$$

where the second line follows from our previous manipulations of the density–density correlation function, and the third results from a substitution $\mathbf{R} = \mathbf{r} - \mathbf{r}'$. Thus we see that up to an additive constant, the structure factor is the Fourier transform of the radial distribution function, and measurable from a scattering experiment.

3.6 Debye–Hückel theory

Having shown the connection between structure and thermodynamics, we are in a position to see how to employ an assumption of Gaussian density statistics to deduce the pair correlation function of a simple system. As a warm-up, we will consider Debye–Hückel theory, a theory for a dilute collection of ions in solution. We will see that an assumption of Gaussian fluctuations, or the linear response of the solution, provides a way to determine an approximation to the radial distribution function.

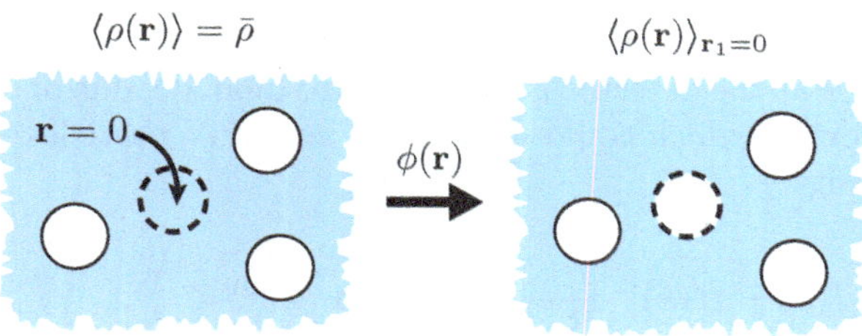

Fig. 3.5 Illustration of the perturbation underlying Debye–Hückel theory.

Our starting point is to imagine a system with a uniform density of ions of one type in an otherwise structureless environment. We will take that density to be $\bar{\rho}$. Such a system is referred to as a one-component plasma, and would physically require a uniform background charge to ensure charge neutrality. Imagine turning on a test ion located at the origin, as in Figure 3.5. This ion exerts forces on particles in a fluid.

Let the field $-\phi(\mathbf{r})$ denote the potential energy field associated with those forces. In the absence of the ion $\phi(\mathbf{r}) = 0$. In the presence of the test ion, the potential can be written as $\phi(\mathbf{r}) = -u(\mathbf{r})$, where $u(\mathbf{r})$ is the pair potential between any two ions. When $\phi(\mathbf{r})$ is changed from 0 to $u(\mathbf{r})$, the density of the ions will change in response. More precisely, from the definition of the radial distribution function, $g(r)$, we expect that as

$$0 \xrightarrow{\phi(r)} u(r)$$

then the corresponding change in average density is

$$\bar{\rho} \xrightarrow{\langle \rho(r) \rangle} \bar{\rho} g(r) \, .$$

Given the uniform reference system and perturbation, we are ready to develop a linear field theory of the ionic density distribution. The theory would be accurate if ions interacted with weak potentials and as a result exhibited only linear response. Physically, this is reasonable if we are in a dilute solution so that any two ions are unlikely to be very close together.

As before, we can write an equation for the density distribution in the presence of the field by Taylor expanding about the reference,

$$\langle \rho(\mathbf{r}) \rangle_\phi \approx \langle \rho(\mathbf{r}) \rangle_0 + \int d\mathbf{r}' \left(\frac{\delta \langle \rho(\mathbf{r}) \rangle}{\delta \phi(\mathbf{r}')} \right)_0 \phi(\mathbf{r}') + \mathcal{O}(\phi^2)$$

$$= \langle \rho(\mathbf{r}) \rangle_0 + \beta \int d\mathbf{r}' \, \chi(\mathbf{r}, \mathbf{r}') \phi(\mathbf{r}')$$

where the subscript zeros means evaluated at $\phi(r) = 0$. However, if the density fluctuations are Gaussian the correlation function is independent of the perturbation. Equating the response function in the presence and absence of the perturbation is equivalent to summing up a certain kind of infinite series. Recognizing perturbed density as related to the $g(r)$, we have

$$\bar{\rho} g(\mathbf{r}) = \bar{\rho} - \beta \int d\mathbf{r}' \, \chi(|\mathbf{r} - \mathbf{r}'|) u(\mathbf{r}')$$

where we can substitute $\chi(|\mathbf{r}|) = \bar{\rho} \delta(\mathbf{r}) + \bar{\rho}^2 \left[g(\mathbf{r}) - 1 \right]$, and simplify the equation,

$$g(\mathbf{r}) - 1 = -\beta u(\mathbf{r}) - \beta \bar{\rho} \int d\mathbf{r}' \, \left[g(\mathbf{r} - \mathbf{r}') - 1 \right] u(\mathbf{r}') \, .$$

Introducing $h(\mathbf{r}) = g(\mathbf{r}) - 1$, we have finally a self-consistent expression,

$$h(\mathbf{r}) = -\beta u(\mathbf{r}) - \beta \bar{\rho} \int d\mathbf{r}' \, h(\mathbf{r} - \mathbf{r}') u(\mathbf{r}')$$

known as the *random phase approximation*. A consequence of Gaussian statistics is that $h(\mathbf{r})$ appears on both sides of the equation, more generally such a relation would require knowledge of a higher bodied moment of the density field. Only Gaussian theories are specified completely by their first two cumulants.

The expression for $h(\mathbf{r})$ is a self-consistent linear integral equation. To solve it, we introduce a Fourier transform, a useful technique whenever you are faced with convolution integrals like the two terms under the integral on the right. Let us assume the pair potential can be Fourier transformed, which is often not possible, but we are all right in the current case because we are assuming here that potentials are weak and not singular. We define our Fourier transforms

$$h(\mathbf{r}) = \int \frac{d\mathbf{k}}{(2\pi)^3}\, \hat{h}(\mathbf{k})e^{-i\mathbf{k}\cdot\mathbf{r}} \qquad u(\mathbf{r}) = \int \frac{d\mathbf{k}}{(2\pi)^3}\, \hat{u}(\mathbf{k})e^{-i\mathbf{k}\cdot\mathbf{r}}$$

and using the convolution theorem, we find that our integral equation has become an algebraic equation,

$$\hat{h}(\mathbf{k}) = -\beta\hat{u}(\mathbf{k}) - \beta\bar{\rho}\hat{h}(\mathbf{k})\hat{u}(\mathbf{k})$$
$$\Rightarrow \quad \hat{h}(\mathbf{k}) = \frac{-\beta\hat{u}(\mathbf{k})}{1 + \beta\bar{\rho}\hat{u}(\mathbf{k})} \ .$$

Until now, the details of the system have not been entered into our analysis. The analysis is general within the assumption of linear response.

Pairs of ions interact with a Coulomb potential of the form $u(\mathbf{r}) = z^2/r$ in an appropriate unity system. Its Fourier transform,

$$\hat{u}(\mathbf{k}) = \frac{4\pi z^2}{k^2}$$

can be substituted into our equation for $\hat{h}(\mathbf{k})$, yielding

$$\hat{h}(\mathbf{k}) = \frac{-\beta 4\pi z^2/k^2}{1 + \beta\bar{\rho}4\pi z^2/k^2} = \frac{-\beta 4\pi z^2}{k^2 + \kappa^2}$$

where,

$$\kappa^{-1} = 1/\sqrt{\beta\bar{\rho}4\pi z^2}$$

is a natural lengthscale that enters into the analysis, known as the *Debye length*.

Exercise 3.7: Demonstrate that the Fourier transform of $1/r$ in $3d$ is $4\pi/k^2$. Hint: You could use the fact that the Coulomb potential is the solution of Poisson's equation $-\nabla^2\psi(\mathbf{r}) = \delta(\mathbf{r})$ where ψ is the electrostatic potential.

In order to evaluate the pair correlation function in real space, we need to take an inverse Fourier transform. This is defined as,

$$h(r) = \int \frac{d\mathbf{k}}{(2\pi)^3}\, \hat{h}(\mathbf{k})e^{-i\mathbf{k}\cdot\mathbf{r}}$$
$$= \frac{-\beta z^2}{2\pi^2} \int d\mathbf{k}\, \frac{1}{k^2 + \kappa^2}e^{-i\mathbf{k}\cdot\mathbf{r}}$$

where in order to evaluate these integrals we define the angle between $\mathbf{k}$ and $\mathbf{r}$ as θ, $\mathbf{k} \cdot \mathbf{r} = kr \cos(\theta)$. It follows in that spherical coordinate system,

$$
\begin{aligned}
h(r) &= \frac{-\beta z^2}{\pi} \int_0^\pi d\theta \int_0^\infty dk\, k^2 \, \sin(\theta) \frac{1}{k^2 + \kappa^2} \exp\left[-ikr\cos\theta\right] \\
&= \frac{-2\beta z^2}{\pi r} \int_0^\infty dk\, k \, \sin(kr) \frac{1}{k^2 + \kappa^2} \\
&= \frac{-\beta z^2}{r} e^{-r\kappa}
\end{aligned}
$$

or in terms of the radial distribution function,

$$
g(r) = 1 - \frac{\beta z^2}{r} e^{-r\kappa} .
$$

This equation is valid for low $\bar{\rho}$, where typical interactions are small and the typical distance between ions is large, as the small r limit of the theory is unphysical. Notice that according to this equation, fluids for which $u(r)$ is long-ranged, decaying as $1/r$, have pair correlations that decay exponentially with a *screening length* $1/\kappa$. Notice too that in the current context we have considered charges of only one sign, and yet correlations are still short ranged. Had we evaluated the correlation between unlike charges we would have found $g(r) = 1 + \beta z^2 \exp[-r\kappa]/r$, identical to the case of like charges except exhibiting an enhancement rather then a suppression at small r. The natural interpretation of these short-ranged correlations is that an ion interacts with many others in its vicinity, all with about the same strength due to the slowly varying potential. The net effect is that no single ion pair becomes strongly correlated. An extension to a solution of multiple types of ions, cations, and anions, is straightforward. Using the pair distribution function, a host of thermodynamic quantities can be evaluated using the relations in the previous section.

Exercise 3.8: A small generalization of the relationship between the Helmholtz free energy and $g(r)$ for a collection of monovalent cations and anions in equal number that interact through a Coulomb potential $u_{ij}(r) = (2\delta_{ij} - 1)z^2/r$ for $i, j = c, a$ for cations and anions, respectively, is

$$
A = k_{\mathrm{B}} T N \ln \bar{\rho} \lambda^3 + \bar{\rho} N \int d\mathbf{r} [g_{cc}(r) u_{cc}(r) + g_{ca}(r) u_{ca}(r)]
$$

where λ is the thermal wavelength, assumed here to be the same for both species, and $g_{ij}(r)$ refers to the pair correlation between ions of type i and j. Using this generalization, compute the chemical potential within the Debye–Hückel theory approximation for the correlation functions. You should find a correction to the ideal form that depends on $\sqrt{z^2 \bar{\rho}}$.

3.7 Gaussian density field theory

Our aim in this section is to construct a theory appropriate for a dense solution. As we will see, this is a bit more subtle than the Debye–Hückel theory we just used, since the excluded volume interactions that dictate liquid structure are not weak, and so it is not clear that a linear theory should work. To construct such a theory, we return to our previous model of density fluctuations, with effective Gaussian Hamiltonian,

$$\beta\mathcal{H}[\delta\rho(\mathbf{r})] = \frac{1}{2}\int d\mathbf{r}\int d\mathbf{r}'\,\delta\rho(\mathbf{r})\chi^{-1}(\mathbf{r},\mathbf{r}')\delta\rho(\mathbf{r}')$$

with correlation function $\chi(\mathbf{r},\mathbf{r}')$. The partition function for this model is,

$$Q = \int \mathcal{D}[\rho(\mathbf{r})]\,e^{-\beta\mathcal{H}[\delta\rho(\mathbf{r})]}$$

where the integral is over all density fields $\rho(\mathbf{r})$ with suitable path measure $\mathcal{D}[\rho(\mathbf{r})]$.

As usual, it will be useful for constructing a cumulant generating function by introducing an external field that acts linearly on the density,

$$Q[\phi] = \int \mathcal{D}[\rho(\mathbf{r})]\,e^{-\beta\mathcal{H}[\delta\rho(\mathbf{r})]}e^{\int d\mathbf{r}\,\phi(\mathbf{r})\rho(\mathbf{r})}$$

$$= Q[0]\exp\left[1/2\int d\mathbf{r}\int d\mathbf{r}'\,\phi(\mathbf{r})\chi(\mathbf{r},\mathbf{r}')\phi(\mathbf{r}') + \int d\mathbf{r}\,\phi(\mathbf{r})\bar{\rho}\right]$$

where the second line has evaluated the functional integral. Note we have absorbed β into ϕ. The second line results from rewriting $\beta\mathcal{H}[\delta\rho(\mathbf{r})]$ as a set of independent Gaussian variables, each of which can be done independently. In Fourier space,

$$\beta\mathcal{H}_0 = \frac{1}{2}\int_{\mathbf{r}}\int_{\mathbf{r}'}\delta\rho(\mathbf{r})\chi^{-1}(\mathbf{r},\mathbf{r}')\delta\rho(\mathbf{r}')$$

$$= \frac{1}{2}\int d\mathbf{r}\int d\mathbf{r}'\int_{\mathbf{k}}\int_{\mathbf{k}'}\int_{\mathbf{k}''}\delta\hat{\rho}(\mathbf{k})e^{-i\mathbf{k}\cdot\mathbf{r}}\hat{\chi}^{-1}(\mathbf{k}')e^{-i\mathbf{k}'\cdot(\mathbf{r}-\mathbf{r}')}\delta\hat{\rho}(\mathbf{k}'')e^{-i\mathbf{k}''\cdot\mathbf{r}'}$$

$$= \frac{1}{2}\int_{\mathbf{k}}|\delta\hat{\rho}(\mathbf{k})|^2\hat{\chi}^{-1}(\mathbf{k})$$

where we have uncoupled the density fluctuations, so that each $\hat{\rho}(\mathbf{k})$ fluctuates independently, like normal modes. In this basis, the matrix inverse is simple,

$$\langle|\delta\hat{\rho}(\mathbf{k})|^2\rangle = \hat{\chi}(\mathbf{k}) = \frac{1}{\hat{\chi}^{-1}(\mathbf{k})}$$

which means that in this way there is a simple analogy that can be made with harmonic oscillators, and thus the Gaussian integral we employed in the case of the forced harmonic oscillator.

From the generating function, we can compute the cumulants of the density field. For example, the first cumulant is just the average density,

$$\left.\frac{\delta \ln Q}{\delta \phi(\mathbf{r})}\right|_{\phi=0} = \langle \rho(\mathbf{r})\rangle_0$$

$$= \bar{\rho}$$

which is found by taking the first derivative and setting the field to $\phi = 0$. Similarly, the second derivative yields the correlation function

$$\left.\frac{\delta^2 \ln Q}{\delta \phi(\mathbf{r})\delta \phi(\mathbf{r}')}\right|_{\phi=0} = \left.\frac{\delta \langle \rho(\mathbf{r})\rangle_0}{\delta \phi(\mathbf{r}')}\right|_{\phi=0}$$

$$= \langle \delta\rho(\mathbf{r})\delta\rho(\mathbf{r}')\rangle_0 = \chi(\mathbf{r},\mathbf{r}')$$

or alternatively, the response function for a perturbation to the density.

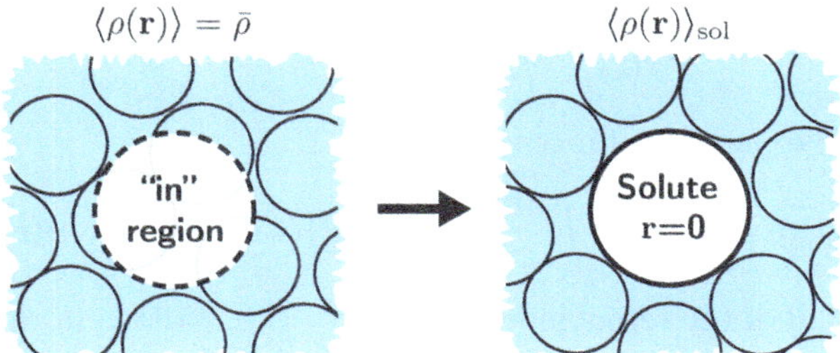

Fig. 3.6 Excluded volume as a constraint on the density field in the *in* region of space.

In order to use this model to compute the structure of a dense liquid, we need to do something more than what we did in the Debye–Hückel theory. If we push our simple linear response with strong repulsive potentials to exclude volume, the density would artificially go negative. This is a consequence of adding a large potential, one that is outside the scope of linear response. Instead of turning on a strong potential to affect an excluded volume, we can instead formulate it as a constraint. What we imagine is shown in Figure 3.6, namely we will tag a region of space, centered at the origin, and grow in a hard object—a solute— that will not act on the surrounding fluid other than to exclude its density. Mathematically, we can implement that constraint as a delta function that sets the density in some region, "in" to zero. It will be advantageous to use a Fourier representation of the delta function,

$$\prod_{\mathbf{r}\in\mathrm{in}} \delta[\rho(\mathbf{r})] = \int \mathcal{D}\psi(\mathbf{r}) \exp\left[i \int_{\mathbf{r}\in\mathrm{in}} d\mathbf{r}\,\psi(\mathbf{r})\rho(\mathbf{r})\right]$$

which is the functional generalization of a Fourier transform, with auxiliary field $\psi(r)$. The integral on the right-hand side is limited to a domain of the region "in."

The Gaussian model with the excluded volume constraint has a partition function, labeled Q_{sol}, that can be written as,

$$Q_{\text{sol}}[\phi] = \int \mathcal{D}[\psi(\mathbf{r})] \int \mathcal{D}[\rho(\mathbf{r})] \, e^{-\beta\mathcal{H}[\delta\rho(\mathbf{r})]} e^{\int_{\mathbf{r}} \phi(\mathbf{r})\rho(\mathbf{r}) + i\int_{\mathbf{r}\in\text{in}} \psi(\mathbf{r})\rho(\mathbf{r})} \, .$$

Because the density fluctuations are Gaussian distributed and the constraint enters linearly, we can integrate the density out, yielding,

$$\frac{Q_{\text{sol}}[\phi]}{Q_0[0]} = \exp\left[\frac{1}{2}\int_{\mathbf{r}}\int_{\mathbf{r}'} \phi(\mathbf{r})\chi(\mathbf{r},\mathbf{r}')\phi(\mathbf{r}') + \int_{\mathbf{r}} \phi(\mathbf{r})\bar{\rho}\right] \int \mathcal{D}[\psi(\mathbf{r})] \exp\left[i\int_{\mathbf{r}\in\text{in}} \psi(\mathbf{r})\bar{\rho}\right.$$
$$\left. - \frac{1}{2}\int_{\mathbf{r}\in\text{in}}\int_{\mathbf{r}'\in\text{in}} \psi(\mathbf{r})\chi(\mathbf{r},\mathbf{r}')\psi(\mathbf{r}') + i\int_{\mathbf{r}\in\text{in}}\int_{\mathbf{r}'} \psi(\mathbf{r})\chi(\mathbf{r},\mathbf{r}')\phi(\mathbf{r}')\right]$$

where we notice that just as in computing the generating function, we have generated bilinear terms in ϕ and in ψ, and a mixed term proportional to $\phi\psi$. In order to compute the generating function and extract information on the density field, we make a substitution,

$$f(\mathbf{r}) = \int_{\mathbf{r}'} \chi(\mathbf{r},\mathbf{r}')\phi(\mathbf{r}')$$

and inserting it into the partition function yields,

$$\frac{Q_{\text{sol}}[\phi]}{Q_0[\phi]} = \int \mathcal{D}[\psi(\mathbf{r})]\exp\left[-\frac{1}{2}\int_{\mathbf{r}\in\text{in}}\int_{\mathbf{r}'\in\text{in}} \psi(\mathbf{r})\chi(\mathbf{r},\mathbf{r}')\psi(\mathbf{r}') + i\int_{\mathbf{r}\in\text{in}} \psi(\mathbf{r})\{\bar{\rho} + f(\mathbf{r})\}\right]$$

where $Q_0[\phi]$ contains all of the terms independent of $\psi(r)$. Written in this way, we see that this too is a Gaussian field, now in the auxiliary field introduced to impose the excluded volume constraint. In order to integrate this field out we need to confront the fact that those bilinear integrals are restricted to a domain, "in." We can anticipate that in integrating them out, we will need to take a functional inverse of χ, but restricted to that domain. To do so we define this inverse as $\chi_{\text{in}}^{-1}(\mathbf{r},\mathbf{r}'')$. It satisfies the property,

$$\int_{\mathbf{r}''\in\text{in}} \chi_{\text{in}}^{-1}(\mathbf{r},\mathbf{r}'')\chi(\mathbf{r}'',\mathbf{r}') = \delta(\mathbf{r}-\mathbf{r}') \qquad \mathbf{r},\mathbf{r}' \in \text{in}$$

Integrating out the auxiliary field, we obtain

$$Q_{\text{sol}}[\phi] \propto Q_0[\phi] \exp\left[-\frac{1}{2}\int_{\mathbf{r}\in\text{in}}\int_{\mathbf{r}'\in\text{in}} \{\bar{\rho} + f(\mathbf{r})\}\chi_{\text{in}}^{-1}(\mathbf{r},\mathbf{r}')\{\bar{\rho} + f(\mathbf{r}')\}\right]$$

which includes the unconstrained part, $Q_0[\phi]$, and an additional Gaussian weight with correlation function $\chi_{\text{in}}^{-1}(\bar{\mathbf{r}},\bar{\mathbf{r}}')$ that exists over a finite domain.

The free energy change for evacuating the solvent, equivalently solvating a particle the size of the cavity, is given by the ratio of partition functions. This standard-state solute chemical potential or solvation free energy is given for $\phi = 0$, by

$$\beta\mu^{(\text{o})} = -\ln\left(Q_{\text{sol}}[0]/Q_0[0]\right)$$
$$= \frac{\bar{\rho}^2}{2}\int_{\mathbf{r}\in\text{in}}\int_{\mathbf{r}'\in\text{in}} \chi_{\text{in}}^{-1}(\mathbf{r},\mathbf{r}') + \frac{1}{2}\ln\left[2\pi\int_{\mathbf{r}\in\text{in}}\int_{\mathbf{r}'\in\text{in}} \chi_{\text{in}}(\mathbf{r},\mathbf{r}')\right]$$

which is the reversible work to remove density from the "in" region. Since this reversible work is identical to the probability that a spontaneous fluctuation empties out the cavity, the first term is identified as the mean density squared divided by an effective microscopic compressibility. Indeed if χ_{in} were just a number, this first term would reduce to $\bar{N}^2/2\chi_{\text{in}}$ where $\bar{N}$ is the average number of particles in the cavity. The second term approximately accounts for the change in the normalization due to the presence of the constraint.

From the generating function, we can compute the mean density around the cavity,

$$\frac{\delta \ln Q_{\text{sol}}}{\delta \phi(\mathbf{r})}\bigg|_{\phi=0} = \langle \rho(\mathbf{r}) \rangle_{\text{sol}} = \bar{\rho} - \bar{\rho} \int_{\mathbf{r}' \in \text{in}} \int_{\mathbf{r}'' \in \text{in}} \chi(\mathbf{r}, \mathbf{r}') \chi_{\text{in}}^{-1}(\mathbf{r}'', \mathbf{r}')$$

and find that, at first blush, it looks very different from the linear response equation we had previously. Making a substitution that defines a function $c(r)$,

$$c(\mathbf{r}') = -\bar{\rho} \int_{\mathbf{r}'' \in \text{in}} \chi_{\text{in}}^{-1}(\mathbf{r}'', \mathbf{r}') \qquad \mathbf{r}' \in \text{in}$$
$$= 0 \quad \mathbf{r}' \notin \text{in}$$

and inserting it into the mean density we find,

$$\langle \rho(\mathbf{r}) \rangle_{\text{sol}} = \bar{\rho} + \int_{\mathbf{r}'} \chi(\mathbf{r}, \mathbf{r}') c(\mathbf{r}')$$

which looks just like Debye–Hückel theory. If $\phi(\mathbf{r}) = -k_{\text{B}}Tc(\mathbf{r})$ then we have a form that is isomorphic to our old linear response. However, hiding under the hood of those definitions is the real difference between that previous calculation and this one. If we compute the fluctuations in the density,

$$\langle \delta\rho(\mathbf{r})\delta\rho(\mathbf{r}') \rangle_{\text{sol}} = \chi(\mathbf{r}, \mathbf{r}') - \int_{\bar{\mathbf{r}} \in \text{in}} \int_{\bar{\mathbf{r}}' \in \text{in}} \chi(\mathbf{r}, \bar{\mathbf{r}}) \chi_{\text{in}}^{-1}(\bar{\mathbf{r}}, \bar{\mathbf{r}}') \chi(\bar{\mathbf{r}}', \mathbf{r}')$$

we find that they have changed! So this is not a simple linear response at all. The condition we have imposed on the density in the region of the solute is akin to

$$\int_{\mathbf{r}'} \chi(\mathbf{r}, \mathbf{r}') c(\mathbf{r}) = -\bar{\rho} \qquad \mathbf{r} \in \text{in}$$

so, $c(\mathbf{r})$ is a field due to the solute that for $\mathbf{r} \notin$ "in" is zero, but causes the density to go to zero for $\mathbf{r} \in$ in. It is known as the *direct correlation function*. The change in response is akin to a change in the normal modes. By constraining the density field at a point in space, we have broken translational symmetry, and this changes the modes.

3.8 Integral equations for liquid structure

We have a theory for the structure of a dense liquid, but by itself the equation we have derived is not closed. Specifically,

$$\langle \rho(\mathbf{r}) \rangle_{\text{sol}} = \bar{\rho} + \int_{\mathbf{r}'} \chi(\mathbf{r} - \mathbf{r}') c(\mathbf{r}')$$

needs further input in order for the density to be determined. One way to solve this equation is to consider the solvent as a collection of hard spheres and set the solute also as a hard sphere with the same size as the solvent. In that case, the density around the tagged particle is

$$\langle \rho(\mathbf{r}) \rangle_{\text{sol}} = \bar{\rho} + \int_{\mathbf{r}'} \left[\bar{\rho}\delta(\mathbf{r} - \mathbf{r}') + \bar{\rho}^2 g(\mathbf{r} - \mathbf{r}') - \bar{\rho}^2 \right] c(\mathbf{r}')$$

and using $\langle \rho(\mathbf{r}) \rangle_{\text{sol}} = \bar{\rho}g(\mathbf{r})$ with $h(\mathbf{r}) = g(\mathbf{r}) - 1$,

$$h(\mathbf{r}) = c(\mathbf{r}) + \bar{\rho} \int_{\mathbf{r}'} h(\mathbf{r} - \mathbf{r}')c(\mathbf{r}')$$

we find a self-consistent equation much like we had in Debye–Hückel theory, with the same $h(r)$ entering on both sides of the equation. This equation is known as the Ornstein–Zernike equation. In order to solve it we need boundary conditions. It can be solved using the following closure due to Percus and Yevick

$$c(r) = 0 \qquad r \geq \sigma$$
$$h(r) = -1 \qquad r \leq \sigma$$

which we can understand from the action of the constraint. With these conditions, the solution for $c(r)$ has a simple form, though one tedious to derive,

$$c(r) = \lambda_1 + \lambda_2 \frac{r}{\sigma} + \lambda_2 \left(\frac{r}{\sigma} \right)^3 \qquad r \leq \sigma$$

where the coefficients λ_i's depend on the macroscopic density $\bar{\rho}$. This dependence is more compactly expressed through the packing fraction $\eta = (\pi/6)\bar{\rho}\sigma^3$

$$\lambda_1 = -\frac{(1 + 2\eta)^2}{(1 - \eta)^4} \qquad \lambda_2 = 6\eta\frac{(1 + \eta/2)^2}{(1 - \eta)^4} \qquad \lambda_3 = \frac{\eta}{2}\lambda_1 \, .$$

The solution for $h(r)$ cannot be written in such a simple analytical form, but can be evaluated numerically. However, by rewriting, $g(r) = \exp[-\beta u(r)]y(r)$, where $y(r)$ is a smooth function and $u(r)$ the hard sphere potential, our previous relationship between the pressure and the $g(r)$ can be used to deduce

$$\beta p = \bar{\rho} + \frac{2\pi\bar{\rho}^2}{3}\sigma^3 g(\sigma)$$

and noting $g(\sigma) = -c(\sigma)$, we arrive at the *virial equation of state,*

$$\beta p/\bar{\rho} = \frac{1 + 2\eta + 3\eta^2}{(1 - \eta)^2}$$

which is accurate for hard spheres over a reasonable range of densities.

In the early days of liquid state theory, there was a tremendous effort trying to find appropriate closures for more complicated fluids, ones not just determined by packing. One particularly useful closure adds a weak potential to the constraint by imposing,

$$c(r) = -\beta u(r) \qquad r \geq \sigma$$
$$h(r) = -1 \qquad r \leq \sigma$$

where $u(r)$ is some weak long-ranged attraction. This is known as the *mean spherical approximation* and is useful for going beyond Debye–Hückel theory to incorporate excluded volume effects into ionic solutions. Alternatively, if the $g(r)$ is known from experiment then it could be used to solve for the density field around a solute. This is the basis of the *Pratt–Chandler theory* of the hydrophobic effect. Molecular extensions and generalizations to polymers are also possible and referred to as reference interaction site models.

3.9 Dielectric continuum theory

The tools we have used to understand liquid structure and packing can also be applied to polar solvation, and the thermodynamics of electrolytes. Empirically, molecular dynamics simulations indicate that the distribution of electrostatic potentials in a cavity embedded in a polar environment is to a very good approximation Gaussian distributed. As such we expect that a linear response or Gaussian field theory may be highly accurate in predicting solvation energies for small charged solutes. Essentially, we will postulate an analogous theory as we did for the density fluctuations of a dense fluid, but here the fluctuation variables are $\mathbf{m}(\mathbf{r})$, the dipole density at $\mathbf{r}$. As before, the normal modes of the liquid can be expressed as the Fourier components of these correlated Gaussian random variables, $\hat{\mathbf{m}}(\mathbf{k})$. To construct this theory, we will define a continuous dipole field by imagining that the polar solvent can be smeared out over a length scale comparable to their molecular diameter. We will not resolve the discreteness of the atoms and molecules, but rather construct a continuous field, which may be justified by the long-ranged nature of the electrostatic interactions we are considering. Specifically, we assume that the variation of the electrostatic potential is slow relative to changes in the density so that the latter is constant. A cartoon of this coarse-graining is illustrated in Figure 3.7.

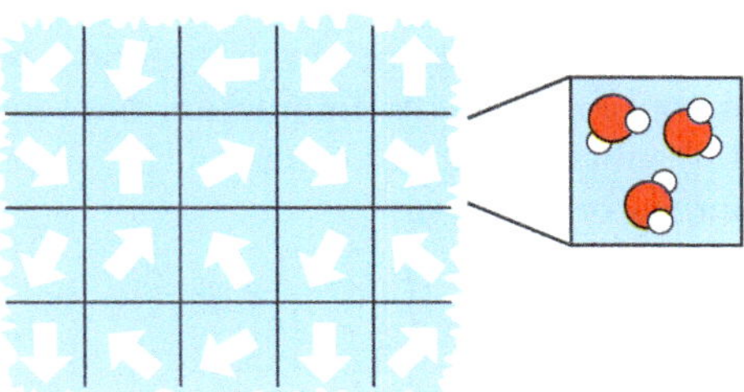

Fig. 3.7 Coarse-grained dipolar field, where the molecular details are integrated out.

Let $\bar{\rho}\mathbf{m}(\mathbf{r})$ be the dipole density at $\mathbf{r}$. The effective Hamiltonian determining the fluctuations of the dipole field in the absence of any perturbation, $\mathcal{H}_0$, can be considered as being the sum of two parts

$$\mathcal{H}_0 = \mathcal{H}_{\text{local}} + \mathcal{H}_{\text{int}}$$

where $\mathcal{H}_{\text{local}}$ is the free energy to create a local dipole and $\mathcal{H}_{\text{int}}$ reflects the interactions between dipoles at different points in space. While the first term reflects an instantaneous local polarization, the second reports on the electrostatic coupling of that polarization to the surroundings. We will assume that this effective Hamiltonian has a Gaussian form,

$$\beta \mathcal{H}_0 = \frac{1}{2} \int_{\mathbf{r}} \int_{\mathbf{r}'} \mathbf{m}(\mathbf{r}) \cdot \chi^{-1}(\mathbf{r} - \mathbf{r}') \cdot \mathbf{m}(\mathbf{r}')$$

where as in our discussions of density fluctuations, $\chi(\mathbf{r}-\mathbf{r}') = \langle \mathbf{m}(\mathbf{r}) \otimes \mathbf{m}(\mathbf{r}') \rangle$, where $\otimes$ denotes outer or tensor product. This correlation function is a little more complicated than in the case of density fluctuations. As $\mathbf{m}(\mathbf{r})$ is a vector quantity, the correlation function contains information about the magnitude of correlations as well as relative orientations. This means it is a tensorial quantity.

The two parts of the effective Hamiltonian can be easily intuited. The local part can be written as

$$\mathcal{H}_{\text{local}} = \frac{1}{2} \bar{\rho} \int_{\mathbf{r}} |\mathbf{m}(\mathbf{r})|^2 / \alpha$$

where $\bar{\rho}$ is the molecular density and α is the bare polarizability, which relates to the energetic cost of developing a local dipole moment. The interaction part follows from a particular writing of the dipole–dipole interaction,

$$\mathcal{H}_{\text{int}} = \frac{1}{2} \bar{\rho}^2 \int_{\mathbf{r}} \int_{\mathbf{r}'} \mathbf{m}(\mathbf{r}) \cdot \nabla\nabla' \frac{1}{|\mathbf{r} - \mathbf{r}'|} \cdot \mathbf{m}(\mathbf{r}')$$

which is just a continuum representation of the dipole–dipole interaction. By collecting terms, this leads us to suspect that the inverse correlation function takes the form in Gaussian units,

$$\chi^{-1}(\mathbf{r} - \mathbf{r}') \overset{?}{=} \beta \left[\frac{\bar{\rho}}{\alpha} \mathbf{1} \delta(\mathbf{r} - \mathbf{r}') + \bar{\rho}^2 \nabla\nabla' \frac{1}{|\mathbf{r} - \mathbf{r}'|} \right]$$

however, we should note that as written, the interaction part of the Hamiltonian contains a singular contribution from the interaction between two dipoles on top of each other that we need to subtract out. To compute the contribution of the local part of $\mathcal{H}_{\text{int}}$, let's define an indicator function,

$$h_R(\mathbf{r}) = \begin{cases} 1, & r < R \\ 0, & r > R \end{cases}$$

which reports on whether a region overlaps with the origin on a scale R. We want to compute

$$\mathcal{H}_{\text{self}} = \lim_{R \to 0} \frac{1}{2} \bar{\rho}^2 \int_{\mathbf{r}} \int_{\mathbf{r}'} \mathbf{m}(\mathbf{r}) \cdot \nabla\nabla' \frac{1}{|\mathbf{r} - \mathbf{r}'|} \cdot \mathbf{m}(\mathbf{r}') h_R(\mathbf{r} - \mathbf{r}')$$

in the limit for $R \to 0$. It will make the calculation much simpler if we do part of it in Fourier space, as there will be a natural need to employ a delta function in the small R limit. For small R, $\mathbf{m}(\mathbf{r}') = \mathbf{m}(\mathbf{r})$, and rewriting the interaction piece

$$\mathcal{H}_{\text{self}} = \frac{1}{2} \bar{\rho}^2 \int d\mathbf{r} \int d\mathbf{r}' \mathbf{m}(\mathbf{r}) \cdot \nabla\nabla \frac{1}{|\mathbf{r} - \mathbf{r}'|} \cdot \mathbf{m}(\mathbf{r}) h_R(\mathbf{r} - \mathbf{r}')$$

$$= \frac{2\pi}{3} \bar{\rho}^2 \int d\mathbf{r} \, |\mathbf{m}(\mathbf{r})|^2$$

where to prove, one uses Fourier transforms and employs the fact that a spherically symmetric function in real space is also spherically symmetric in Fourier space. Collecting this additional term we find that the correlator must be

$$\chi^{-1}(\mathbf{r} - \mathbf{r}') = \beta\bar{\rho}\left(\alpha^{-1} - \frac{4\pi\bar{\rho}}{3}\right)\mathbf{1}\delta(\mathbf{r} - \mathbf{r}') + \beta\bar{\rho}^2\nabla\nabla'\frac{1}{|\mathbf{r} - \mathbf{r}'|}$$

which now cleanly separates into a self-term and a correlation term.

Exercise 3.10: Confirm the calculation of $\mathcal{H}_{\text{self}}$ following the suggested expression in Fourier space.

3.10 Response to an external field

Now that we have the inverse of the correlation function, we can use the same relationships developed in other contexts to write down the response of the dipole field to an external perturbation that couples to it. Let us take as an example, the response to an external electric field. An electric field couples to a dipole linearly, so our full Hamiltonian can be written as

$$\mathcal{H} = \mathcal{H}_0 - \bar{\rho}\int d\mathbf{r}\,\mathbf{m}(\mathbf{r}) \cdot \mathcal{E}_{\text{ext}}(\mathbf{r})$$

where the dot product of the electric and dipole fields, integrated over space, is subtracted off from our original effective Hamiltonian. We know immediately from linear response how this changes the average dipole at a point $\mathbf{r}$,

$$\langle\mathbf{m}(\mathbf{r})\rangle = \beta\bar{\rho}\int d\mathbf{r}'\,\chi(\mathbf{r} - \mathbf{r}') \cdot \mathcal{E}_{\text{ext}}(\mathbf{r}')$$

namely that it is proportional to the field, with a proportionality constant that depends on the fluctuations of the dipoles.

The total electric field generated from the application of the external field has a contribution also from induced polarization, $\mathcal{E}(\mathbf{r}) = \mathcal{E}_{\text{ext}}(\mathbf{r}) + \mathcal{E}_{\text{ind}}(\mathbf{r})$, where the inductive part on average is

$$\langle\mathcal{E}_{\text{ind}}(\mathbf{r})\rangle = -\bar{\rho}\int d\mathbf{r}'\,\nabla\nabla'\frac{1}{|\mathbf{r} - \mathbf{r}'|} \cdot \langle\mathbf{m}(\mathbf{r}')\rangle\,.$$

The average induced field is proportional to the averaged dipole, and by substituting our expression for the average dipole and writing out the inverse correlation function we find,

$$\begin{aligned}
\langle\mathcal{E}_{\text{ind}}(\mathbf{r})\rangle &= -\bar{\rho}\int_{\mathbf{r}'}\frac{1}{\beta\bar{\rho}^2}\left[\chi^{-1}(\mathbf{r} - \mathbf{r}') - \beta\bar{\rho}\left(\frac{1}{\alpha} - \frac{4\pi\rho}{3}\right)\mathbf{1}\delta(\mathbf{r} - \mathbf{r}')\right] \cdot \langle\mathbf{m}(\mathbf{r})\rangle \\
&= -\int_{\mathbf{r}''}\left[\delta(\mathbf{r} - \mathbf{r}'')\mathcal{E}_{\text{ext}}(\mathbf{r}'') - \beta\bar{\rho}\left(\frac{1}{\alpha} - \frac{4\pi\bar{\rho}}{3}\right)\chi(\mathbf{r} - \mathbf{r}'') \cdot \mathcal{E}_{\text{ext}}(\mathbf{r}'')\right] \\
&= -\mathcal{E}_{\text{ext}}(\mathbf{r}) + \left(\frac{1}{\alpha} - \frac{4\pi\bar{\rho}}{3}\right)\langle\mathbf{m}(\mathbf{r})\rangle
\end{aligned}$$

or rearranging,

$$\langle \mathbf{m}(\mathbf{r}) \rangle = \left(\alpha^{-1} - \frac{4\pi\bar{\rho}}{3} \right)^{-1} \langle \mathcal{E}(\mathbf{r}) \rangle$$

there is a simple relationship between the total average electric field and the average dipole. Note the linearity of the contributions to the total electric field from the external and the induced field allowed us to find this relationship without inverting $\chi(\mathbf{r} - \mathbf{r}')$.

The relationship between the average electric field and the average dipole bears a striking relationship to an old result in electrostatics from the theory of continuous dielectric media, or *dielectric continuum theory*. In dielectric continuum theory, it is postulated that there is a linear relationship between the induced polarization field, $\mathbf{P}(\mathbf{r})$, and the electric field,

$$\mathbf{P}(\mathbf{r}) \propto \mathcal{E}(\mathbf{r})$$

where the proportionality is a constant

$$\mathbf{P}(\mathbf{r}) = \chi_{\mathrm{D}} \mathcal{E}(\mathbf{r}) \qquad \chi_{\mathrm{D}} = \frac{\epsilon - 1}{4\pi}$$

which defines ϵ as the well-known *dielectric constant*. To understand what the dielectric constant means, consider the electrostatic potential generated by a polarization field constrained to a volume V of dielectric media,

$$\phi_{\mathrm{p}}(\mathbf{r}) = \int_V d\mathbf{r}' \, \mathbf{P}(\mathbf{r}') \cdot \nabla' \frac{1}{|\mathbf{r} - \mathbf{r}'|}$$
$$= \int_S dS' \, \hat{\mathbf{n}} \cdot \mathbf{P}(\mathbf{r}') \frac{1}{|\mathbf{r} - \mathbf{r}'|} - \int_V d\mathbf{r}' \, \nabla' \cdot \mathbf{P}(\mathbf{r}') \frac{1}{|\mathbf{r} - \mathbf{r}'|}$$

where the first line integrates over the volume of the dielectric material, and the second rewrites that integral using the divergence theorem as a surface integral with surface normal $\hat{\mathbf{n}}$. Taking the Laplacian of the potential

$$\nabla^2 \phi_{\mathrm{p}}(\mathbf{r}) = \int_S dS' \, \hat{\mathbf{n}} \cdot \mathbf{P}(\mathbf{r}') \delta(\mathbf{r} - \mathbf{r}') + 4\pi \nabla \mathbf{P}(\mathbf{r})$$
$$= 4\pi \nabla \mathbf{P}(\mathbf{r})$$

where we note that the polarization is 0 inside of V, we find an expression of Poisson's equation in terms of the polarization field. With the addition of a charge density $\rho_c(\mathbf{r})$ there is an additional contribution to the potential,

$$\nabla^2 \phi(\mathbf{r}) = 4\pi \nabla \mathbf{P}(\mathbf{r}) - 4\pi \rho_c(\mathbf{r})$$
$$= -4\pi \chi_{\mathrm{D}} \nabla^2 \phi(\mathbf{r}) - 4\pi \rho_c(\mathbf{r})$$
$$= -\frac{4\pi}{1 + 4\pi\chi_{\mathrm{D}}} \rho_c(\mathbf{r})$$

where we have utilized the supposed relationship between the polarization and electric field. Since $\chi_{\mathrm{D}} = (\epsilon - 1)/4\pi$ or $4\pi\chi_{\mathrm{D}} + 1 = \epsilon$ we have finally

$$\nabla^2 \phi(\mathbf{r}) = -\frac{4\pi}{\epsilon} \rho_c(\mathbf{r})$$

which shows that the dielectric constant attenuates the effective strength of electrostatic interactions. This is where our notion of screening comes from. It is worth noting the typical size of ϵ. For the vacuum $\epsilon = 1$ by definition. For an ideal conductor, $\epsilon = \infty$ as no electric fields can be generated. For typical nonpolar solvents $\epsilon \approx 2$, while for a polar solvent like water $\epsilon = 80$.

Microscopically, we have a relationship between the average dipole and the average electric field,

$$\langle \mathbf{m}(\mathbf{r}) \rangle = \left(\alpha^{-1} - \frac{4\pi\bar{\rho}}{3} \right)^{-1} \langle \mathcal{E}(\mathbf{r}) \rangle$$

since $\mathbf{P}(\mathbf{r}') = \bar{\rho} \langle \mathbf{m}(\mathbf{r}) \rangle$, in the macroscopic limit, it must be that

$$\chi_{\mathrm{D}} = \frac{\bar{\rho}}{\alpha^{-1} - 4\pi\bar{\rho}/3} = \frac{\epsilon - 1}{4\pi}$$

or rearranging

$$\frac{4\pi\bar{\rho}\alpha}{3} = \frac{\epsilon - 1}{\epsilon + 2}$$

we find a relationship between the microscopic polarizability and the macroscopic dielectric response constant. This is known as the *Clausius–Mossotti relation* and is an example of a fluctuation–dissipation relationship. This lets us express the correlator as

$$\chi^{-1}(\mathbf{r} - \mathbf{r}') = \beta\bar{\rho}^2 \left(\frac{4\pi}{\epsilon - 1} \mathbf{1}\delta(\mathbf{r} - \mathbf{r}') + \nabla\nabla' \frac{1}{|\mathbf{r} - \mathbf{r}'|} \right)$$

in terms of an easily measurable quantity.

To go further with this theory, we will need to invert χ^{-1}. The first step is to express it in normal modes

$$\hat{\chi}^{-1}(\mathbf{k}) = \beta\bar{\rho}^2 \left(\frac{4\pi}{\epsilon - 1} \mathbf{1} + 4\pi\hat{k}\hat{k}^T \right)$$

using a Fourier representation. Additionally, we need a basis for the tensorial elements of $\hat{\chi}^{-1}(\mathbf{k})$. Let's introduce a basis set for working with tensors,

$$\mathbf{J}_+ = \hat{k}\hat{k}^T , \qquad \mathbf{J}_- = 1 - \hat{k}\hat{k}^T , \qquad \mathbf{J}_+ + \mathbf{J}_- = 1$$

where $\hat{k}$ are unit vectors, and the $\mathbf{J}$ tensors satisfy the orthogonality relationships

$$\mathbf{J}_+\mathbf{J}_+ = \mathbf{J}_+ , \qquad \mathbf{J}_-\mathbf{J}_- = \mathbf{J}_- , \qquad \mathbf{J}_+\mathbf{J}_- = 0$$

so that

$$(a\mathbf{J}_+ + b\mathbf{J}_-)^{-1} = \frac{1}{a}\mathbf{J}_+ + \frac{1}{b}\mathbf{J}_-$$

is simply invertible. Expressing $\hat{\chi}^{-1}$ in this basis, we have

$$\hat{\chi}^{-1}(\mathbf{k}) = 4\pi\beta\bar{\rho}^2 \left(\frac{1}{\epsilon - 1}\mathbf{J}_- + \frac{\epsilon}{\epsilon - 1}\mathbf{J}_+ \right)$$

or inverting using this convenient basis

$$\hat{\chi}(\mathbf{k}) = \frac{1}{4\pi\beta\bar{\rho}^2}\left[(\epsilon - 1)\mathbf{J}_- + \frac{\epsilon - 1}{\epsilon}\mathbf{J}_+ \right]$$

$$= \frac{\epsilon - 1}{4\pi\beta\bar{\rho}^2}\left[\mathbf{1} - \frac{\epsilon - 1}{\epsilon}\hat{k}\hat{k}^T \right]$$

we have a simple expression of the normal modes of the dipole field. In real space this becomes,

$$\chi(\mathbf{r} - \mathbf{r}') = \frac{\epsilon - 1}{4\pi\beta\bar{\rho}^2}\left[\mathbf{1}\delta(\mathbf{r} - \mathbf{r}') - \frac{\epsilon - 1}{4\pi\epsilon}\nabla\nabla'\frac{1}{|\mathbf{r} - \mathbf{r}'|} \right]$$

$$= \langle \mathbf{m}(\mathbf{r}) \otimes \mathbf{m}(\mathbf{r}') \rangle$$

where we find the dipole–dipole correlation function.

Because the dipole field is a vector field, its Fourier transform inherits this vector quality. A useful way to decompose the Fourier-transformed vector field is to consider the resultant field parallel and perpendicular to the wave-vector direction,

$$\hat{\mathbf{m}}(\mathbf{k}) = \int d\mathbf{r}\, e^{i\mathbf{k}\cdot\mathbf{r}}\mathbf{m}(\mathbf{r})$$

$$= \hat{\mathbf{m}}_{||}(\mathbf{k}) + \hat{\mathbf{m}}_\perp(\mathbf{k})$$

where $\hat{\mathbf{m}}_{||}(\mathbf{k})$ is the Fourier-transformed dipole field in the direction of the wave-vector, the so-called *longitudinal component*, and $\hat{\mathbf{m}}_\perp(\mathbf{k})$ is the Fourier-transformed dipole field perpendicular to the wavevector, the so-called *transverse component*. The longitudinal component is defined as the component in the direction of the wavevector,

$$\hat{\mathbf{m}}_{||}(\mathbf{k}) = \hat{k}\left(\hat{k} \cdot \hat{\mathbf{m}}(\mathbf{k}) \right) = \mathbf{J}_+\hat{\mathbf{m}}(\mathbf{k})$$

where we find the operator $\mathbf{J}_+$ corresponds directly to that direction. Analogously, we can construct the transverse component through the sum rule

$$\hat{\mathbf{m}}_\perp(\mathbf{k}) = \hat{\mathbf{m}}(\mathbf{k}) - \hat{\mathbf{m}}_{||}(\mathbf{k}) = \mathbf{J}_-\hat{\mathbf{m}}(\mathbf{k})$$

where $\mathbf{J}_-$ is found to project out the component orthogonal to the wave vector.

The structure of $\hat{\chi}(\mathbf{k})$ is block diagonal. It is diagonal in that it is a function of only a single wave vector, but along the diagonal are 3×3 matrixes mixing each spatial component. Remember in real space χ is a dense matrix, and it is the Fourier transform that transforms it into this block diagonal form as the reciprocal space is the eigenbasis for translationally invariant functions. Essentially the basis $\mathbf{J}_\pm$ further diagonalizes this block diagonal structure, which enabled us to invert it,

$$\hat{\chi}(\mathbf{k}) = \frac{1}{4\pi\beta\bar{\rho}^2}\left[(\epsilon - 1)\mathbf{J}_- + \frac{\epsilon - 1}{\epsilon}\mathbf{J}_+ \right]$$

$$= \frac{\epsilon - 1}{4\pi\beta\bar{\rho}^2}\left[\mathbf{1} - \frac{\epsilon - 1}{\epsilon}\hat{k}\hat{k}^T \right]$$

by writing it in terms of $\mathbf{J}_\pm$. That tensor basis also allows us to decompose the correlation function into transverse and longitudinal components,

$$\langle \hat{\mathbf{m}}_\perp \otimes \hat{\mathbf{m}}_\perp^* \rangle = \mathbf{J}_- \langle \hat{\mathbf{m}} \otimes \hat{\mathbf{m}}^* \rangle \, \mathbf{J}_- = \frac{1}{4\pi\beta\bar{\rho}^2}(\epsilon - 1)\mathbf{1}$$

and

$$\left\langle \hat{\mathbf{m}}_{\parallel} \otimes \hat{\mathbf{m}}_{\parallel}^* \right\rangle = \mathbf{J}_+ \langle \hat{\mathbf{m}} \otimes \hat{\mathbf{m}}^* \rangle \, \mathbf{J}_+ = \frac{1}{4\pi\beta\bar{\rho}^2}\frac{\epsilon - 1}{\epsilon}\mathbf{1}$$

respectively, where $\hat{\mathbf{m}}^*$ denotes complex conjugate of $\hat{\mathbf{m}}$. It is noteworthy that there are no inherent lengthscales associated with these correlation functions. As should be expected, the correlational function between these two components

$$\left\langle \hat{\mathbf{m}}_\perp \hat{\mathbf{m}}_{\parallel}^* \right\rangle = 0$$

is zero. For polar fluid, the dielectric constant is large, $\epsilon > 1$. For water, for example $\epsilon \approx 80$. Therefore, the transverse components of the polarization fluctuations are much larger than those associated with the longitudinal direction.

3.11 Polar solvation of an excess charge

Now that we have a better understanding of the form of the correlation function, we are in a position to compute the response of the dipole field to placing a charged cavity inside it, and in so doing deduce the solvation free energy. The setup in the system is shown in Figure 3.8.

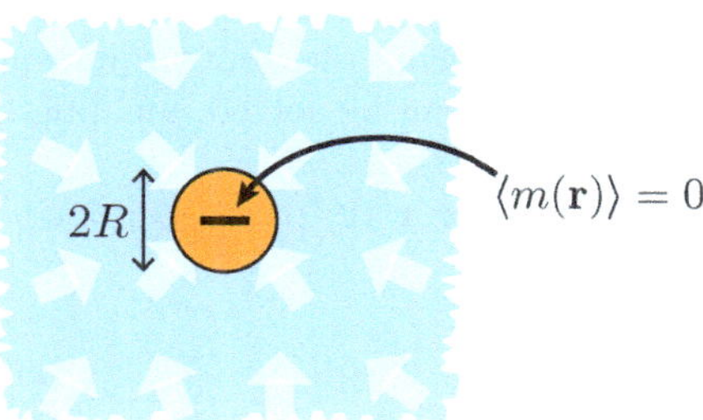

Fig. 3.8 Solvation of a charged cavity of radius R by a polar medium.

We can proceed microscopically by using the same trick we used for the density field calculation. Specifically, we imagine that the action of placing a charged cavity inside the dipole field is to constrain the value of the dipole field

$$\prod_{r\in\,\text{"in"}} \delta[\mathbf{m}(\mathbf{r})]$$

so that the dipole in the cavity is 0. A lesson we learned in doing this to the density field was that adding such a constraint will modify the fluctuations of the field. In this case, the modified correlation function, $\chi^{(m)}(\mathbf{r}, \mathbf{r}')$, is given by

$$\chi^{(m)}(\mathbf{r}, \mathbf{r}') = \chi - \int_{\text{in}} d\mathbf{r}_1 \int_{\text{in}} d\mathbf{r}_2 \, \chi(\mathbf{r} - \mathbf{r}_1)\chi_{\text{in}}^{-1}(\mathbf{r}_1, \mathbf{r}_2)\chi(\mathbf{r}_2 - \mathbf{r}')$$

where as before, we find that we must invert the correlation function inside the region of the cavity (the "in" region). To arrive at a closed-form expression for the solvation energy, we must try to compute this explicitly.

It is possible to perform this inversion for the case of a spherical cavity by asserting that $\chi_{\text{in}}^{-1}(\mathbf{r}, \mathbf{r}')$ could be written as

$$\chi_{\text{in}}^{-1}(\mathbf{r}, \mathbf{r}') = \beta\bar{\rho}^2 \left[\frac{4\pi}{\epsilon - 1}\mathbf{1}\delta(\mathbf{r} - \mathbf{r}') + \nabla\nabla'\frac{1}{|\mathbf{r} - \mathbf{r}'|} + \nabla\nabla' f(\mathbf{r}, \mathbf{r}') \right]$$

where $f(\mathbf{r}, \mathbf{r}')$ was some unknown function. From the definition of the functional inverse,

$$\int_{\text{in}} d\mathbf{r}'' \, \chi(\mathbf{r} - \mathbf{r}'')\chi_{\text{in}}^{-1}(\mathbf{r}'', \mathbf{r}') = \mathbf{1}\delta(\mathbf{r} - \mathbf{r}')$$

plugging this ansatz in

$$\int_{\text{in}} d\mathbf{r}'' \left[\nabla''\frac{1}{|\mathbf{r} - \mathbf{r}''|} \cdot \nabla''\frac{1}{|\mathbf{r}' - \mathbf{r}''|} + \nabla''\frac{1}{|\mathbf{r} - \mathbf{r}''|} \cdot \nabla'' f(\mathbf{r}', \mathbf{r}'') \right]$$
$$= -\frac{\epsilon}{\epsilon - 1}f(\mathbf{r}, \mathbf{r}') + \frac{1}{|\mathbf{r} - \mathbf{r}'|}$$

we arrive at an integro-differential equation. While it looks intimidating, the spherical symmetry of the cavity and the Coulomb potential admits an expansion of $f(\mathbf{r}, \mathbf{r}')$ in terms of spherical harmonics. By integrating the above equation by parts a number of times, the coefficients of the expansion can be solved, yielding

$$f(\mathbf{r}, \mathbf{r}') = \sum_{\ell,j} B_\ell(\mathbf{r}, \mathbf{r}')Y_{\ell,j}^*(\theta, \phi)Y_{\ell,j}(\theta, \phi)$$

$$B_\ell(\mathbf{r}, \mathbf{r}') = -4\pi(\epsilon - 1)\frac{\ell + 1}{2\ell + 1}\frac{1}{(\ell + 1)\epsilon + \ell}\frac{(\mathbf{r}\mathbf{r}')^\ell}{R^{2\ell+1}}$$

where $Y_{\ell,j}$ is the ℓ, j'th spherical harmonic and R is the radius of the cavity.

With the modified response function, the change in the average dipole in the presence of the field generated by the charged cavity can be written down

$$\langle \mathbf{m}(\mathbf{r}) \rangle = \beta\bar{\rho} \int d\mathbf{r}' \, \chi^{(m)}(\mathbf{r} - \mathbf{r}')\mathcal{E}(\mathbf{r}')$$

$$= -q\beta\bar{\rho} \int d\mathbf{r}' \, \chi^{(m)}(\mathbf{r} - \mathbf{r}')\nabla'\frac{1}{|\mathbf{r}'|}$$

where we have explicitly written the resultant electric field as the gradient of the Coulomb potential for a charge q. The average change in the internal energy of the system is given by the interaction of the average dipole field with the electric field,

$$\langle \Delta U \rangle = -\frac{1}{2}\bar{\rho} \int d\mathbf{r} \, \langle \mathbf{m}(\mathbf{r}) \rangle \, \mathcal{E}(\mathbf{r})$$

$$= \frac{1}{2}\beta\bar{\rho}^2 q^2 \int d\mathbf{r} \int d\mathbf{r}' \, \nabla\frac{1}{|\mathbf{r}|}\chi^{(m)}(\mathbf{r}-\mathbf{r}')\nabla'\frac{1}{|\mathbf{r}'|}$$

where we have input the linear response form for the average dipole in terms of the field and the modified correlation function. One immediate consequence of this is that the change in the energy scales like q^2, which is a consequence of the charge inversion symmetry of our dipole field.

To proceed we need to input the modified correlation function and perform a bunch of integrals. While this is pretty tedious, all of the integrals can be done using tricks like

$$\int d\mathbf{r}' \, \nabla\nabla'f(\mathbf{r},\mathbf{r}') \cdot \nabla'\frac{1}{|\mathbf{r}'|} = \int_S dS' \, \hat{n}\,(\dots) - \int d\mathbf{r}' \, \nabla f(\mathbf{r},\mathbf{r}') \cdot \nabla'^2\frac{1}{|\mathbf{r}'|}$$

$$= 4\pi \int d\mathbf{r}' \, \nabla f(\mathbf{r},\mathbf{r}')\delta(\mathbf{r}')$$

$$= 4\pi\nabla f(\mathbf{r},0)$$

where we make judicious use of the divergence theorem. Going through many such integrals one finds that only $\ell = 0$ terms can contribute, which is a consequence of the radial symmetry of the problem. The internal energy can be written as,

$$\langle \Delta U \rangle = \frac{1}{2}q^2 f(0,0)$$

Evaluating f for $\mathbf{r} = \mathbf{r}' = 0$, we find

$$f(0,0) = -\frac{1}{R}\frac{\epsilon-1}{\epsilon}$$

and putting this together we arrive at

$$\langle \Delta U \rangle = -\frac{1}{2}\frac{q^2}{R}\left(1 - \frac{1}{\epsilon}\right)$$

which is a result known as the *Born solvation model*. It has an intuitive interpretation, which is that a charge polarizes its surroundings, accumulating a build-up of counter charge around its radius. The amount of counter charge that is accumulated depends on how polarizable the surroundings are. The limit of $\epsilon \to \infty$ results in a full charge of the opposite sign, and finite ϵ is less than 1. This energy change is a good approximation to the reversible work to solvate the ion, or the excess part of its chemical potential.

We can generalize this result to a point charge not located in the center of the cavity, but off center by a distance b. In this case, in general, all the ℓ components of the spherical harmonic expansion are needed,

$$\langle \Delta U \rangle = -\frac{1}{2}\frac{q^2}{R}\left(1 - \frac{1}{\epsilon}\right)\sum_{\ell=0}\frac{(\ell+1)\epsilon}{(\ell+1)\epsilon+\ell}\left(\frac{b}{R}\right)^{2\ell}$$

which while still proportional to q^2 and inversely proportional to R, the magnitude of the accumulated charge from the polarization field can be dramatically reduced as $b < R$. Note that this is different from an analogous result if the unmodified, or macroscopic, dielectric constant is used.

> **Exercise 3.11:** Use the above result for a spherical cavity of radius R with a point charge located off center to evaluate the solvation energy of a point dipole
>
> $$\langle \Delta U \rangle = -\frac{m_{\text{sol}}^2}{R^3} \left(\frac{\epsilon - 1}{2\epsilon + 1} \right)$$
>
> with magnitude $m_{\text{sol}} = qb$, centered in the cavity.

3.12 Excess electron in a polar lattice

Gaussian field theories are not restricted to classical systems, and can be used to understand the reorganization of a quantum mechanical environment. For example, an electron placed in a polar lattice will induce a polarization around it, breaking the symmetry of the electrostatic potential of the material. It was suspected by Landau that the back reaction of that field on the charge could act to localize it spatially, decohering the plane wave state prescribed by Bloch's theory. This is illustrated in Figure 3.9. At finite temperature, the charge can still move through the lattice, and thus it is expected that the polarization will follow the charge. The charge carrier together with the induced polarization can thus be considered as one dynamical entity, dubbed a *polaron* by Pekar. The physical properties of a polaron differ from those of a band-carrier, as the charge is dressed by the polarization field. The effect is typically characterized by an effective mass of the charge plus field, and the binding energy associated with the interaction. Here we will construct a Gaussian field theory to describe the quantum solvation of an electron in a polar lattice.

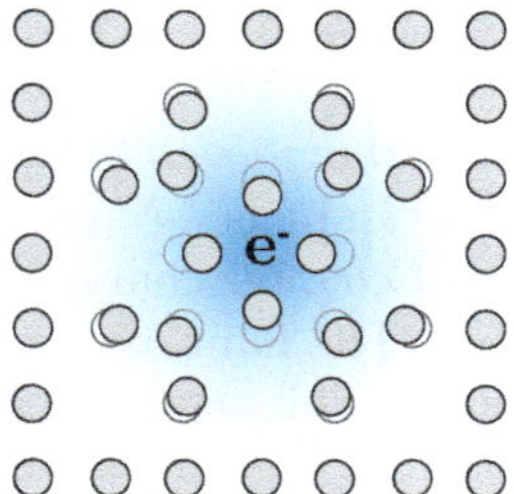

Fig. 3.9 A free charge (blue) polarizes its surrounding lattice (grey) displacing the phonons from their equilibrium positions.

Fröhlich proposed a model Hamiltonian for a polaron, appropriate in a limit where it is delocalized over a number of unit cells. The polarization, carried by the longitudinal optical (LO) phonons, is represented by a set of quantum oscillators with frequency ω_{LO}, the long-wavelength LO-phonon frequency, and the interaction between the charge and the polarization field is linear in the displacement, reflecting the charge–dipole interaction within a harmonic limit,

$$\hat{\mathcal{H}} = \frac{1}{2m_e}\hat{\mathbf{P}}^2 + \sum_{\mathbf{k}} \frac{1}{2}\hat{\mathbf{p}}_{\mathbf{k}}^2 + \frac{\omega_{\mathbf{k}}^2}{2}\hat{\mathbf{r}}_{\mathbf{k}}^2 + \sqrt{8\pi\alpha}\sum_{\mathbf{k}} V_{\mathbf{k}}\frac{\hat{\mathbf{r}}_{\mathbf{k}}}{k}e^{i\mathbf{k}\cdot\hat{\mathbf{R}}}$$

where uppercase bold variables refer to the position and momentum of the electron, and lowercase bold variables to the mass-weighted phonon degrees of freedom. We have adopted hats for operators and the notation

$$\sum_{\mathbf{k}} = \int \frac{d\mathbf{k}}{(2\pi)^3}.$$

The mass m_e of the electron is envisioned to be the band mass, computable from the curvature of the conduction band of the semiconductor, accounting for the influence of the other electrons and the lattice in their rest positions. While in principle optical phonons are dispersive, near zero frequency it is reasonable to neglect this $\omega_{\mathbf{k}} \approx \omega_{\mathrm{LO}}$. The coupling constant, $V_{\mathbf{k}} = \hbar\omega_{\mathbf{k}}\left(\omega_{\mathbf{k}}/2\hbar m_e\right)^{1/4}$, results from the electrostatics of charge-dipole coupling. The strength of the electron–phonon interaction is expressed by a dimensionless constant α, which is defined as

$$\alpha = \frac{e^2}{\hbar}\sqrt{\frac{m_e}{2\hbar\omega_{\mathrm{LO}}}}\left(\frac{1}{\epsilon_\infty} - \frac{1}{\epsilon}\right)$$

where ϵ_∞ and ϵ are the electronic and the static dielectric constant of the polar crystal, respectively. The coupling constant takes values between 0 and 10, for example for GaAs $\alpha =0.068$ while for $SrTiO_3$ $\alpha =3.77$. In deriving the form of electron–phonon coupling it is assumed that (1) the spatial extension of the polaron is large compared to the lattice parameter of the solid so that continuum electrostatics is valid, (2) the band-electron has parabolic dispersion, (3) in line with the first approximation it is also assumed that the LO-phonons are not dispersive.

3.13 Quantum classical correspondence

In order to frame the polaron problem in the form of a Gaussian field theory, we need to transform this quantum mechanical system into a classical one. At first blush, this sounds impossible, and it is in general! However, it turns out that if we confine our attention to the static properties, there exists a so-called quantum-classical isomorphism that translates the calculation of a partition function of a quantum system into the calculation of a partition function in a classical system with extra degrees of freedom. We will first sketch out how this works generally using so-called imaginary time path integrals before returning to the polaron problem.

To uncover the quantum-classical isomorphism, let us consider a single particle Hamiltonian whose eigenvalues E_i and eigenfunctions $|\epsilon_i\rangle$ satisfy the time independent Schrodinger equation

$$\hat{\mathcal{H}}|\epsilon_i\rangle = E_i|\epsilon_i\rangle$$

where $\hat{\mathcal{H}}$ is an operator on this Hilbert space. Provided the eigenvalues, the form of a quantum partition function is identical to a classical one as a weighted sum over energy levels. However, we can rewrite it in a form independent of the way we count states by re-expressing the eigenvalues using the Schrodinger equation

$$Q = \sum_i e^{-\beta E_i}$$

$$= \sum_i \langle \epsilon_i | e^{-\beta \hat{\mathcal{H}}} | \epsilon_i \rangle = \text{Tr} \left[e^{-\beta \hat{\mathcal{H}}} \right]$$

where in the second line, we replace the eigenvalues with the action of the Hamiltonian on its eigenfunctions, and in the third we recognize the form of a matrix trace operation that is conveniently independent of the choice of basis.

Let us now assume that the Hamiltonian takes the form $\hat{\mathcal{H}} = \hat{K}(\mathbf{p}) + \hat{V}(\mathbf{r})$ which is a sum of a kinetic energy operator $\hat{K}(\mathbf{p})$ dependent on momentum $\mathbf{p}$ and a potential energy $\hat{V}(\mathbf{r})$ dependent on position $\mathbf{r}$. As the partition function Q is a trace, we can carry out the sum in the space of real space functions,

$$Q = \int d\mathbf{r} \, \langle \mathbf{r} | e^{-\beta \hat{\mathcal{H}}} | \mathbf{r} \rangle$$

which are complete and orthogonal. The exponential of the Hamiltonian entering into the partition function can be rewritten using a Trotter expansion

$$e^{-\beta \hat{\mathcal{H}}} = \lim_{n \to \infty} \left[e^{-\beta \hat{V}(\mathbf{r})/2n} e^{-\beta \hat{K}(\mathbf{p})/n} e^{-\beta \hat{V}(\mathbf{r})/2n} \right]^n$$

$$= \lim_{n \to \infty} \left[I_n \right]^n$$

where we define I_n as the kernel in the first line. The Trotterization is necessary as the operators $\hat{K}(\mathbf{p}) = \hat{\mathbf{p}}^2/2m$ and $\hat{V}(\mathbf{r})$ do not commute, $e^{K+V} \neq e^K e^V$. Introducing resolutions of the identity in real space,

$$\int d\mathbf{r} \, |\mathbf{r}\rangle \langle \mathbf{r}| = 1 \qquad \langle \mathbf{r}|\mathbf{r}'\rangle = \delta(\mathbf{r} - \mathbf{r}')$$

and in Fourier space,

$$\int d\mathbf{p} \, |\mathbf{p}\rangle \langle \mathbf{p}| = 1 \qquad \langle \mathbf{p}|\mathbf{p}'\rangle = \delta(\mathbf{p} - \mathbf{p}') \qquad \langle \mathbf{r}|\mathbf{p}\rangle = \frac{e^{i\mathbf{p}\cdot\mathbf{r}/\hbar}}{(2\pi\hbar)^{d/2}}$$

we can integrate out the momentum as a series of Gaussian integrals. The product of all n exponents in the initial factorization, results in integrals over dynamic degrees of freedom $\{\mathbf{r}_i\}$, or a form of the partition function

$$Q = \lim_{n \to \infty} \left(\frac{nm}{2\pi\beta\hbar^2} \right)^{nd/2} \int d\mathbf{r}_0 \int d\mathbf{r}_1 \dots \int d\mathbf{r}_{n-1} e^{-\mathcal{S}[\{\mathbf{r}_i\}]}$$

where we refer to $\mathcal{S}$ as an action an define it as

$$\mathcal{S} = \beta \sum_{i=0}^{n-1} \frac{1}{n} \left[\frac{n^2 m (\mathbf{r}_i - \mathbf{r}_{i+1})^2}{2\beta^2 \hbar^2} + V(\mathbf{r}_i) \right]$$

with periodic boundary conditions $\mathbf{r}_0 = \mathbf{r}_n$ from the initial trace operation. This partition function for a quantum particle in a potential $V(\mathbf{r})$ shares the same thermal statistics as a collection of classical particles tethered together by springs moving in a potential with reduced strength $V(\mathbf{r})/n$. Defining $\tau = \beta\hbar$ as a basic unit of time and $d\tau = \beta\hbar/n$, we can take a continuum limit $n \to \infty$

$$S = \frac{1}{\hbar} \int_0^{\beta\hbar} d\tau \, \frac{m\dot{\mathbf{r}}^2(\tau)}{2} + V[\mathbf{r}(\tau)]$$

and defining the integral over all fictitious degrees of freedom as a path integral

$$\int \mathcal{D}[\mathbf{r}(\tau)] = \lim_{n \to \infty} \left(\frac{nm}{2\pi\beta\hbar^2} \right)^{nd/2} \prod_{i=1}^{n} \int d\mathbf{r}_i$$

we arrive at a compact representation of the partition function

$$Q = \int \mathcal{D}[\mathbf{r}(\tau)] e^{-S[\mathbf{r}(\tau)]}$$

where the springs that tether Trotter slices together take the form of a classical kinetic energy, albeit one where time is imaginary, and $\mathbf{r}(0) = \mathbf{r}(\beta\hbar)$. Indeed, one can perform the same calculation on the quantum mechanical propagator and find an analogous answer except for a factor of i in the exponential. In that case the partition function is a real-time path integral, and its evaluation provides the probability amplitude of a transition. For these reasons, the expression above for Q is referred to as an imaginary time path integral. Generalization to many particles is possible, but introduces additional complications if they are indistinguishable.

3.14   Imaginary time influence functional

With the quantum-classical isomorphism clarified, we can now put the polaron problem into a Gaussian field theory form. From the Frohlich polaron Hamiltonian, we find three distinct contributions to the imaginary time action. From the kinetic energy operator of the electron we have,

$$S_e = \frac{1}{\hbar} \int_0^{\beta\hbar} d\tau \frac{m_e}{2} \dot{\mathbf{R}}^2$$

from the independent, distinguishable harmonic oscillators,

$$S_p = \frac{1}{2\hbar} \int_0^{\beta\hbar} d\tau \sum_{\mathbf{k}} \dot{\mathbf{r}}_{\mathbf{k}}^2 + \omega_{\mathbf{k}}^2 \mathbf{r}_{\mathbf{k}}^2$$

and finally the coupling provides a potential for the electron and phonons

$$S_{ep} = \omega_{\mathrm{LO}} \sqrt{8\pi\alpha} \left(\frac{\omega_{\mathrm{LO}}}{2\hbar m_e} \right)^{1/4} \int_0^{\beta\hbar} d\tau \sum_{\mathbf{k}} \frac{\mathbf{r}_{\mathbf{k}}}{k} e^{i\mathbf{k}\cdot\mathbf{R}}.$$

While the contribution to the action from the electron is nonlinear, due to the complex exponential function of its position, the contributions from the phonons are quadratic with a linear coupling to the electron variable. This means from the perspective of the phonons, we have a Gaussian field theory, though one in imaginary time rather than space as we have encountered previously.

> **Exercise 3.12:** Using the path integral framework employed above, evaluate the phonon partition function Q_p and demonstrate that it is equivalent to that determined by the factorization approaches used in Chapter 2.

Introducing the Fourier transform of the phonon positions, $\mathcal{S}_p$ becomes

$$\mathcal{S}_p = \frac{1}{2\hbar} \int \frac{d\omega}{2\pi} \sum_{\mathbf{k}} |\hat{\mathbf{r}}_{\mathbf{k}}(\omega)|^2 (\omega^2 + \omega_{\mathrm{LO}}^2)$$

where we can identify the Fourier transform of the position-position correlation function as

$$\hat{\chi}_{\mathbf{k}}(\omega)^{-1} = \frac{1}{\hbar}(\omega^2 + \omega_{\mathrm{LO}}^2) \quad \chi_{\mathbf{k}}(\tau) = \int \frac{d\omega}{2\pi} \hat{\chi}_{\mathbf{k}}(\omega) e^{-i\omega\tau}.$$

Because the partition function is Gaussian in the harmonic oscillators we can integrate them out

$$Q = \int \mathcal{D}[\mathbf{R}(\tau)] \int \mathcal{D}[\{\mathbf{r}_{\mathbf{k}}(\tau)\}] e^{-\mathcal{S}_e[\mathbf{R}(\tau)] - \mathcal{S}_p[\mathbf{r}_{\mathbf{k}}(\tau)] - \mathcal{S}_{ep}[\mathbf{R}(\tau), \mathbf{r}_{\mathbf{k}}(\tau)]}$$

$$= Q_p \int \mathcal{D}[\mathbf{R}(\tau)] e^{-\mathcal{S}_e} \times$$

$$\exp\left[\alpha\omega_{\mathrm{LO}}^2 \sqrt{\frac{2\omega_{\mathrm{LO}}}{\hbar m_e}} \sum_{\mathbf{k}} \int_{\tau}\int_{\tau'} \frac{4\pi}{k^2} \chi_{\mathbf{k}}(\tau - \tau') e^{i\mathbf{k}\cdot[\mathbf{R}(\tau) - \mathbf{R}(\tau')]}\right]$$

resulting in a new effective action for the electron. We have introduced the quantum partition function for the phonons as Q_p. To evaluate this action we need the harmonic oscillator correlation function in imaginary time,

$$\chi_{\mathbf{k}}(\tau) = \hbar \int \frac{d\omega}{2\pi} \frac{1}{(\omega^2 + \omega_{\mathrm{LO}}^2)} e^{-i\omega\tau} = \frac{\hbar}{2\omega_{\mathrm{LO}}} e^{-\omega_{\mathrm{LO}}|\tau|}$$

which is just a simple decaying exponential. Note the use of continuous Fourier transforms here assumes that the temperature is very low. Otherwise, the frequencies would be discrete, $\omega_j = 2\pi j/\beta\hbar$ with $j = \{0, \pm 1, \pm 2 \dots\}$. Collecting all of the terms in the exponential, we can define a new effective action of the electron, $\tilde{\mathcal{S}}_e$,

$$\tilde{\mathcal{S}}_e = \frac{1}{\hbar} \int_0^{\beta\hbar} d\tau \frac{m_e}{2} \dot{\mathbf{R}}^2 - \alpha\omega_{\mathrm{LO}} \sqrt{\frac{\hbar\omega_{\mathrm{LO}}}{2m_e}} \int_{\tau}\int_{\tau'} \sum_{\mathbf{k}} \frac{4\pi}{k^2} e^{i\mathbf{k}\cdot|\mathbf{R}(\tau) - \mathbf{R}(\tau')|} e^{-\omega_{\mathrm{LO}}|\tau - \tau'|}$$

where we have a nonlocal function of the electron position in imaginary time. Recognizing the Fourier transform of the Coulomb potential,

$$\sum_{\mathbf{k}} \frac{4\pi}{k^2} e^{-i\mathbf{k}\cdot\mathbf{r}} = \frac{1}{r}$$

we can arrive at a relatively simple functional,

$$\tilde{\mathcal{S}}_e = \frac{1}{\hbar}\int_0^{\beta\hbar} d\tau\, \frac{m_e}{2}\dot{\mathbf{R}}^2(\tau) - \alpha\omega_{\mathrm{LO}}\sqrt{\frac{\hbar\omega_{\mathrm{LO}}}{2m_e}}\int_0^{\beta\hbar} d\tau \int_0^{\beta\hbar} d\tau'\, \frac{e^{-\omega_{\mathrm{LO}}|\tau-\tau'|}}{|\mathbf{R}(\tau)-\mathbf{R}(\tau')|}$$

where in addition to the kinetic energy of the electron, we find a screened attractive Coulomb self-interaction. This self-interaction results from the scattering of the electron with a phonon that subsequently back-reacts on the electron. Its non-locality reflects the time it takes for the phonon to back-react, here with rate ω_{LO}. A picture of the effective interactions within an imaginary time path is shown in Figure 3.10. Provided a large-enough α, the attractive interaction can act to localize the electron. In real time, the phonon contribution to $\tilde{\mathcal{S}}_e$ is called the influence functional, so some refer to its analog for the partition function as the imaginary time influence functional.

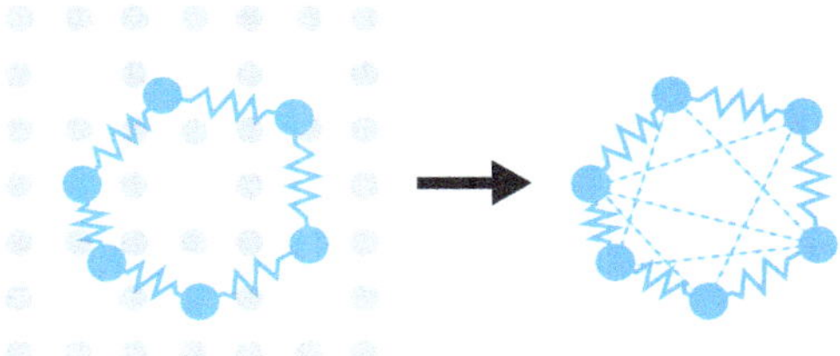

Fig. 3.10 Integrating out the phonons, the electron acquires effective interactions along its imaginary time path.

Unfortunately, integrating out the phonons leaves us with a path integral that we can not evaluate exactly due to the Coulomb potential between the different imaginary time slices. However, when Feynman originally applied this approach to the polaron problem, he used a clever approximation. Specifically, imagine there exists a quadratic action $\mathcal{S}_0$ which is exactly computable, then one can bound the full partition function

$$Q = Q_p Q_0 \frac{\int \mathcal{D}[\mathbf{R}(\tau)] e^{-\mathcal{S}_0[\mathbf{R}(\tau)] - (\tilde{\mathcal{S}}_e[\mathbf{r}(\tau)] - \mathcal{S}_0[\mathbf{R}(\tau)])}}{\int \mathcal{D}[\mathbf{R}(\tau)] e^{-\mathcal{S}_0[\mathbf{R}(\tau)]}}$$

$$= Q_p Q_0 \left\langle e^{-(\tilde{\mathcal{S}}_e - \mathcal{S}_0)} \right\rangle_0 \geq Q_p Q_0 e^{-\langle(\tilde{\mathcal{S}}_e - \mathcal{S}_0)\rangle_0}$$

where Q_0 has been introduced as the partition function for $\mathcal{S}_0$, and we have used Jensen's inequality to move the expectation value of the exponential to the exponential of the expectation value. This bound relies on the convexity of the exponential, and is very useful. We will see it employed throughout this textbook. Employing the bound on Q, the free energy can be similarly bounded from above,

$$\Delta F = -k_{\mathrm{B}}T\ln(Q/Q_p Q_0) \leq F_0 + k_{\mathrm{B}}T\left\langle(\tilde{\mathcal{S}}_e - \mathcal{S}_0)\right\rangle_0$$

where $\beta F_0 = -\ln(Q_0 Q_p)$. This expression is very general for any pair of partition functions, and is known as the Feynman–Bogoliubov inequality. We could employ this bound by parameterizing $\mathcal{S}_0$ and solving for the set of parameters that minimized the estimate of ΔF, or in the low-temperature limit, the corresponding ΔE. Feynman was able to find a reference $\mathcal{S}_0$ that gave nearly quantitative accuracy across the range of coupling constants α spanned by typical materials. However, the final answer is not completely closed form and requires an integral to be evaluated numerically.

Instead of considering an optimal reference for all α, we can consider a $\mathcal{S}_0$ which is accurate for small α. In the small coupling limit, you might expect that a free electron reference is a good approximation,

$$\mathcal{S}_0[\mathbf{R}(\tau)] = \frac{1}{\hbar} \int_0^{\beta\hbar} d\tau \, \frac{m_e}{2} \dot{\mathbf{R}}^2$$

for which the-low temperature limit provides an estimate of the so-called *polaron binding energy* or the interaction energy between the electron and surrounding phonon cloud. This is analogous to our calculation in the previous section for the solvation energy of a charge in a dielectric, only here the solvent is quantum mechanical. The energy is expressible as a single expectation value,

$$\Delta E \leq \lim_{T \to 0} -k_\mathrm{B} T \left\langle \alpha\omega_\mathrm{LO} \sqrt{\frac{\hbar\omega_\mathrm{LO}}{2m_e}} \int_0^{\beta\hbar} d\tau \int_0^{\beta\hbar} d\tau' \frac{e^{-\omega_\mathrm{LO}|\tau-\tau'|}}{|\mathbf{R}(\tau) - \mathbf{R}(\tau')|} \right\rangle_0$$

where the fluctuating quantity is $\mathbf{R}(\tau')$ and it is averaged under the action $\mathcal{S}_0$. To evaluate this expectation value it is easiest to go back to the Fourier space representation of the Coulomb potential,

$$\left\langle \frac{1}{|\mathbf{R}(\tau) - \mathbf{R}(\tau')|} \right\rangle_0 = \sum_k \frac{4\pi}{k^2} \left\langle e^{ik|\mathbf{R}(\tau)-\mathbf{R}(\tau')|} \right\rangle_0$$

$$= \sum_\mathbf{k} \frac{4\pi}{k^2} \left\langle e^{ik \int du \, g(u)\mathbf{R}(u)} \right\rangle_0 = \sum_\mathbf{k} \frac{4\pi}{k^2} e^{-\hbar k^2 |\tau-\tau'|/2m_e}$$

where we have introduced $g(u) = \delta(u - \tau) - \delta(u - \tau')$ and taken the $\omega_\mathrm{LO} \to 0$ limit of the correlation function in the previous section to find $\chi(\tau - \tau') = -\hbar|\tau - \tau'|/2m_e$ as the correlation function of a free electron in $\mathcal{S}_0$. Putting the expectation value into our expression for the energy and integrating over imaginary time,

$$\Delta E \leq \lim_{T \to 0} -k_\mathrm{B} T \left(\alpha\omega_\mathrm{LO} \sqrt{\frac{\hbar\omega_\mathrm{LO}}{2m_e}} \sum_\mathbf{k} \frac{4\pi}{k^2} \int_\tau \int_{\tau'} e^{-(\omega_\mathrm{LO}+\hbar k^2/2m_e)|\tau-\tau'|} \right)$$

$$\leq -\alpha\hbar\omega_\mathrm{LO}$$

which is an upper bound to the polaron binding energy, saturated in the weak coupling limit. This result can be derived independently from first-order perturbation theory, however demonstrating it is a bound from that calculation is difficult. As a reminder α is a dimensionless parameter encoding the strength of the electron-phonon coupling and $\hbar\omega_\mathrm{LO}$ is the energy to excite a phonon mode. Therefore, the interpretation of the binding energy in the weak coupling limit is that through its coupling to the lattice, the electron excites α phonons.

In this example, as with the solvation of an ion in a classical dielectric or a solute by a hard sphere fluid, we have rendered a difficult many-body problem tractable using an assumption of Gaussian statistics and the machinery of linear response. In each, the fact that the Gaussian field is determined entirely by a two-point correlation function, either in space or imaginary time, allows us to formulate expressions to solve for these correlation functions in a way that would not otherwise be possible if non-Gaussian fluctuations would need to be confronted. With these models of correlation functions, the thermodynamics of interacting systems were obtained.

Further reading

The material in this chapter is touched upon in Mehran Kardar's *Statistical physics of fields*, Jean-Pierre Hansen and Ian McDonald's *Theory of simple liquids*, Mark Tuckerman's *Statistical mechanics: theory and molecular simulation*, and Richard Feynman's *Statistical mechanics*. However much of it exists only in the primary literature, and in particular David Chandler's *Gaussian field model of fluids with an application to polymeric fluids*, Physical Review E 1993.

Additional exercises

Exercise 3.13: A set of random variables $\{x_1, x_2, \ldots, x_N\}$, each with zero mean, is Gaussian distributed according to the energy function

$$\beta U_0[\{x_j\}] = \frac{1}{2} \sum_{j,k}^{N} x_j K_{jk} x_k$$

where $\mathbf{K}$ is a real, symmetric, non-singular matrix whose off-diagonal terms couple fluctuations of different variables. In the following problems, you may find it useful to consider fluctuations of the normal modes

$$\eta^{(n)} = \sum_{j=1} \xi_j^{(n)} x_j \,,$$

where $\boldsymbol{\xi}^{(n)}$ is the nth eigenvector with components $\xi_j^{(n)}$ of $\mathbf{K}$ with corresponding eigenvalue $\lambda^{(n)}$:

$$\sum_k K_{jk} \xi_k^{(n)} = \lambda^{(n)} \xi_j^{(n)} \,.$$

Notice that we are using subscripts to index the $\{x_i\}$ variables, and superscripts to index normal modes. Recall that $\mathbf{K}$ is diagonal in the basis of its eigenvectors, and therefore can be inverted trivially in that basis:

$$K_{jk} = \sum_n^{N} \xi_j^{(n)} \lambda^{(n)} \xi_k^{(n)} \,, \qquad K_{jk}^{-1} = \sum_n^{N} \xi_j^{(n)} \frac{1}{\lambda^{(n)}} \xi_k^{(n)}$$

Finally, note that the set of eigenvectors is orthonormal and complete:

$$\sum_{k=1}^{N} \xi_k^{(n)} \xi_k^{(m)} = \delta_{nm} , \qquad \sum_{n=1}^{N} \xi_j^{(n)} \xi_k^{(n)} = \delta_{jk}$$

where δ_{jk} is the Kronecker delta, which vanishes when $j \neq k$ and is 1 for $j = k$.

1. Show that the $\{x_i\}$ variables can be expressed as linear combinations of the normal mode coordinates:

$$x_i = \sum_n \xi_i^{(n)} \eta^{(n)}$$

2. For a single normal mode η, which like x is a continuous variable, calculate the equilibrium probability distribution $p(\eta^{(n)})$. (You could begin by writing βU_0 in terms of the $\{\eta^{(n)}\}$ variables.)

3. Show that

$$\left\langle e^{b\eta^{(n)}} \right\rangle = \exp\left[\frac{b^2}{2\lambda^{(n)}} \right]$$

where b is a constant. Using this result, calculate

$$\left\langle \exp\left[\sum_{n=1}^{N} c^{(n)} \eta^{(n)} \right] \right\rangle$$

where $\{c^{(1)}, c^{(2)}, \ldots, c^{(N)}\}$ is an array of constants.

4. Show that

$$\left\langle \exp\left[\sum_{j=1}^{N} a_j x_j \right] \right\rangle = \exp\left[\frac{1}{2} \sum_{j,k} a_j K_{jk}^{-1} a_k \right] ,$$

where $\langle \cdots \rangle$ denotes an average over the $\{x_i\}$ variables, and K_{ij}^{-1} is the (i,j) element of the inverse of the matrix $\mathbf{K}$. You may find it useful to define $c^{(n)} = \sum_j a_j \xi_j^{(n)}$.

5. The average you computed in the previous part can be viewed as a function of the coefficients $\{a_j\}$. By differentiating with respect to these coefficients, show that the elements of K^{-1} determine correlations between the $\{x_i\}$ variables:

$$K_{ij}^{-1} = \langle \delta x_i \delta x_j \rangle$$

6. Now imagine that the $\{x_i\}$ variables are subjected to a set of external forces $\{f_1, f_2, \ldots, f_N\}$, giving a total energy

$$U[\{x_j\}] = U_0 - \sum_{j=1}^{N} f_j x_j .$$

By first calculating the partition function

$$Q(\mathbf{f}) = \int dx_1 \int dx_2 \ldots \int dx_N\, e^{-\beta U} ,$$

and relating it to $Q(0)$, show that the induced response $\langle x_i \rangle_{\mathbf{f}}$ of variable i is given by

$$\langle x_i \rangle_{\mathbf{f}} = \beta \sum_{j=1}^{N} \langle \delta x_i \delta x_j \rangle f_j$$

Exercise 3.14: The Ornstein-Zernicke equation,

$$h(r) = c(r) + \bar{\rho} \int d\mathbf{r}' \, h(|\mathbf{r} - \mathbf{r}'|)c(r')$$

can be viewed as a definition of the direct correlation function $c(r)$ in terms of $h(r) = g(r) - 1$. In practice it often forms the basis for approximate theories of $h(r)$, which make physically motivated assumptions about the form of $c(r)$. Percus–Yevick theory is one such theory that has an analytical solution shown in the main text.

$$c(r) = \left[\lambda_1 + \lambda_2 \left(\frac{r}{\sigma} \right) + \lambda_3 \left(\frac{r}{\sigma} \right)^3 \right] \Theta(\sigma - r), \tag{3.1}$$

where $\Theta(x)$ is the Heaviside step function, which is 1 for $x > 0$ and 0 for $x < 0$ and λ_i are functions of the density. The solution for $h(r)$ cannot be written in such a simple analytical form. Here you will determine it numerically through the Ornstein–Zernicke equation.

1. For hard spheres, the functions $h(r)$ and $c(r)$ are discontinuous at $r = \sigma$. Their Fourier transforms are therefore highly oscillatory and troublesome to contend with numerically. Their difference, $I(r) = h(r) - c(r)$, however, is a continuous function of r. Write the Fourier transform of this difference, $\hat{I}(k)$, solely in terms of $c(k)$ and the density $\bar{\rho}$. As usual, we define spatial Fourier transforms according to

$$\hat{f}(\mathbf{k}) = \int d\mathbf{r} \, e^{i\mathbf{k}\cdot\mathbf{r}} f(\mathbf{r}) \qquad f(\mathbf{r}) = \int \frac{d\mathbf{k}}{(2\pi)^3} e^{-i\mathbf{k}\cdot\mathbf{r}} \hat{f}(\mathbf{k})$$

   As in the previous assignment, you should find it useful to exploit the convolution theorem. Specifically, for a convolution $z(\mathbf{r})$ of two functions $x(\mathbf{r})$ and $y(\mathbf{r})$,

$$z(\mathbf{r}) = \int d\mathbf{r}' \, x(\mathbf{r} - \mathbf{r}')y(\mathbf{r}')$$

   Fourier integration yields $\hat{z}(\mathbf{k}) = \hat{x}(\mathbf{k})\hat{y}(\mathbf{k})$.

2. By writing $h(r)$, rather than $h(\mathbf{r})$, we have already acknowledged that correlations in an isotropic fluid do not depend upon direction, only on distance $r = |\mathbf{r}|$. For such rotationally symmetric functions, the angular part of Fourier transform integrals can be performed straightforwardly. For a function $f(r)$ of distance r only, show that

$$\hat{f}(k) = \int d\mathbf{r} \, e^{i\mathbf{k}\cdot\mathbf{r}} f(r) = \frac{4\pi}{k} \int dr \, r \, \sin(kr) f(r)$$

   indicating that the Fourier transform of $f(r)$ depends only on the magnitude $k = |\mathbf{k}|$ of wavevector k. Similarly, show that

$$f(r) = (2\pi^2 r)^{-1} \int dk \, k \, \sin(kr) \hat{f}(k)$$

   When integrating over the position r, it is convenient to take the z-axis to point in the same direction as k.

3. Compute $I(r)$ by numerical Fourier inversion of your result from part (1). To do so, you will need to construct $\hat{c}(k)$. The following integrals should come in handy:

$$\int d\mathbf{r}\, e^{i\mathbf{k}\cdot\mathbf{r}}\Theta(\sigma - r) = 4\pi\sigma^3 \left[-\frac{\cos k\sigma}{(k\sigma)^2} + \frac{\sin k\sigma}{(k\sigma)^3} \right]$$

$$\int d\mathbf{r}\, e^{i\mathbf{k}\cdot\mathbf{r}}\Theta(\sigma - r)\left(\frac{r}{\sigma}\right) = 4\pi\sigma^3 \left[-\frac{\cos k\sigma}{(k\sigma)^2} + 2\frac{\sin k\sigma}{(k\sigma)^3} + 2\frac{\cos k\sigma - 1}{(k\sigma)^4} \right]$$

$$\int d\mathbf{r}\, e^{i\mathbf{k}\cdot\mathbf{r}}\Theta(\sigma - r)\left(\frac{r}{\sigma}\right)^3 =$$

$$4\pi\sigma^3 \left[-\frac{\cos k\sigma}{(k\sigma)^2} + 4\frac{\sin k\sigma}{(k\sigma)^3} + 12\frac{\cos k\sigma}{(k\sigma)^4} - 24\frac{\sin k\sigma}{(k\sigma)^5} - 24\frac{\cos k\sigma - 1}{(k\sigma)^6} \right]$$

Plot your results for $h(r)$ for densities $\rho\sigma^3 = 0.02$, 0.1, 0.2, 0.3, 0.4, 0.5, 0.6, 0.7, 0.8, and 0.9. How do they compare to the figure of the $g(r)$ in the main text.

Exercise 3.15: Solutions of long polymers are prone to phase separation. One way to stabilize a polymer blend is to covalently bond together two different types of polymers in large domains. Such a polymer is known as a block copolymer, and due to its chemical constraint, it cannot undergo macroscopic demixing. A coarse-grained model for block copolymers with equal composition of type 1 and type 2 can be derived. The effective Hamiltonian for the well-mixed phase becomes

$$\beta\mathcal{H}[\phi(\mathbf{r})] = \int d\mathbf{r}\, a\phi^2(\mathbf{r}) + m|\nabla\phi(\mathbf{r})|^2 + c\int d\mathbf{r}\int d\mathbf{r}'\, \phi(\mathbf{r})G(\mathbf{r},\mathbf{r}')\phi(\mathbf{r}')$$

where $\phi(\mathbf{r}) = (\rho_1(\mathbf{r}) - \rho_2(\mathbf{r}))/\rho$ is the local composition and a and m are functions of the polymer length and Flory's χ parameter. The second term proportional to c enforces the covalent bond constraint. Using the random phase approximation, Liebler found that this function is equal to

$$G(\mathbf{r},\mathbf{r}') = \frac{1}{|\mathbf{r} - \mathbf{r}'|}$$

We will use this model to compute the composition-composition correlation function, $\chi(\mathbf{r}) = \langle\phi(0)\phi(\mathbf{r})\rangle$.

1. Fourier transform the Hamiltonian, expressing it as a sole function of $\mathbf{k}$.
2. Express the Fourier transform of the correlation function $\hat{\chi}(\mathbf{k})$ from your result in part 1. Note that for a translational symmetric function $\hat{\chi}(\mathbf{k})$, the inverse can be performed trivially in the Fourier space.
3. Back-transform $\hat{\chi}(\mathbf{k})$ to real space to solve for $\chi(\mathbf{r})$ as a function of a, m, and c. Note that in this region of the phase diagram, it is safe to assume that a is small and positive, $m > 0$ and $c > 0$. *Hint*: You may need to use Cauchy's residue theorem to evaluate the complex integral

$$\oint_C f(z)\,dz = 2\pi i \sum_{k=1}^{n} \text{Res}_f(z_k)$$

where z_k are the poles of $f(z)$.

Exercise 3.16: The Helfrich Hamiltonian is a canonical model for membrane mechanics. For a tensionless membrane, it is given as a potential energy that acts on a height field $h(\mathbf{r})$, which is the distance of the two-dimensional surface of area L^2 from a reference plane. Taking the plane to be located at $h(\mathbf{r}) = 0$, the resultant potential is a sum of two terms

$$U[h(\mathbf{r})] = \frac{1}{2} \int d\mathbf{r} \left\{ a_0 h^2(\mathbf{r}) + \kappa |\nabla^2 h(\mathbf{r})|^2 \right\}$$

an external potential with strength a_0 and a bending rigidity with stiffness κ. As the height field enters bilinearly into the potential, this is a Gaussian theory. Here we will work out what happens when two mobile protein receptors pin the membrane at specific heights and in particular, work out a membrane-mediated interaction that is generated.

1. Like other Gaussian theories, the Helfrich Hamiltonian can be decomposed into independent modes using Fourier transforms. Rewrite the potential in Fourier space

$$\beta U[h(\mathbf{k})] = \frac{1}{2} \sum_{\mathbf{k}} |h(\mathbf{k})|^2 \chi_{\mathbf{k}}^{-1}$$

   and identify the Fourier transformed height correlation function $\chi_{\mathbf{k}}$.

2. Imagine the receptors pin the membrane at $\mathbf{r}_1$ and $\mathbf{r}_2$, to $h(\mathbf{r}_1)$ and $h(\mathbf{r}_2)$. Using a Fourier representation of the delta function

$$\delta[h(\mathbf{r}_j)] = \int d\lambda_j\, e^{-i\lambda_j h(\mathbf{r}_j)} = \int d\lambda_j\, e^{-i\lambda_j \sum_{\mathbf{k}} h_{\mathbf{k}} e^{-i\mathbf{k}\cdot\mathbf{r}_j}}$$

   show that the partition function with the constraint, Q_c, is equal to

$$\frac{Q_c}{Q_0} = \int_{\lambda_1} \int_{\lambda_2} \exp\left[-\frac{(\lambda_1^2 + \lambda_2^2)}{2} \sum_{\mathbf{k}>0} \chi_{\mathbf{k}} - \frac{\lambda_1 \lambda_2}{2} \sum_{\mathbf{k}>0} \chi_{\mathbf{k}} \cos\left(\mathbf{k}\cdot|\mathbf{r}_1 - \mathbf{r}_2|\right) \right]$$

   where Q_0 is the partition function without the constraint.

3. By carrying out the integrals over λ_1 and λ_2 and sum over $\mathbf{k}$, compute $\beta \Delta A(|\mathbf{r}_1 - \mathbf{r}_2|) = -\ln Q_c/Q_0$, an effective potential between the two pinning sites due to the constraint on the membrane. The evaluation of the sum over $\mathbf{k}$ will require knowledge of a special function, the Kelvin function $\mathrm{kei}_0(r)$

$$\mathrm{kei}_0(r/b) = -\frac{b^2}{2\pi} \int_0^{2\pi} d\theta \int_0^\infty dk\, k \frac{1}{1 + (bk)^4} \cos\left(kr\cos(\theta)\right)$$

   for positive b. Comment on its shape and sign.

4

Phase transitions and symmetry breaking

The most exotic thermodynamic phenomena are those associated with the transitions between different states of matter. The manner by which the stability of one state of matter is lost in favor of another is provocative, occurring sharply at particular thermodynamic conditions. As we will see in this chapter, the singular changes accompanying a phase transition require the existence of macroscopic correlations. As a consequence, phase transitions mark a stark departure from the noninteracting and linear responding systems we have considered thus far. Understanding phase transitions requires that we study strongly interacting models, pushed far into regimes of nonlinear response. This complication renders detailed studies largely intractable analytically, relegating them to the domain of numerical simulations like those discussed in the next chapter. However, some simplifications exist that allow for a general theory of phase transitions to be developed. These are the notions of universality that accompany phase transition phenomenology, and the ability to classify phase transitions simply using ideas of spontaneous symmetry breaking. The former allows us to postulate simple models bereft of molecular detail and yet still provide quantitative predictions for observable behavior. The latter allows us to predict what phases can exist, and understand why phase transitions occur.

4.1 Singular phenomenology

A phase transition is marked by a singular, or discontinuous, change in an observable property of a physical system. Some examples of phase transitions are shown in Figure 4.1. For example, the density of a fluid will discontinuously change from a large value to a small value upon heating at a specific temperature, in a process called boiling. The change upon boiling is so distinct that we differentiate the material on either side of the transition, with the dense phase called a liquid and the dilute phase called a vapor. Likewise, a fluid at a low enough temperature upon compression can be made to transition abruptly to a crystal, at a specific value of the applied pressure in a process known as freezing. Accompanying a freezing transition is a release of heat, resulting from an abrupt decrease in the energy of the system. A magnet if exposed to a magnetic field can have its magnetic moment change abruptly from one pointing in one direction to a moment pointing in another direction.

The locations of these abrupt transitions, in which one state of matter, or phase, transitions into another, depends on the thermodynamic variables that can be con-

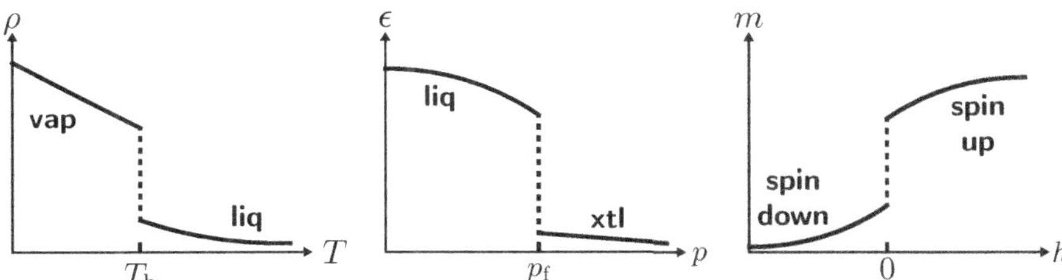

Fig. 4.1 Phase transitions are marked by discontinuous changes of observables like the density of a fluid, ρ, the energy density of a solid, ϵ, or the net magnetization per spin of a magnet, m, with changes in external fields, like temperature, T, pressure p, or magnetic field, h.

trolled. The locus of points at which transitions occur can be summarized into phase diagrams that relay which phase under thermodynamic conditions is most stable. Examples of canonical phase diagrams for simple molecules consisting of regions of liquid and vapors, and for multicomponent mixtures that have miscibility transitions between solutions and phase-separated states are shown in Figure 4.2. These two phase diagrams, encoding distinct physical behavior, bear a striking resemblance. In each, below a critical temperature, T_c, two distinct phases can be realized, or put into coexistence. By coexistence, we mean that both phases are present with an interface separating them. The boundaries of the coexistence region denote the values of the densities or compositions residing on either side of the discontinuous change, as observable by manipulating a conjugate field like pressure or chemical potential.

The manner by which the two phases can be distinguished is through an *order parameter*. For a boiling or its reverse condensation transition, the order parameter is the molecular density. It is the density that is the fundamental difference between liquids and vapors, as molecularly both are disordered fluids. Far below the critical point, the coexistence densities of the liquid and vapor are very different, and depend on details of the fluid. Near the critical point, independent of the molecular system, the difference between the coexistence densities and its value at the critical point, ρ_c, disappears like a power law,

$$|\rho - \rho_c| \propto \left(\frac{|T - T_c|}{T_c} \right)^{\hat{\beta}}$$

with a critical exponent $\hat{\beta}$ empirically determined to be $\hat{\beta} \approx 0.32$.

Analogously, while a binary mixture at high enough temperature will become miscible, below T_c it will phase separate into two fluids with more of one component in each phase. The order parameter for a miscibility transition is thus the molar fraction of one component in the other. Near the critical point located at a critical molar fraction, x_c, the coexistence concentration scales like a power law in precisely the same manner as the liquid-vapor transition,

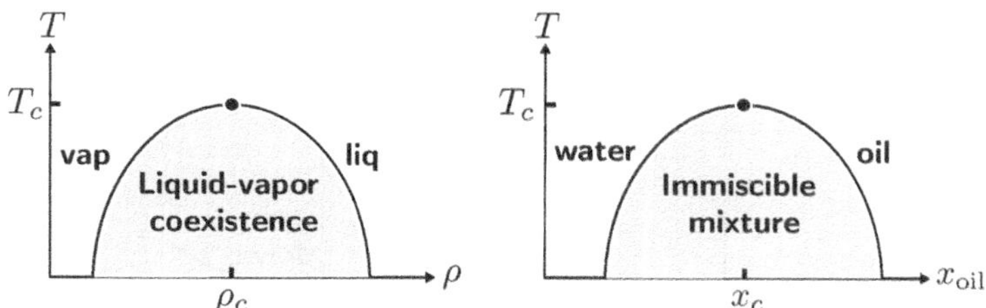

Fig. 4.2 Typical phase diagrams for liquid-vapor systems, and binary mixtures.

$$|x - x_c| \propto \left(\frac{|T - T_c|}{T_c} \right)^{\hat{\beta}}$$

with the same empirically determined exponent for a host of binary mixtures. Not shown, but a magnet which can have domains of net magnetization that point in one of two directions below the so-called *Curie–Weiss transition*, the order parameter distinguishing those phases is the magnetization, m. Near the critical point, where the net magnetization disappears,

$$|\langle m \rangle| \sim \left(\frac{|T - T_c|}{T_c} \right)^{\hat{\beta}}$$

where $\hat{\beta}$ again, for a wide array of magnets, is 0.32, just like the phase diagrams for the liquid-vapor and binary mixtures.

Exercise 4.1: What are possible order parameters for distinguishing a liquid and a solid? or a metal and an insulator?

This quantitatively common behavior, despite being molecularly very distinct phenomena, is known as *universality*. It extends beyond the behavior of order parameters, to other thermodynamic quantities. For example, shown in Figure 4.3, the heat capacity of a system passing through a critical point obeys another power law

$$C_V \propto |T - T_c|^{-\hat{\alpha}}$$

with critical exponent quantifying the strength of its divergence $\hat{\alpha} = 0.11$ for condensation, miscibility, and Curie–Weiss transitions. Analogously, the susceptibility of the order parameter to an external field generically diverges upon approaching a critical point. For example, for the magnet in applied field h

$$\chi = \frac{d \langle m \rangle}{d\beta h} \sim |T - T_c|^{-\hat{\gamma}}$$

where $\hat{\gamma} = 1.2$. Analogous susceptibilities can be defined for the change in the density with pressure or composition with chemical potential. This independence of molecular detail or even of the specifics of the phase transition considered is remarkable. A primary objective of this chapter is to explain it.

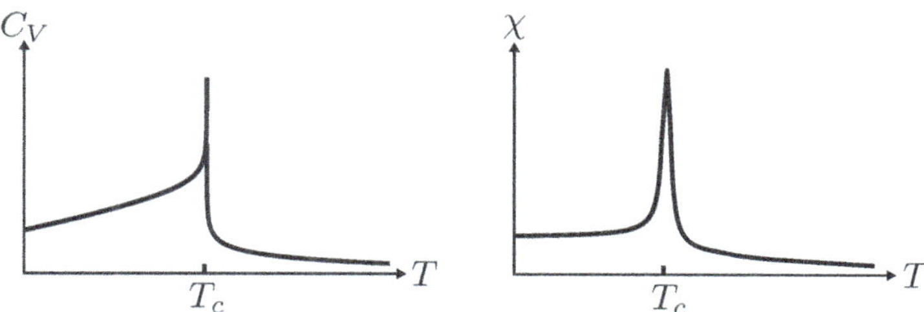

Fig. 4.3 The heat capacity and susceptibility are examples of other universal properties for systems near critical points.

4.2   Thermodynamics of phase coexistence

In order to understand the thermodynamic constraints on phase transitions, we can use results from Chapter 2 where we discussed the notion of mass equilibrium. If phases α and γ are in contact with each other sharing a stable interface, for that coexisting state to be in thermal equilibrium we require that

$$T^{(\alpha)} = T^{(\gamma)} \qquad p^{(\alpha)} = p^{(\gamma)} \qquad \mu^{(\alpha)} = \mu^{(\gamma)}$$

so that there are no persistent heat, volume, or mass fluxes. If the two phases consist of multiple components, we require $\mu_i^{(\alpha)} = \mu_i^{(\gamma)}$, for all species i that can transfer between the two phases. From the Gibbs–Duhem relationship back in Chapter 1, the chemical potential is a function of T and p, so these constraints can be summarized by $\mu_i^{(\alpha)}(T,p) = \mu_i^{(\gamma)}(T,p)$. In the temperature-pressure plane, phase boundaries therefore identify regions of changing relationship between the chemical potentials of two phases. For example, as shown in Figure 4.4, regions of T and p for which $\mu_i^{(\alpha)}(T,p) > \mu_i^{(\gamma)}(T,p)$ will find γ as the thermodynamically stable phase. Alternatively, regions for which $\mu_i^{(\alpha)}(T,p) < \mu_i^{(\gamma)}(T,p)$ will find a system prepared in γ spontaneously transforming into α. The line separating these two regions of the phase diagram denotes where $\mu_i^{(\alpha)}(T,p)$ equals $\mu_i^{(\gamma)}(T,p)$.

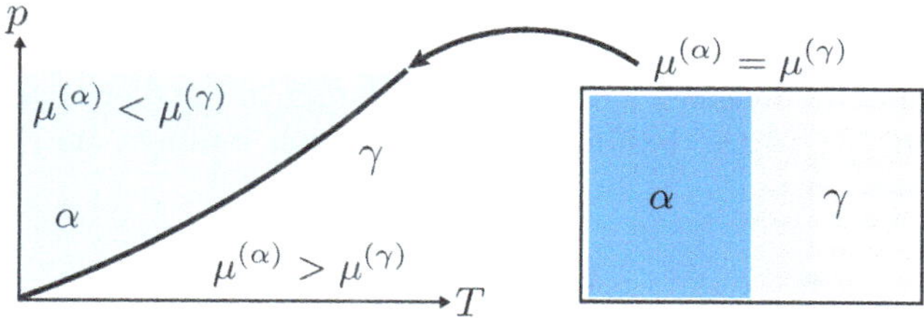

Fig. 4.4 Information contained within a phase diagram, like that separating regions of stability for phases α and γ.

Moving along the phase boundary in such a way as to maintain coexistence by changing the temperature and at the same time the pressure, requires that $d\mu^{(\alpha)} = d\mu^{(\gamma)}$. This means that the Gibbs–Duhem relationship,

$$d\mu^{(\alpha,\gamma)} = -s^{(\alpha,\gamma)}dT + v^{(\alpha,\gamma)}dP$$

must be equal for both phases, where $s^{(\alpha,\gamma)} = (S/N)^{(\alpha,\gamma)}$ is the entropy per particle and $v^{(\alpha,\gamma)} = (V/N)^{(\alpha,\gamma)}$ is the volume per particle. Equating $d\mu^{(\alpha)}$ and $d\mu^{(\gamma)}$,

$$-s^{(\gamma)}dT + v^{(\gamma)}dP = -s^{(\alpha)}dT + v^{(\alpha)}dP$$

$$\implies \left(\frac{\partial p}{\partial T}\right)_{\text{coex}} = \frac{s^{(\gamma)} - s^{(\alpha)}}{v^{(\gamma)} - v^{(\alpha)}} = \frac{\Delta s}{\Delta v}$$

we find a relationship for the coexistence line in terms of changes to the volume and entropy per particle. This equation is known as the *Clausius–Clapeyron relation* and helps us rationalize why phase diagrams appear so similarly for many different molecular systems.

Exercise 4.2: It is often said that pressure-induced melting is the reason ice skating is possible. Calculate how heavy a person would need to be to melt ice under their skates at $T = -12°\text{C}$ using the Clausius–Clapeyron relation. The blade edge of a typical skate is about one-eighth of an inch wide and about 11 inches long, and the volume change and heat of fusion are easy to look up.

Take, for example, a freezing transition. The expected change in the entropy and volume upon freezing is

$$s^{(\text{xtl})} - s^{(\text{liq})} < 0 \qquad v^{(\text{xtl})} - v^{(\text{liq})} \lesssim 0$$

where the crystal nearly always has a lower entropy, while the change in the volume can be positive or negative but is almost always very small. This is because both liquids and crystals are not very compressible. As a consequence, the slope of the freezing line in the temperature pressure plane is very nearly vertical, due to being inversely proportional to a small Δv. The vertical line might tilt positively or negatively depending on whether the density of the crystal is larger or smaller than that of the liquid. This is in contrast to the boiling transition. Upon boiling, the change in the entropy and volume are

$$s^{(\text{vap})} - s^{(\text{liq})} > 0 \qquad v^{(\text{vap})} - v^{(\text{liq})} \gg 0$$

which while both are positive, the volume change is expected to be very large. As a consequence, the slope of the boiling line is positive though very nearly flat due to Δv being large. Examples of these phase boundaries are shown in Figure 4.5.

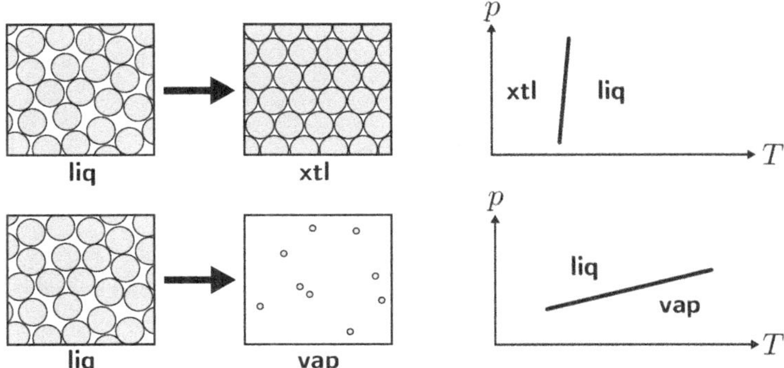

Fig. 4.5 Consequences of the Clausius–Clapeyron relation for freezing and boiling.

Thermodynamically, predicting the location of phase boundaries or regions of coexistence can be accomplished by employing a so-called *Maxwell construction*. Consider two phases α and γ, whose equations of state are known. Integrating those equations of state can allow one to parameterize the functional dependence of the Helmholtz free energy, $A^{(\alpha)}(T, V, N)$ and $A^{(\gamma)}(T, V, N)$, within each phase as a function of T, V, and N. Let $a = A/N$ be the free energy density. As a function of the volume per particle, v, stability requires that

$$\left(\frac{\partial a}{\partial v}\right)_T = \left(\frac{\partial A}{\partial V}\right)_{T,N} = -p \leq 0 \qquad \left(\frac{\partial^2 a}{\partial v^2}\right)_T = -\left(\frac{\partial p}{\partial v}\right)_T \geq 0$$

or that $a(v)$ is a decreasing function of v with positive curvature. Given these free energy densities, we can ask at a given v, which phase do we expect to observe? At equilibrium $T^{(\alpha)} = T^{(\gamma)} = T$, $p^{(\alpha)} = p^{(\gamma)} = p$ and $\mu^{(\alpha)} = \mu^{(\gamma)}$ for coexistence between the two phases. Since the pressure is the tangent of a with respect to v, the equal pressure constraint requires that $a^{(\alpha)}(T, v)$ and $a^{(\gamma)}(T, v)$ have the same slope at coexistence. Since, $E = TS - pV + \mu N$ and $A = -pV + \mu N$, due to the extensivity of thermodynamic systems, $a = -pv + \mu$ or $\mu = a + pv$, therefore,

$$a^{(\gamma)} + pv^{(\gamma)} = a^{(\alpha)} + pv^{(\alpha)} \implies p = -\frac{\Delta a}{\Delta v}$$

or that at coexistence, the slope is actually a common tangent between the two curves $a^{(\alpha)}(T, v)$ and $a^{(\gamma)}(T, v)$. Take, for example, the free energy densities plotted in Figure 4.6. If for $v < v^{(\alpha)}$, $\mu^{(\alpha)} < \mu^{(\gamma)}$ and thus phase α is stable, and for $v > v^{(\gamma)}$, $\mu^{(\alpha)} > \mu^{(\gamma)}$ and thus phase γ is stable, then these coexistence arguments imply that for $v^{(\alpha)} < v < v^{(\gamma)}$ a mixture of α and γ is stable.

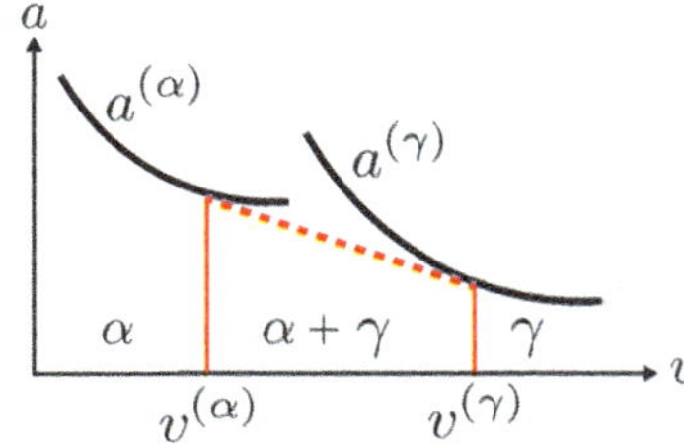

Fig. 4.6 Example of a Maxwell construction between two phases α and γ.

Exercise 4.3: Imagine that the equation of state of a liquid and its crystal can be measured, yielding expressions for the Helmholtz free energy over a range of densities and temperatures

$$A^{(\text{liq})}/V = \frac{b}{2T}\rho^2 \qquad A^{(\text{xtl})}/V = \frac{c}{3T}\rho^3$$

where b and c are positive constants. Using a Maxwell construction, determine the phase diagrams in ρ versus T and p versus T, by evaluating the coexistence densities ρ_{liq} and ρ_{xtl} and the melting pressure p_m as a function of temperature.

4.3   Necessary conditions for a phase transition

Microscopically, the ability of a thermal system to exhibit a discontinuity, or infinity sharp change, is peculiar. The partition function for an equilibrium system in the canonical ensemble is a sum of Boltzmann factors,

$$Q = \sum_{\nu} e^{-\beta E(\nu)}$$

where each Boltzmann factor $\exp\left[-\beta E(\mathbf{r}^N)\right]$ is a smooth function of the temperature, away from $T = 0$. The partition function, Q, is thus a sum of smooth functions, and consequently must itself be smooth. This means that its first derivatives, like the average energy, are smooth

$$\left(\frac{\partial \ln Q}{\partial \beta}\right)_{V,N} = -\langle E \rangle, \qquad \left(\frac{\partial^2 \ln Q}{\partial \beta^2}\right)_{V,N} \propto C_V$$

and its second derivative, proportional to the heat capacity, is finite. So then, how can a system whose properties are determined by this partition function support a phase transition? This would require that Q is non-analytic. The answer lies in a loophole, which is if the number of terms in the sum is infinite, in other words $N \to \infty$, then Q need not be smooth. However, if the macroscopic system can be broken up into subregions, each of which are statistically independent, then the partition function of the whole system would factorize,

$$Q = q^N, \qquad \left(\frac{\partial \ln Q}{\partial \beta}\right)_{N,V} = N \left(\frac{\partial \ln q}{\partial \beta}\right)_{N,V}$$

and each subregion partition function, not being in the thermodynamic limit, must be smooth. In other words, the thermodynamic limit is a necessary but not sufficient condition for a phase transition, the other being that correlations span the entire system so regions cannot be factorized. Indeed, a hallmark of a phase transition is the collective reorganization of a material, where correlations are established over macroscopic scales.

The necessity of the thermodynamic limit and macroscopic ranged correlations can similarly be understood by considering the divergence of the heat capacity. As shown in Chapter 1, the heat capacity is proportional to fluctuations in the energy. If particles interact over a finite distance, we can define the total energy of the system as the sum over local energies, ε_i in each of N regions, $E = \sum_{i=1}^{N} \varepsilon_i$. The molar heat capacity is given by

$$C_V = \frac{1}{Nk_{\mathrm{B}}T^2} \left\langle \delta E^2 \right\rangle = \frac{1}{Nk_{\mathrm{B}}T^2} \sum_{i,j=1}^{N} \left\langle \varepsilon_i \varepsilon_j \right\rangle - \left\langle \varepsilon_i \right\rangle \left\langle \varepsilon_j \right\rangle$$

where inside the double sum is $\langle \delta\varepsilon_i \delta\varepsilon_j \rangle$, the correlation function between local energies. In a system with translational invariance, the correlation function should only depend on the separation between regions i and j, and so one of those indices can be summed over leaving

$$C_V = \frac{1}{k_{\mathrm{B}}T^2} \sum_{j=1}^{N} \left\langle \varepsilon_1 \varepsilon_j \right\rangle - \left\langle \varepsilon_1 \right\rangle \left\langle \varepsilon_j \right\rangle$$

which is intensive provided cell 1 is correlated over a domain that does not scale with N. If correlations are microscopic, the molar heat capacity will be finite even if $N \to \infty$. Only if cells are correlated over a macroscopic domain, such that all terms in that sum are nonzero, in the thermodynamic limit $C_V \to \infty$. So the singularity in the heat capacity that accompanies a phase transition requires diverging correlation lengths. Associated with a diverging correlation length is a divergence in the characteristic time to decorrelate fluctuations. Correspondingly, critical phenomena occur only in stable, not metastable, states.

Note that our notions of ensemble equivalence, used pervasively throughout the early chapters of this book, relied on being able to take the thermodynamic limit and employ the resultant smoothness of thermodynamic auxiliary functions. This was a consequence of the large deviation forms that result from finite correlations and statistical independence. For example, the definition of the Gibbs free energy as the Legendre transform of the Helmholtz free energy

$$G(T, p, N) = A(T, V, N) - \left(\frac{\partial A}{\partial V}\right)_{T,N} V$$

allowed us to equate systems held at fixed volume to those at fixed pressure. However, the uniqueness of the Legendre transform is predicated on the smoothness of

$A(T, V, N)$. At a phase transition $A(T, V, N)$ is singular, and so this transformation is not invertible and consequently ensembles are not equivalent. One can picture a fluid near its liquid–vapor transition, where at a fixed volume and number of particles, coexistence can be enforced, provided the density is chosen in between the equilibrium densities of the liquid and vapor. However, if one were to control the pressure, enforcing coexistence is not possible. Without a constraint of the density, the excess free energy associated with forming the interface would drive a transformation to a pure phase. So while a two-phase system can be stable at constant volume, one would only ever find a one-phase system at fixed pressure. This manifests the breaking of the large deviation scaling when for long-ranged correlations.

The singular nature of thermodynamic quantities at a phase transition allows one a means of classifying transitions. As thermodynamic quantities are all related through derivatives of the partition function, or equivalently the free energy, it is standard to refer to the order of a phase transition as the derivative that first becomes non-analytic. For example, freezing or boiling, in which quantities like E or V exhibit a jump discontinuity with changes of T or p are examples of non-analytic behavior of the first derivative of the free energy, $V = \partial G/\partial p$. As a consequence, they are referred to as first-order transitions. A system passing through its critical point exhibits a smooth change of E and V but typically has an infinite infection point, rendering second derivatives of the free energy, quantities like the heat capacity or susceptibility, divergent. As a consequence, critical points are referred to as second-order transitions.

4.4 Simple lattice models

Because phase transitions arise from the collective behavior of many interacting degrees of freedom, even the simplest models that support them are difficult to study analytically. However, the universality associated with phase transitions, especially critical points, suggests that simple models may be formulated and studied. Here we will introduce two lattice models that form the bedrock of our understanding of critical phenomena. These minimal models do not attempt to describe all behaviors quantitatively across the phase diagram, but will capture universal properties near critical points.

First is the *Ising model*, a minimal model for ferromagnetization. Its Hamiltonian is given by two terms. The first is an interaction term with strength J, which favors nearby spins to point in the same direction. This interaction is meant to reflect exchange, and J referred to as an exchange coupling. The interaction is expected to be relatively short-ranged. The term second is an external field with strength h that tries to align spins in its direction. If $s_i = \pm 1$ denotes the local direction of the spin at lattice site i, then the Hamiltonian and partition functions are

$$\mathcal{H} = -J \sum_{i,j}{}' s_i s_j - h \sum_i s_i \qquad Q = \sum_{\{s_i^N\}} e^{-\beta \mathcal{H}}$$

where the $'$ in the sum denotes a restriction to nearest neighbors, and the sum in the partition function is over all configurations of the N spins. The Ising model is motivated by the short-ranged nature of the exchange interaction, and an assumption that there is a definite axis along which the magnetization is generated. A picture of such a spin system is shown in Figure 4.7.

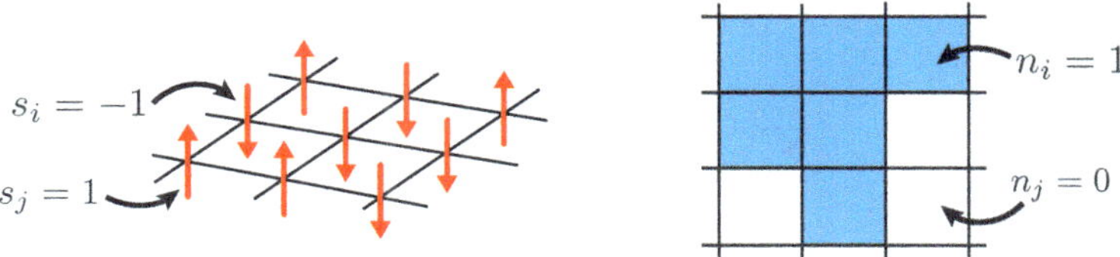

Fig. 4.7 Configurations of the Ising model (left) and the interacting lattice gas (right).

In the absence of an applied external field, $h = 0$, the Hamiltonian exhibits an up/down symmetry. Specifically, the Hamiltonian and therefore the statistical likelihood of a configuration is invariant to a global inversion of all of the spins,

$$\mathcal{H}[\{s_i^N\}] = -J \sum_{i,j}{}' s_i s_j$$

$$= -J \sum_{i,j}{}' (-s_i)(-s_j) = \mathcal{H}[\{-s_i^N\}]$$

which is true of any Hamiltonian that is even in the spin variables. This symmetry at $h = 0$ implies that the ground state is two-fold degenerate, with an equal likelihood of forming a fully magnetized state in either the up or down direction. If these states persist at finite temperatures, the system must spontaneously break the symmetry of the Hamiltonian to form them. Indeed, we will find that microscopically a phase transition is a point at which an underlying symmetry of the Hamiltonian is broken, where the nature of the broken symmetry dictates much of the macroscopic consequences of the phase transition.

A similarly simple model can be formulated for a liquid–vapor transition by generalizing a lattice gas model to include interactions. Specifically, envisioning density fluctuations on a grid, the Hamiltonian is given by an interaction term with strength ϵ that favors nearby cells to be occupied. We can take $n_i = 0,1$ to denote the local value of the density at lattice site i. As n_i fluctuates, so does the total density of the system, as the number of particles is given by $N = \sum_i n_i$, so it will be natural to consider fluctuations within the grand canonical ensemble with chemical potential μ. The Hamiltonian and partition function are

$$\mathcal{H} = -\epsilon \sum_{i,j}{}' n_i n_j \qquad \Xi = \sum_{\{n_i^N\}} e^{-\beta \mathcal{H} + \beta \mu N}$$

whereas with the Ising model, we restrict interactions to nearest neighbors, envisioning that the cohesive energy of the liquid results from short-ranged interactions. A picture of the lattice gas is shown in Figure 4.7.

At first blush, the Ising model and lattice gas would seem to describe very different phenomena. However, both are defined as a set of dynamic variables that can take two different values on a lattice, and whose statistical weights are determined by a sum of one- and two-body interactions. A mapping between the two models can be constructed to make this connection concrete. Let $s_i = 2n_i - 1$, then the energy of the Ising model can be written as

$$\mathcal{H}_{\text{Ising}}[\{s_i^N\}] = -J \sum_{i,j}' s_i s_j - h \sum_i s_i$$

$$= -4J \sum_{i,j}' n_i n_j + 4J \sum_{i,j}' n_i - 2h \sum_i n_i + \text{const.}$$

$$= -4J \sum_{i,j}' n_i n_j - 2(h - Jz) \sum_i n_i + \text{const.} = \mathcal{H}_{\text{LG}}[\{n_i^N\}] - \mu N + \text{const.}$$

where we have performed one of the sums over nearest neighbors in going from the second to third line by defining the coordination number of the lattice, z. If we identify $\epsilon = 4J$ and $\mu = 2(h - Jz)$, we find that the statistical weight of the Ising model in the canonical ensemble is equal to the lattice gas in the grand canonical ensemble up to a multiplicative constant. This means that these two models are exactly equivalent, or isomorphic, and expectation values in one directly relate to expectation values in the other. This notion of exact equivalence is distinct from the approximate equivalence of systems near their critical points.

Exercise 4.4: Consider a lattice model for a binary mixture,

$$\mathcal{H} = -\sum_{i,j}' \epsilon_{11} n_i n_j + \epsilon_{22}(1 - n_i)(1 - n_j) + \epsilon_{12} n_i (1 - n_j) + \epsilon_{12}(1 - n_i)n_j,$$

which is also isomorphic to the Ising model. Here, $n_i = 1$ represents the presence of species 1 at site i, and $1 - n_i$ represents the presence of species 2 at site i. Establish the isomorphism of the binary mixture model with the Ising model and determine the corresponding relations to $4J$ and to $2(h - Jz)$ in terms of the interaction energies between like and dissimilar species, $\epsilon_{11}, \epsilon_{22}$, and ϵ_{12}, and the chemical potentials μ_1 and μ_2.

Noting that the Ising model exhibited an up/down symmetry for $h = 0$, then the lattice gas must also have a symmetry line, which from the mapping occurs at $\mu = -\epsilon z/2$. Along that line, the statistical weight of a configuration of lattice gas variables in the grand canonical ensemble can be written as

$$\mathcal{H} - \mu N = \frac{\epsilon}{2} \sum_{i,j}' (n_i - n_j)^2$$

$$= \frac{\epsilon}{2} \sum_{i,j}' [(1 - n_i) - (1 - n_j)]^2 = \frac{\epsilon}{2} \sum_{i,j}' (n_i - n_j)^2$$

which is invariant to changing all filled cells to empty and all empty to filled, $\{n_i^N\} \rightarrow \{1 - n_i^N\}$. This is an example of particle-hole symmetry. Along the line $\mu = -\epsilon z/2$ configurations of all filled and all empty cells are equally likely. A succinct way to refer to both it and the up/down symmetry of the Ising model is to refer to the binary group, Z_2.

4.5   Role of dimensionality

The Ising model or lattice gas is difficult to solve in general, but in one spatial dimension it is possible using the *transfer matrix approach*. Let's consider the Ising model on a ring, for which the energy can be written

$$\mathcal{H} = -J \sum_{i=1}^{N} s_i s_{i+1} - h \sum_i s_i$$

where s_N and $s_1 = s_{N+1}$ are nearest neighbors on the ring. The partition function is

$$Q = \sum_{\{s_i^N\}} e^{\beta J \sum_{i=1}^{N} s_i s_{i+1} + \beta h \sum_i s_i}$$

$$= \sum_{\{s_i^N\}} \prod_{i=1}^{N} T(s_i, s_{i+1})$$

where we have introduced a symmetrized function of s_i and s_{i+1},

$$T(s_i, s_{i+1}) = e^{\beta J s_i s_{i+1} + \beta h s_i/2 + \beta h s_{i+1}/2}$$

which is an exact rewriting of the product of Boltzmann factors. However, this is suggestive of a particular mathematical structure for the partition function. Grouping two of the T functions with a sum of a spin variable

$$\sum_{s_{i+1}=\pm 1} T(s_i, s_{i+1}) T(s_{i+1}, s_{i+2}) = (\mathbf{T} \cdot \mathbf{T})_{s_i, s_{i+2}}$$

we recognize a matrix product form. Carrying this out for all of the spin variables, we are left with

$$Q = \sum_{s_1 = \pm 1} \left(\mathbf{T}^N\right)_{s_1, s_1} = \mathrm{Tr}(\mathbf{T}^N)$$

a relation between the partition function and the trace of N factors of a *transfer matrix* $\mathbf{T}$. The elements of $\mathbf{T}$, $T(1, 1), T(-1, 1), T(1, -1), T(-1, 1)$ are

$$\mathbf{T} = \begin{pmatrix} e^{\beta J + \beta h} & e^{-\beta J} \\ e^{-\beta J} & e^{\beta J - \beta h} \end{pmatrix}$$

as spanned by $(s_i = \pm 1, s_{i+1} = \pm 1)$. This matrix is diagonalizable, and in the eigenbasis,

$$\mathbf{T}^N = \begin{pmatrix} \lambda_+^N & 0 \\ 0 & \lambda_-^N \end{pmatrix}$$

and providing $\lambda_+ > \lambda_-$ and the number of spins N is large, reduces the partition function to $Q = \lambda_+^N$, or N factors of the dominant eigenvalue of the transfer matrix. Diagonalizing $\mathbf{T}$ for the Ising model yields,

$$\lambda_\pm = e^{\beta J} \left[\cosh(\beta h) \pm \sqrt{\sinh^2(\beta h) - e^{-4\beta J}} \right]$$

and the resultant free energy for the one dimensional Ising model is

$$-\beta A/N = \beta J + \ln \left[\cosh(\beta h) + \sqrt{\sinh^2(\beta h) - e^{-4\beta J}} \right]$$

which is a smooth function of h and β away from $T = 0$. This is suggestive of the lack of a phase transition, which indeed can be verified by computing the average magnetization,

$$\begin{aligned}
\frac{\langle M \rangle}{N} &= \frac{1}{N} \left(\frac{\partial \ln Q}{\partial \beta h} \right)_{T,N} \\
&= \frac{\sinh(\beta h)}{\sqrt{\sinh^2(\beta h) - e^{-4\beta J}}}
\end{aligned}$$

which goes to zero as $h \to 0^+$ from above, unless $T = 0$, in which case $M/N = 1$. So there is no finite temperature critical point in the one-dimensional Ising model, or spontaneous symmetry breaking. At $T = 0$ there is a first order transition reflecting the degeneracy in the ground state that can be lifted by changing h away from 0.

Exercise 4.5: The transfer matrix approach can also be used to evaluate the spin-spin correlation function $\langle s_i s_{i+r} \rangle$. For $h = 0$ this average can be written as,

$$\langle s_i s_{i+r} \rangle = \frac{1}{Q} \mathrm{Tr}(\boldsymbol{\sigma}_x \mathbf{T}^r \boldsymbol{\sigma}_x \mathbf{T}^{N-r})$$

where $\boldsymbol{\sigma}_x$ encodes s_i in the eigenbasis of $\mathbf{T}$

$$\boldsymbol{\sigma}_x = \begin{pmatrix} 0 & 1 \\ 1 & 0 \end{pmatrix}$$

Show that in the large N limit, $\langle s_i s_{i+r} \rangle \approx (\lambda_-/\lambda_+)^r \equiv \exp(-r/\xi)$, where the correlation length is $\xi = -1/\ln \tanh(\beta J)$.

The lack of a phase transition is a peculiarity of $1d$, not the Ising model itself. A system that exists in the thermodynamic limit in only $1d$, in which only one direction can be taken macroscopically large, cannot support a phase transition away from $T = 0$

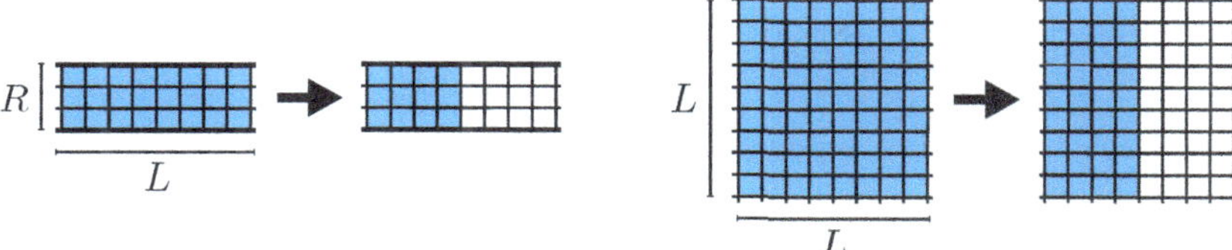

Fig. 4.8 A thermodynamically-$1d$ system, where $L \to \infty$ with fixed R, cannot generally support a phase transition. Exceptions are systems that interact over long ranges or systems at $T = 0$. A $2d$ system however can sustain an interface between two phases in the limit that $L \to \infty$.

if the interactions are short ranged. This point was originally articulated by Landau, and follows from a simple argument. Consider a thermodynamically $1d$ system, of width R, and length L, like that shown in Figure 4.8 for the lattice gas. If the two different states of the system are at phase coexistence, like liquid and vapor, then their chemical potentials are equivalent. Therefore, the free energy, ΔG, to bring these two states into contact with each other is given by a sum of two interfacial terms,

$$\Delta G = \gamma \Delta E R - k_{\mathrm{B}} T \ln(L/a)$$

where $\gamma \Delta E R$ is the energetic penalty for establishing an interface and $k_{\mathrm{B}} T \ln(L/a)$ is an estimate of the entropy gain for placing an interface at any degenerate place along the length of the system in units of the lattice constant a. As long as $\gamma \Delta E$ does not change with L, which is reasonable for systems interacting over short ranges, then in the limit of $L \gg R$, the free energy to form an interface will be negative for any finite temperature. In this limit, interfaces will proliferate, and the two domains of each phase will cease being well defined. Thus multiple phases, or phase coexistence, is not possible at finite temperature. Replacing the $1d$ system with one in $2d$, the interfacial energy would scale like L, and thus in the limit L is large it could overcome the entropy term, which will still scales logarithmically. Whether or not phase coexistence is stable would depend on the temperature, but at least in principle a phase transition could occur.

Onsager solved the Ising model in $2d$ for $h = 0$ exactly using a generalization of the transfer matrix method, demonstrating definitively that there is a phase transition at finite temperature in the thermodynamic limit with $k_{\mathrm{B}} T_c = 2J/\ln(1 + \sqrt{2})$. In three and higher dimensions, numerical solutions are possible. As summarized in Table 4.1, the Ising model is capable of producing quantitatively accurate results for the critical exponents of systems, which undergo a critical point separating states with broken Z_2 symmetry like the magnet, fluid, or binary mixture. In fact, measurements on binary mixtures in $2d$, realized in lipid membranes by Keller, agree with the exponents predicted from Onsager's solution to the Ising model. For the Ising model, there is no phase transition below $2d$, so we call it the lower critical dimension. For $4d$ and above, the critical exponents cease changing, so we call it the upper critical dimension. Generally, dimensionality plays an important role in universality, and measured critical exponents in $3d$ are distinct from those in other d.

d	1	2	3	4	5	∞
$\hat{\beta}$	-	1/8	0.32	1/2	1/2	1/2
$\hat{\gamma}$	-	7/4	1.2	1	1	1

Table 4.1 Summary of the critical exponents for systems in the Ising universality class as a function of dimensionality.

It turns out that the infinite dimensional case can also be solved exactly using another specialized trick known as the *Hubbard–Stratonovich transformation.* An infinite dimensional lattice is one in which all spins interact with all other spins. In order for the energy to remain extensive, the coupling strength needs to be scaled by the system size so that the energy is given by

$$\mathcal{H} = -\frac{J}{2N} \sum_{i \neq j=1}^{N} s_i s_j - h \sum_i s_i$$

where we note the sum over i and j is no longer restricted to nearest neighbors. The Hubbard–Stratonovich transformation is essentially the Gaussian integral tricks employed so often in the previous chapter, in reverse. Specifically, we can use a Gaussian integral identity to rewrite the Boltzmann factor,

$$e^{\beta \frac{J}{2N} \sum_{i,j=1}^{N} s_i s_j + \beta h \sum_i s_i} = \sqrt{\frac{\beta J N}{2\pi}} \int_{-\infty}^{\infty} d\phi \, e^{-\beta J N \phi^2/2 + \beta J \phi \sum_i s_i + \beta h \sum_i s_i}$$

which equates a pairwise interacting model to a model of independent spins subject to a Gaussian random field ϕ. As a collection of independent spins, the usual factorization techniques work to evaluate the partition function

$$Q = \sqrt{\frac{\beta J N}{2\pi}} \int_{-\infty}^{\infty} d\phi \, e^{-N\beta J \phi^2/2} \left(\sum_{\{s_i = \pm 1\}} e^{\beta J \phi s_i + \beta h s_i} \right)^N$$

$$= \sqrt{\frac{\beta J N}{2\pi}} \int_{-\infty}^{\infty} d\phi \, e^{-N\beta J \phi^2/2 + N \ln[2 \cosh(\beta J \phi + \beta h)]}$$

where we have solved for Q up to an integral over the field. In the large N limit, this integral can be evaluated using a saddle point integration. As $N \to \infty$, the integral will be dominated by the value of ϕ that minimizes the argument of the exponential,

$$\frac{d}{d\phi} \left\{ \beta J \phi^2/2 - \ln[2 \cosh(\beta J \phi + \beta h)] \right\} \bigg|_{\phi = \bar{\phi}} = 0$$

$$\implies \bar{\phi} = \tanh\left(\beta J \bar{\phi} + \beta h\right)$$

where we have denoted $\bar{\phi}$ the extremum, which satisfies a self-consistent expression. In terms of $\bar{\phi}$, the free energy up to a constant is

$$A = NJ\bar{\phi}^2/2 - k_{\mathrm{B}}TN \ln[2 \cosh\left(\beta J \bar{\phi} + \beta h\right)].$$

The self-consistent condition on $\bar{\phi}$ has a simple interpretation. If we note that the mean value of the magnetization per site, $m = \langle M \rangle / N$, is given by

$$m = \frac{1}{N} \left(\frac{\partial A}{\partial h} \right)_{T,N}$$
$$= \tanh\left(\beta J \bar{\phi} + \beta h\right) = \tanh(\beta J m + \beta h)$$

we find the same self-consistent equation as $\bar{\phi}$, so in fact $\bar{\phi} = m$. With this relation, the infinite dimensional Ising model can be shown to spontaneously break symmetry at a critical temperature $T_c = J/k_B$. Near the critical point, $h = 0$ and m is small, so expanding the self-consistent expression to cubic order

$$m = \beta J m - (\beta J m)^3 / 3$$

and solving for m, yields $m \propto \pm \sqrt{J/k_B - T}$ if $T < J/k_B$, two degenerate values and a critical exponent, $\hat{\beta} = 1/2$. A similar calculation leads to $\hat{\gamma} = 1$, consistent with the exponents found for $d > 3$.

Exercise 4.6: Demonstrate the earlier claims of the location of the critical temperature and critical exponents are correct for the infinite dimensional Ising model. To do this you should employ the self-consistent expression for the mean magnetization $h = 0$. Above T_c you should find $m = 0$ while below you should find $m = \pm\sqrt{3(T_c - T)/T_c}$. The susceptibility can be computed by $\tanh(\beta J m + \beta h) \approx \beta J m + \beta h$ and recalling that $\chi = dm/d\beta h \sim |T - T_c|^{-\hat{\gamma}}$.

4.6   Landau's theory of phases

The exact techniques suitable for low and very high dimensional systems are not easily generalizable for $d = 3$, nor for models not defined on a lattice. An approximate theory developed by Landau is general and widely employed. Landau theory presents a unified framework for classifying and understanding phase transitions based on their order, dimensionality, and symmetry. It is based on a power series expansion of the local free energy in terms of an order parameter field that distinguishes an ordered phase from a disordered phase. It typically assumes that the order parameter is small so that only the lowest order terms required by symmetry are kept. This approximation is reasonable near a critical point, but must be done with care near a first-order transition. The utility of Landau theory is in its simplicity. When evaluated approximately, properties of a phase transition can be determined by solving simple algebraic equations. Going beyond simple approximations field is possible, but becomes increasingly difficult.

To develop Landau theory, we will consider for concreteness the Curie–Weiss transition for magnetization, but the arguments we will use are quite general. We need to first construct a field representation of an appropriate order parameter. We can do this by coarse-graining the original spin variables from an Ising model, over a length scale, ℓ, which is larger than the lattice size, a. Specifically, the local magnetization density at $\mathbf{r}$ can be constructed from an indicator $\Theta(x) = 1/\ell^d$ if $x < 0$ and 0 else,

$$\bar{m}(\mathbf{r}) = \sum_i s_i \Theta(|\mathbf{r} - \mathbf{r}_i| - \ell)$$

where the sum is taken over all of the spins within a coarse-graining volume, ℓ^d. The probability of observing a specific pattern of magnetization can be represented as

$$P[m(\mathbf{r})] = \left\langle \prod_{\mathbf{r}} \delta \left(m(\mathbf{r}) - \sum_i s_i \Theta(|\mathbf{r} - \mathbf{r}_i| - \ell) \right) \right\rangle$$

where the inner product constrains the magnetization for each $\mathbf{r}$ in the system. We know that the free energy, or the reversible work surface, is simply related to the probability, so we can construct an effective potential by,

$$\mathcal{H}[m(\mathbf{r})] = -k_{\mathrm{B}}T \ln \left\langle \prod_{\mathbf{r}} \delta \left[m(\mathbf{r}) - \sum_i s_i \Theta(|\mathbf{r} - \mathbf{r}_i| - \ell) \right] \right\rangle$$

where $\mathcal{H}[m(\mathbf{r})]$ is an effective Hamiltonian for changes in $m(\mathbf{r})$. Because we have integrated over some short-ranged degrees of freedom, $\mathcal{H}[m(\mathbf{r})]$ is strictly a free energy including energetic as well as entropic components and as such may have an explicit thermodynamic state dependence.

Assuming that the order parameter is small, and $\langle m(\mathbf{r}) \rangle = 0$, we can represent this probability distribution with a cumulant expansion. Expanding the fluctuations in the field we obtain the following series,

$$\beta \mathcal{H}[m(\mathbf{r})] \approx \int_{\mathbf{r}} \Gamma_1(\mathbf{r}) m(\mathbf{r}) + \frac{1}{2} \int_{\mathbf{r}} \int_{\mathbf{r}'} m(\mathbf{r}) \Gamma_2(\mathbf{r}, \mathbf{r}') m(\mathbf{r}')$$
$$+ \frac{1}{3!} \int_{\mathbf{r}} \int_{\mathbf{r}'} \int_{\mathbf{r}''} m(\mathbf{r}) m(\mathbf{r}') m(\mathbf{r}'') \Gamma_3(\mathbf{r}, \mathbf{r}', \mathbf{r}'')$$
$$+ \frac{1}{4!} \int_{\mathbf{r}} \int_{\mathbf{r}'} \int_{\mathbf{r}''} \int_{\mathbf{r}'''} m(\mathbf{r}) m(\mathbf{r}') m(\mathbf{r}'') m(\mathbf{r}''') \Gamma_4(\mathbf{r}, \mathbf{r}', \mathbf{r}'', \mathbf{r}''')$$

where we have truncated it at the 4'th order. The expansion coefficients, which in general depend on all of the field points $(\mathbf{r}, \mathbf{r}', \dots)$, are known as vertex functions. We have encountered the second-order vertex function previously, though we used a different notation,

$$\Gamma_2(\mathbf{r}, \mathbf{r}') = \chi^{-1}(\mathbf{r}, \mathbf{r}') \qquad \chi(\mathbf{r}, \mathbf{r}') = \langle m(\mathbf{r}) m(\mathbf{r}') \rangle$$

where we identify the vertex functions are inverses of higher-order correlation functions. In order for this effective Hamiltonian to obey the same symmetries as the underlying molecular degrees of freedom, in this case up/down symmetry, we can identify the odd order coefficients as 0,

$$\Gamma_1(\mathbf{r}) = 0 \qquad \Gamma_3(\mathbf{r}, \mathbf{r}', \mathbf{r}'') = 0$$

and also set the constant term to 0 since it is irrelevant. Doing this we are left with only two terms,

$$\beta\mathcal{H}[m(\mathbf{r})] = \frac{1}{2}\int_{\mathbf{r}}\int_{\mathbf{r}'} m(\mathbf{r})\Gamma_2(\mathbf{r},\mathbf{r}')m(\mathbf{r}')$$
$$+ \frac{1}{4!}\int_{\mathbf{r}}\int_{\mathbf{r}'}\int_{\mathbf{r}''}\int_{\mathbf{r}'''} m(\mathbf{r})m(\mathbf{r}')m(\mathbf{r}'')m(\mathbf{r}''')\Gamma_4(\mathbf{r},\mathbf{r}',\mathbf{r}'',\mathbf{r}''')$$

but to evaluate this theory requires knowing in advance the two-point and four-point correlation functions.

Here we will make an approximation. Because our aim is to understand the phase transition, which is a phenomenon that occurs on large lengthscales, we imagine that we can approximate these correlation functions by their longest wavelength components. To do this, it is easier to consider the effective Hamiltonian in Fourier space,

$$\beta\mathcal{H}[\hat{m}(\mathbf{k})] = \frac{1}{2}\int_{\mathbf{k}} \hat{m}(\mathbf{k})\hat{m}(-\mathbf{k})\hat{\Gamma}_2(|\mathbf{k}|)$$
$$+ \frac{1}{4!}\int_{\mathbf{k}}\int_{\mathbf{k}'}\int_{\mathbf{k}''} \hat{m}(\mathbf{k})\hat{m}(\mathbf{k}')\hat{m}(\mathbf{k}'')\hat{m}(-\mathbf{k}-\mathbf{k}'-\mathbf{k}'')\hat{\Gamma}_4(\mathbf{k},\mathbf{k}',\mathbf{k}'')$$

where $\hat{m}(\mathbf{k})$ is the Fourier component of the order parameter field and we have used the delta function definition of the Fourier transform to get rid of a few of the integrals. Each of the vertex functions can now be expanded about $k = 0$,

$$\hat{\Gamma}_2(|\mathbf{k}|) \approx a + \gamma k^2 + \mathcal{O}(q^4) \qquad \hat{\Gamma}_4(\mathbf{k},\mathbf{k}',\mathbf{k}'') \approx 6u$$

where to 4'th order due to inversion symmetry, we are left with only constants, and one quadratic term. Putting this back together we have,

$$\beta\mathcal{H}[\hat{m}(\mathbf{k})] = \int_{\mathbf{k}} \frac{(a + \gamma k^2)}{2}|\hat{m}(\mathbf{k})|^2 + \frac{u}{4}\int_{\mathbf{k}}\int_{\mathbf{k}'}\int_{\mathbf{k}''} \hat{m}(\mathbf{k})\hat{m}(\mathbf{k}')\hat{m}(\mathbf{k}'')\hat{m}(-\mathbf{k}-\mathbf{k}'-\mathbf{k}'')$$

which includes only single integrals over k. This is the Landau theory for a second-order phase transition with a scalar order parameter field. Transforming back to real space, and invoking a boundary condition that $\nabla m(\mathbf{r}) = 0$, we find,

$$\beta\mathcal{H}[m(\mathbf{r})] = \int_{\mathbf{r}} \frac{1}{2}a\,m^2(\mathbf{r}) + \frac{1}{4}u\,m^4(\mathbf{r}) + \frac{1}{2}\gamma|\nabla m(\mathbf{r})|^2$$

which shows that the low wavevector expansion is the same as a gradient expansion in real space. This clarifies the nature of this approximation, namely that it contains only those slowest varying spatial fluctuations. Apart from the gradient term, the theory has spatial locality. If solved numerically this effective Hamiltonian exhibits the same critical exponents as the original Ising model.

4.7 Mean field theory

While seemingly simpler than the original Ising model, the Landau theory is still not exactly solvable. The effective Hamiltonian,

$$\beta\mathcal{H}[m(\mathbf{r})] = \int d\mathbf{r}\, \frac{a}{2}m^2(\mathbf{r}) + \frac{u}{4}m^4(\mathbf{r}) + \frac{1}{2}\gamma|\nabla m(\mathbf{r})|^2$$

includes quartic and gradient terms of a fluctuating field, $m(\mathbf{r})$, that must be averaged over to yield a partition function

$$Q = \int \mathcal{D}[m(\mathbf{r})]e^{-\beta\mathcal{H}[m(\mathbf{r})] + \beta \int_{\mathbf{r}} h(\mathbf{r})m(\mathbf{r})}$$

which now has a compact representation as a path integral and in general could have an additional external field, $h(\mathbf{r})$. The simplest approximate treatment of this theory is to evaluate the partition function by a saddle point integration. The saddle point is given by minimizing the argument of the exponential, by finding the most typical configuration of the field. This can be done by evaluating the functional derivative,

$$\frac{\delta}{\delta m(\mathbf{r})}\left[\beta\mathcal{H}[m(\mathbf{r})] - \beta\int_{\mathbf{r}} h(\mathbf{r})m(\mathbf{r})\right] = 0$$

and finding the field $m(\mathbf{r})$ that solves the above equation. The functional derivative can be done simply for the first two terms,

$$am(\mathbf{r}) + um^3(\mathbf{r}) + \frac{1}{2}\gamma\frac{\delta}{\delta m(\mathbf{r})}\int_{\mathbf{r}}|\nabla m(\mathbf{r})|^2 = \beta h(\mathbf{r})$$

but the functional derivative of the squared gradient is a little tricky to do. A simple way to understand its action is to remind ourselves that the functional derivative is the response to an infinitesimal change,

$$\int_{\mathbf{r}}\nabla\left(m(\mathbf{r}) + \delta m(\mathbf{r})\right)\nabla\left(m(\mathbf{r}) + \delta m(\mathbf{r})\right) = \int_{\mathbf{r}}|\nabla m(\mathbf{r})|^2 + 2\int_{\mathbf{r}}\nabla\delta m(\mathbf{r})\nabla m(\mathbf{r})$$

where we have kept only the first order term in the infinitesimal change in the field, $\delta m(\mathbf{r})$. The second term can be evaluated by integration by parts yielding,

$$2\int_{\mathbf{r}}\nabla\delta m(\mathbf{r})\nabla m(\mathbf{r}) = -2\int_{\mathbf{r}}\delta m(\mathbf{r})\nabla^2 m(\mathbf{r})$$

which is now linear in $\delta m(\mathbf{r})$, allowing us to evaluate the functional derivative straightforwardly. This yields

$$a\langle m(\mathbf{r})\rangle + u\langle m(\mathbf{r})\rangle^3 - \gamma\nabla^2\langle m(\mathbf{r})\rangle = \beta h(\mathbf{r})$$

which in general for a spatially dependent external field, or with inhomogeneous boundary conditions, requires us to solve a second-order differential equation. This is known as a *Euler–Lagrange equation*, and the approximation to the partition function that results from considering only the most likely behavior is mean field theory.

If we neglect the external field, $h(\mathbf{r}) = 0$, then the lowest energy solution will always be a homogenous state, $\langle m(\mathbf{r}) \rangle = m$ because the squared gradient is strictly positive. The magnitude of the order parameter is determined by the algebraic equation,

$$am + um^3 = 0$$

which, as a cubic polynomial, has three solutions. If we assume that we can expand a to linear order in the temperature near T_c, $t = (T - T_c)/T_c$,

$$a(t) = ct + \mathcal{O}(t^2)$$

then the three solutions are given by

$$m = \begin{cases} 0 & t > 0 \\ \pm\sqrt{\frac{-ct}{u}} & t < 0 \end{cases}$$

or we have found that above the critical temperature, $t > 0$, there is one disordered phase, characterized by an average magnetization that is zero. Below the critical temperature, $t < 0$, we have two degenerate phases, characterized by an average magnetization that is nonzero, with equal magnitude and opposite sign. Landau theory within its mean field approximation successfully predicts the spontaneous symmetry breaking of the Ising magnet. Within the mean field theory approximation, the free energy to change the magnetization is determined by the polynomial expansion in m, $A \approx \mathcal{H}_{\mathrm{MF}}[m] = V(am^2/2 + um^4/4)$. When the temperature dependence is incorporated into the quadratic coefficient, the progression in temperature is illustrated in Figure 4.9.

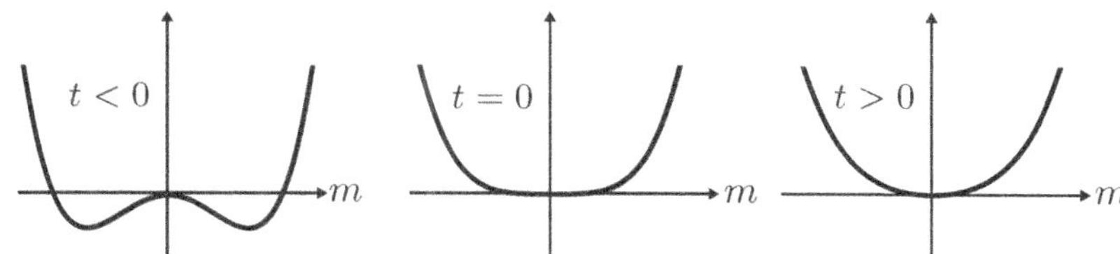

Fig. 4.9 Free energy as a function of magnetization above (right), at (middle), and below (left) the critical temperature. Spontaneous symmetry breaking is evident as a bimodal free energy surface.

As we found for the infinite dimensional Ising model,

$$m \sim t^{1/2}$$

or that the critical exponent $\hat{\beta} = 1/2$. We can analogously compute the critical exponent associated with the susceptibility,

$$\chi(t) = \left(\frac{dm}{d\beta h}\right)_T \bigg|_{h=0}$$

by adding back a small homogeneous magnetic field,

$$\frac{d}{d\beta h}\left(ctm + um^3\right) = \frac{d}{d\beta h}\beta h$$

$$\chi(t)\left(ct + 3um^2\right) = 1$$

$$\chi(t) = \frac{1}{ct + 3um^2}$$

and setting it zero. Rearranging this and evaluating it with our mean magnetization we find

$$\chi(t) = \begin{cases} \frac{1}{ct} & \text{if} \quad t > 0 \\ \frac{1}{2c|t|} & \text{if} \quad t < 0 \end{cases}$$

or that the critical exponent, $\hat{\gamma} = 1$, again just as we found previously in infinite dimensions. The critical behavior predicted within mean field theory is shown in Figure 4.10. Notice two short-comings. The critical exponents are not those associated with a $3d$ system, and further, within mean field theory, they do not depend on d.

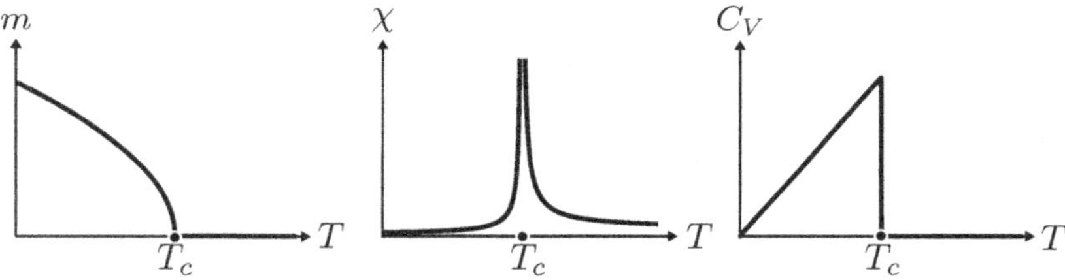

Fig. 4.10 Critical behavior predicted by mean field theory. On the left is the predicted magnetization as a function of temperature. In the middle is the predicted susceptibility, which diverges at the critical point. A failing of mean field theory is the prediction of a discontinuity of the heat capacity, which experimentally is weakly divergent.

Exercise 4.7: Show that the heat capacity $C_V = -T(d^2A/dT^2)$ within the mean field approximation $A = \mathcal{H}_{\mathrm{MF}}[m]$ is equal to 0 for $t > 0$ and $c^2 k_{\mathrm{B}} T / 2 u T_c$ for $t < 0$.

4.8 Divergent correlations and fluctuations

We have motivated an effective free energy functional to describe a second-order phase transition like that associated with a magnet. This functional came from a coarse-graining of the molecular degrees of freedom into a continuum field. The subsequent procedure was to take a general order parameter field, $\phi(\mathbf{r})$, and expand the Hamiltonian in a power series, which assuming its magnitude is small, can be truncated at some order and simplified by invoking symmetry. In order to retain some amount of spatial information, a gradient expansion is made to approximate the correlation function expansion coefficients. The resulting free energy functional for a field with $\phi \to -\phi$ symmetry is

$$\beta \mathcal{H}[\phi(\mathbf{r})] = \int d\mathbf{r}\, a\phi^2(\mathbf{r})/2 + u\phi^4(\mathbf{r})/4 + \gamma\, |\nabla\phi(\mathbf{r})|^2/2$$

where for consistency with the truncation and to allow for a temperature-driven transition we set,

$$a(t) = ct \qquad u, \gamma > 0$$

where t is the reduced temperature, $t = T/T_c - 1$, and u and γ are positive constants. From this effective Hamiltonian, the partition function is given by

$$Q = \int \mathcal{D}[\phi(\mathbf{r})] e^{-\beta\mathcal{H}[\phi(\mathbf{r})]}$$

which is a path integral over all fields, $\phi(\mathbf{r})$. This model is equivalent to that deduced from the Ising model. This specific theory is often referred to as a "$\phi-4$" field theory, as is applicable to any of the Z_2 symmetry-breaking phase transitions we have introduced, provided a suitable interpretation of the order parameter field, ϕ.

Previously we worked out a mean field approximation to the partition function. Such an approximation neglected all fluctuations of the field about its average value. We can evaluate the scale of those fluctuations by computing the correlation function of field fluctuations in different regions of space. To proceed, let's separate out the fluctuations of the field from its average by defining

$$\delta\phi(\mathbf{r}) = \phi(\mathbf{r}) - \langle\phi(\mathbf{r})\rangle$$

where $\langle\phi(\mathbf{r})\rangle$ is the average value of the field. Inserting this definition into the Hamiltonian, we naturally find that it separates into two pieces, a part that depends on just the average of the field,

$$\beta\mathcal{H}[\phi(\mathbf{r})] = \beta\mathcal{H}_{\mathrm{MF}}[\langle\phi(\mathbf{r})\rangle] + \beta\Delta\mathcal{H}[\delta\phi(\mathbf{r})]$$

which we label the mean field part of the Hamiltonian, and a piece that depends on the fluctuations. The mean field part is given by

$$\beta\mathcal{H}_{\mathrm{MF}}[\langle\phi(\mathbf{r})\rangle] = \int d\mathbf{r}\, a(t)\langle\phi(\mathbf{r})\rangle^2/2 + u\langle\phi(\mathbf{r})\rangle^4/4 + \gamma\, |\nabla\langle\phi(\mathbf{r})\rangle|^2/2$$

and the corresponding fluctuation contribution

$$\beta\Delta\mathcal{H}[\delta\phi(\mathbf{r})] = \int d\mathbf{r}\, \frac{1}{2}\left(a(t) + 3u\langle\phi(\mathbf{r})\rangle^2\right)\delta\phi^2(\mathbf{r}) + \frac{1}{4}u\delta\phi^4(\mathbf{r}) + \frac{1}{2}\gamma\, |\nabla\delta\phi(\mathbf{r})|^2$$

which we label with a delta since it's the correction to mean field theory. Inserting these two pieces into the partition function,

$$Q = Q_{\mathrm{MF}}[\langle \phi(\mathbf{r}) \rangle] \int \mathcal{D}[\delta\phi(\mathbf{r})] e^{-\beta \Delta \mathcal{H}[\delta\phi(\mathbf{r})]}$$

we find that it factorizes in a mean field contribution that comes outside of the integration over the field since it is just a constant and a piece that must be evaluated over all field fluctuations.

Assuming $\beta \Delta \mathcal{H}$ is small,

$$Q \approx Q_{\mathrm{MF}}[\langle \phi(\mathbf{r}) \rangle] \qquad \beta A \approx - \ln Q_{\mathrm{MF}}[\langle \phi(\mathbf{r}) \rangle] \qquad \left. \frac{\delta \beta A}{\delta \phi(\mathbf{r})} \right|_{\langle \phi(\mathbf{r}) \rangle} = 0$$

we can proceed as we did previously, neglecting a spatially dependent external field or inhomogenious boundary conditions, we find the average field is constant through space,

$$\langle \phi(\mathbf{r}) \rangle = \begin{cases} 0 & t > 0 \\ \pm \sqrt{\frac{-ct}{u}} & t < 0 \end{cases}$$

and is exactly given by our previous mean field solution.

With the decomposition earlier, we can also ask questions about fluctuations away from the mean field, or correspondingly, about spatial correlations. This is most easily done in Fourier space. Transforming $\Delta \mathcal{H}$ we find,

$$\beta \Delta \mathcal{H}[\delta\hat{\phi}(\mathbf{k})] = \int \frac{d\mathbf{k}}{(2\pi)^d} \frac{1}{2} \left(a(t) + 3u \langle \phi \rangle^2 \right) |\delta\hat{\phi}(\mathbf{k})|^2 + \frac{1}{2} m \mathbf{k}^2 |\delta\hat{\phi}(\mathbf{k})|^2 + \ldots$$

which would include a quartic term in the field, making it difficult to preform any sort of direct integration. We can proceed only by assuming that the fluctuations are small and truncating $\Delta \mathcal{H}$ to terms bilinear in the field,

$$\beta \Delta \mathcal{H}[\delta\hat{\phi}(\mathbf{k})] \approx \int \frac{d\mathbf{k}}{(2\pi)^d} \frac{1}{2} \left(a(t) + 3u \langle \phi \rangle^2 + m \mathbf{k}^2 \right) |\delta\hat{\phi}(\mathbf{k})|^2$$
$$= \int \frac{d\mathbf{k}}{(2\pi)^d} \frac{1}{2} \chi^{-1}(\mathbf{k}) |\delta\hat{\phi}(\mathbf{k})|^2$$

which is just like the effective Hamiltonians studied in Chapter 3. This is a Gaussian field, so we know how to read off the correlation function, which in Fourier space is

$$\chi^{-1}(\mathbf{k}) = \left\langle \delta\hat{\phi}(-\mathbf{k}) \delta\hat{\phi}(\mathbf{k}) \right\rangle$$
$$= \left(a(t) + 3u \langle \phi \rangle^2 + \gamma \mathbf{k}^2 \right)$$

and retains some wave-vector dependence. We can rewrite this by non-dimensionalizing the wave-vector by introducing a length scale ξ,

$$\chi(\mathbf{k}) = \frac{\chi_0}{\xi^{-2} + \mathbf{k}^2}$$

where $\chi_0^{-1} = \gamma$ and ξ is

$$\xi = \begin{cases} \sqrt{\frac{\gamma}{ct}} & t > 0 \\ \sqrt{\frac{\gamma}{-2ct}} & t < 0 \end{cases}$$

and reflects a correlation length for field fluctuations that diverge at the critical point as

$$\xi(t) \sim t^{-1/2}$$

with a critical exponent of $1/2$. This exponent dictating the divergence of the correlation length is typically denoted $\hat{\nu}$. Like other exponents in the Ising universality class, $\hat{\nu}$ is independent of molecular details but depends on dimensionality below $4d$. Experimentally in $3d$, $\hat{\nu}$ is around 0.63. Accompanying the divergence of the correlation length is a divergent correlation time, known as critical slowing down, with an associated critical dynamical exponent dependent on whether the order parameter is globally conserved or not.

We can interrogate the correlations in real space by inverse Fourier transforming $\chi(\mathbf{k})$. Defining $C(r)$ as the real space correlation function,

$$C(r) = \langle \delta\phi(0)\delta\phi(r) \rangle = \mathcal{F}^{-1}\left[\chi(\mathbf{k})\right]$$

with generic d dimensional inverse Fourier transform defined as

$$C(r) = \frac{1}{(2\pi)^d} \int d\Omega_d \int dk\, k^{d-1} e^{i\mathbf{k}\cdot\mathbf{r}} \chi(\mathbf{k})$$

where Ω_d denotes the d dimensional solid angle. Specializing for the moment to $3d$, we can evaluate the correlation function,

$$\begin{aligned} C(r) &= \frac{\chi_0}{(2\pi)^2} \int dk \int d\theta\, \sin(\theta) k^2 e^{ikr\cos(\theta)} \frac{1}{\xi^{-2} + k^2} \\ &= \frac{2\chi_0}{(2\pi)^2 r} \int dk\, k \frac{\sin(kr)}{\xi^{-2} + k^2} \\ &= \frac{\chi_0}{4\pi r} e^{-r/\xi} \end{aligned}$$

where we find a familiar form of a screened Coulomb interaction. At the critical point when $\xi \to \infty$, correlations become *scale-free*, in the sense that correlations decay algebraically like $1/r$. This power law decay results in patterns at the critical point that are fractal or self-similar. More generally, in arbitrary dimension,

$$C(r) = \frac{\chi_0}{|r|^{d-2}} Y(r/\xi) \qquad Y(\eta) = \frac{1}{(2\pi)^d} \int_0^\infty dz\, z^{d-1} \int d\Omega_d e^{iz\cos(\theta)} \frac{1}{z^2 + \eta^2}$$

the correlation function has a generalized Coulomb part, $1/r$ in $3d$ or logarithmic decay in $2d$, and a screening part that is inconsequential at the critical point, but given by a generalized Bessel function.

Exercise 4.8: In the presence of a spatially varying external field $h(\mathbf{r})$, the partition function within a Landau theory can be written as

$$Q_h = \int \mathcal{D}[\phi(\mathbf{r})] e^{-\beta \mathcal{H}[\phi(\mathbf{r})] + \beta \int d\mathbf{r}\, h(\mathbf{r})\phi(\mathbf{r})}$$

$$= Q_0 \left\langle e^{\beta \int d\mathbf{r}\, h(\mathbf{r})\phi(\mathbf{r})} \right\rangle_0$$

where the exponential average is identified as a generating function. Show that taking a two functional derivatives of $\ln Q_h$ with respect to $\beta h(\mathbf{r})$ and $\beta h(\mathbf{r}')$, provides an equation for the correlation function when $h = 0$,

$$a(t)\chi(\mathbf{r} - \mathbf{r}') + 3u \langle \phi \rangle^2 \chi(\mathbf{r} - \mathbf{r}') - \gamma \nabla^2 \chi(\mathbf{r} - \mathbf{r}') = \delta(\mathbf{r} - \mathbf{r}')$$

and demonstrate that this equation is satisfied in Fourier space by the $\chi(\mathbf{k})$ deduced above.

4.9   Fluctuation corrections

With the proceeding calculation of an approximate treatment of fluctuations, we are in a position to evaluate how well the mean field approximation works. Specifically, for mean field theory to be accurate, we require

$$\langle \delta\phi^2 \rangle \ll \langle \phi \rangle^2$$

namely that the fluctuations of the field are much smaller than the average. While to evaluate this inequality exactly would be to exactly solve the problem, we can make an estimate of the inequality evaluated in the mean field approximation to check if our result is self-consistent,

$$\langle \delta\phi^2 \rangle_{\mathrm{MF}} \ll \langle \phi \rangle^2_{\mathrm{MF}}$$

and therefore check when we are permitted to throw away fluctuation corrections. In the mean field approximation, the average field for $t < 0$ is,

$$\langle \phi \rangle^2_{\mathrm{MF}} = \frac{-a}{u}$$

and we can estimate the average size of fluctuations by integrating our correlation function over a region defined by the correlation length,

$$\langle \delta\phi^2 \rangle_{\mathrm{MF}} = \frac{1}{v_\xi} \int_{v_\xi} d\mathbf{r}\, C(r)$$

$$= \frac{\chi_0 \Omega_d}{\xi^{d-2}}$$

which upon inserting our previous expression for $C(r)$ and using $v_\xi \sim \xi^d$, we find that the scale of fluctuations is inversely proportional to ξ^{2-d} with a proportionality constant that depends on the solid angle and integral of the Bessel function $Y(r)$ that we denote for simplicity as Ω_d. The correlation function we worked out previously is given by

$$\xi(t) = \xi_\mathrm{o} t^{-1/2} \qquad \xi_\mathrm{o} = (\gamma/2c)^{-1/2}$$

Inserting this into the inequality, known as the *Ginzburg criteria*, and rearranging

$$\frac{-a}{u} \gg \frac{\Omega_d}{\gamma \xi^{d-2}}$$

$$\frac{-2a}{\gamma} \gg \frac{2u\Omega_d}{\gamma^2 \xi^{d-2}}$$

$$\left(\frac{\xi}{\xi_\mathrm{o}}\right)^{d-4} \gg \frac{\Omega_d k_\mathrm{B}}{4C_V \xi_\mathrm{o}^d}$$

we find a particular dependence on the dimensionality. Since as $t \to 0$, $\xi \to \infty$, we can infer that our mean field theory will break down close to the critical point, as fluctuations increase unboundedly, but only if $d \leq 4$! If $d > 4$ as the critical point is approached the relative scale of fluctuations does not compete with the growing scale of the average. Indeed, it is found that for $d > 4$ mean field theory is capable of accurately predicting critical exponents, as fluctuation corrections are not important.

Another way to view this inequality is to restore the explicit temperature dependence of ξ,

$$\left(\frac{T - T_c}{T_c}\right)^{(4-d)/2} \gg \frac{\Omega_d k_\mathrm{B}}{4C_V \xi_\mathrm{o}^d}$$

which for a given dimension and temperature away from the critical point may be satisfied. We can define a temperature, known as the Ginzburg temperature, where the two contributions are equal,

$$t_G = \left(\frac{T_G - T_c}{T_c}\right) = \left(\frac{\Omega_d k_\mathrm{B}}{4C_V \xi_\mathrm{o}^d}\right)^{2/(4-d)}$$

which represents how close we can come to the critical point and describe the system with mean field theory. This temperature is inversely proportional to ξ_o the *bare correlation length* and the change in the heat capacity between the ordered and disordered states C_V. If these are large, we will see mean field exponents very close to the critical point. Indeed, in superconductivity the bare correlation length is large, and consequently it is hard to measure non-mean field behavior.

With the above analysis, we can work out fluctuation corrections to observables, such as the location of the critical point. To do this we go back to our formally exact expression for the partition function,

$$Q = Q_\mathrm{MF}[\langle\phi(\mathbf{r})\rangle] \int \mathcal{D}[\delta\phi(\mathbf{r})] e^{-\beta\Delta\mathcal{H}[\delta\phi(\mathbf{r})]}$$

separated out into its mean field contribution and the remainder. If we truncate that remainder at $\delta\phi^2(\mathbf{r})$, we can perform the Gaussian function integral. Denoting that approximation to the partition function as Q_c, we find

$$Q_c = \int \mathcal{D}[\delta\phi(\mathbf{r})]e^{-\frac{1}{2}\int d\mathbf{k}\,\delta\phi^2(\mathbf{k})\chi^{-1}(\mathbf{k})}$$

$$= \prod_\mathbf{k} \sqrt{2\pi/\chi^{-1}(\mathbf{k})}$$

which is just the generalization of the norm of a Gaussian integral made easy by working in the normal modes of the field. The corresponding free energy is

$$\beta A - \beta A_\mathrm{MF} = \frac{1}{2}V \int \frac{d\mathbf{k}}{(2\pi)^d} \ln[\chi^{-1}(\mathbf{k})]$$

where the log of the product over modes has been approximated as an integral. While we could evaluate this directly, in order to understand how the critical temperature changes, we can use the fact that the susceptibility diverges at the critical point and compute that divergence within a random phase approximation. Reminding ourselves that the susceptibility is given by

$$\chi^{-1}(\mathbf{r} - \mathbf{r}') = \frac{\delta^2 \beta A}{\delta\langle\phi(\mathbf{r})\rangle\,\delta\langle\phi(\mathbf{r}')\rangle}$$

we can compute it by taking these derivatives. We find

$$\chi^{-1}(\mathbf{k}) = \chi_\mathrm{MF}^{-1}(\mathbf{k}) + \frac{1}{2} \int \frac{d\mathbf{k}'}{(2\pi)^d} \chi(\mathbf{k}') \frac{\delta^2\chi^{-1}(\mathbf{k}')}{\delta\langle\phi(k')\rangle\,\delta\langle\phi(-k')\rangle}$$

that the susceptibility contains a mean field term and a correction proportional to the susceptibility under the integral. In order to find out what happens to the macroscopic phase transition with the addition of fluctuations, we consider χ at zero wavevector. Defining, $\tau = \chi^{-1}(0)$, our expression simplifies to

$$\tau = a + 3u \int \frac{d\mathbf{k}}{(2\pi)^d} \frac{1}{\tau + \gamma k^2}$$

where the transition is located where the susceptibility is infinite, or where $\tau = 0$. Rearranging, and performing the integral in arbitrary d dimensions we find

$$a = -\frac{3u}{\gamma} \int \frac{d\mathbf{k}}{(2\pi)^d} \frac{1}{k^2}$$

$$= -\frac{3u}{\gamma} \frac{\Omega_d}{(2\pi)^d} \int_0^\Lambda dk\, k^{d-1} \frac{1}{k^2}$$

$$= -\frac{3u}{\gamma(d-2)} \frac{\Omega_d}{(2\pi)^d} \Lambda^{d-2}$$

where we have introduced an upper wavevector (or ultraviolet) cutoff, Λ to regularize the integral. While within the mean field approximation, the susceptibility diverged at

$a = 0$, we now find a correction that shifts this point. Inserting $a = c(T_c - T_c^{\mathrm{MF}})/T_c^{\mathrm{MF}}$ and rewriting in terms of the new critical temperature,

$$T_c = \left(1 - \frac{3u}{\gamma c(d-2)} \frac{\Omega_d}{(2\pi)^d} \Lambda^{d-2}\right) T_c^{\mathrm{MF}}$$

we find that the new critical temperature is necessarily depressed due to fluctuations. The extent depends on d, and is dependent on the high wavevector cutoff. This latter point expresses a sentiment that while near the critical point some properties are universal, others like the location of the critical point are intrinsically dependent on molecular details. There are further a number of useful limits to consider. In the limit that $d \to \infty$, the critical temperature is exactly the mean field estimate. In the opposite limit, for $d \to 2$, the critical temperature is pushed to $T_c \to 0$.

These two examples serve to illustrate that arbitrarily close to T_c, for dimensions lower than the upper critical dimension, simple fluctuation corrections to mean field theory will fail to provide quantitatively accurate predictions. To understand the precise values of critical exponents, or to predict the locations of critical points, a dramatically different approach is needed. Historically, it was Wilson, Kadanoff, and Fisher who developed such an approach, known as the renormalization group. This theory leverages the self-similar structure of a system at its critical point directly, and is able to make accurate estimates of critical exponents. However it is not easy to generalize, and as a consequence currently more often numerical approaches are used.

4.10 Interfaces between coexisting phases

The real utility of Landau theory is its generality. In enables a characterization of arbitrary phase transitions in terms of the spontaneous symmetry breaking of their order parameter fields. To begin to demonstrate that generality, let us shift our attention to the liquid–vapor transition. As discussed previously, the order parameter that distinguishes a liquid from its vapor is the number density ρ. If we really wanted to, we could begin with a microscopic Hamiltonian, coarse-grain it to find an effective theory for its density fluctuations. Such care is needed if we wish to ask about microscopic features, however that detail comes at the cost of making the theory largely analytically intractable. If we are concerned with the behavior of a fluid near its critical point, we can propose an effective theory that includes a local free energy function of the density and a gradient term that penalizes slowly varying density fields. Such a theory would look like,

$$\beta \mathcal{H}[\rho(\mathbf{r})] = \int d\mathbf{r}\, f[\rho(\mathbf{r})] + \frac{\gamma}{2} |\nabla \rho(\mathbf{r})|^2 \qquad f(\rho) = \frac{a}{2}\rho^2 - w\rho^3 + u\rho^4 - h\rho$$

where unlike the magnetization, the density is not up/down symmetric and thus there is no reason in particular that odd order terms in the local free energy $f(\rho)$ need be zero. However, consider rewriting the density as its value at the liquid-vapor critical point ρ_c and some temperature dependence difference $\Delta\rho$, where $\rho = \rho_c + \Delta\rho$. Such a change of variables let us rewrite the local free energy as

$$f(\rho) = \frac{a}{2}(\rho_c + \Delta\rho)^2 - \frac{w}{3}(\rho_c + \Delta\rho)^3 + \frac{u}{4}(\rho_c + \Delta\rho)^4 - h(\rho_c + \Delta\rho)$$

$$= C(\rho_c) + \frac{a(\rho_c)}{2}\Delta\rho^2 - \frac{w(\rho_c)}{3}\Delta\rho^3 + \frac{u(\rho_c)}{4}\Delta\rho^4 - h(\rho_c)\Delta\rho$$

where we can rename our expansion coefficients,

$$h(\rho_c) = h - a\rho_c + w\rho_c^2 - u\rho_c^3$$
$$a(\rho_c) = a - 2w\rho_c + 3u\rho_c^2$$
$$w(\rho_c) = 3u\rho_c - w$$
$$u(\rho_c) = u$$

while constant $C(\rho_c)$ is neglected as it just resets the zero of energy. Inspecting the above equations, we can find a symmetry line in which deviations of the density $\Delta\rho$ are symmetric. Choosing, $\rho_c = w/3u$, and $h = a\rho_c + w\rho_c^2 - u\rho_c^3$, and renaming $\tilde{a} = a - w^2/3u$, we have

$$f(\Delta\rho) = \frac{\tilde{a}}{2}\Delta\rho^2 + u\Delta\rho^4$$

which is identical to our $\phi - 4$ field theory for the Ising model. This transformation clarifies an isomorphism between the up/down symmetry of a simple magnet and the particle/hole symmetry of a fluid, now in the language of fields. It also follows that provided the same form of effective free energy, all of the universal properties of a demagnetization transition are shared by a liquid-vapor transition. While in the former, the symmetry breaking is apparent along the line of zero applied field, the latter requires us to move along a more complicated symmetry line, one which you might suspect is of equal chemical potential between the liquid and its vapor.

Given the isomorphism between our theory of magnets and now that for the liquid–vapor transition, a number of results simply follows from our previous analysis. Specifically, at the mean field theory level of approximation, the mean deviation of the density is

$$\langle\Delta\rho\rangle = \begin{cases} 0 & \text{if} \quad \tilde{a} > 0 \\ \pm\sqrt{\frac{-\tilde{a}(t)}{u}} & \text{if} \quad \tilde{a} < 0 \end{cases}$$

elucidating the spontaneous symmetry breaking between dense and dilute states. The specific densities of the vapor, ρ_v and its liquid, ρ_l are

$$\rho_v = \rho_c - \sqrt{\frac{-\tilde{a}(t)}{u}} \qquad \rho_l = \rho_c + \sqrt{\frac{-\tilde{a}(t)}{u}}$$

below the critical point. Analogous results for the heat capacity and susceptibility would follow provided a phenomenological parameterization of the expansion coefficients a, w, u, and h on temperature and pressure.

Within the context of the liquid–vapor transition, it is natural to ask questions concerning the behavior of the field at coexistence, under conditions of constant density where the liquid is in contact with its vapor. Consider for example the situation in

Figure 4.11 whereby below the critical point, but along the symmetry line, we expect to be able to prepare an interface. In the absence of an applied field, but with boundary conditions that dictate that the liquid and vapor exist far away from the interface placed perpendicular to x, the mean field solution will be given by

$$a(t)\,\langle\Delta\rho(x)\rangle + u\,\langle\Delta\rho(x)\rangle^3 = \gamma\frac{d^2}{dx^2}\,\langle\Delta\rho(x)\rangle$$

where the boundary conditions will enforce an inhomogeneous density distribution. This is a second-order nonlinear ordinary differential equation, which is generally difficult to solve. However in this case, we can guess a solution informed by our physical intuition and underlying assumption that the density varies smoothly. Specifically a function that could satisfy the boundary conditions that $\rho(-\infty) = \rho_l$ and $\rho(\infty) = \rho_v$ could be

$$-\,\langle\Delta\rho(x)\rangle \overset{?}{=} \langle\Delta\rho\rangle\tanh(x/\delta)$$

a simple hyperbolic tangent with characteristic length δ and mean interface position $x = 0$. Plugging this into the Euler–Lagrange equation for the mean profile, the right-hand side yields

$$\gamma\frac{d^2}{dx^2}\,\langle\Delta\rho\rangle\tanh(x/\delta) = \frac{\gamma\,\langle\Delta\rho\rangle}{\delta}\frac{d}{dx}\left(1 - \tanh^2(x/\delta)\right)$$

$$= -\frac{2\gamma\,\langle\Delta\rho\rangle}{\delta^2}\tanh(x/\delta)\left(1 - \tanh^2(x/\delta)\right)$$

which must be equal to the left-hand side

$$\langle\Delta\rho\rangle\tanh(x/\delta)[a(t) + u\,\langle\Delta\rho\rangle^2\tanh^2(x/\delta)] =$$
$$a(t)\,\langle\Delta\rho\rangle\tanh(x/\delta)[1 - \tanh^2(x/\delta)]$$

which is a valid solution provided

$$\delta = \sqrt{-\frac{2\gamma}{a(t)}}$$

which identifies the width of the interface as simply twice the correlation length of the field $\delta = 2\xi$. Just as the correlation length diverges near the critical point, so to must the width of the interface. This should make intuitive sense, as at the critical point there is no meaningful interface between the two phases. This characteristic form of the interface is also shown in Figure 4.11.

The squared-gradient term in the effective field theory penalizes spatially inhomogeneous states, like the interface we just studied. Thus, the additional free energy over the homogeneous phase due to this interface has a thermodynamic meaning. It is precisely the surface tension, σ, or the reversible work to create an interface. Within our Landau theory, we can evaluate the surface tension given the interfacial solution

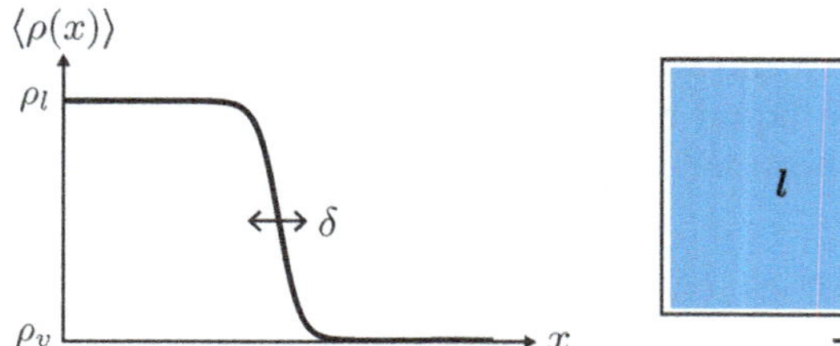

Fig. 4.11 Liquid–vapor coexistence below the critical point results in interface formation with a characteristic interfacial profile predicted by Landau theory.

we just derived. Specifically, at coexistence the excess free energy due to the interface per area L^2 is

$$\sigma = \frac{\gamma}{2L^2} \int d\mathbf{r}\, |\nabla \phi(\mathbf{r})|^2$$

$$= \frac{\gamma}{2} \langle \Delta\rho \rangle^2 \int dx\, |\nabla \tanh(x/\delta))|^2$$

$$= \frac{2\gamma}{3\delta} \langle \Delta\rho \rangle^2 = \frac{\sqrt{2\gamma}}{3u} a^{3/2}(t)$$

where in the last line we have substituted the width of the interface and the mean $\Delta\rho$. From the first equality in the last line, we find a relationship between the width of the interface and the surface tension. This reflects the fact that interfaces with high surface tensions are more resistant to deformation from thermal fluctuations. Upon approaching the critical point, the width diverges, and as a consequence the surface tension goes to 0. Generally, such interfacial fluctuations are known as *capillary waves*.

Exercise 4.9: Two alternative expressions for the surface tension are

$$\sigma = \int_{-\infty}^{\infty} dx\, f[\Delta\rho(x)] \qquad \sigma = \int_{\rho_v}^{\rho_l} d\rho\, \sqrt{\gamma f(\rho)/2}$$

where the integral in the second relation is between the two coexistence densities. Show these relations follow from the Euler–Lagrange equation for the mean profile.

4.11 First order transitions

We have thus far only really discussed second-order transitions that are thermally driven. However, the analysis of the liquid–vapor transition above and specifically the calculation of the surface tension, points to the ability to analyze first-order transitions as well. In the following, we will go over the standard Landau theory for thermally driven first-order transitions. For the moment we will abstract away the precise nature of the order parameter ϕ.

Lacking in specific symmetry, an effective theory for the fluctuations of ϕ would be

$$\beta\mathcal{H}[\phi(\mathbf{r})] = \int d\mathbf{r}\, f[\phi(\mathbf{r})] + \frac{\gamma}{2}|\nabla\phi(\mathbf{r})|^2 \qquad f(\phi) = \frac{a}{2}\phi^2 - w\phi^3 + u\phi^4$$

whereas before we have a local free energy density $f(\phi)$ expanded to low-order in ϕ and truncated our treatment of spatial variation at the squared gradient term. In the local free energy we have neglected a linear field, which can always be incorporated into the cubic order term w, with an appropriate change of variables. As before, we can phenomenologically parameterize the expansion coefficients, and will consider only adding a temperature dependence to a such that

$$a = c(T - T_s) \qquad c, w, u, \gamma > 0$$

where c multiplies the difference between the working temperature T and a reference T_s. We can apply our standard mean field theory approximation to computing the average value of the field through the solution of the Euler–Lagrange equation.

$$\frac{\delta}{\delta\phi(\mathbf{r})}\int d\mathbf{r}\, f[\phi(\mathbf{r})] + \frac{\gamma}{2}|\nabla\phi(\mathbf{r})|^2 \bigg|_{\langle\phi\rangle} = 0$$

which absent boundary conditions reduces to the algebraic equation

$$a(t)\langle\phi\rangle - 3w\langle\phi\rangle^2 + 4u\langle\phi\rangle^3 = 0$$

for $\langle\phi\rangle$. This cubic equation has three roots,

$$\langle\phi\rangle = 0, \quad \frac{3w}{8u} \pm \sqrt{\frac{9w^2}{64u^2} - \frac{a}{2u}}$$

whose stability depends on the discriminate $D = 9w^2 - 16au$, or the factors under the square root and the conditions for the field to be real. Assuming that w and u are fixed positive constants, the existence of multiple solutions depends on a, which phenomenologically depends on temperature.

The evolution of the mean field free energy, which is the local free energy function evaluated at the mean field ϕ is shown in Figure 4.12. When the discriminate $D =$

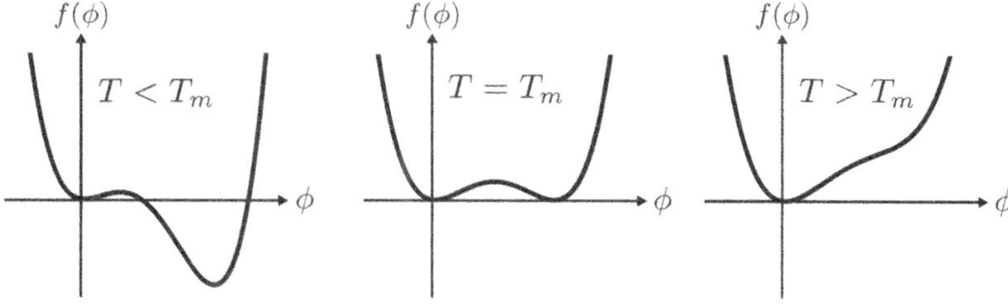

Fig. 4.12 Progression of the mean field free energy density for a first-order transition as described by Landau theory.

$9w - 4au$ is less than 0, there is a single solution at $\phi = 0$. This is the high-temperature solution found when a is large and corresponds to the disordered state. When D is very large, there is a single stable solution at finite ϕ. This is the low-temperature solution found when a is small and corresponds to the ordered state. At a special value of $D = D^*$ there are two degenerate solutions, and thus coexistence between the ordered and disordered state. This occurs at a finite value of ϕ in the ordered state, and thus the mean value of ϕ changes discontinuously as a changes with w and u fixed, or as the temperature is varied. The reference temperature T_s corresponds to the temperature where the disordered state becomes unstable, as a vanishing second derivative.

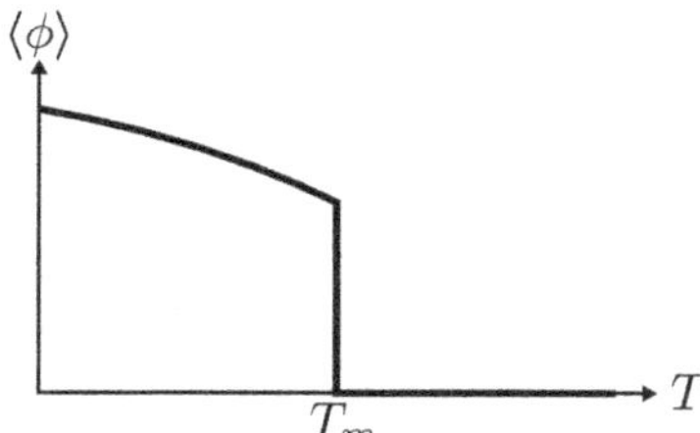

Fig. 4.13 The discontinuous change of $\langle\phi\rangle$ for a first-order phase transition.

We can use this mean value with the coexistence condition that the free energy in the disordered state $\phi = 0$ is equal to the free energy of the ordered state, $f(\langle\phi\rangle) = f(0) = 0$ where we have acknowledged the free energy of the disordered state is 0. This coexistence condition yields a first-order transition temperature, T_m, and associated discontinuous increase of the order parameter,

$$T_m = T_s + \frac{w^2}{4cu} \qquad \langle\phi\rangle_{T_m} = \frac{w}{2u}$$

where for $T > T_m$ there is a single stable solution at $\phi = 0$ and for $T < T_m$ there is a single stable solution at $\phi = \langle\phi\rangle$. Exactly at $T = T_m$ the two solutions are degenerate. The discontinuous nature of the order parameter is a hallmark of a first-order transition and is illustrated in Figure 4.13. Mathematically, this discontinuity is a consequence of the non-vanishing cubic order term in the free energy density that will persist for specific order parameter symmetry groups. This theory in particular has been useful in the analysis of symmetry breaking associated with liquid crystals, in which an order parameter is constructed from the global orientation of elongated molecules.

Exercise 4.10: Evaluate the latent heat released q_{lat} accompanying the first order transition within mean field theory by

$$q_{\text{lat}} = -T_m \left.\frac{df}{dT}\right|_{\langle\phi\rangle, T_m}$$

which is equal to the ordering temperature times the change in entropy.

4.12 Goldstone modes

Most first-order transitions we commonly experience involve the spontaneous breaking of a continuous symmetry. Consider for example freezing a liquid into a solid. Exactly at the melting temperature a crystal forms, propagating a repeating unit, the unit cell, in a specific direction. While a liquid is both translationally and rotationally invariant with particles equally likely to be at any point in space and their neighbors arranged around them isotropically, the crystalline lattice breaks both by orienting the unit cell. A crystal retains some symmetry in rotations and translations, but these are discrete rather than continuous. Suitable order parameters must encode these symmetries. For example, an order parameter for translational symmetry breaking could be the Fourier components of the density field, or for rotational symmetry could be the spherical harmonics of the local environment.

Accompanying the breaking of a continuous symmetry is a so-called Goldstone mode. This is a low-energy excitation occurring in the ordered or symmetry-broken state. This reflects the fact that a continuous symmetry can be broken in any of an infinite number of ways. For example, rotational symmetry can be broken by aligning in any direction. As a consequence, we expect low energy excitations or deformations can act to slowly vary the order parameter over a long range. Indeed, a global rotation leaves the system thermodynamically invariant, so if the energetic price of a slowly varying modulation is just due to the square gradient term, then it is not clear that such modulations would ever allow for a continuous symmetry breaking to be stable.

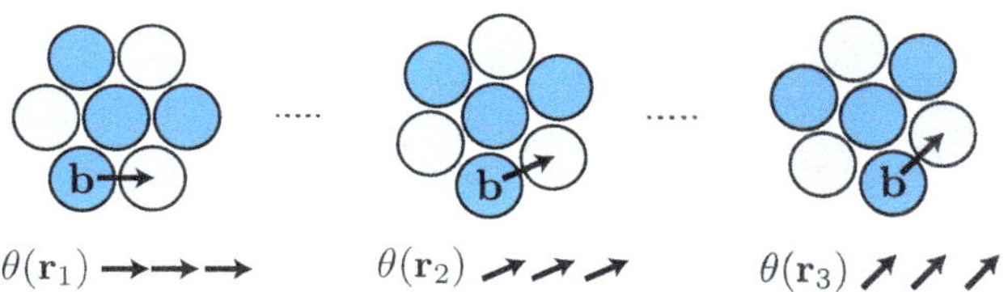

Fig. 4.14 Illustration of the gradual change of a continuous order parameter.

Let us explore this idea concretely by asking for how local orientations are correlated in a state with broken rotational symmetry. We can evaluate the decay of orientations with a correlation function of the form

$$G(r) = \langle \mathbf{b}(0) \cdot \mathbf{b}(r) \rangle$$

where $\mathbf{b}(r)$ is a local direction of our field at position r. Concretely imagine $\mathbf{b}(r)$ as a lattice vector like that shown in Figure 4.14. The stability of broken symmetry is concerned with the asymptotic behavior of $G(r)$. If $G(r) \to 0$ order is destroyed, however if $G(r) > 0$ at large r, order is stable. Within our Landau theory, we can assign an energetic penalty for spatial variations through a squared gradient of the local orientation $\theta(r)$

$$\beta\mathcal{H} = \frac{\gamma}{2} \int d\mathbf{r} |\nabla\theta(\mathbf{r})|^2 = \frac{\gamma}{2} \int d\mathbf{k}\, k^2 |\hat{\theta}(\mathbf{k})|^2$$

where the second line Fourier transforms the squared gradient. Reading off the correlation function for this Gaussian function, we find $\langle\hat{\theta}(\mathbf{k})|^2\rangle = 1/\gamma k^2$. In order to use this correlation in θ, we can rewrite $G(r)$ as

$$G(r) = \langle\mathbf{b}(0)\mathbf{b}(r)\rangle$$
$$= \langle\cos[\theta(0) - \theta(r)]\rangle = \mathrm{Re}\left\langle e^{i[\theta(0)-\theta(r)]}\right\rangle$$

employing standard trigonometry identities. By first writing the argument of the exponential in $G(r)$ in Fourier space, we can evaluate the Gaussian functional integral

$$G(r) = \mathrm{Re}\left\langle \exp\left[i\int d\mathbf{k}\,\theta(\mathbf{k})(1 - e^{-i\mathbf{k}\cdot\mathbf{r}})\right]\right\rangle$$
$$= \mathrm{Re}\exp\left[-\int d\mathbf{k}\,\frac{1}{2\gamma k^2}(1 - e^{i\mathbf{k}\cdot\mathbf{r}})(1 - e^{-i\mathbf{k}\cdot\mathbf{r}})\right]$$
$$\propto \exp\left[\int d\mathbf{k}\,\frac{1}{\gamma k^2}e^{i\mathbf{k}\cdot\mathbf{r}}\right]$$

where the first line is just the Fourier transform of $\theta(0) - \theta(r)$ and the second follows from standard Gaussian integral relations. If we define write $G(r) = \exp[g(r)]$ then

$$g(r) = \int d\mathbf{k}\,\frac{1}{\gamma k^2}e^{i\mathbf{k}\cdot\mathbf{r}}$$

is just an inverse Fourier transform of the Green's function associated with the Laplacian. We have seen these integrals before, and noted that they depend on dimensionality. For a three-dimensional system,

$$g_{3d}(r) = \frac{1}{4\pi\gamma r} \qquad G(r) \propto e^{1/4\pi\gamma r}$$

the Green's function is just a $3d$ Coulomb potential and the correlation function is finite at infinite r, i.e. $G(r) \to 1$. However in $2d$,

$$g_{2d}(r) = -\frac{2\pi}{\gamma}\ln r \qquad G(r) \propto r^{-\gamma}$$

which follows from the $2d$ Coulomb potential, and subsequently $G(r)$ decays as r gets large! This implies that Goldstone modes destroy long ranged order in $2d$ systems. This realization is known as the *Hohenberg–Mermin–Wagner theorem*, and is strictly a no-go theorem for the existence of broken continuous symmetry in low dimensions. So formally, there is no such thing as a $2d$ crystal. However, the algebraic decay of $G(r)$ is very slow, so in practice order can persist over very large lengthscales.

> **Exercise 4.11:** For a two-dimensional crystal like graphene, it can be shown that $\gamma = \beta\Sigma a^2$ where Σ is the shear modulus and a the lattice constant. Determine the value of r for which $G(r)$ will have decayed to $1/e$ of its initial value, using the known material properties of graphene.

4.13 Ordering at a finite wavevector

We will conclude our discussion of Landau theory by discussing the possibility of ordering at a finite wavevector. Physical examples of orders forming with a special, non-infinite, spatial scale include systems like block-copolymers, micelles, or charged fluids where constraints prohibit macroscopic phase separation from occurring. In the context of block copolymers this might occur for example in cases where the individual monomers would phase separate, but the covalent bond linking segments of the two monomers prohibits this. In the context of charged fluids, like ionic liquids, while the individual anion and cation might be prone to demixing in the absence of charges, the requirement of macroscopic neutrality prohibits large domains from forming.

Our typical Landau framework for an order parameter $\phi(\mathbf{r})$ has included spatial degrees of freedom only in the squared gradient term

$$\beta \mathcal{H} = \int d\mathbf{r}\, f[\phi(\mathbf{r})] + \frac{\gamma}{2} |\nabla \phi(\mathbf{r})|^2$$
$$= \int d\mathbf{k}\, |\phi(\mathbf{k})|^2 \left(a + \gamma k^2\right)^2$$

which is simple in Fourier space, as shown in the second line. Reading off the correlation function we again find,

$$\hat{\chi}(k) = \frac{\hat{\chi}(0)}{1 + \xi^2 k^2}$$

where the strongest fluctuations occur in the ordered state around $k = 0$. Remembering the relationship between fluctuations and response, this implies that ordering occurs in the most permission mode, or that with the largest fluctuations, i.e., near $k = 0$. This results in macroscopic domain formation. If we wanted to describe periodic order at wavelength $2\pi/k_0$, we could try

$$\beta \mathcal{H} = \int d\mathbf{k}\, |\phi(\mathbf{k})|^2 \left[a(t) + \gamma^2 (k - k_{\mathrm{o}})^2\right] / 2 + \int d\mathbf{r}\, u\phi^4(\mathbf{r})/4$$

yielding a correlation function

$$\hat{\chi}(k) = \frac{\hat{\chi}(0)}{1 + \xi^2 (k - k_0)^2}$$

which for large ξ is sharply peaked at $k = k_0$. Within our mean field theory, parameterizing $a(t) = c(T - T_c)/T_c$, we would find that ordering occurs at $T = T_c$ and is a second-order transition, just as in the theory of magnetization. However, ordering in this case can be in any direction as long as $|\mathbf{k}_0| = k_0$. This large degeneracy results in fundamentally different behavior when fluctuations are included. In fact, upon including fluctuations, the second-order transition is transformed into a first-order transition, a so-called *fluctuation-induced first-order transition*.

To understand how fluctuations alter the phase transition behavior, we will consider fluctuations at Gaussian level of approximation, using a variational theory to find the best Gaussian field that approximates the full Landau theory. Specifically, we will consider a Gaussian field theory of the form

$$\beta \mathcal{H}_{\mathrm{o}} = \int d\mathbf{k}\, |\phi(\mathbf{k})|^2 \chi_{\mathrm{o}}^{-1}(k)/2$$

where $\chi_{\mathrm{o}}^{-1}(k)$ is optimized to best approximate the full partition function of $\beta \mathcal{H}$. This follows from the Feynman–Bogoliubov inequality we have seen in Chapter 3

$$Q = \int \mathcal{D}[\phi(\mathbf{r})] e^{-\beta \mathcal{H}_{\mathrm{o}}[\phi(\mathbf{r})] + \beta \Delta \mathcal{H}[\phi(\mathbf{r})]}$$
$$= Q_{\mathrm{o}} \langle e^{\beta \Delta \mathcal{H}} \rangle_{\mathrm{o}} \geq Q_{\mathrm{o}} e^{\beta \langle \Delta \mathcal{H} \rangle_{\mathrm{o}}}$$

where $\Delta \mathcal{H} = \mathcal{H}_{\mathrm{o}} - \mathcal{H}$. To optimize $\chi_{\mathrm{o}}^{-1}(k)$ we need to maximize the full right-hand side of the last equation. This includes two contributions, one from Q_{o} and the other from $\langle \Delta \mathcal{H} \rangle_{\mathrm{o}}$. The first is simple,

$$\frac{\delta \ln Q_{\mathrm{o}}}{\delta \chi_{\mathrm{o}}^{-1}(k)} = -\frac{1}{2} \chi_{\mathrm{o}}(k)$$

which follows from the normalization of a Gaussian integral. The second follows from evaluating the average,

$$\langle \Delta \mathcal{H} \rangle_{\mathrm{o}} = \frac{1}{2} \int d\mathbf{k}\, \chi^{-1}(k) \chi_{\mathrm{o}}(k) + \frac{3u}{4} \int d\mathbf{k} \int d\mathbf{k}'\, \chi_{\mathrm{o}}(k) \chi_{\mathrm{o}}(k')$$

which follows from the fact that

$$\langle \phi(\mathbf{k}) \phi(\mathbf{k}') \phi(\mathbf{k}'') \phi(-\mathbf{k} - \mathbf{k}' - \mathbf{k}'') \rangle_{\mathrm{o}} = 3 \delta_{k,-k''} \chi_{\mathrm{o}}(k) \chi_{\mathrm{o}}(k')$$

for a system with Gaussian statistics.

Exercise 4.12: Confirm this result, $\langle \phi(\mathbf{k}) \phi(\mathbf{k}') \phi(\mathbf{k}'') \phi(-\mathbf{k} - \mathbf{k}' - \mathbf{k}'') \rangle_{\mathrm{o}} = 3 \delta_{k,-k''} \chi_{\mathrm{o}}(k) \chi_{\mathrm{o}}(k')$, an example of Wick's theorem. You can perform an explicit integration over the Gaussian field, with $\beta \mathcal{H}_{\mathrm{o}} = \int d\mathbf{k}\, |\phi(\mathbf{k})|^2 \chi_{\mathrm{o}}^{-1}(k)/2$.

Taking the functional derivative of both terms with respect to $\chi_{\mathrm{o}}(k)$ we find

$$\chi_{\mathrm{o}}^{-1}(k) = \chi^{-1}(k) + 3u \int d\mathbf{k}'\, \chi_{\mathrm{o}}(k')$$

in which we note that the last term has no dependence on k, and so $\chi_{\mathrm{o}}^{-1}(k)$ must have the same dependence on k as $\chi^{-1}(k)$. Writing $\chi_{\mathrm{o}}^{-1}(k)$ as

$$\chi_{\mathrm{o}}^{-1}(k) = \bar{a} + \gamma(k - k_0)^2$$

we have

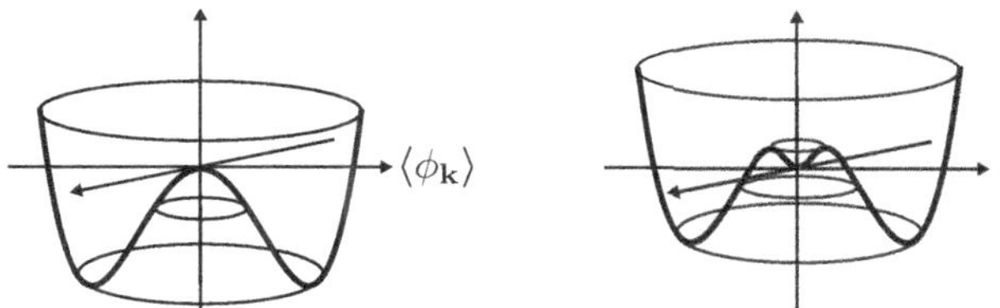

Fig. 4.15 Fluctuation-induced stability of the disordered state, as encoded in the free energy in mean field (left) and with fluctuation corrections (right).

$$\bar{a} = a + 3u \int d\mathbf{k}' \frac{1}{\bar{a} + \gamma(k' - k_0)^2} \, .$$

Evaluating the integral in the vicinity of $k' \approx k_0$, we find

$$\bar{a} = a + \frac{3k_0^2}{2\pi\sqrt{\gamma}} \frac{1}{\sqrt{\bar{a}}}$$

a self-consistent expression for $\bar{a}$, which encodes the transition temperature. Graphically, it is clear that $\bar{a}$ is positive for all a, implying that the disordered state never loses stability. However, at sufficiently small a it becomes metastable, leading to a first-order transition and abrupt ordering along a direction that is chosen spotaneously. The impact on the free energy as a function of ϕ_k is illustrated in Figure 4.15.

In this chapter, we have mostly described a means for understanding the phenomenology of phase transitions. While in a few special cases we could develop sharp quantitative statements, most of our calculations were only semi-quantitative or qualitative. This is because when mean field theory breaks down, computing corrections is difficult and mostly impossible. Solutions can be found in two distinct avenues. The quantitative evaluation of critical exponents in $3d$ is provided by the theory of renormalization group, which we do not address here, but which curious students are encouraged to read about. This theory employs a completely different perspective routed in the observation that a system at its critical point is scale invariant. The alternative avenue is to study systems numerically. The tools with which this can be done are the subject of the next chapter.

Further reading

There are many great texts that present the modern theory of phase transitions. These include Mehran Kardar's *Statistical physics of fields* and Nigel Goldenfeld's *Lectures on phase transitions and the renormalization group*. The presentation here is most heavily influenced by Paul Chaikin and Tom Lubensky's *Principles of condensed matter physics*, who expand on Landau theory particularly lucidly.

Additional exercises

Exercise 4.13: Consider a polymer blend— a liquid mixture of two different types of polymers. Creating a blend requires a remarkable balance between the two species because the effects of any differences in interactions are multiplied by the number of segments on a chain, and as that number is large, there is typically a large driving force to de-mix the blend. This problem examines that driving force within Flory–Huggins theory. For this purpose, write the total potential energy as

$$U(\{\mathbf{r}_{i,P}^{(\alpha)}\}) = U_0(\{\mathbf{r}_{i,P}^{(\alpha)}\}) + U_1(\{\mathbf{r}_{i,P}^{(\alpha)}\})$$

where $\mathbf{r}_{i,P}^{(\alpha)}$ is the position of the αth segment of the ith polymer of type P. There are two types of polymers, so that $P = 1$ or 2, and a polymer of type P has N_P segments or monomers, so that α extends from 1 to N_P. The reference potential energy, $U_0(\{\mathbf{r}_{i,P}^{(\alpha)}\})$, contains all the intra-chain interactions and all the repulsive potentials between different polymer chains. The remaining contribution to the potential energy contains all the interactions that make it energetically unfavorable to mix the two polymer species.

For simplicity, assume that the reference system is that of identical threads: both species have the same number of segments, $N_1 = N_2 = N$, and the segment types are identical, too. The only difference between the two types of polymers is reflected in

$$U_1(\{\mathbf{r}_{i,P}^{(\alpha)}\}) = \sum_{i=1}^{M_1}\sum_{j=1}^{M_2}\sum_{\alpha=1}^{N}\sum_{\gamma=1}^{N} \Delta u(|\mathbf{r}_{i,1}^{(\alpha)} - \mathbf{r}_{j,2}^{(\gamma)}|)$$

where M_P is the number of polymers of type P in a container of volume V, and $\Delta u(r)$ is short-ranged and positive. In what follows, you will consider the thermodynamic consequences of this perturbation potential.

1. In the absence of any inter-molecular interactions, show that for a dilute solution of M_1 type 1 molecules and M_2 type 2 molecules, the Helmholtz free energy per unit volume can be written as

$$\beta A_{\text{ideal}}/V = \frac{1}{N}\rho_1\left[\ln(\rho_1 a^3) - 1\right] + \frac{1}{N}\rho_2\left[\ln(\rho_2 a^3) - 1\right],$$

   where $\rho_1 = N \times M_1/V$ is the mean monomer density associated with polymers of type 1, ρ_2 is defined similarly for species 2, and a is a length, the choice of which defines the standard state of the mixture.

2. Show that the reference free energy per unit volume can be written as

$$A_0/V = \frac{k_{\text{B}}T}{N}\rho\left[x\ln x + (1 - x)\ln(1 - x)\right] + f_0(\rho, N),$$

   where x is the mole-fraction of species 1, $\rho = \rho_1 + \rho_2$, and $f_0(\rho, N)$ is independent of x.

3. Accounting for $\Delta u(r)$, show that the free energy of the system is given by

$$\beta A/V = \beta A_0/V + \beta \rho^2 x(1 - x)\int d\mathbf{r} g_0(r)\Delta u(r) + \dots$$

where $g_0(r)$ is the segment-segment radial distribution function in the reference system, and the omitted terms are $\mathcal{O}[(\beta \Delta u)^2]$.

4. If the higher-order terms are neglected, the resulting free energy expression predicts that the blend exhibits a de-mixing transition. In this case, show that the predicted critical temperature for de-mixing is

$$T_c = N \Delta \bar{u}/2k_{\mathrm{B}},$$

where

$$\Delta \bar{u} = \rho \int d\mathbf{r} \, g_0(r) \Delta u(r).$$

Establish the existence of the phase transition by applying the thermodynamic conditions for phase coexistence, such as equating chemical potentials of like species between two distinct phases. You can assume that the total density of monomers of type 1 and type 2 is the same, so near the critical point any given region of space likely has a composition of $x \approx 1/2$. The critical temperature is the highest temperature at which equilibrium between distinct phases can be established. In polymer physics, the quantity $\Delta \bar{u}$ is called Flory's χ parameter. It is independent of polymer mass. As such, the formula you have derived indicates that T_c grows linearly with polymer mass. A very small unbalancing potential is thus sufficient to cause a de-mixing transition.

5. It has been claimed that the formula for T_c derived above is an exact result. Justify this claim. In particular, argue why the terms omitted make no contribution to T_c.

Exercise 4.14: In this problem, you will explore a mean field theory introduced by Maier and Saupe for the spontaneous rotational symmetry breaking of a collection of long, rod-like molecules. In particular you will study the transition between a high-temperature *isotropic* state, where the liquid possesses continuous translational and rotational symmetry, to a low-temperature *nematic* state, where the molecules have a tendency to align parallel to each other, while maintaining translational symmetry. The axis along which the molecules align is called the *director*, and it is denoted by the unit vector $\hat{n}$. Here you will use mean field theory to explore the isotropic-nematic phase transition. A parameter which distinguishes the nematic phase from the isotropic phase is

$$S = \frac{1}{N} \sum_{j=1}^{N} \frac{3}{2} \cos^2 \theta_j - \frac{1}{2},$$

where $\cos \theta_j = \hat{n} \cdot \hat{\Omega}_j$, where $\hat{\Omega}_j$ is the orientation of molecule j.

1. What are the values of S when all the molecules are perfectly aligned with the director ($\hat{n}$), and when molecules' orientations are distributed uniformly? To compute the latter average, you will need to take care to work in spherical coordinates.

2. In general, the molecules comprising a liquid crystal will interact with one another via a complicated intermolecular potential. Maier and Saupe made the simplifying, mean field assumption that all the molecules are subject to the same average potential, which has the form

$$v_{MF}(\theta) = -v_0 \langle S \rangle \cos^2 \theta.$$

   which depends on the average value of the order parameter, $\langle S \rangle$. Given this form for the potential energy, determine the probability distribution for a molecule's orientation, $p(\theta)$. The normalization constant can be evaluated analytically, but this is unnecessary here.

3. By recalling the definition of S, and noting that the ensemble average involves $\langle S \rangle$, establish a self-consistent equation for $\langle S \rangle$. Your answer should involve the following integral identities valid for $a > 0$

$$I_1(a) = \int_{-1}^{1} du \, e^{au^2} = \sqrt{\frac{\pi}{a}} \mathrm{Erfi}[\sqrt{a}]$$

   where $\mathrm{Erfi}[x]$ is the imaginary error function, and

$$I_2(a) = \int_{-1}^{1} du \, u^2 \, e^{au^2} = \frac{1}{a} e^a - \frac{1}{2a} I_1(a)$$

4. Using your answers from parts (1) to define the range of possible $\langle S \rangle$ values, numerically evaluate and plot both sides of this self-consistent equation as a function of $\langle S \rangle$ for values of $\beta v_0 = \{1, 4, 5, 8, 10\}$. Comment on the nature of the solutions to the self-consistent equation over this range of parameters.

5. Adopting a unit system where $v_0/k_{\mathrm{B}} = 1$, solve this self-consistent equation numerically to determine $\langle S \rangle$ as a function of T. Make a plot of $\langle S \rangle$ as a function of T and label the location of the transition temperature.

Exercise 4.15: In this problem you will work through a theory due to Weeks for the roughening transition, a transition between distinct surface phases that is responsible for faceting of crystals and the temperature dependence of their growth rates. In particular we will consider the transition between an interface whose fluctuations in a height field $h(\mathbf{r})$ above some fixed plane are macroscopically large, where the interface width diverges due to its ability to move freely, and an interface whose position is pinned, where the interface width remains finite. The Hamiltonian for the system will be taken as

$$\mathcal{H}[h(\mathbf{r})] = \frac{1}{2} \int d\mathbf{r} \, \gamma |\nabla h(\mathbf{r})|^2 + u h^2(\mathbf{r})$$

where γ is related to the surface tension and u is an external potential, and $\mathbf{r}$ denotes a position in two dimensions. In the Hamiltonian, we will imagine that $h(\mathbf{r})$ can take any value, $-\infty < h(\mathbf{r}) < \infty$. However, for a real crystal, $h(\mathbf{r})$ should be limited to discrete values, integer multiples of the lattice constant. In order to incorporate that constraint we will write the partition function

$$Q = \int \mathcal{D}[h(\mathbf{r})] W[h(\mathbf{r})] e^{-\beta \mathcal{H}[h(\mathbf{r})]}$$

as an integral over height fluctuations with a weighting function $W[h(\mathbf{r})]$ that can be used to impose discrete constraint of $h(\mathbf{r})$.

1. Starting with the free Gaussian field, where $W = 1$, show that the partition function is given by

$$Q_G = \int \mathcal{D}[\hat{h}(\mathbf{k})] e^{-\beta \frac{1}{2} \int d\mathbf{k} \, (\gamma k^2 + u)|\hat{h}(k)|^2} = \prod_{\mathbf{k}} \sqrt{\frac{2\pi k_B T}{\gamma k^2 + u}}$$

   when the Hamiltonian is diagonalized using a Fourier basis.

2. Demonstrate that the mean squared fluctuations of the height diverge logarithmically with the system size as $u \to 0$,

$$\langle \delta h^2 \rangle = k_B T \ln(L/a)/2\pi\gamma$$

   for any finite temperature, where L is the system size and a a microscopic length. This behavior is a signature of the rough phase of the interface, where interfacial fluctuations, and thus the width of the interface, diverge. To show this, you will need to regularize the integral over k using cutoffs of $k = 2\pi/L$ and $k = 2\pi/a$ where a is the lattice constant.

3. A way to impose a periodic discreteness in the height field is to use a weighting function of the form
$$W[h(\mathbf{r})] = e^{-\int d\mathbf{r} \, \cos[qh(\mathbf{r})]}$$

   where $q = 2\pi/a$. However, with it the direct evaluation of the partition function is no longer possible. To proceed, we will use Feynman's variational theory and try to find an optimal value of the harmonic potential u to mimic the effect of W. First, evaluate the variational free energy

$$\beta A \leq \beta A_G + \int d\mathbf{r} \langle \cos[qh(\mathbf{r})] - \beta u h^2(\mathbf{r})/2 \rangle_G$$

   where $\beta A_G = -\ln Q_G$ and A is an estimate of the free energy of the free interface, $u = 0$, in the presence of $W[h(\mathbf{r})]$.

4. By taking the derivative of A with respect to u, show that the optimal value of the potential u^* satisfies a self-consistent equation of the form

$$u^* = k_B T q^2 \left(1 + q^2 \frac{\gamma}{u^*}\right)^{-k_B T q^2/8\gamma\pi}$$

   and demonstrate that at a temperature T_R, $u^* > 0$ resulting in an interface with $\langle \delta h^2 \rangle$ that remains finite as $L \to \infty$.

5
Monte Carlo methods

Any modern presentation of the statistical mechanics of equilibrium phenomena would be incomplete if it did not discuss the numerical and computational methods that have enabled the study of increasingly sophisticated models of matter. Of the many numerical tools available, Monte Carlo sampling has played a particularly important role in benchmarking approximate treatments of nontrivial systems composed of interacting particles, and in establishing the phase diagrams of model systems. In this chapter, we will focus exclusively on Monte Carlo approaches employed to study the statistical mechanics of equilibrium systems, and relegate alternative approaches based on solving specific equations of motion, so-called molecular dynamics simulations, to a later chapter. The basic idea of Monte Carlo is to estimate ensemble averages by sampling configurations from the Boltzmann distribution, which can be done by evolving the system using a stochastic process known as a Markov chain. In many cases, this can be done straightforwardly and the limitations of such an approach are only in the accuracy with which a system's Hamiltonian is modeled, and the finite capacity of a computer that will restrict the system sizes and number of samples collected. Here, we will introduce both basic and more advanced algorithms, as well as discuss some practical aspects associated with simulating efficiently. We will render our presentation concrete by considering a few paradigmatic models related to both classical and quantum mechanical systems.

5.1 Sampling not summing

We have shown that the partition function encodes all of the observable properties of a system. However, is not generally possible to evaluate it for interacting systems. While we have reviewed a number of techniques to approximate it and specific expectation values, like the field theories and perturbative expansions discussed in previous chapters, a natural question to ponder is how one could use a computer to solve such problems. The most straightforward means of employing a computer would be to try to evaluate the partition function directly,

$$Q = \sum_{\nu} e^{-\beta E_\nu} \overset{?}{=} e^{-\beta E_1} + e^{-\beta E_2} + \ldots$$

as it is expressible as a sum. This is easiest to imagine for a discrete system, but even for a system in a continuum, one could in principle express the integrals on a grid. Performing the sum would be equivalent to quadrature, albeit in a very high dimensional space. However, that dimensionality is what makes this straightforward approach intractable. Consider a 10×10 Ising model in two dimensions. The number of microstates, or the number of terms in the sum for the partition function, is $2^{100} \approx 10^{30}$! To sum that many numbers using a 1 GHz computer processor would take 10^{12} *years*. Even then, the properties of the Ising model would not be a very accurate representation of a macroscopic system, which would require studying ever larger lattices resulting in exponentially more configurations. The symmetry of the lattice could be invoked, but even so, this direct method is only possible for very small systems.

There is, however, a simplification. Most of the terms in the partition function sum, at most thermodynamic conditions of interest, contribute very little to Q. Indeed at $T = 0$ there are only two configurations of an Ising model that matter, and at finite temperatures most configurations are still irrelevant. With this insight, the question then becomes how to find the states that contribute the most. This question can be sharpened by considering not the evaluation of Q, but the estimation of an expectation value. For an observable O, its thermal average is given by

$$\langle O \rangle = \frac{\sum_\nu O_\nu e^{-\beta E_\nu}}{\sum_\nu e^{-\beta E_\nu}}$$

which is just a sum over the normalized Boltzmann distribution. The question of finding the configurations that matter is thus equivalent to finding those that have the largest statistical weight or that are most probable.

Monte Carlo sampling attempts to move through possible microstates of a system with a bias toward the most important states, those that will contribute the most to an ensemble average. In general, if M samples are drawn from a distribution $\tilde{P}_\nu$, then a thermal average can be estimated from

$$\langle O \rangle \approx \frac{\sum_{i=1}^{M} O_{\nu_i} \tilde{P}_{\nu_i}^{-1} e^{-\beta E_{\nu_i}}}{\sum_{i=1}^{M} \tilde{P}_{\nu_i}^{-1} e^{-\beta E_{\nu_i}}}$$

which in the limit of $M \to \infty$ is expected to converge to the ensemble average. Notice if we sample from a uniform distribution, $\tilde{P}_\nu \propto 1$, then we return to the previous problem that most microstates will not contribute significantly to the average. However, if we can sample directly from a Boltzmann distribution, $\tilde{P}_\nu \propto \exp[-\beta E_\nu]$ then the expectation value will reduce to

$$\langle O \rangle \approx \frac{1}{M} \sum_{i=1}^{M} O_{\nu_i}$$

just a regular average. This basic idea of tuning a sampling distribution to effectively evaluate expectation values is known as *importance sampling*, and most Monte Carlo approaches implement it by constructing a stochastic process known as a Markov chain.

5.2   Metropolis Monte Carlo

In Monte Carlo approaches, we desire to sample the Boltzmann distribution directly in order to efficiently estimate ensemble averages. But how can we generate microstates in proportion to their Boltzmann weights? The trick is to design an effective, fictitious dynamics that will evolve the system randomly but with a bias toward more probable microstates. If done carefully, we can construct the dynamics so as to ensure that the Boltzmann distribution is its stationary state. The basic idea is illustrated in Figure 5.1, in which the system evolves between microstates ν_i and ν_j with a rate k_{ij}.

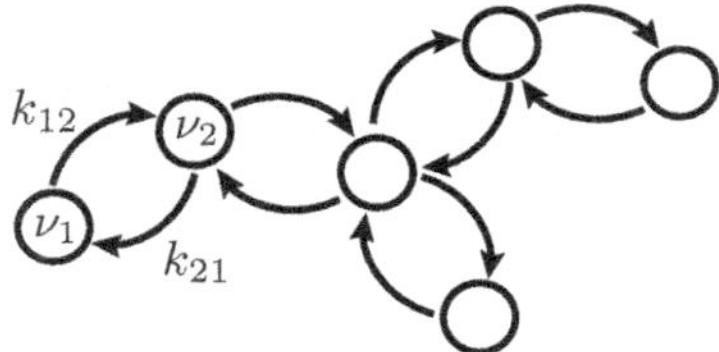

Fig. 5.1 Example of a Monte Carlo walk through configuration space with rates k_{12} and k_{21} between states ν_1 and ν_2.

In order to determine a form for the k_{ij}'s, let us denote the probability of finding the system at microstate ν_j at a time t as $P_j(t)$. In the limit that $t \to \infty$, the system reaches a steady state, $\bar{P}_j$, independent of time. If the dynamics are such that the probability to evolve between any two states just depends on the current state, a so-called *Markov process*, then the evolution of the probability $P_j(t)$ is given by

$$\frac{dP_j(t)}{dt} = \sum_{i \neq j} k_{ij} P_i(t) - k_{ji} P_j(t)$$

where the first term on the right-hand side sums up all of the ways state j can gain probability, while the second term sums over all of the ways state j can lose probability. The stationary distribution satisfies, $d\bar{P}_j/dt = 0$, or in other words within the steady state there are no net fluxes of probability. The necessary condition for the steady state is $\sum_i k_{ij} \bar{P}_i = \sum_i k_{ji} \bar{P}_j$. This reduces to

$$\bar{P}_j \sum_{i \neq j} k_{ji} = \sum_{i \neq j} k_{ij} \bar{P}_i$$

which is known as the global balance condition. However, this condition is difficult to solve, as it is a function of all of the transition rates. A sufficient condition to guarantee a specific steady state is to ensure that for $\bar{P}_j$,

$$k_{ij}\bar{P}_i = k_{ji}\bar{P}_j \quad \forall i, j$$

or that there are no net fluxes between any two states. Rearranging this expression, and plugging in the desired Boltzmann distribution, results in a condition for the relative rates between any two states,

$$\frac{k_{ij}}{k_{ji}} = \frac{\bar{P}_j}{\bar{P}_i} = e^{-\beta(E_j - E_i)}$$

known as the principle of *detailed balance*. We will see later on that this relationship holds for physical dynamics, and is a design constraint that is relatively easy to satisfy. This procedure for generating a sampling distribution is known as Markov chain Monte Carlo, as we chain together a sequence of states using a Markov process to estimate averages. Note that this process does not let us evaluate the partition function, as we have no way of representing the normalization of the probability distribution. The random walk through configuration space is a normalized process.

Exercise 5.2: Consider a two-level system, consisting of a lower energy state labeled 1 with energy E_1 and a higher energy state labeled 2 with energy E_2. If P_i denotes the probability of finding the system in state i, and satisfies the equations

$$\frac{dP_1}{dt} = -k_{12}P_1 + k_{21}P_2 \qquad \frac{dP_2}{dt} = -k_{21}P_2 + k_{12}P_1$$

demonstrate that if detailed balance holds that the steady state probability of finding the system in state i is $\bar{P}_i = e^{-\beta E_i}/(e^{-\beta E_2} + e^{-\beta E_1})$.

With the detailed balance principle, we have a relation that ratios of rates have to satisfy. However, there are many choices of k_{ij} that can satisfy this constraint and therefore evolve a Boltzmann distribution as a steady state. A very common choice is known as Metropolis Monte Carlo, and breaks each transition rate into a product of distinct terms. For a transition between microstates v_i and v_j, the rate is taken as

$$k_{ij} = k^{(o)} P_{\mathrm{att}}(v_i \rightarrow v_j) P_{\mathrm{acc}}(v_i \rightarrow v_j)$$

where $k^{(o)}$ is an arbitrary prefactor that is independent of the microstate. The two other terms on the right encode probabilities associated with the transition that aid in bookkeeping. The first, $P_{\mathrm{att}}(v_i \rightarrow v_j)$, is the *attempt probability*. It is the probability associated with proposing a specific move, or update to the microstate of the system. This could be the probability of proposing to move a particle or flip a spin. In the original Metropolis algorithm and most common current implementations, this is chosen to be symmetric, $P_{\mathrm{att}}(v_i \rightarrow v_j) = P_{\mathrm{att}}(v_j \rightarrow v_i)$. Each transition is equally likely to be generated in the forward or backward direction. That is, a particle is equally likely to move forward or backward, or a spin is equally likely to be flipped up or down.

The last term in k_{ij}, $P_{\text{acc}}(v_i \to v_j)$, is the *acceptance probability*. As it is all that is left to differentiate between the two microstates, so it must be chosen to satisfy detailed balance. This implies that the ratio of acceptance probabilities must equal

$$\frac{P_{\text{acc}}(v_i \to v_j)}{P_{\text{acc}}(v_j \to v_i)} = e^{-\beta(E_j - E_i)}$$

where the right-hand side is the ratio of equilibrium probabilities associated with the initial, ν_i, and final, ν_j, states. Thus, to evaluate the acceptance probability requires knowing how the energy of the system changes over the transition. Again, there are many specific forms that would satisfy this ratio, and the one most often used, which proves to be rather efficient, is

$$P_{\text{acc}}(v_i \to v_j) = \min\left[1, e^{-\beta(E_j - E_i)}\right]$$

which takes the minimum of 1 or the ratio of Boltzmann factors of the final and initial states. As the exponential is strictly less than one for the negative values of its argument and strictly greater than one for the positive values of its argument, this minimum returns 1 if $E_j - E_i < 0$, and $\exp[-\beta(E_j - E_i)]$ if $E_j - E_i > 0$.

Exercise 5.3: Show that the Metropolis acceptance criteria, $P_{\text{acc}}(v_i \to v_j) = \min\left[1, e^{-\beta(E_j - E_i)}\right]$ satisfies detailed balance.

The Metropolis acceptance criteria imply that proposed moves that decrease the energy of the system are accepted. Moves that result in an increase in energy must be accepted with a probability equal to the ratio of Boltzmann factors for the final and initial state. In order to decide whether to accept a move that increases in energy, we can draw a uniform random number. If η is a sample from a uniform distribution on the interval $[0, 1]$, then the probability that η is less than $\exp[-\beta(E_j - E_i)]$ is

$$P\left(\eta < e^{-\beta \Delta E}\right) = e^{-\beta \Delta E}$$

which implies that we should accept the proposed move if $\eta < \exp[-\beta(E_j - E_i)]$. In practice, random numbers are approximated computationally by constructing pseudo-random number generators, whose quality is instrumental in an accurate Monte Carlo program. If the pseudo-random number generator is biased, then the algorithm will also be biased.

A way to represent and summarize the Metropolis Monte Carlo approach is to write it down in *pseudocode*. Pseudocode is a plain language description of the steps in an algorithm, which for the Metropolis Monte Carlo algorithm is presented in Algorithm 1. In order to evaluate averages, the number of Monte Carlo updates, denoted n_{steps} in Algorithm 1, needs to be large. This is in order to ensure that statistical expectation values are converged. Asymptotically, we expect from the central limit theorem that for local observables that have finite variance, the statistical error bar will scale as $\sim 1/\sqrt{n_{\text{steps}}}$ for large n_{steps}. The constant of proportionality will depend on the intrinsic variance of the observable and the number of Monte Carlo updates over which the observable is correlated.

Algorithm 1 Metropolis Monte Carlo

1: **for** $n = 1$, $n \leq n_{\text{steps}}$ **do**
2:     MC Move($\nu \to \nu'$) $\triangleright$ Propose an update to the system
3:     Compute $\Delta E = E(\nu') - E(\nu)$
4:     **if** $\Delta E < 0$ **then** $\triangleright$ Accept if energy change is negative
5:         $\nu \to \nu'$
6:     **else if** Rand()$<$ exp$[-\beta\Delta E]$ **then** $\triangleright$ Accept if random number is small
7:         $\nu \to \nu'$
8:     **end if**
9:     **if** mod(n, n_{sample}) $= 0$ **then** Compute Properties(ν)
10:     **end if**
11: **end for**

The details of the specific proposal move, denoted as the function *MC Move*($\nu \to \nu'$), will vary depending on the details of the system studied. Similarly, the details of the function *Compute Properties*(ν) that will evaluate specific expectation values of interest will vary, as will the periodicity with which such properties are computed, here implemented using the *mod*(a, b) function that returns the remainder of the quotient a/b. Finally, we have denoted the uniform random number generator as *Rand*().

Exercise 5.4: Show that if the attempt probabilities are not symmetric that the Metropolis acceptance criteria becomes,

$$P_{\text{acc}}(v_i \to v_j) = \min\left[1, \frac{e^{-\beta E_j} P_{\text{att}}(\nu_j \to \nu_i)}{e^{-\beta E_i} P_{\text{att}}(\nu_i \to \nu_j)}\right]$$

which depends on ΔE as well as the ratio of attempt probabilities.

5.3   Trial moves in discrete space

With the general procedure outlined, let us consider two specific cases that represent the two typical kinds of systems studied–those that are defined on a lattice and have a discrete set of internal states, and those that are defined off-lattice and have a continuum set of internal states. As the means of sampling the distribution is not physical, we have significant flexibility in deciding what type of Monte Carlo moves we can make. These two cases will help us understand what sorts of moves might be effective.

For the discrete model, we can revisit the minimal model of a magnet introduced in Chapter 4. The Ising model is defined by a set of spins, $\{s^N\}$, on a lattice where each spin can take a value of $s = \pm 1$. A reasonable Monte Carlo move that obeys the assumed symmetric attempt probability would be to select a spin at random and attempt to flip it. For concreteness, let's consider a two-dimensional model on a square lattice, such that if there are N spins, the number of spins on each side is $\sqrt{N}$. This requires that N is a perfect square. We could choose a random spin by drawing two random integers between 1 and $\sqrt{N}$ for the x and y coordinates of the randomly selected spin. Attempting to flip the spin takes its original value and $s_{x,y} \to -s_{x,y}$, where we index the spin with two variables, rather than a single index as we have done previously. An illustration of the single spin flip move is illustrated in Figure 5.2.

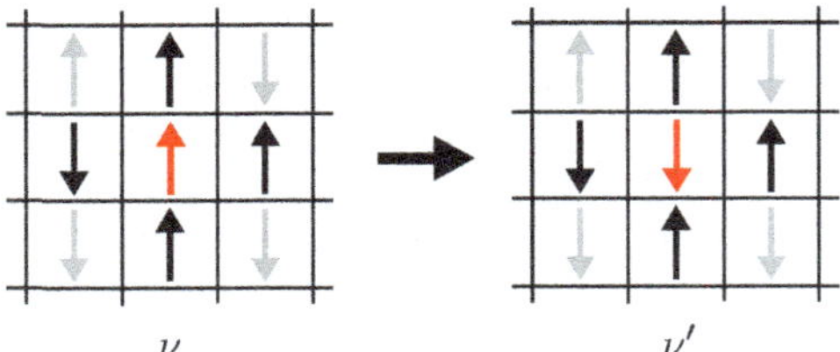

Fig. 5.2 Example spin flip move for studying the Ising model with Metropolis Monte Carlo, where the red spin is the randomly selected spin to be flipped, which interacts directly only with the spins in black not the remaining spins in gray.

To decide whether or not to accept the trial move, we must evaluate the accompanying change in energy, ΔE. For the single spin flip, this takes a particularly simple form

$$\Delta E = E(-s_{x,y}) - E(s_{x,y}) = 2s_{x,y} \left(h + J \sum_{i,j \in n_{x,y}} s_{i,j} \right)$$

where the sum is over nearest neighbors, denoted by the set $n_{x,y}$. Notice first that since the interactions between Ising model spins are restricted to nearest neighbors, the information required to evaluate the change in the energy is local. The calculation requires only knowledge of the state of the neighboring spins, not the full configuration. Because the model is defined on a lattice, the location of the nearest neighbors is straightforward to evaluate. Notice also, however, that for a finite number of spins the evaluation of nearest neighbors is subtle. If $x = \sqrt{N}$ on our square lattice, what should be taken as its neighbor in the position at $x = \sqrt{N} + 1$? To answer this requires establishing boundary conditions. In principle, we could assert fixed boundary conditions, in which a spin at the end of the lattice simply does not have the same number of neighbors as a spin in the interior. However, that would render some spins statistically distinct. The properties of the spins at the boundaries would be different in general from those in the bulk. As we are usually interested in macroscopic systems, we would need to simulate large lattices in order for the collective properties to not be significantly influenced by the boundary spins. An alternative to more rapidly approach a simulation of the bulk is to employ *periodic boundary conditions*, which envisions the N spin lattice as the primitive cell of an infinite periodic lattice of identical cells.

Under such boundary conditions for our square lattice, $x_{\sqrt{N}+1,y} = x_{1,y}$, with the same periodicity invoked in the y direction. This will restore translational invariance within the lattice, as all cells will now be statistically equivalent again, each having 4 nearest neighbors. While periodic boundary conditions render each spin statistically identical, a finite lattice is still distinct from an infinite lattice, as the former is capable or representing only certain Fourier modes, $k > 2\pi/\sqrt{N}$. As such, care should be taken to ensure that properties are converged with respect to system size.

The pseudocode of the single spin flip Monte Carlo algorithm is displayed in Algorithm 2. As before, it uses Rand() as a uniform pseudorandom number generator. The evaluation of the change in energy follows from the discussion earlier, and assumes some means of invoking the periodic boundary condition. A simple means of doing this on a square lattice is to replace $x \pm 1$ with $\mathrm{mod}(x \pm 1, \sqrt{N}) + 1$, which will return 1 if $x + 1 > \sqrt{N}$ and if $x - 1 < 1$ will return $\sqrt{N}$.

Algorithm 2 Spin flip Monte Carlo

1: **for** $i = 1,\ i \leq N$ **do**
2: $\quad$ $x=\mathrm{Int}[\sqrt{N}\times \mathrm{Rand}() + 1]$ $\hfill$ ▷ Pick random x position
3: $\quad$ $y=\mathrm{Int}[\sqrt{N}\times \mathrm{Rand}() + 1]$ $\hfill$ ▷ Pick random y position
4: $\quad$ $s_{\mathrm{old}} = s_{x,y}$
5: $\quad$ $\Delta E = 2s_{x,y}\left[J(s_{x+1,y} + s_{x-1,y} + s_{x,y+1} + s_{x,y-1}) + h\right]$
6: $\quad$ **if** $\Delta E < 0$ **then**
7: $\quad\quad$ $s_{x,y} = -s_{\mathrm{old}}$ $\hfill$ ▷ Replace with flipped spin
8: $\quad$ **else if** $\mathrm{Rand}() \leq \exp[-\beta\Delta E]$ **then**
9: $\quad\quad$ $s_{x,y} = -s_{\mathrm{old}}$
10: $\quad$ **end if**
11: **end for**

The Markov chain we have constructed to generate a Boltzmann distribution is guaranteed to relax to a steady state independent of the initial condition provided enough time; however, the number of Monte Carlo updates required to do so and, thus, the computational resources needed can depend strongly on that initial condition. This is because a system started in a state that is atypical under the thermodynamic conditions studied will need to relax toward a more typical state and doing so may require passing through many improbable states. Take, for example, a system like that pictured in Figure 5.3 in which two domains of primarily up and down spins are separated by an interface. The presence of the extended interface increases the free energy of the system relative to a homogeneous system of primarily up or down spins, and as such without a constraint on the total magnetization this state is unstable. However, to relax away the interface can require many Monte Carlo updates. Thus, if a typical state of the system is known under specific thermodynamic conditions it is advantageous to initialize the system in it, in order to avoid wasting time collecting statistics of the system that are not representative of thermal ensemble. For example, for the Ising model, if the low temperature behavior is of interest, $T < T_c$, it makes sense to start the system with a net magnetization, either all up or all down. If the high temperature behavior is more of interest, $T > T_c$, each spin orientation may be selected randomly with equal likelihood of an up or down spin.

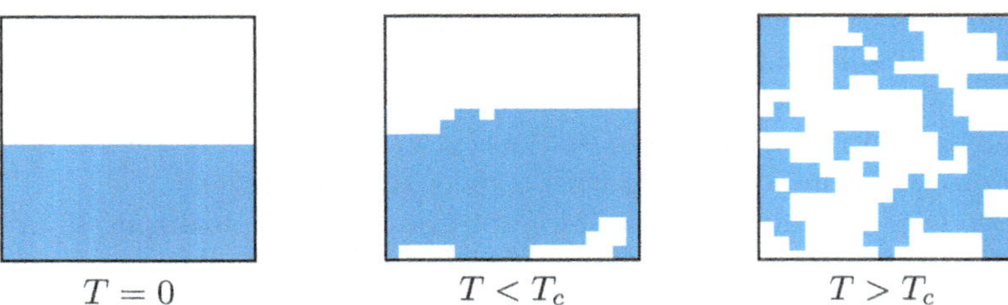

Fig. 5.3 Typical configurations of the Ising model initialized in the left-most configuration and sampled at various temperatures for a fixed small number of steps.

The single spin flip move is not the only Monte Carlo move that is possible. Indeed, in principle multiple spin flips could be attempted at once. However, the single spin flip move is often found to be most efficient. To understand this, we can consider the statistics associated with proposing moves. The acceptance probability depends on the average value of the Boltzmann factor resulting from a proposed move, $\langle \exp[-\beta \Delta E] \rangle$. This average obeys a sum rule resulting from detailed balance, when the attempt probabilities are symmetric

$$\langle e^{-\beta \Delta E} \rangle = \int d\nu_i \int d\nu_j P(\nu_i) P_{\text{att}}(\nu_i \to \nu_j) e^{-\beta \Delta E}$$

$$= \int d\nu_i \int d\nu_j P(\nu_j) P_{\text{att}}(\nu_j \to \nu_i) e^{\beta \Delta E} e^{-\beta \Delta E} = 1$$

where the second equality follows from substituting the symmetric proposal probability and the third follows from normalization. From this equality, Jensen's inequality implies

$$e^{-\beta \langle \Delta E \rangle} \leq 1$$

or $\langle \Delta E \rangle > 0$, which means that on average we will propose moves that increase the energy of the system. This does not mean that the change in the energy after a move will be positive on average. Within a stationary state, the mean change will be zero, but this results from a combination of the proposal and the acceptance parts of the Monte Carlo procedure. Take, for example, a low temperature configuration. If most spins are pointed parallel to each other, it is more likely than not to propose a spin flip that will render the spin antiparallel to its neighbors.

This increase on average is a reflection of the second law of thermodynamics, a point that will be clarified in subsequent chapters in the context of dynamical nonequilibrium processes. Because the average energy for an attempted move is positive, if we attempt to flip many spins in an independent way, the energy will increase extensively, resulting in a larger ΔE on average, and thus an exponentially smaller probability to accept a move. Without a clever means of flipping many spins at once, taking into account likely correlations between spins for example, the tradeoff between relaxing an extensive number of spins and a likely extensive energy increase does not usually work out to be in our favor. Clever methods do exist, known as cluster moves, that flip multiple spins at a time and work particularly well near the critical point where the fractal correlation functions are well described within Landau theory.

> **Exercise 5.5:** The spin flip Monte Carlo discussed does not preserve the total magnetization, $M = \sum_i^N s_i$. Construct a move that conserves magnetization by changing two spins, and evaluate its acceptance probability.

5.4 Trial moves in continuous space

A canonical continuum model studied with Monte Carlo is a system of hard disks, a two-dimensional model of a fluid with purely repulsive interactions. The three-dimensional version of this system was introduced in Chapter 3, in the context of our discussion of liquid structure. Here we will consider the configurational properties of a collection of hard disks, as the fluctuations associated with the momentum can be trivially integrated out. The remaining potential energy function is given by a sum over pair potentials, $U(\mathbf{r}^N) = \sum_{i<j} u_{\mathrm{HS}}(r_{ij})$, with

$$
u_{\mathrm{HS}}(r) = \begin{cases} 0 & \text{if } r \geq \sigma \\ \infty & \text{if } r < \sigma \end{cases}
$$

where σ denotes the diameter of the hard disk. For this system, the Boltzmann factor reduces to $\exp(-\beta \Delta E) = 0$ or 1, depending only on whether there are particle overlaps or not. As a consequence, the configurational partition function is independent of β, retaining only its dependence on N and V.

Since the disks are point particles that have no internal structure, sampling the Boltzmann distribution reduces to sampling different relative positions of the particles. A reasonable single particle move is thus to perturb a particle's position by some small random amount. If we define a parameter Δ as the largest possible displacement, then the proposal takes the original position for particle i, $\{x_i, y_i\}$, and creates a new position, $\{x_i', y_i'\}$, as

$$
x' \rightarrow x + 2\Delta(\eta - 1/2)
$$
$$
y' \rightarrow y + 2\Delta(\eta - 1/2)
$$

where η is a uniform random number. This move will sample a displacement that is uniformly distributed in a box of area Δ^2. An illustration of this move is in Figure 5.4. The attempt probability is manifestly symmetric, and so the acceptance probability will involve only a ratio of Boltzmann factors. In the case of hard disks, the infinite repulsion means that if the proposed position results in the particle coming closer than σ to any other particle, the Boltzmann factor would become 0 and the move would be rejected. If there are no overlaps, then the change in the energy is 0 and the move will be accepted.

The resulting single translational move algorithm is presented in pseudocode in Algorithm 3. The first step is to randomly select a particle, and then draw a random

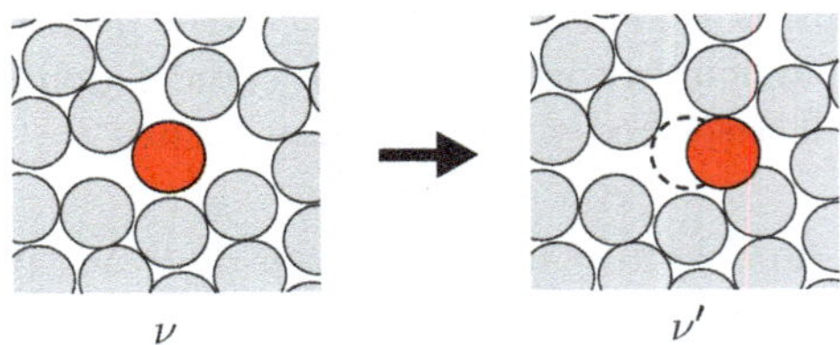

Fig. 5.4 Single particle translational move for hard disk Monte Carlo in which a move is rejected if the particles overlap.

displacement to its position, which for the hard disks consists of two random numbers. Because of the infinite repulsive force, the acceptance step is transformed into a query concerning particle overlaps, implemented in the function *Overlaps*(). Finally, if the new position of the particle is such that the distance between it and any other particle is less than σ, employing periodic boundary conditions, then the move is rejected.

Algorithm 3 Hard disk Monte Carlo

1:	**for** $i = 1, i \leq N$ **do**	
2:	$j = \text{Int}[N \times \text{Rand}() + 1]$	$\triangleright$ Pick random particle
3:	$x_{\text{old}} = x_j$	
4:	$y_{\text{old}} = y_j$	
5:	$x_{\text{new}} = x_{\text{old}} + 2\Delta[\text{Rand}() - 1/2]$	$\triangleright$ Draw random displacement
6:	$y_{\text{new}} = y_{\text{old}} + 2\Delta[\text{Rand}() - 1/2]$	$\triangleright$ Draw random displacement
7:	$\text{Flag} = \text{Overlaps}(\mathbf{r}^N)$	$\triangleright$ Check for overlaps
8:	**if** Flag$= 0$ **then**	
9:	$x_j = x_{\text{new}}$	$\triangleright$ Replace with new position
10:	$y_j = y_{\text{new}}$	$\triangleright$ Replace with new position
11:	**end if**	
12:	**end for**	

Exercise 5.6: Write out the pseudo-code for computing distances between particles employing periodic boundary conditions if the length of the box is L. What does this imply about interactions that are larger then $L/2$?

For implementing this algorithm, an immediate question concerns how large Δ should be. A large value will create large changes to the configuration, but may have a small probability of acceptance if the density of disks is high. In general, Δ should be chosen to optimize the convergence of expectation values. This means that a reasonable measure of efficiency is how quickly uncorrelated samples can be generated, as for a given total number of Monte Carlo steps, those with the most uncorrelated samples will have the smallest statistical error bars. While there is no single simple measure of correlation, because all observables are related to the particles' configuration, a sensible metric is the mean squared displacement averaged over accepted moves, $\langle |\mathbf{r}(n) - \mathbf{r}(0)|^2 \rangle_{\text{accepted}} /n$, where n counts the number of accepted moves. This is a number that remains finite even if the system is not diffusive, and will convey how much change there is in the configuration space.

For the system of hard disks, how should we initialize the Monte Carlo calculation? At the beginning of a simulation, all of the microscopic degrees of freedom need to be assigned, in this case we need to determine the positions of all of the hard disks. As before, it makes sense therefore to start the system with as much forethought as possible. For the collection of hard disks, the crystalline state, a hexagonal closed-packed arrangement, is the stable state at high densities. Therefore, it is advisable to start the simulation in such an arrangement if the properties of the crystal are of interest, or to avoid it and its potential metastability if the fluid is of interest. For a fluid, it might be more useful to set up an unstable square lattice and allow it to melt to generate an initial disordered arrangement. Typical configurations of hard disks relaxed from an initial square lattice are shown in Figure 5.5.

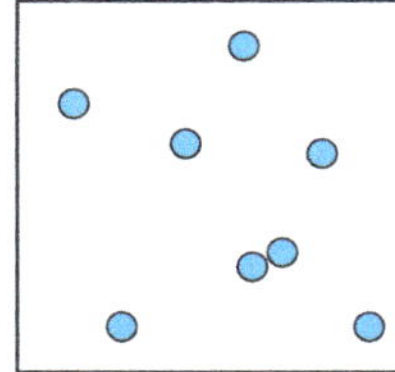 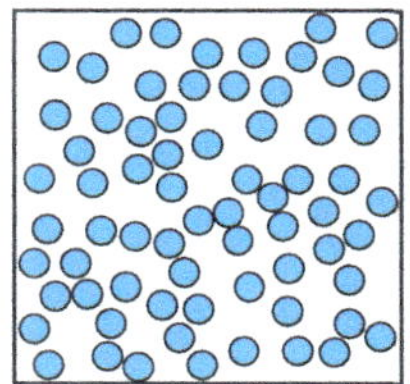 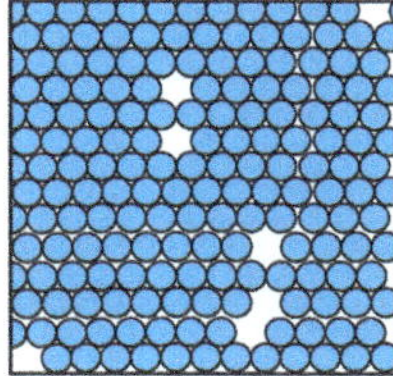

Fig. 5.5 Typical configurations of the hard disk model initialized in a square lattice at various densities and evolved for a small number of steps.

In order to attempt an update of N hard disks, in principle $N - 1$ distances must be computed to check for potential overlaps. This means that a full system update scales like N^2. This can be tamed somewhat by acknowledging that it only requires one overlap to result in a rejected move, and thus the calculation of distances may be aborted once a single overlap is found. However, this is not helpful when a move is accepted. This scale is at odds with the physical notion that hard disks interact only with the small number of particles in their vicinity, a number independent of N.

It is possible to create a linearly scaling update using a *neighbor list*. First introduced by Verlet, this bookkeeping device allows us to check for overlaps by evaluating distances over a local set. This is done by introducing a length scale δ that, when added to σ, comprises the region around a tagged particle encompassing all of the pairs that need to be computed, as illustrated in Figure 5.6. Before the overlaps are evaluated, a list of all the particles within $\sigma + \delta$ of a tagged particle is constructed. Only the overlap with these particles needs to be checked. While the construction of the list requires all pair distances to be evaluated, as long as no particle moves more than $\delta/2$, the list can be saved and reused. The choice of δ is a practical one, chosen small enough that the lists are much smaller than N but large enough that the list

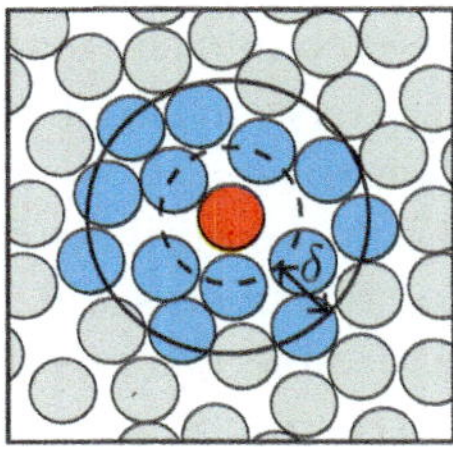

Fig. 5.6 Illustration of a neighbor list, where only blue particles within $\delta + \sigma$ of the tagged red particle need to be checked for overlaps.

does not have to be reconstructed very often. This choice depends sensitively on the density. For a density of $\rho\sigma^2 > 0.6$, like that of a liquid, δ can be taken quite small, $\delta \sim \sigma/5$.

Algorithm 4 MC with Verlet list

1: **for** $i = 1$, $i \leq N$ **do**
2:     $j = \text{Int}[N \times \text{Rand}() + 1]$ $\triangleright$ Pick random particle
3:     $\mathbf{r}_{\text{new}} = \mathbf{r}_{\text{old}} + 2\Delta[\text{Rand}() - 1/2]$ $\triangleright$ Draw random displacement
4:     **if** $|\mathbf{r}_{\text{new}} - \mathbf{r}_j^v| > \delta/2$ **then** MakeList()
5:     **end if**
6:     Flag=Overlaps($\mathbf{r}^N$) $\triangleright$ Check for overlaps
7:     **if** Flag$= 0$ **then**
8:         $\mathbf{r}_j = \mathbf{r}_{\text{new}}$ $\triangleright$ Replace with new position
9:     **end if**
10: **end for**

The incorporation of the Verlet list into a Monte Carlo simulation of the form of our random translational move algorithm is shown in Algorithm 4, with the list construction in Algorithm 5. Within the main Monte Carlo loop, the only addition is to check that the proposed position is not so far from the old position that a new list needs to be constructed. The major computational savings is hidden in the *Overlap* function that will now only loop over the particles in the neighbor list of the particle whose position we are changing. The construction of the list is straightforward, proceeding first by reinitializing all of the variables and then checking whether or not each unique pair is within $\sigma + \delta$ of each other. Here we have made explicit the use of periodic boundary conditions assuming the linear dimension of the box is L and the function $int()$ implements a rounding function.

5.5 Ensemble reweighting

The Markov chain Monte Carlo algorithm we have discussed concentrates sampling to the microstates that have the highest probability by construction. While this aids the convergence of simple expectation values, it can render studying others difficult.

Algorithm 5 Construction of the Verlet list

1: **for** $i = 1$, $i \leq N$ **do**   ▷ Clear old list
2:      Nlist(i)=0
3:      $\mathbf{r}_i^v = \mathbf{r}_i$ ▷ Store current position
4: **end for**
5: **for** $i = 1$, $i \leq N - 1$ **do**
6:      **for** $j = i + 1$, $j \leq N$ **do**
7:          $R = |\mathbf{r}_i - \mathbf{r}_j|$ ▷ Compute distance
8:          $R = R - L\operatorname{int}(R/L)$ ▷ Employ periodic boundary conditions
9:          **if** $R < \sigma + \delta$ **then**   ▷ If within cutoff, add to list
10:              Nlist(i)=Nlist(i)+1
11:              Nlist(j)=Nlist(j)+1
12:              list$[i,$Nlist$(i)]$=j
13:              list$[j,$Nlist$(j)]$=i
14:          **end if**
15:      **end for**
16: **end for**

Take for example the distribution of magnetization in the Ising model, $P(M) = \langle \delta(M - \sum_i s_i) \rangle$. Above the critical temperature, $P(M)$ is peaked near $M = 0$, and sampling in the vicinity of $M = 0$ is easy. However, below the critical temperature, $P(M)$ is bimodal, having peaks at $\pm \langle M \rangle$. In between those two peaks is a region of $P(M)$ that has vanishing probability, yet to transition between configurations with $M \approx \langle M \rangle$ and those with $M \approx - \langle M \rangle$ requires passing through that rare region. Thus an accurate estimation of $P(M)$ is made difficult because transitions between the two typical states is a rare event. Indeed, many systems exhibit activation barriers that separate distinct typical states of the system. It would be useful therefore to devise a means of sampling rare events efficiently, but in such a way that we can evaluate their likelihood relative to the typical behavior of the system.

In order to evaluate rare events, we can employ reweighing principles introduced previously. Specifically, let us consider the free energy $A(M) = -k_\mathrm{B} T \ln P(M)$, which by virtue of the probability being bimodal at low temperatures will have two minima separated by a high barrier. We can rewrite it by introducing an extra potential $U_B(M)$ dependent on the magnetization,

$$
\begin{aligned}
e^{-\beta A(M)} &= \frac{1}{Q} \sum_{\{\mathbf{s}^N\}} \delta\left(M - \sum_i s_i\right) e^{-\beta E} \\
&= \frac{Q_B}{Q} \frac{1}{Q_B} \sum_{\{\mathbf{s}^N\}} \delta\left(M - \sum_i s_i\right) e^{-\beta E - \beta U_B(M) + \beta U_B(M)} \\
&= e^{-\beta A_B(M) - \beta U_B(M) + \beta \Delta A}
\end{aligned}
$$

where we find that the free energy of the system we are interested in is equal to the free energy $A_B(M)$ evaluated under the action of a biasing potential, less the biasing potential and an additive constant, $\beta\Delta A = \ln Q_B/Q$. The constant depends on the ratio of partition functions

$$Q = \sum_{\{\mathbf{s}^N\}} e^{-\beta E} \qquad Q_B = \sum_{\{\mathbf{s}^N\}} e^{-\beta E - \beta U_B(M)}$$

for the reference and auxiliary systems. These relations are true for any potential $U_B(M)$. However, if we choose $U_B(M)$ such that it is small over a range of M we would like to sample, and very large outside of that particular range, then a Monte Carlo procedure using the energy $E + U_B(M)$ will focus the sampling onto specific configurations of a prescribed range of M. This sampling technique is called *umbrella sampling*.

Exercise 5.7: Adding an umbrella potential is just one way to enhancing the sampling of rare events. An alternative is simulate the system at a higher temperature and try to infer the properties at a lower temperature. Work out an expression relating averages of an arbitrary observable O with energy E at temperature T_1 and a weighted average of O at another temperature T_2.

The method of umbrella sampling aims to enhance the sampling of specific microstates by adding a bias potential to localize the Monte Carlo sampling, and then employs the ensemble reweighting equation to relate the statistics measured under the bias to those of the system of interest. Take, for example, a bias potential of the form

$$U_B(M) = \frac{k}{2}\left(M - M_{\mathrm{o}}\right)^2$$

parameterized by a stiffness k and a minimum M_{o}. In the limit that k is large, this potential will ensure that sampling occurs in the vicinity of M_{o}. Employing many simulations with different values of M_{o} will enable the full range of M to be sampled. With the reweighting expression, $A(M)$ can then be reconstructed provided a set of normalization constants, additive shifts to $A(M)$ can be found in order to render the free energy a smooth function of M. The procedure is schematically shown in

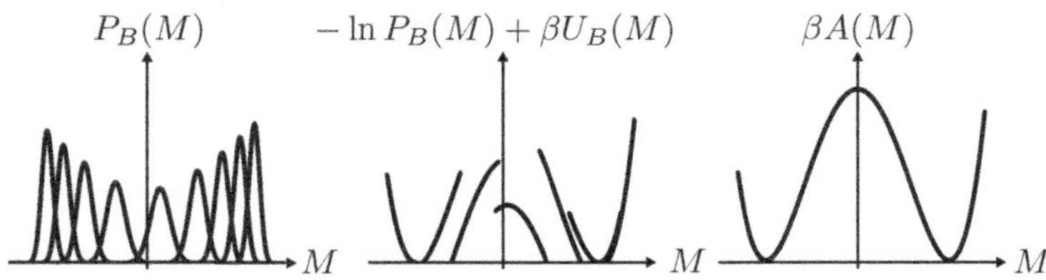

Fig. 5.7 Schematic of umbrella sampling procedure in which a set of biased histograms are generated (left), and reweighed to provide a piecewise estimate of the unbiased free energy (middle). By requiring that the curve is continuous, a set of constant shifts can be determined to allow for the complete free energy to be reconstructed (right).

Figure 5.7, where a set of simulations are performed with M_o's chosen to span the range of M of interest. The resultant histograms can be unweighted using the known form of the biasing potential, and the constant shifts applied to produce a continuous curve. The evaluation of the normalization constants can be accomplished in many ways. Provided two estimates of $A(M)$ taken from different biasing potentials have a region of M for which they overlap, a naive estimate is to just average the difference along the range of overlapping M. A slightly more robust way is to fit the different $A(M)$'s to a polynomial and then solve for the constant offset. More sophisticate ways go by acronyms WHAM and BAR.

The free energy surfaces computed from umbrella sampling are routinely used to demonstrate the existence of phase transitions and their conditions of coexistence in detailed models. This is done by studying the system size dependence of the free energy as a function of an order parameter. Taking the Ising model as an example, below its critical point the system undergoes a first-order phase transition between states with predominately up spins and predominately down spins. If a first-order transition exists, there must also be a surface that would separate the two phases. Otherwise, lacking a finite surface tension the two states would mix, resulting in a single phase. The surface free energy is encoded in $A(M)$, illustrated in Figure 5.8, as the reversible work to start in one stable phase and create an interface. This is found by the difference between the free energy at the barrier and that at its minimum. Reflecting the interface scaling, this barrier height should scale like $N^{(d-1)/d}$ or $N^{1/2}$ in $2d$. This is distinct from the $1/\sqrt{N}$ scaling of the width of the basin. Computing free energy profiles and studying the finite system size scaling of the barrier provides a means of demonstrating the stability of the two phases in the thermodynamic limit.

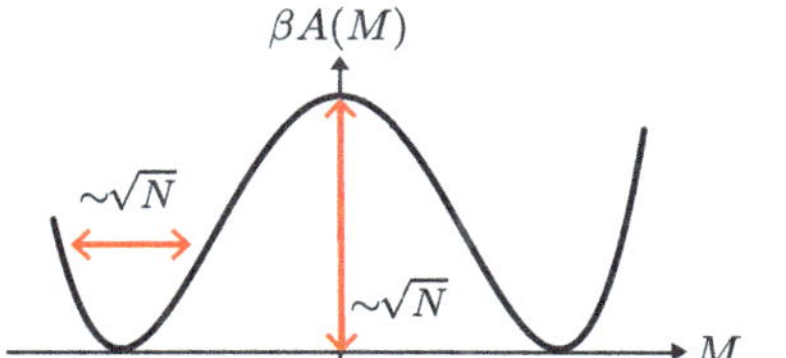

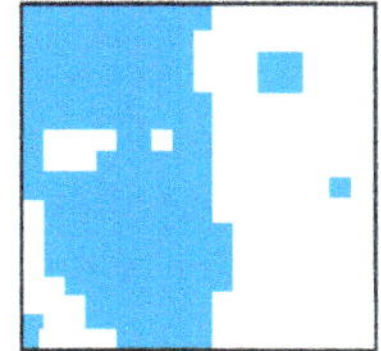

Fig. 5.8 Expected size scaling of the free energy of an order parameter (left) for a system with a first-order transition, for which an interface (right) would separate the two phases.

5.6　Path integral Monte Carlo

In Chapter 3, we introduced the quantum-classical isomorphism that related the partition function of a quantum particle to that of a classical particle in an extended space. For a system whose quantum mechanical Hamiltonian is $\mathcal{H} = \mathbf{p}^2/2m + U(\mathbf{r})$, the partition function becomes

$$Q = \lim_{n\to\infty} \left(\frac{nm}{2\pi\beta\hbar^2}\right)^{nd/2} \int d\mathbf{r}_0 \int d\mathbf{r}_1 \dots \int d\mathbf{r}_{n-1} e^{-\beta E[\{\mathbf{r}_i\}]}$$

an integral over a set of n fictitious classical particle positions, often referred to as beads along the imaginary time path. Each configuration of the path is weighted by an energy function of the form

$$E = \sum_{i=0}^{n-1} \frac{1}{n} \left[\frac{n^2 m (\mathbf{r}_i - \mathbf{r}_{i+1})^2}{2\beta^2 \hbar^2} + U(\mathbf{r}_i) \right]$$

with periodic boundary conditions $\mathbf{r}_0 = \mathbf{r}_n$. This energy function is isomorphic to a collection of particles, tethered together by springs into a ring, moving in the original potential attenuated by the number of particles n, and is exact in the limit that n is large. This formulation of quantum statistical mechanics together, with Monte Carlo sampling, admits a computationally efficient way of evaluating the thermal properties of quantum mechanical systems.

As in the other Monte Carlo approaches considered there are many different types of moves one could adopt to sample the paths. An additional complexity in this case arises because the system now has both internal configurations, different arrangements of beads relative to one another, as well as distinct translational states whose symmetry is broken by the external potential. As a consequence, we can imagine developing Monte Carlo moves that sample each, like those illustrated in Figure 5.9.

Fig. 5.9 Illustration of two path integral Monte Carlo moves. On the left is a move that attempts a change to the internal structure of the path, and on the right is a move that attempts a change to the center of mass of the path.

For example, consider a move that attempts an update to a single bead along the path. If a random bead index is drawn, then a new displacement to that bead can be randomly selected as in the hard disk Monte Carlo, $\mathbf{r}_i' = \mathbf{r}_i + 2\Delta(\eta - 1/2)$ with maximum displacement Δ. The corresponding acceptance probability will be determined by the change in the effective energy of the path, given by a sum of two terms

$$\Delta E = \frac{nm}{\beta^2 \hbar^2} \left(\mathbf{r}_i' - \mathbf{r}_i \right) \left(\mathbf{r}_i' + \mathbf{r}_i - \mathbf{r}_{i-1} - \mathbf{r}_{i+1} \right) + \frac{1}{n} \left[U(\mathbf{r}_i') - U(\mathbf{r}_i) \right]$$

where the first set of terms corresponds to the change in the springs that tether the path together, while the last two terms correspond to the change in the energy due to the external potential. While in principle repeated single bead moves will translate the whole center of mass, such motion will be slow for a large number of beads. As a consequence, it could be useful in addition to attempt moves that translate the center of mass of the path. In this case, a random displacement is drawn and added to each bead, $\{\mathbf{r}'_i\} = \{\mathbf{r}_i + 2\Delta(\eta - 1/2)\}$ for all i. The change in effective energy is

$$\Delta E = \frac{1}{n} \sum_{i=0}^{n-1} [U(\mathbf{r}'_i) - U(\mathbf{r}_i)]$$

where all of the terms from the springs between particles vanish as the relative internal positions are the same. Note in the case of a single quantum particle in an external potential there is no need for a neighbor list. This methodology can also be applied to many particles, however care must be take in order to obey the quantum statistics of the particles.

The path integral Monte Carlo algorithm is illustrated in Algorithm 6. It is distinct from the previous two algorithms in that there are two different potential moves that can be performed, and thus for the attempt probability to be symmetric we must also randomly decide which of the two to perform at any given step. This can be done compactly by drawing a uniform random number between 1 and $n + 1$, where if the random number returns $n + 1$ a center of mass move is attempted, while other values result in an attempted change to the n'th bead.

Algorithm 6 Path integral Monte Carlo

1: **for** $i = 1$, $i \leq n + 1$ **do**
2:      $j = \mathrm{Int}[n \times \mathrm{Rand}() + 2]$ $\triangleright$ Pick a move randomly
3:      **if** $j < n$ **then** $\triangleright$ Random single bead displacement
4:          $\mathbf{r}_{\mathrm{old}} = \mathbf{r}_j$
5:          $\mathbf{r}_{\mathrm{new}} = \mathbf{r}_{\mathrm{old}} + 2\Delta[\mathrm{Rand}() - 1/2]$
6:          Compute ΔE
7:      **else if** $j = n$ **then** $\triangleright$ Random path displacement
8:          $\mathbf{r}_{\mathrm{old}} = \mathbf{r}$
9:          $\mathbf{r}_{\mathrm{new}} = \mathbf{r}_{\mathrm{old}} + 2\Delta[\mathrm{Rand}() - 1/2]$
10:          Compute ΔE
11:      **end if**
12:      **if** $\Delta E < 0$ **then**
13:          $\mathbf{r} = \mathbf{r}_{\mathrm{old}}$ $\triangleright$ Replace position
14:      **else if** $\mathrm{Rand}() \leq \exp[-\beta \Delta E]$ **then**
15:          $\mathbf{r} = \mathbf{r}_{\mathrm{old}}$
16:      **end if**
17: **end for**

With a path integral Monte Carlo algorithm, one has to be mindful of the calculation of expectation values. This is because the real system, described quantum mechanically, has been mapped to an effective classical system through a equivalence of the partition functions. There is not a corresponding equivalence between arbitrary observables. Take, for example, the energy. The correct quantum mechanical average energy can be computed from the β dependence of the partition function

$$\langle E \rangle = -\left(\frac{\partial \ln Q}{\partial \beta}\right)_{N,V} = \frac{nd}{2\beta} - \frac{mn}{2\beta^2\hbar^2}\sum_{i=0}^{n-1}\left\langle(\mathbf{r}_i - \mathbf{r}_{i+1})^2\right\rangle + \frac{1}{n}\sum_{i=0}^{n-1}\langle U(\mathbf{r}_i)\rangle$$

which is not equivalent to averaging the effective energy of the path used in sampling. The average energy above consists of a simple average of the potential energy, a consequence of expressing the path integral in a real space basis, but a nontrivial function for the kinetic energy. The first two terms that estimate the kinetic energy are in fact pathological, as both scale as n, making it difficult to use in practice. Indeed, the situation is even worse, as it can be shown that the variance of the kinetic energy increases with n, while the average kinetic energy does not, it converges to its true quantum mechanical value. An alternative estimator to the so-called primitive one earlier is to use the virial theorem,

$$\langle E \rangle = \frac{1}{n}\sum_{i=0}^{n-1}\left(\langle U(\mathbf{r}_i)\rangle + \frac{1}{2}\langle \mathbf{r}_i \cdot \nabla_i U(\mathbf{r}_i)\rangle\right)$$

where the average energy is transformed into averages solely over the potential. This virial estimator behaves well in the large n limit and, as a consequence, is the standard means for evaluating the energy. Generically, observables that are simple functions of position can be straightforwardly evaluated, like the density $\rho(x) = \sum_{0=1}^{n-1}\delta(x-x_i)/n$, while those that depend on the momentum take some consideration.

Exercise 5.8: Derive the virial relation

$$\frac{1}{2m}\langle \mathbf{p}_i^2\rangle = \frac{1}{2}\langle \mathbf{r}_i \cdot \nabla_i U\rangle$$

using the identity that within an equilibrium state, $d\langle \mathbf{r}_i \cdot \mathbf{v}_i\rangle /dt = 0$ and Heisenberg's equations of motion.

Further reading

Monte Carlo techniques are widely used and many textbooks exist that provide ample background, and additional practical techniques. Of these, Daan Frenkel and Berend Smit's *Understanding molecular simulation: From algorithms to applications*, and Michael Allen and Dominic Tildesley's *Computer simulations of liquids* are the most related to this presentation. Mark Tuckerman's *Statistical mechanics: Theory and molecular simulation* includes discussion on path integral Monte Carlo, for which David Ceperley's *Path integrals in the theory of condensed helium* in *Reviews of modern physics* 1995 in a canonical source.

Additional exercises

Exercise 5.9: This rexercise centers on Monte Carlo simulations of a two-dimensional Ising magnet. In each of the problems, study a lattice no smaller than $N = 15\times15$ spins unless otherwise indicated. It is wise to examine trajectories of the system, the net magnetization as a function of the number of steps, as you perform these calculations.

1. In the absence of an external magnetic field, $h = 0$, the transition to a ferromagnetic state upon decreasing temperature is second order. In other words, the magnetization per spin $\langle m \rangle = N^{-1}\langle \sum_{i=1}^{N} s_i \rangle$ changes continuously but not smoothly at $T = T_c$. Try to confirm this fact by computing $\langle m \rangle$ from several Monte Carlo trajectories above and below T_c and plot your results, along with snapshots of the system at a temperature below T_c, a temperature very close to T_c, and a temperature above T_c.

2. Your results will differ from the exact behavior of this system determined by Onsager in several respects. For example, the apparent transition in your simulated system should not occur precisely at $k_B T_c = 2.269J$. The transition should also appear smooth rather than exhibiting a sharp kink. Explain these deviations.

3. Singularities at the critical point arise from correlated fluctuations of spins separated by macroscopically large distances. Confirm the existence of long-ranged correlations at the critical point by computing $c(r_{ij}) = \langle s_i s_j \rangle - \langle s_i \rangle \langle s_j \rangle$ as a function of $r_{ij} = |\mathbf{r}_i - \mathbf{r}_j|$, the distance between spins i and j, at several temperatures above and below the critical point. Plot your results.

4. A key concept of statistical mechanics is that fluctuations encode response. In the context of the Ising model, fluctuations in the magnetization report on the susceptibility of the magnetization to an applied magnetic field. Specifically, we showed the relationship $\chi = d\langle m \rangle / d\beta h|_{h=0} = N\langle \delta m^2 \rangle$. Demonstrate this relation numerically by first computing the average magnetization at some temperature above T_c, and a few values of h around 0 so that that χ can be evaluated by a finite difference. Then compute χ by averaging the mean squared magnetization fluctuations at that same temperature and $h = 0$.

Exercise 5.10: In this problem you will continue studying the Ising model, and use some of the enhanced sampling methods to evaluate the reservable work surface for changing the magnetization.

1. Using the umbrella sampling method described in this chapter, compute the Helmholtz free energy $A(M)$ for a 20×20 Ising magnet constrained to have net magnetization $M = \sum_{i=1}^{N} s_i$. Perform your calculation with a small external magnetic field $h/J = 0.1$ at two temperatures, one above T_c and the other below T_c. Plot the probability distributions of M in each window used in your calculations as well as your reconstruction of the quantity $A(M)$.

2. Repeat your calculations of $A(M)$ in the absence of a magnetic field, $h = 0$, for two system sizes, $N = 10 \times 10$ and $N = 20 \times 20$. Provide representative snapshots of the $N = 20 \times 20$ system at $M = -Nm$, $M = 0$, and $M = +Nm$.

   (a) For the case $T < T_c$, extract from your calculation the reversible work required to change the net magnetization M from its equilibrium value $M = Nm$ to $M = 0$. Does your result change with system size as expected? Explain.

   (b) For the case $T > T_c$, show that the magnetic susceptibility $\chi = d\langle m \rangle / d\beta h|_{h=0}$ can be deduced from the curvature of $A(M)$, and then carry out the numerical calculation. Does your result change with system size as expected? Explain.

Exercise 5.11: In this problem, you will study density fluctuations in a fluid of hard disks with diameter σ. We will focus on a density of $\rho^* = 0.5$, where $\rho^* = (N/L_x^2)\sigma^2$ is the dimensionless number of particles, N, per unit area, L_x^2. The Monte Carlo sampling strategy presented requires an initial configuration that is free of overlapping particles. At $\rho\sigma^2 = 0.5$, this constraint can be satisfied very simply by placing particles on a square lattice. The magnitude of particle displacements is an important parameter to optimize in Monte Carlo simulations. By performing short simulations with a variety of displacement sizes (and computing the fraction of moves that are accepted), you should be able to quickly identify a useful value.

1. Perform a Monte Carlo simulation of $N = 100$ hard disks, tracking the position of a particular particle over $1000N$ steps. Plot the x, y coordinates of this particle as functions of the number of sweeps elapsed (i.e., 'MC time' , or N steps).

2. This problem concerns the occupation statistics of a small circular probe area a in a fluid of hard disks. We will take the diameter, $D = 0.9\sigma$, of the probe area to be so small that at most one particle's center could lie within a. In other words, since $D < \sigma$, the number N_a of hard disks whose center lies inside the probe area can only take values of $N_a = 0$ or $N_a = 1$.

   (a) Note that for such a small probe area, $P(N_a) = 0$ for all $N_a > 1$. Using this fact, write a relationship between $P(N_a = 0)$ and $P(N_a = 1)$.

   (b) Exploiting the small size of the probe area, write a simple relationship between the average number N_a of particles in a and the probability $P(N_a = 1)$ of single occupancy.

   (c) Combining your results, together with the expression for $\langle N_a \rangle$ calculate $P(N_a = 0)$ in terms of the bulk density N/L_x^2 and probe area a.

   (d) Using a Monte Carlo simulation of $N = 100$ hard disks at density $\rho\sigma^2 = 0.5$, compute the probability $P(N_a = 0)$ that a probe area with diameter $D = 0.9\sigma$ is vacant. As with any stochastic expectation value, the error in your estimate should asymptote to $\sim 1/\sqrt{T}$ where T is the number of samples of the expectation value. Provide a plot of the running average of $P(N_a = 0)$ to demonstrate statistical convergence. Compare your simulation result with the theoretical prediction from part (c).

3. In this problem you will explore the occupation statistics of somewhat larger probe areas, again using Monte Carlo simulations of $N = 100$ hard disks at density $\rho\sigma^2 = 0.5$. Perform each of the calculations described next for spherical probe area of diameter $D = 2\sigma, 3\sigma$, and 4σ.

   (a) Compute the average number of hard disks whose center lies inside the probe area. Plot your results for $\langle N_a \rangle$ as a function of probe size D, along with the expected result, ρa.

   (b) Compute the variance of N_a, i.e., $\langle (\delta N_a)^2 \rangle = \langle N_a^2 \rangle - \langle N_a \rangle^2$. Plot $\langle (\delta N_a)^2 \rangle$ as a function of probe size D.

   (c) Compute a normalized histogram $P(N_a)$ of the occupation number N_a. Plot $\ln P(N_a)$ as a function of N_a. Values of N_a for which you have gathered fewer than 100 counts are not statistically well represented. Do not include them in your plots. Compare your result with a Gaussian distribution,

$$p_G(N_a) = \exp\left[-(N_a - \langle N_a \rangle)^2/2\langle(\delta N_a)^2\rangle\right]/\sqrt{2\pi\langle(\delta N_a)^2\rangle}$$

with the same mean and variance as $P(N_a)$. Include plots of $\ln p_G(N_a)$ alongside your simulation results, and comment on the quality of agreement.

Exercise 5.12: We have previously defined the pair distribution function $g(r) = \langle\rho(r)\rangle_{r_1=0}/\rho$ of a liquid in terms of the average density field $\langle\rho(r)\rangle_{r_1=0}$ surrounding a tagged particle at the origin, divided by the bulk density ρ. Here you will use Monte Carlo simulations of a hard disk fluid to obtain numerical results for $g(r)$.

1. Let N_{R_1,R_2} be the fluctuating number of particles in a spherical shell, with inner radius R_1 and outer radius R_2, centered on a tagged particle. Show that

$$\langle N_{R,R+dR}\rangle \approx 4\pi\rho R^2 g(R)dR$$

for a very thin shell of width dR.

2. Show that the average number of particles in a thin shell can be written in terms of the probability distribution $P(R)$ for the distance R between pairs,

$$\langle N_{R,R+dR}\rangle \approx (N-1)P(R)dR$$

3. Adapt your hard disk Monte Carlo simulation program to calculate $g(r)$. You might find the results from parts 1 and 2 useful for this purpose, since they relate $g(r)$ to a histogram that can be computed very straightforwardly. Note that parts 1 and 2 provide relationships for 3 spatial dimensions and your code is operating in only 2, so take care to use the proper volume element. Make sure to take advantage of the particles' statistical equivalence. Perform calculations for systems of at least $N = 100$ particles, at densities $N/L^2 = 0.9\sigma^{-2}, 0.8\sigma^{-2}, 0.7\ \sigma^{-2}, 0.6\sigma^{-2}, 0.5\sigma^{-2}, 0.4\sigma^{-2}, 0.3\sigma^{-2}, 0.2\sigma^{-2}, 0.1\sigma^{-2}$ and $0.02\sigma^{-2}$. Plot your results.

Exercise 5.13: In this problem, you will explore path integral Monte Carlo approaches to evaluate the quantum thermodynamics of a particle in an external potential. Throughout you should run calculations with 20 and 40 imaginary time slices and compare your results in order to check for convergence. You should use the virial estimator for all energy calculations.

1. Consider a quantum harmonic oscillator, with $\hat{\mathcal{H}} = \hat{p}^2/2m + m\omega^2\hat{x}^2/2$. Setting $\beta\hbar\omega = 10$, $m = 1$, compute $\langle E\rangle$ and make a histogram of the bead position. Compare these to expectations from the ground state of the harmonic oscillator.

2. Compute the average energy for the harmonic oscillator in part 1 over a range of temperatures $\beta\hbar\omega = 10$ to $\beta\hbar\omega = 0.5$. Make a plot and compare your results to the exact $\langle E\rangle$ derived from the canonical partition function. At what temperature does equipartition hold?

3. Consider next a particle in a double well potential $\hat{\mathcal{H}} = \hat{p}^2/2m - a_1\hat{x}^2/2 + a_4\hat{x}^4$. For $\beta a_1 = 4$, $\beta a_4 = 1$ and a range of masses $m = 0.5, 1, 5, 10$, compute a histogram of the bead position and compare your results to the expectations for a classical particle using a Boltzmann distribution. At what mass is the classical approximation valid? Explain.

6

Time dependence near equilibrium

Thermal equilibrium is characterized macroscopically by the absence of change. Microscopically, equilibrium is a statement concerning typical behavior, and the absence of change *on average*. At a particular equilibrium state, atoms and molecules are constantly moving, engaged in a complex molecular dynamics. In this chapter, we will develop the formalism to quantify dynamical evolution statistically. The dynamical view of how an equilibrium distribution emerges supplements the static perspective we have thus far built. In contrast with our discussion of static equilibrium phenomena, the development will be concerned with the probability distribution associated with trajectories that microstates trace out over time, rather than the particular microstates themselves. Initially, the basis of our development will be Hamiltonian mechanics, and from it we will deduce theorems concerning the conservation of probability, and by association, the equations governing the evolution of collections of particles or their collective properties like mass and momentum. In later chapters, we will consider trajectories that are not solely determined by a set of initial conditions, but evolve stochastically due to our imperfect information. The statements we will develop here are sufficient to clarify principles of reversibility alluded to macroscopically in early chapters of this book, and state general theorems restricting the behavior of systems evolving irreversibly. Of overarching importance will be the perspective we develop on time, both its extent and its direction. As with our development in the early chapters, these exact relationships are difficult to use directly, away from idealized cases. Nevertheless they serve as a foundation from which to construct useful approximations.

6.1 Ergodic hypothesis

We begin this section on time dependence by first considering a few points regarding how time enters into thermal equilibrium. We will begin classically by denoting the equilibrium Boltzmann distribution, as

$$f_{\mathrm{eq}}(\mathbf{r}^N, \mathbf{p}^N) = \exp\left[-\beta \mathcal{H}(\mathbf{r}^N, \mathbf{p}^N)\right]/Q$$

where

$$\mathcal{H}(\mathbf{r}^N, \mathbf{p}^N) = \sum_{i=1}^{N} \mathbf{p}_i^2/2m + U(\mathbf{r}^N)$$

is the Hamiltonian of the system. It has two parts, a kinetic energy that is the sum of single particle momentum squared divided by twice the mass, and a potential energy that in principle depends on the position of all N particles. In equilibrium, we can in principle talk about two different kinds of averages. The first, denoted by an overbar,

$$\bar{\mathbf{r}}_i = \frac{1}{\tau} \int_0^\tau dt\, \mathbf{r}_i(t)$$

is a time average. It is an average over the trajectory that a tagged observable, in this case, $\mathbf{r}_i$, traces out over an observation time τ. The trajectory can be computed by noting the equations of motion the observable follows. For a Hamiltonian system, this is given by the set of equations for the time dependence of the positions and momenta of all N particles. These are,

$$\frac{d\mathbf{p}_i}{dt} = -\frac{\partial \mathcal{H}}{\partial \mathbf{r}_i} \qquad \frac{d\mathbf{r}_i}{dt} = \frac{\partial \mathcal{H}}{\partial \mathbf{p}_i}$$

where substituting in the Hamiltonian,

$$\frac{d\mathbf{p}_i}{dt} = -\frac{\partial}{\partial \mathbf{r}_i} U(\mathbf{r}^N) \qquad \frac{d\mathbf{r}_i}{dt} = \mathbf{p}_i/m$$

we find perhaps the more familiar Newtonian equations of motion. These equations are first-order, so they are uniquely determined by a single boundary condition. Most often it is easiest to consider solving these equations with an initial condition, in which case there exists a unique mapping

$$\mathbf{r}_i(t) = \mathbf{r}_i[\mathbf{r}^N(t), \mathbf{p}^N(t)] \to \mathbf{r}_i[t; \mathbf{r}^N(0), \mathbf{p}^N(0)]$$

between the time evolved observable, and the initial conditions of all N particles. As a formal function of a set of initial conditions, one can use this dependence to define a separate type of average,

$$\langle \mathbf{r}_i \rangle = \int d\mathbf{r}^N \int d\mathbf{p}^N\, f_{\text{eq}}(\mathbf{r}^N, \mathbf{p}^N) \mathbf{r}_i[t; \mathbf{r}^N, \mathbf{p}^N]$$
$$= \langle \mathbf{r}_i(t) \rangle = \langle \mathbf{r}_i(0) \rangle$$

where $\langle \ldots \rangle$ denotes ensemble average, and for a system at thermal equilibrium, the likelihood of a given set of initial conditions is just given by the Boltzmann distribution. In addition, a Boltzmann distribution is stationary, so averages at one time are equivalent to another time. This is a property we will define more carefully shortly. In cases where the single-particle trajectory evolves such that it generates a Boltzmann distribution, then as $\tau \to \infty$, one expects

$$\langle \mathbf{r}_i \rangle = \bar{\mathbf{r}}_i$$

an equivalence that is often difficult to prove, and it is typically regarded as an assumption. Formally, it is known as the *ergodic hypothesis* and relies on the dynamics of the molecules being chaotic. This is a dynamical restatement of our fundamental postulate of equal probabilities for microstates of a closed system.

Exercise 6.1: Evaluate the equations of motion for a harmonic oscillator Hamiltonian, $\mathcal{H} = p^2/2m + m\omega^2 r^2/2$ using Hamiltonian mechanics. By evaluating the total time derivative of $\mathcal{H}$,

$$\frac{d\mathcal{H}}{dt} = \dot{r}\frac{\partial \mathcal{H}}{\partial r} + \dot{p}\frac{\partial \mathcal{H}}{\partial p}$$

show that it is a constant of motion.

Exercise 6.2: The equation of motion for the harmonic oscillator can be integrated yielding

$$r(t) = r_0 \cos(\omega t) + \frac{p_0}{m\omega}\sin(\omega t)$$

where r_0 is the initial position and p_0 is the initial momentum. Compute the time average of $r^2(t)$ using thermal averages for the necessary initial conditions $p_0^2 = \langle p^2 \rangle$, $r_0^2 = \langle r^2 \rangle$, and $p_0 = r_0 = \langle p \rangle = \langle r \rangle = 0$. Show that it is equal to the thermal result of $\langle r^2 \rangle = k_{\mathrm{B}}T/m\omega^2$.

6.2 Liouville's theorem

In order to give a plausible argument for why one might expect a system to evolve a Boltzmann distribution, we need to remind ourselves of some of the structure of classical mechanics. In particular, we need to understand the mapping,

$$\{\mathbf{r}^N(0), \mathbf{p}^N(0)\} \rightarrow \{\mathbf{r}^N(t), \mathbf{p}^N(t)\}$$

which relates two sets of points in a $6N$ dimensional space. To understand this mapping, we can consider how an arbitrary density of points in this space evolves. We denote a distribution of points, f, in phase space, $\mathbf{x} = \{\mathbf{r}^N, \mathbf{p}^N\}$. The total time derivative of this distribution is,

$$\frac{df(\mathbf{x},t)}{dt} = \frac{\partial f}{\partial t} + \dot{\mathbf{x}} \cdot \nabla_{\mathbf{x}} f$$

$$= \frac{\partial f}{\partial t} + \sum_i^N \left(\dot{\mathbf{p}}_i \cdot \frac{\partial}{\partial \mathbf{p}_i} + \dot{\mathbf{r}}_i \cdot \frac{\partial}{\partial \mathbf{r}_i} \right) f$$

where the total, or *convective* derivative, includes a time derivative that acts on the density directly, and a term that depends on the variation of f in phase space. The time dependence of the density can be understood in the following way. Consider a region of phase space, V, bounded by a surface S with surface normal $\mathbf{n}$. Such a

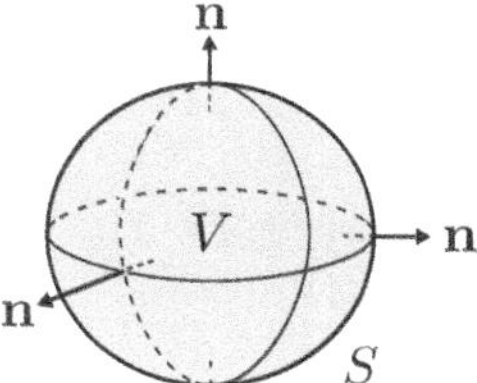

Fig. 6.1 A region of volume V, bounded by the surface S, with the surface normal $\mathbf{n}$.

geometry is illustrated in Figure 6.1. The total accumulation of points in that region can be computed by integrating the time derivative over the region,

$$\int_V d\mathbf{x}\,\frac{\partial f}{\partial t} = -\int_S dS\,\mathbf{n}\cdot(\dot{\mathbf{x}}f) = -\int_V d\mathbf{x}\,\nabla_{\mathbf{x}}\cdot(\dot{\mathbf{x}}f)$$

where the second equality follows from the divergence theorem.

For this equation to be true for all V,

$$\frac{\partial f}{\partial t} = -\nabla_{\mathbf{x}}\cdot(\dot{\mathbf{x}}f) = -\dot{\mathbf{x}}\cdot(\nabla_{\mathbf{x}}f) - (\nabla_{\mathbf{x}}\cdot\dot{\mathbf{x}})f$$

$$= -\sum_i^N \dot{\mathbf{p}}_i\cdot\frac{\partial f}{\partial\mathbf{p}_i} + \dot{\mathbf{r}}_i\cdot\frac{\partial f}{\partial\mathbf{r}_i} + \left(\frac{\partial\dot{\mathbf{r}}_i}{\partial\mathbf{r}_i} + \frac{\partial\dot{\mathbf{p}}_i}{\partial\mathbf{p}_i}\right)f\,.$$

Substituting in Hamilton's equation of motion for the two time derivatives on the right side,

$$\frac{\partial f}{\partial t} = -\sum_i^N \dot{\mathbf{p}}_i\cdot\frac{\partial f}{\partial\mathbf{p}_i} + \dot{\mathbf{r}}_i\cdot\frac{\partial f}{\partial\mathbf{r}_i} + \left(\frac{\partial^2\mathcal{H}}{\partial\mathbf{r}_i\partial\mathbf{p}_i} - \frac{\partial^2\mathcal{H}_i}{\partial\mathbf{r}_i\partial\mathbf{p}_i}\right)f$$

$$= -\sum_i^N \dot{\mathbf{p}}_i\cdot\frac{\partial f}{\partial\mathbf{p}_i} + \dot{\mathbf{r}}_i\cdot\frac{\partial f}{\partial\mathbf{r}_i}$$

we find that the second term on the right-hand side vanishes. The leftover terms are identical to $-\dot{\mathbf{x}}\cdot\nabla_{\mathbf{x}}$, which implies from the total derivative of f that

$$\frac{df(\mathbf{x},t)}{dt} = 0\,.$$

This is known as the *Liouville's theorem*, and implies that the phase-space distribution function is constant along the trajectories of the system. It is a statement of conservation of probability, and is an example of a continuity equation. Liouville's equation can be written as a linear differential equation for the motion of phase points,

$$\frac{\partial f}{\partial t} = -\mathcal{L}f \qquad \mathcal{L}f \equiv \dot{\mathbf{x}}\cdot\nabla_{\mathbf{x}}f = \sum_i^N\left(\dot{\mathbf{p}}_i\cdot\frac{\partial}{\partial\mathbf{p}_i} + \dot{\mathbf{r}}_i\cdot\frac{\partial}{\partial\mathbf{r}_i}\right)f$$

where $\mathcal{L}$ is the Liouvillian, or the linear operator that generates dynamics in phase space.

Liouville's equation can be formally integrated

$$f(\mathbf{x},t) = e^{-\mathcal{L}t} f(\mathbf{x},0)$$

where in the reference frame of the trajectory, the density of the phase points is invariant. Operating on a Boltzmann distribution,

$$\mathcal{L}f_{\mathrm{eq}} = 0$$

we find that it is stationary under the action of the Liouvillian, so Hamilton's equation of motion will preserve a Boltzmann distribution. Note too that a uniform density in phase space would also be conserved. Thus, Liouville's theorem does not supply a uniqueness of the stationary distribution. Such uniqueness comes from phenomenological considerations like the extensiveness of the entropy or free energy, or other constraints imposed on the system.

> **Exercise 6.3:** Demonstrate explicitly that the Boltzmann distribution is a stationary state of the Liouville operator, $\mathcal{L}f_{\mathrm{eq}} = 0$ for $f_{\mathrm{eq}}(\mathbf{r}^N, \mathbf{p}^N) = \exp\left[-\beta \mathcal{H}(\mathbf{r}^N, \mathbf{p}^N)\right]/Q$.

Generally, the time dependence of any function of phase space is determined by the Liouvillian. Consider a variable $A[\mathbf{x}(t)]$. Its time dependence is computed as

$$\frac{d}{dt} A[\mathbf{x}(t)] = \dot{\mathbf{x}} \cdot \nabla_{\mathbf{x}} A[\mathbf{x}(t)]$$

$$= \sum_i^N \dot{\mathbf{p}}_i \cdot \frac{\partial A}{\partial \mathbf{p}_i} + \dot{\mathbf{r}}_i \cdot \frac{\partial A}{\partial \mathbf{r}_i} = \mathcal{L}A[\mathbf{x}]$$

which is identical to the Liouville equation with the exception of the minus sign in front of the operator $\mathcal{L}$. Thus at some time t, a phase space variable is evaluated as the action of the Liouvillian on any function of phase space,

$$A[\mathbf{x}(t)] = e^{\mathcal{L}t} A[\mathbf{x}(0)]$$

provided its initial value $A[\mathbf{x}(0)]$. Associating time dependence to the average of an observable or to the probability distribution is a choice we can make, analogously with the complementary Heisenberg and Schrodinger pictures in quantum mechanics. The change of sign when operating on the phase space distribution or an observable is a signature of the anti-Hermitian or skew-Hermitian nature of $\mathcal{L}$, from which we can deduce that the eigenvalues of this linear operator are purely imaginary. Because of this, often the Liouvillian is defined with the imaginary i in front, which we forego here. Note that if $A[\mathbf{x}(t)]$ is the Hamiltonian, $\mathcal{H}$, the action of $\mathcal{L}$ on it returns 0, implying that the Hamiltonian is conserved. We will see that the existence of a conserved quantity implies, via Noether's theorem, the existence of a symmetry. The symmetry is invariance under time translations, and the generator of the symmetry is the Hamiltonian.

Aside: Functions of differential operators. The action of the exponential differential operators can be understood by Taylor expanding the exponential

$$e^{a\frac{d}{dy}}g(y) = \left(1 + a\frac{d}{dy} + \frac{a^2}{2}\frac{d^2}{dy^2} + \dots\right)g(y) = g(y + a)$$

where the second equality follows from identifying the portion in parenthesis as a Taylor series expansion of the function $g(y)$ around $y = a$. Using this rule, for example we could show that for a free particle in the z direction

$$e^{\mathcal{L}t}z(0) = z(0) + tp_z/m$$

the action of the Liouville operator returns the expected solution of Hamilton's equations of motion.

6.3 Microreversibility and Jarzynski's equality

Liouville's theorem provides the necessary conditions to connect molecular dynamics to irreversible behavior in thermodynamics. The nature of reversibility is manifest in Hamiltonian mechanics by the acknowledgement that the solutions of these equations come in pairs. For every trajectory of a system described by a set of initial conditions and integrated forward in time, there is an equivalent trajectory described by a final condition specified by a set of positions and reversed momenta integrated backward in time. This is the notion of *microscopic reversibility*. We will denote a configuration of the system at time t in the forward direction of time as $\mathbf{x}(t) = \{\mathbf{r}^N(t), \mathbf{p}^N(t)\}$, and the time reversed configuration of the system $\mathbf{x}^*(t) = \{\mathbf{r}^N(\tau - t), -\mathbf{p}^N(\tau - t)\}$, for a trajectory of length τ. A representative pair of trajectories are shown in Figure 6.2, where $\mathbf{p}$ is reflected about the origin.

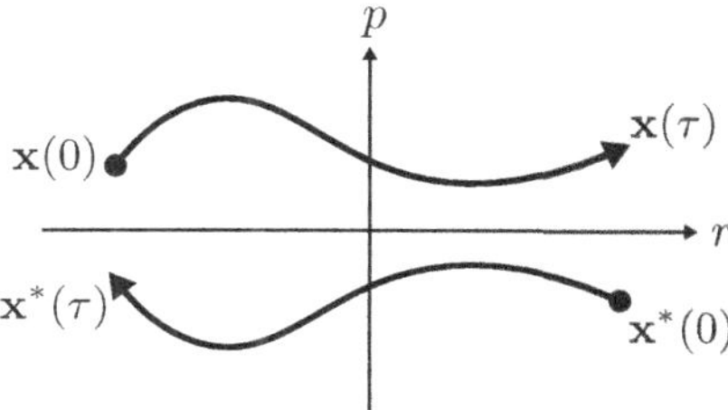

Fig. 6.2 Example of a pair of trajectories related by time reversal symmetry.

Consider a Hamiltonian dependent on a control parameter $\Lambda(t)$. It is invariant under time reversal so that $\mathcal{H}[\mathbf{x}, \Lambda] = \mathcal{H}[\mathbf{x}^*, \Lambda]$. From our understanding of equilibrium thermodynamics, we know that if we change $\Lambda(t)$, we will change the internal energy, which for a closed system is identifiable as work $\mathcal{W}$,

$$\mathcal{W} = \int dt\, \dot{\Lambda} \frac{\partial \mathcal{H}}{\partial \Lambda}.$$

If we change $\Lambda(t)$ slowly, that work will approach its reversible limit determined by the change in free energy of the system $\mathcal{W} \to \Delta A$. What happens when we change the control parameter Λ at a finite rate, or irreversibly? Imagine preparing the system with $\Lambda(0)$, such that its initial configuration is Boltzmann distributed,

$$f_{\Lambda(0)}[\mathbf{x}(0)] = \frac{e^{-\beta \mathcal{H}[\mathbf{x}(0), \Lambda(0)]}}{Q_{\Lambda(0)}}$$

with the initial value of the control parameter $\Lambda(0)$. The associated partition function is $Q_\Lambda = \int d\mathbf{x}\, \exp\left[-\beta \mathcal{H}(\mathbf{x}, \Lambda)\right]$, which has the usual connection to a free energy $A_\lambda = -k_\mathrm{B} T \ln Q_\Lambda$.

Because the dynamics are deterministic, the likelihood of a trajectory of a system acted upon by a prescribed protocol $\Lambda(t)$ is completely determined by this initial condition. As $\Lambda(t)$ is varied between its initial value and its final value, the work that is accumulated is equal to the change the internal energy

$$\mathcal{W}[\mathbf{x}(t)] = \mathcal{H}[\mathbf{x}(t), \Lambda(t)] - \mathcal{H}[\mathbf{x}(0), \Lambda(0)]$$

which will fluctuate through different realizations of the process due to the randomness associated with the initial condition. We can characterize those fluctuations by evaluating the generating function of the work done on the system by the external protocol,

$$\begin{aligned}
\left\langle e^{-\beta \mathcal{W}[\mathbf{x}(\tau)]} \right\rangle &= \int d\mathbf{x}(0)\, f_{\Lambda(0)}[\mathbf{x}(0)] e^{-\beta \mathcal{W}[\mathbf{x}(\tau)]} \\
&= \int d\mathbf{x}(0)\, \frac{e^{-\beta \mathcal{H}[\mathbf{x}(0), \Lambda(0)]}}{Q_{\Lambda(0)}} e^{-\beta(\mathcal{H}[\mathbf{x}(\tau), \Lambda(\tau)] - \mathcal{H}[\mathbf{x}(0), \Lambda(0)])} \\
&= \int d\mathbf{x}(0)\, \frac{e^{-\beta \mathcal{H}[\mathbf{x}(\tau), \Lambda(\tau)]}}{Q_{\Lambda(0)}}
\end{aligned}$$

which we can rewrite by inserting the initial equilibrium distribution and the definition of the work. Note the argument of the Hamiltonian in the last line is $\mathbf{x}(\tau)$. To integrate it we can perform a change of measure

$$\int d\mathbf{x}(0) = \int d\mathbf{x}(\tau) \left| \frac{d\mathbf{x}(\tau)}{d\mathbf{x}(0)} \right|^{-1}$$

where the Jacobian $|d\mathbf{x}(\tau)/d\mathbf{x}(0)|$ is equal to 1 from Liouville's theorem, as the evolution of phase space points is conserved along a trajectory. Microscopically this is analogous to the fact that the end point $\mathbf{x}(\tau)$ can be reversed in time to regenerate the initial condition. With the change of measure

$$\left\langle e^{-\beta \mathcal{W}[\mathbf{x}(\tau)]} \right\rangle = \frac{1}{Q_{\Lambda(0)}} \int d\mathbf{x}(\tau) e^{-\beta \mathcal{H}[\mathbf{x}(\tau),\lambda(\tau)]}$$

$$= \frac{Q_{\Lambda(\tau)}}{Q_{\Lambda(0)}} = e^{-\beta \Delta A}$$

we can identify the integral as the partition function of a system prepared initially under a fixed value of the control parameter, $\Lambda(\tau)$.

Exercise 6.4: Using the same arguments as above, prove the Crooks work fluctuation theorem

$$\frac{P(\mathcal{W})}{P(-\mathcal{W})} = e^{\beta \mathcal{W} - \beta \Delta A}$$

using $P(\mathcal{W}) = \langle \delta(\mathcal{W} - \mathcal{W}[\mathbf{X}(\tau)] \rangle$ and noting that $\mathcal{W}(\mathbf{x}) = -\mathcal{W}(\mathbf{x}^*)$.

This incredible result is known as the *Jarzynski equality*. It relates the work done on a system during an arbitrary nonequilibrium process to the change in the system's free energy. This is significant. In traditional thermodynamics free energy differences are determined only in an idealized limit of a reversible process, which can be difficult or even impossible to achieve in practice, especially for small systems. Indeed, the convexity of the exponential implies

$$\langle \mathcal{W} \rangle \geq \Delta A$$

from Jensen's inequality, which is a previous statement of the second law made in Chapter 1 known as the reversible work theorem. By virtue of the microscopic reversibility of Hamiltonian dynamics, the Jarzynski equality implies that free energies are encoded in microscopic dynamics even when perturbed greatly from their equilibrium state by changing a control parameter at a finite rate. This derivation also emphasizes that work is a fluctuating quantity and has a distribution of values, like those illustrated in Figure 6.3. While typically a finite time process will do *more* work

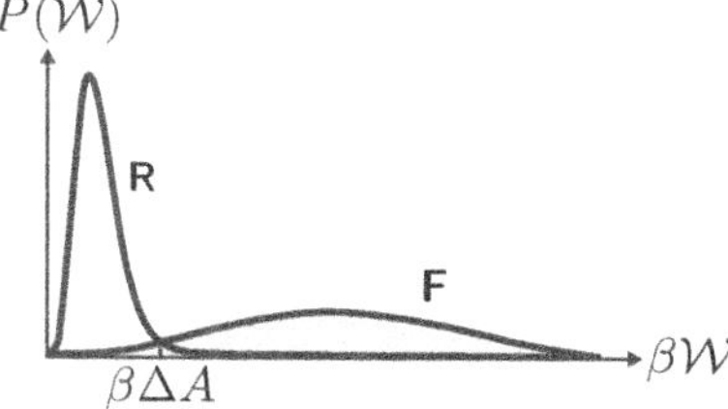

Fig. 6.3 Example of the work distribution for a forward (F) and reverse (R) process.

than its reversible limit, there are necessarily fluctuations in the statistics of the work that result in a process that does *less*. These latter realizations must be very rare however. As the dynamics are deterministic, the source of these fluctuations are encoded in the uncertainty of the initial condition.

An analogous result to the Jarzynski equality can be derived for heat flows between two systems held initially at equilibrium, but at different temperatures. Consider a system illustrated in Figure 6.4 that is partitioned into initially non-interacting regions A and B, such that the full configuration of the system is given by $\mathbf{x} = \{\mathbf{x}_A, \mathbf{x}_B\}$. The total Hamiltonian of the system decomposes into $\mathcal{H}[\mathbf{x}] = \mathcal{H}_A[\mathbf{x}_A] + \mathcal{H}_B[\mathbf{x}_B] + h_{\text{int}}[\mathbf{x}_A, \mathbf{x}_B, t]$ where $\mathcal{H}_i$ is a Hamiltonian that acts only on the configuration for the ith region, while h_{int} is an interaction term between the two partitions. We will take h_{int} as initially zero, will turn on for a small amount of time τ, and then turn back off. We expect that throughout the process h_{int} is small. This is physically motivated by the expectation that the interactions are mediated by a common interface between the two regions, and this manifests an infinitesimal contribution to the total energy.

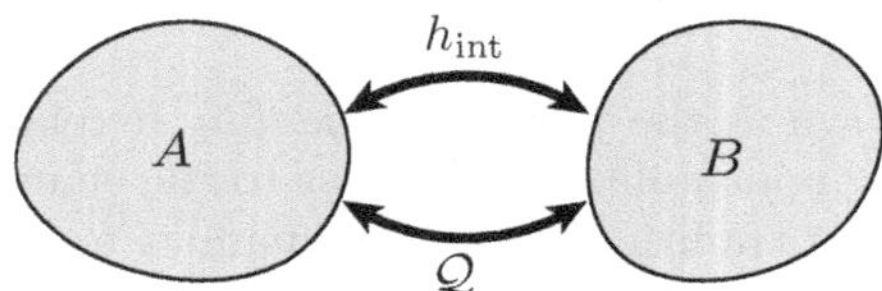

Fig. 6.4 Two isolated systems A and B are temporarily allowed to interact through h_{int}, through which heat $\mathcal{Q}$ can be transferred.

Imagine the total system is initially prepared in equilibrium with $h_{\text{int}} = 0$,

$$f[\mathbf{x}(0)] = \frac{1}{Q_A Q_B} e^{-\beta_A \mathcal{H}_A[\mathbf{x}_A(0)]} e^{-\beta_B \mathcal{H}_B[\mathbf{x}_B(0)]}$$

where region i is at fixed temperature β_i. Imagine now integrating the system forward in time by an amount τ, with h_{int} finite, to a final state $\mathbf{x}(\tau)$ and then turning off h_{int}. Because the Hamiltonian is invariant under time reversal, we can construct the equilibrium likelihood of sampling the end point of the forward time trajectory $\mathbf{x}(\tau) = \mathbf{x}^*(0)$ as an initial condition in a virtual time-reversed process

$$f[\mathbf{x}^*(0)] = \frac{1}{Q_A Q_B} e^{-\beta_A \mathcal{H}_A[\mathbf{x}_A^*(0)]} e^{-\beta_B \mathcal{H}_B[\mathbf{x}_B^*(0)]}$$

and can define the change in energy between the initial condition $\mathbf{x}(0)$ and $\mathbf{x}^*(0)$ for each partition as $\Delta E_A = \mathcal{H}_A[\mathbf{x}_A^*(0)] - \mathcal{H}_A[\mathbf{x}_A(0)]$ and analogously $\Delta E_B = \mathcal{H}_B[\mathbf{x}_B^*(0)] - \mathcal{H}_B[\mathbf{x}_B(0)]$. In the limit that we can neglect the small amount of work performed by turning h_{int} on and back off, then $\Delta E_B = -\Delta E_A$ and it is natural to identify the change as heat, $\mathcal{Q} \approx \Delta E_B = -\Delta E_A$ that is transferred spontaneously from one region to another.

As a consequence of microscopic reversibility and our definitions, the ratio of the initial conditions of the forward and backward trajectories is

$$\frac{f[\mathbf{x}(0)]}{f[\mathbf{x}^*(0)]} = e^{-(\beta_A - \beta_B)\mathcal{Q}}$$

which is just given by the difference in temperatures of the two systems times the heat exchanged between them. In the absence of a heat flow, the forward and backward directions are indistinguishable, but not so in its presence. Note that the heat in the forward direction of time is opposite to that in the backward direction of time, $\mathcal{Q}(\mathbf{x}) = -\mathcal{Q}(\mathbf{x}^*)$. In different realizations of this process, different amounts of heat can flow and we can consider the statistics of heat exchange by evaluating the probability $P(\mathcal{Q})$ that a given amount of heat is exchanged. Importantly this distribution obeys a deep symmetry known as the *fluctuation theorem*,

$$\begin{aligned}
P(\mathcal{Q}) &= \int d\mathbf{x}(0) f[\mathbf{x}(0)] \delta(\mathcal{Q} - \mathcal{Q}[\mathbf{x}(0)]) \\
&= \int d\mathbf{x}^*(0) f[\mathbf{x}^*(0)] e^{-(\beta_A - \beta_B)\mathcal{Q}} \delta(\mathcal{Q} + \mathcal{Q}[\mathbf{x}^*(0)]) \\
&= e^{-(\beta_A - \beta_B)\mathcal{Q}} P(-\mathcal{Q})
\end{aligned}$$

where in the second line we have used Liouville's theorem to change the measure of integration, and employed the ratio of equilibrium probabilities in the forward and backward directions of time. The result is that the ratio of the likelihood of observing a positive heat flow, relative to observing a negative heat flow, is exponentially large in the heat times the difference in inverse temperatures of the two initial states. Since heat is an extensive quantity, for a macroscopic system the ratio yields a deterministic direction as the relative likelihoods become extremely small. However, for a microscopic system this result implies that heat can move spontaneously from hot to cold or from cold to hot, just with unequal probabilities. More generally, the ambiguity of the direction of heat flow reflects an ambiguity in the direction of time. As $\mathcal{Q}$ becomes macroscopic, the arrow of time emerges, as heat flow becomes deterministic.

The moment generating function for the heat flow obeys the same sort of integral relation as the Jarzynski equality. Averaging over trajectories

$$\left\langle e^{(\beta_A - \beta_B)\mathcal{Q}} \right\rangle = \int d\mathcal{Q}\, e^{(\beta_A - \beta_B)\mathcal{Q}} P(\mathcal{Q})$$

$$= \int d\mathcal{Q}\, P(-\mathcal{Q}) = 1$$

we find that the moment generating function is equal to 1. As a consequence of Jensen's inequality

$$(\beta_B - \beta_A)\langle \mathcal{Q} \rangle \geq 0$$

implying that if $\beta_B > \beta_A$ or $T_A > T_B$, heat flows from A into B, $Q = \Delta E_B > 0$, just as we found from our fundamental postulate.

Exercise 6.5: Assuming the distribution of heat flows is Gaussian, $P(Q) \propto \exp\left[-(Q - \langle Q \rangle)^2/2 \langle \delta Q^2 \rangle\right]$, relate the mean and variance of the heat exchanged using the fluctuation theorem, $P(Q) = P(-Q) \exp[(\beta_B - \beta_A)Q]$.

6.4   The many body hierarchy

The Liouville equation contains an enormous amount of information. It assigns a likelihood to a point in the $6N$ dimensional phase space of a system, and determines how that likelihood changes in time from some initial distribution. Most macroscopic properties of interest do not require such a detailed description, and as a consequence it is useful to introduce reduced phase space distributions that encode the behavior of a subset of particles of the system. Generally, an *n-body reduced phase space distribution* can be defined as

$$f_n(\mathbf{r}^n, \mathbf{p}^n, t) = \frac{N!}{(N-n)!} \int d\mathbf{r}^{(N-n)} \int d\mathbf{p}^{(N-n)} f(\mathbf{r}^N, \mathbf{p}^N)$$

which is just an integral over $N - n$ phase space variables, with a prefactor that accounts for the permutations of n indistinguishable particles. For $n = 1$ the reduced distribution function is the joint probability of finding a particle at some point in space, with a given momentum, and for $n > 1$ these distribution functions encode correlations between n particles' positions and their corresponding momenta. In equilibrium, each n-body distribution would factorize into separate momentum and position components, though away from equilibrium, these variables may be correlated.

For each n-body distribution function, we can derive an associated equation of motion by integrating out variables from the original Liouville equation. It is particularly constructive to do this in the case where particles interact through a potential U that decomposes as $U = \sum_{i=1}^{N} u_1(\mathbf{r}_i) + \sum_{i<j=1}^{N} u_2(|\mathbf{r}_i - \mathbf{r}_j|)$, where u_1 and u_2 are external and pair potentials. The resulting force on particle i from this potential is $\mathbf{F}_i = \mathbf{F}_i^{(1)} + \sum_{j \neq i} \mathbf{F}_{ij}^{(2)}$, and introducing it into the Liouville equation we find,

$$\frac{\partial f}{\partial t} = -\sum_i^N \left(\mathbf{F}_i^{(1)} \cdot \frac{\partial f}{\partial \mathbf{p}_i} + \frac{\mathbf{p}_i}{m} \cdot \frac{\partial f}{\partial \mathbf{r}_i} \right) - \sum_{i,j \neq i}^N \mathbf{F}_{ij}^{(2)} \cdot \frac{\partial f}{\partial \mathbf{p}_i}$$

where we have dropped the arguments of f. Multiplying the Liouville equation by $N!/(N - n)!$ and integrating over $N - n$ phase space variables, we can construct an evolution equation for f_n,

$$\frac{\partial f_n}{\partial t} = -\sum_i^n \left(\mathbf{F}_i^{(1)} \frac{\partial f_n}{\partial \mathbf{p}_i} + \frac{\mathbf{p}_i}{m} \frac{\partial f_n}{\partial \mathbf{r}_i} \right) - \sum_{i,j}^n \mathbf{F}_{ij}^{(2)} \frac{\partial f_n}{\partial \mathbf{p}_i}$$

$$- \frac{N!}{(N-n)!} \sum_i^n \sum_{j=n+1}^N \int d\mathbf{r}^{(N-n)} \int d\mathbf{p}^{(N-n)} \mathbf{F}_{ij}^{(2)} \frac{\partial f}{\partial \mathbf{p}_i}$$

which can be simplified by noting that particle indistinguishability implies that f_n is invariant to permutation of particle labels. This means that the sum over $N - n$ terms indexed by j can be replaced by any one term times a factor of $N - n$ allowing us to rewrite the last term

$$\frac{\partial f_n}{\partial t} = -\sum_i^n \left(\mathbf{F}_i^{(1)} \frac{\partial f_n}{\partial \mathbf{p}_i} + \frac{\mathbf{p}_i}{m} \frac{\partial f_n}{\partial \mathbf{r}_i} \right) - \sum_{i,j}^n \mathbf{F}_{ij}^{(2)} \frac{\partial f_n}{\partial \mathbf{p}_i}$$

$$- \sum_i^n \int d\mathbf{r}_{n+1} \int d\mathbf{p}_{n+1} \mathbf{F}_{i,n+1}^{(2)} \frac{\partial f_{n+1}}{\partial \mathbf{p}_i}$$

by introducing a higher-bodied distribution function f_{n+1}. This system of coupled equations is known as the BBGKY hierarchy, for Bogolyubov, Born, Green, Kirkwood, and Yvon, and relates an n-bodied reduced phase space distribution to an $(n+1)$-bodied distribution. It is an exact set of equations that while intractable to use without approximation, serves as the starting point for a number of approaches and formal insight.

The conceptually simplest equation in the hierarchy is for $n = 1$,

$$\frac{\partial f_1}{\partial t} = -\mathbf{F}_1^{(1)} \frac{\partial f_1}{\partial \mathbf{p}_1} - \frac{\mathbf{p}_1}{m} \frac{\partial f_1}{\partial \mathbf{r}_1} - \int d\mathbf{r}_2 \int d\mathbf{p}_2 \, \mathbf{F}_{1,2}^{(2)} \frac{\partial f_2}{\partial \mathbf{p}_1}$$

which is expressible in terms of convective terms in the phase space of a single body, and an integral of the two-body distribution function. The last term on the right-hand side can be considered a functional of f_1. Indeed, this expression with the definition

$$\left(\frac{\partial f_1}{\partial t} \right)_c = - \int d\mathbf{r}_2 \int d\mathbf{p}_2 \, \mathbf{F}_{1,2}^{(2)} \frac{\partial f_2}{\partial \mathbf{p}_1}$$

where the subscript c denotes two-body collision, yields the Boltzmann equation

$$\left(\frac{\partial}{\partial t} + \mathbf{F}_1^{(1)} \frac{\partial}{\partial \mathbf{p}_1} + \frac{\mathbf{p}_1}{m} \frac{\partial}{\partial \mathbf{r}_1} \right) f_1 = \left(\frac{\partial f_1}{\partial t} \right)_c$$

used in the earliest development of statistical mechanics. In particular, if we define a one-body entropy as

$$S = -\frac{k_B}{h} \int d\mathbf{r} \int d\mathbf{p} f_1(\mathbf{r}, \mathbf{p}, t) \ln f_1(\mathbf{r}, \mathbf{p}, t)$$

then we can observe that its time derivative

$$\dot{S} = -\frac{k_B}{h} \int d\mathbf{r} \int d\mathbf{p} \left(\frac{\partial f_1}{\partial t} \right)_c \ln f_1(\mathbf{r}, \mathbf{p}, t)$$

is determined directly by the collision term. This relationship follows from substituting the equation of motion for f_1, and integrating the convective terms by parts twice. Thus it is clear that in the absence of collisions, as mediated by two-particle interactions, there is no change in entropy. The system in such a limit is simply a single particle moving in an external potential, which by Hamilton's equations of motion will conserve its energy and be completely reversible. Microscopically, irreversibility results from the loss of inter-particle correlations.

> **Exercise 6.6:** Write out an expression relating the $n = 2$ to $n = 3$ reduced distribution functions within the BBGKY hierarchy in the limit where there are no external forces, $F_i^{(1)} = 0$.

It is difficult to work out a specific form for the collision integral. The original Boltzmann theory assumed a dilute limit, in which particles are initially uncorrelated and become correlated only after an interaction that served to exchange momentum. Using this form, applicable to gases, Boltzmann could demonstrate that the change in entropy was non-negative. A conceptually similar approach applicable to fluids results in the so-called *BGK equation*, for Bhatnagar, Gross, and Krook, that can be proposed by assuming that collisions act to locally equilibrate the fluid with a constant rate of local relaxation. The collision integral is written as

$$\left(\frac{\partial f_1}{\partial t}\right)_c = -\lambda \left[f_1(\mathbf{r}, \mathbf{p}, t) - \bar{f}_1(\mathbf{r}, \mathbf{p}, t) \right]$$

where λ is the rate of local relaxation of the current one-body distribution to a local equilibrium distribution. This closure to the equation for the 1-body reduced distribution function simplifies the solution of the many-body dynamical problem by positing an effective single-particle equation of motion, in a procedure shown schematically in Figure 6.5.

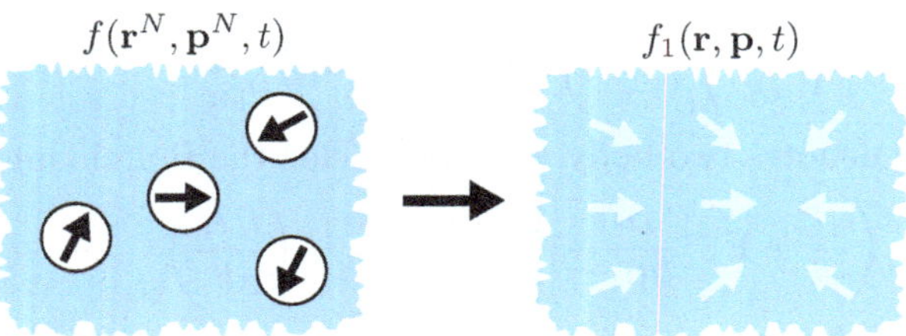

Fig. 6.5 The BGK equation reduces a many particle dynamical problem into an effective equation of motion for the density and momentum fields.

The local equilibrium distribution function, $\bar{f}_1(\mathbf{r}, \mathbf{p}, t)$ is taken to be of the form of a Boltzmann distribution,

$$\bar{f}_1(\mathbf{r}, \mathbf{p}, t) = \rho(\mathbf{r}, t) \left(\frac{1}{2\pi m k_\mathrm{B} T(\mathbf{r}, t)} \right)^{3/2} e^{-[\mathbf{p} - \bar{\mathbf{p}}(\mathbf{r}, t)/\rho(\mathbf{r}, t)]^2 / 2k_\mathrm{B} T(\mathbf{r}, t) m}$$

but with finite, spatially varying momentum, and density. The density is the marginal distribution of the one-body reduced distribution,

$$\rho(\mathbf{r}, t) = \int d\mathbf{p}\, f_1(\mathbf{r}, \mathbf{p}, t)$$

which is normalized to 1 when integrated over the constant spatial volume. The corresponding momentum density is

$$\bar{\mathbf{p}}(\mathbf{r}, t) = \int d\mathbf{p}\, \mathbf{p} f_1(\mathbf{r}, \mathbf{p}, t)$$

which is the first moment of the 1-body distribution. This leaves a local temperature, $T(\mathbf{r}, t)$, or kinetic energy density,

$$\rho(\mathbf{r}, t) T(\mathbf{r}, t) = \rho(\mathbf{r}, t) \frac{1}{3mk_{\mathrm{B}}} \int d\mathbf{p} [\mathbf{p} - \bar{\mathbf{p}}(\mathbf{r}, t)/\rho(\mathbf{r}, t)]^2 f_1(\mathbf{r}, \mathbf{p}, t)$$

as the second centered moment of the momentum. The local equilibrium distribution is defined such that

$$\int d\mathbf{r} \int d\mathbf{p} f_1(\mathbf{r}, \mathbf{p}, t) \ln \bar{f}_1(\mathbf{r}, \mathbf{p}, t) = \int d\mathbf{r} \int d\mathbf{p} \bar{f}_1(\mathbf{r}, \mathbf{p}, t) \ln \bar{f}_1(\mathbf{r}, \mathbf{p}, t)$$

is true. Indeed, any average over $\mathbf{p}$ is equivalent within the full or local equilibrium phase space distribution implying an assumption of facile local momentum relaxation.

Exercise 6.7: Demonstrate that averages of $\mathbf{p}$ and $\mathbf{p}^2$ are equivalent when taken over the full distribution $f_1(\mathbf{r}, \mathbf{p}, t)$ or its local equilibrium limit $\bar{f}_1(\mathbf{r}, \mathbf{p}, t)$.

Within the BGK approximation, the time evolution of the entropy becomes

$$\dot{S} = \frac{\lambda k_{\mathrm{B}}}{h} \int d\mathbf{r} \int d\mathbf{p} \left[f_1(\mathbf{r}, \mathbf{p}, t) - \bar{f}_1(\mathbf{r}, \mathbf{p}, t) \right] \ln f_1(\mathbf{r}, \mathbf{p}, t)$$

$$= \frac{\lambda k_{\mathrm{B}}}{h} \int d\mathbf{r} \int d\mathbf{p} \left[f_1(\mathbf{r}, \mathbf{p}, t) - \bar{f}_1(\mathbf{r}, \mathbf{p}, t) \right] \ln \frac{f_1(\mathbf{r}, \mathbf{p}, t)}{\bar{f}_1(\mathbf{r}, \mathbf{p}, t)} \geq 0$$

where in the second line we have used the equivalence of averages in the local equilibrium distribution. For a model of collisions that acts to locally equilibrate the system, the entropy change is non-negative. It is zero if locally at equilibrium and increases otherwise. The inequality follows from the fact that $(x - y) \ln(x/y) \geq 0$ for any real positive numbers x and y. This is Boltzmann's famous H-theorem that led him to the identification of the microscopic origin of the second law. Though we see that we had to put in by hand a mechanism for relaxing to equilibrium, as a purely Hamiltonian system cannot really produce entropy without explicit loss of information.

6.5   Slow fluctuations and hydrodynamics

The evolution equation for the single particle distribution function serves as a starting point for a macroscopic approach to dynamics for specific variables whose equations of motion can be approximately closed. Take for example the single particle density $\rho(\mathbf{r}, t)$. An equation of motion for $\rho(\mathbf{r}_1, t)$ can be determined by integrating the 1-body equation over $\mathbf{p}_1$,

$$\int d\mathbf{p}_1 \frac{\partial f_1}{\partial t} = -\int d\mathbf{p}_1 \left(\mathbf{F}_1^{(1)} \frac{\partial f_1}{\partial \mathbf{p}_1} + \frac{\mathbf{p}_1}{m} \frac{\partial f_1}{\partial \mathbf{r}_1} + \int d\mathbf{r}_2 \int d\mathbf{p}_2 \, \mathbf{F}_{1,2}^{(2)} \frac{\partial f_2}{\partial \mathbf{p}_1} \right)$$

by noting that the first and third terms on the right-hand side can be integrated by parts. Doing so and assuming that f_1 goes to 0 at $\mathbf{p}_1 \to \pm\infty$ yields

$$\frac{\partial \rho(\mathbf{r}_1, t)}{\partial t} = -\frac{\partial}{\partial \mathbf{r}_1} \int d\mathbf{p}_1 \frac{\mathbf{p}_1}{m} f_1(\mathbf{r}_1, \mathbf{p}_1, t)$$

$$= -\frac{\partial}{\partial \mathbf{r}_1} \bar{\mathbf{p}}(\mathbf{r}_1, t)/m$$

where in the second line we have used our previous definition of the average momentum. This expression demonstrates that the density is conserved, as it takes the form of a continuity equation. It is not closed, as the equation of motion for $\rho(\mathbf{r}, t)$ depends on another field, $\bar{\mathbf{p}}(\mathbf{r}, t)$. However, within traditional hydrodynamic theory, one can postulate a so-called linear law, that relates the two. For small gradients in the density, in an otherwise quiescent fluid, the momentum is assumed to be

$$\bar{\mathbf{p}}(\mathbf{r}, t)/m = -D \frac{\partial \rho(\mathbf{r}, t)}{\partial \mathbf{r}}$$

where D is the diffusion constant. The diffusion constant is a phenomenological parameter defined through this linear expression, known as Fick's first law. Substituting this linearization of the momentum field back into the equation of motion for $\rho(\mathbf{r}, t)$ we find

$$\frac{\partial \rho(\mathbf{r}, t)}{\partial t} = D \frac{\partial^2}{\partial \mathbf{r}^2} \rho(\mathbf{r}, t)$$

which is Fick's second law of diffusion, a closed equation of motion for ρ that depends on a sole material property D and can be solved with appropriate boundary conditions. Fick's laws are only valid when convective terms can be neglected, such as in transport within porous media where the support can quench the momentum, or in multicomponent systems where there is no net velocity, only relative motion.

In Fourier space, the diffusion equation can be written as

$$\frac{\partial \hat{\rho}(\mathbf{k}, t)}{\partial t} = -Dk^2 \hat{\rho}(\mathbf{k}, t)$$

which can be integrated to yield

$$\hat{\rho}(\mathbf{k}, t) = \hat{\rho}(\mathbf{k}, 0) e^{-Dk^2 t}$$

illustrating that each Fourier mode of the density field relaxes independently with a rate that depends on the diffusion constant and magnitude of the wavevector squared.

For small wavevectors, this rate of relaxation becomes very small, and because the density is globally conserved, the rate goes to zero as $k \to 0$. The product structure of the integrated Fourier transformed diffusion equation can be used with the convolution theorem to yield in one dimension

$$\rho(r,t) = \sqrt{\frac{\pi}{Dt}} \int dr' \, \rho(r',0) e^{-(r-r')^2/4Dt}$$

which is shown for a $\rho(r,0) = \bar{\rho}\delta(r)$ in Figure 6.6.

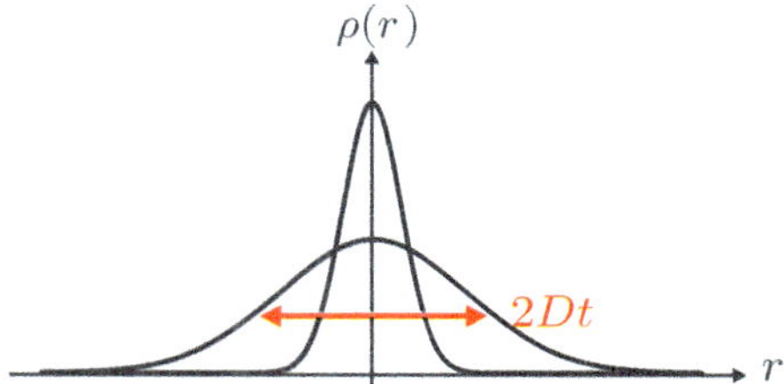

Fig. 6.6 Evolution of the density field from a localized initial condition evolving diffusively.

An analogous approach can be used to express an equation of motion for the momentum field. Multiplying the 1-body equation in the BBGKY hierarchy by $\mathbf{p}_1$ and integrating it over $\mathbf{p}_1$ yields, after some integration by parts,

$$\frac{\partial \bar{\mathbf{p}}}{\partial t} = \mathbf{F}_1^{(1)}(\mathbf{r})\rho(\mathbf{r},t) - \frac{\partial}{\partial \mathbf{r}_1} \int d\mathbf{p}_1 f_1 \frac{\mathbf{p}_1 \otimes \mathbf{p}_1}{m} + \int d\mathbf{r}_2 \int d\mathbf{p}_1 \int d\mathbf{p}_2 \, \mathbf{F}_{1,2}^{(2)} f_2$$

where the first term on the right-hand side follows from the definition of ρ and integration by parts. The second term of the right-hand side is the second moment of the single particle momentum where $\otimes$ denotes a tensor product. We can introduce another field, known as the kinetic stress tensor $\sigma_K(\mathbf{r}_1,t)$,

$$\sigma_K(\mathbf{r}_1,t) = -\int d\mathbf{p}_1 f_1(\mathbf{r}_1,\mathbf{p}_1) \frac{(\mathbf{p}_1 - \bar{\mathbf{p}}/\rho) \otimes (\mathbf{p}_1 - \bar{\mathbf{p}}/\rho)}{m}$$

where we have rewritten it in terms of the mean momentum field $\bar{\mathbf{p}}/\rho$ and its deviations. The last term can be simplified by noting the two integrals over momentum,

$$\rho_2(\mathbf{r}_1,\mathbf{r}_2) = \int d\mathbf{p}_1 \int d\mathbf{p}_2 \, f_2(\mathbf{r}_1,\mathbf{p}_1,\mathbf{r}_2,\mathbf{p}_2)$$

reduce the two-particle reduced density distribution to the two-particle density function. We have seen the second moment of the density field before, and shown in Chapter 3 that in a fluid it decomposes as a self-term and the radial distribution function $\rho_2(\mathbf{r}_1,\mathbf{r}_2) = \rho(\mathbf{r}_1)\delta(\mathbf{r}_1 - \mathbf{r}_2) + \rho(\mathbf{r}_1)\rho(\mathbf{r}_2)g(|\mathbf{r}_1 - \mathbf{r}_2|)$, where here we keep the explicit spatial dependence of the single particle densities for generality. The $g(r)$ is defined as

$g(|\mathbf{r}_1 - \mathbf{r}_2|) = (N - 1) \langle \delta(\mathbf{r}_1 - \mathbf{r}_2) \rangle / \rho(\mathbf{r}_1)$ which in the limit of a slowly varying density field and large N, becomes

$$N \langle \delta(\mathbf{r}_1 - \mathbf{r}_2) \rangle \approx N \langle \delta(\mathbf{r}_2) \rangle + N (\mathbf{r}_1 - \mathbf{r}_2) \frac{\partial}{\partial \mathbf{r}_2} \langle \delta(\mathbf{r}_2) \rangle$$

$$= \rho(\mathbf{r}_2) + (\mathbf{r}_1 - \mathbf{r}_2) \frac{\partial}{\partial \mathbf{r}_2} \rho(\mathbf{r}_2)$$

where we have expanded the mean pair density to first-order in the displacement between the two particles. This locality approximation was introduced by Irving and Kirkwood and allows us to define another field, the virial stress tensor, $\sigma_\mathrm{V}(\mathbf{r}_1)$ as

$$\sigma_\mathrm{V}(\mathbf{r}_1, t) = -\frac{1}{2} \int d\mathbf{r}_2 \, \mathbf{F}_{1,2}^{(2)}(|\mathbf{r}_1 - \mathbf{r}_2|) \rho(\mathbf{r}_1) \rho(\mathbf{r}_2) (\mathbf{r}_1 - \mathbf{r}_2)$$

Combining the two stress tensors $\sigma = \sigma_\mathrm{K} + \sigma_\mathrm{V}$, and inserting them back into the equation of motion for the momentum

$$\frac{\partial \bar{\mathbf{p}}(\mathbf{r}, t)}{\partial t} + \frac{\partial}{\partial \mathbf{r}} \frac{\bar{\mathbf{p}}(\mathbf{r}, t) \otimes \bar{\mathbf{p}}(\mathbf{r}, t)}{m \rho(\mathbf{r}, t)} = \mathbf{F}^{(1)}(\mathbf{r}) \rho(\mathbf{r}, t) + \frac{\partial}{\partial \mathbf{r}} \sigma(\mathbf{r}, t)$$

we find an evolution equation for the momentum field that depends on another time and spatially dependent field, $\sigma(\mathbf{r}, t)$. Note the second term on the left comes from the definition of the kinetic stress tensor. Postulating another linear law, valid in the limit that the average stress varies slowly in space, we can relate $\sigma(\mathbf{r}, t)$ to $\bar{\mathbf{p}}(\mathbf{r}, t)$ through Newton's law of viscosity

$$\sigma(\mathbf{r}, t) = -1 p(\mathbf{r}, t) + \mu \frac{\partial}{\partial \mathbf{r}} \bar{\mathbf{p}}(\mathbf{r}, t)$$

where $p(\mathbf{r}, t)$ is the pressure, the diagonal component of the stress tensor, and μ is the viscosity coefficient of the fluid. Inserting Newton's law of viscosity into the equation for the momentum

$$\frac{\partial \bar{\mathbf{p}}(\mathbf{r}, t)}{\partial t} + \frac{\partial}{\partial \mathbf{r}} \frac{\bar{\mathbf{p}}(\mathbf{r}, t) \otimes \bar{\mathbf{p}}(\mathbf{r}, t)}{m \rho(\mathbf{r}, t)} = -\frac{\partial}{\partial \mathbf{r}} p(\mathbf{r}, t) + \mu \frac{\partial^2}{\partial \mathbf{r}^2} \bar{\mathbf{p}}(\mathbf{r}, t) + \rho(\mathbf{r}, t) \mathbf{F}^{(1)}(\mathbf{r})$$

we arrive at the Navier–Stokes equation, which is a momentum balance equation for a fluid driven by a body force $\mathbf{F}^{(1)}(\mathbf{r})$ and with viscous response.

> **Exercise 6.8:** A common assumption in hydrodynamics is incompressibility, where $\rho(\mathbf{r}, t)$ is assumed constant. A consequence of incompressibility is
>
> $$\frac{\partial \rho(\mathbf{r}, t)}{\partial t} = -\frac{\partial}{\partial \mathbf{r}} \bar{\mathbf{p}}(\mathbf{r}, t) = 0$$
>
> Simplify the Navier–Stokes equation in this limit for the x component of the momentum, $\bar{p}_x(\mathbf{r}, t)$.

The hydrodynamic equations derived earlier, together with the linear laws establish a closed set of dynamical expressions for the mass and momentum at single points in space. Analogous equations can also be derived for other conserved quantities like energy and angular momentum, and supplemented with additional linear laws. The resulting set of equations are of considerable utility when considering the evolution of macroscopic systems where gradients are relatively small, and the linear laws are valid. At this level of description, the linear laws are fully phenomenological. We will see, that they emerge naturally from a dynamical response theory. Understanding these laws from microscopic motion and their domain of validity is explored in Chapter 7.

6.6 Time correlation functions

With an understanding of the time dependence of points in phase space, we are in a position to work out some time-dependent properties in thermal equilibrium. Let's go back to the thermal average we considered before, namely the time dependence of the position of a tagged particle $\mathbf{r}_i$. Writing the expectation value over an arbitrary distribution,

$$\langle \mathbf{r}(t) \rangle = \int d\mathbf{x}\, f(\mathbf{x}, t)\mathbf{r}(0)$$

$$= \int d\mathbf{x}\, e^{-\mathcal{L}t} f(\mathbf{x}, 0)\mathbf{r}(0)$$

$$= \int d\mathbf{x}\, f(\mathbf{x}, 0)e^{\mathcal{L}t}\mathbf{r}(0) = \int d\mathbf{x}\, f(\mathbf{x}, 0)\mathbf{r}(t)$$

where in the second line we have inserted the Liouvillian. Because the Liouvillian is anti-Hermitian, we can switch its order of evaluation from left to right at the cost of an overall negative sign in the argument of exponential— effectively reversing the sign of the time. This follows from integration by parts. Using this property and the stationarity of the Boltzmann distribution,

$$\langle \mathbf{r}(t) \rangle = \int d\mathbf{x}\, \mathbf{r}(0)e^{-\mathcal{L}t} f_{\text{eq}}(\mathbf{x})$$

$$= \int d\mathbf{x}\, \mathbf{r}(0) f_{\text{eq}}(\mathbf{x}) = \langle \mathbf{r}(0) \rangle$$

we find that the expectation value is time independent, a consequence of the stationarity of $f_{\text{eq}}(\mathbf{x})$.

This time translational invariance is general for equilibrium averages of single time quantities. From our discussion of the fluctuation theorems, we can understand this as time having an *extent* but not a *direction* within an equilibrium state. Therefore, nontrivial dynamical information within equilibrium comes from time correlation functions, or thermal averages of operators at two different points in time, an example of which is shown in Figure 6.7. Time correlation functions satisfy a number of simple properties. For example,

$$\langle A(0)B(t) \rangle = \langle A(0)e^{\mathcal{L}t}B(0) \rangle = \langle B(0)e^{-\mathcal{L}t}A(0) \rangle = \langle A(-t)B(0) \rangle$$

which is another statement of stationarity. In the limit that $t \to 0$,

$$\langle A(0)B(t)\rangle \to \langle AB\rangle$$

correlations functions are just static averages. In the opposite limit, $t \to \infty$,

$$\langle A(0)B(t)\rangle \to \langle A\rangle\langle B\rangle$$

provided A and B eventually decorrelate. This statement cements the relationship between relaxation and the loss of correlation.

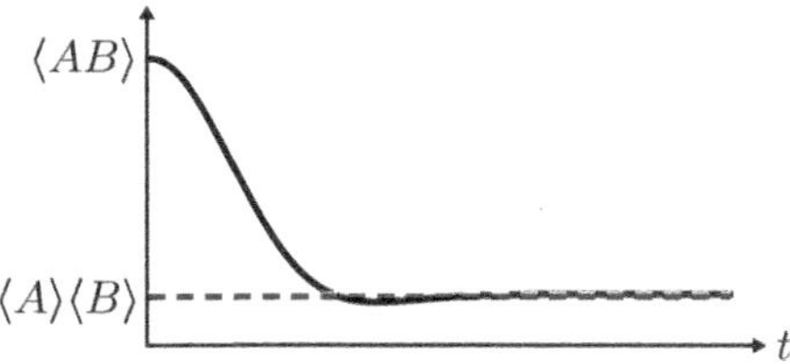

Fig. 6.7 Characteristic behavior of a time correlation function.

For a system with time translational invariance, like an equilibrium system,

$$\langle A(t')B(t'')\rangle = \langle A(0)B(t'' - t')\rangle$$

where the equality comes from stationarity. Additionally, time correlation functions have a number of symmetry properties, such as

$$\frac{d}{d\tau}\langle A(\tau)B(\tau + t)\rangle = \left\langle \dot{A}(\tau)B(\tau + t)\right\rangle + \left\langle A(\tau)\dot{B}(\tau + t)\right\rangle = 0$$

where due to stationarity the total time derivative must be zero. Evaluating this expression for $\tau = 0$, we find,

$$\left\langle \dot{A}(0)B(t)\right\rangle = -\left\langle A(0)\dot{B}(t)\right\rangle .$$

Evaluating it at $t = 0$, for $B = A$, $\left\langle A\dot{A}\right\rangle = -\left\langle \dot{A}A\right\rangle$, which can only be true if both are 0. This means that generically, classical time autocorrelation functions are even functions of time, and therefore have zero slope about the origin. More generically, time correlation functions could have either even or odd symmetry. For example, for observables that are arbitrary functions of phase space,

$$\langle A(0)B(t)\rangle = \langle A^*(0)B^*(-t)\rangle$$

where $A^*[\mathbf{x}] = A[\mathbf{x}^*]$ is the time reversed operator. If A and B are both solely functions of positions, then

$$\langle A(0)B(t)\rangle = \langle A(-t)B(0)\rangle = \langle A(0)B(-t)\rangle$$

is an even function of time. This is also true if A and B are both solely functions of momentum. However

$$\langle r(0)v(t)\rangle = -\langle r(0)v(-t)\rangle$$

and other mixed functions of position and momentum, are odd functions in time.

6.7   Green–Kubo relations

As an example of the information contained in time correlation functions, we can consider the motion of a tagged particle in a dense solution. The equation of motion for its position is

$$\frac{d\mathbf{r}}{dt} = \mathbf{v}$$

which can be integrated,

$$\mathbf{r}(t) - \mathbf{r}(0) = \int_0^t dt' \, \mathbf{v}(t') \,.$$

The equilibrium expectation of the particle's motion, in terms of its position at time t relative to its position at time 0, is

$$\langle \mathbf{r}(t) - \mathbf{r}(0) \rangle = \langle \mathbf{r}(t) \rangle - \langle \mathbf{r}(0) \rangle = 0$$

$$= \int_0^t dt' \, \langle \mathbf{v}(t') \rangle = 0$$

as can be seen from either the difference in the expectation values of the position, which upon thermal averaging become time independent, or the expectation value of the velocity, which is similarly zero.

Let $\Delta \mathbf{r}(t) = \mathbf{r}(t) - \mathbf{r}(0)$. The square of the displacement can be written as a sum $t/\Delta t$ of discrete displacements $\Delta \mathbf{r}_j$ over a time Δt, as

$$\Delta \mathbf{r}(t) = \sum_j^{t/\Delta t} \Delta \mathbf{r}_j$$

$$\Delta \mathbf{r}^2(t) = \sum_{i,j}^{t/\Delta t} \Delta \mathbf{r}_i \Delta \mathbf{r}_j$$

$$= \sum_i^{t/\Delta t} \Delta \mathbf{r}_i^2 + \sum_{i \neq j}^{t/\Delta t} \Delta \mathbf{r}_i \Delta \mathbf{r}_j$$

where we have pulled over the terms where $i = j$. For large Δt, provided an isotropic system, we expect that the cross terms are uncorrelated and average to zero,

$$\langle \Delta \mathbf{r}^2(t) \rangle = \langle \Delta \mathbf{r}_i^2 \rangle \, t/\Delta t$$

and the mean squared displacement should grow linearly with time. The alternative limit, at short times,

$$\mathbf{r}(t) - \mathbf{r}(0) = \int_0^t dt' \, \mathbf{v}(t')$$

$$\approx \mathbf{v}(t)t$$

where we assume that over a short time the velocity does not change. Squaring and averaging it,

$$\langle \Delta \mathbf{r}^2(t) \rangle = \langle \mathbf{v}^2 \rangle t^2$$
$$= d k_{\mathrm{B}} T t^2 / m$$

we find that the mean squared displacement grows quadratically with time. Figure 6.8 provides a characteristic example of the mean squared displacement. Traditionally, the long time limit

$$\lim_{t \to \infty} \frac{\langle \Delta \mathbf{r}^2(t) \rangle}{t} = 2Dd$$

defines a diffusion constant, D, while the short time is inertial.

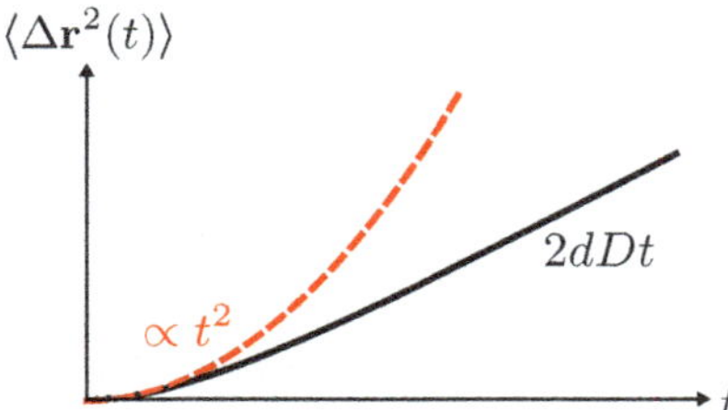

Fig. 6.8 Mean squared displacement as a function of time.

The fact that the slope is taken as the diffusion constant can be understood by remembering Fick's law,

$$\frac{\partial \rho(r,t)}{\partial t} = D \frac{\partial^2 \rho(r,t)}{\partial r^2}$$

where $\rho(r,t)$ is the density field, which from our earlier discussion is identical to the probability of finding a particle at r. Defining an expectation value of $\Delta r^2(t)$

$$\langle \Delta r^2 \rangle = \int dr \, \Delta r^2 \rho(r,t)$$

and taking its time derivative,

$$\frac{\partial}{\partial t} \langle \Delta r^2 \rangle = \int dr \, \Delta r^2 \frac{\partial \rho(r,t)}{\partial t}$$
$$= D \int dr \, \Delta r^2 \frac{\partial^2 \rho(r,t)}{\partial r^2}$$
$$= 2D \int dr \, \rho(r,t) = 2D$$

where the third line follows from integration by parts assuming $\rho(r)$ goes to zero at $\pm\infty$. The fourth line follows from normalization.

An alternative view of the mean squared displacement comes from considering the velocity side of the equation. Specifically, integrating the equation of motion and squaring it,

$$\langle \Delta \mathbf{r}^2(t) \rangle = \int_0^t dt' \int_0^t dt'' \, \langle \mathbf{v}(t')\mathbf{v}(t'') \rangle$$

$$= \int_0^t dt' \int_0^t dt'' \, \langle \mathbf{v}(0)\mathbf{v}(t'' - t') \rangle$$

$$= 2 \int_0^t d\tau \int_0^{t-\tau} d\bar{t} \, \langle \mathbf{v}(0)\mathbf{v}(\tau) \rangle$$

$$= 2 \int_0^t d\tau \, (t - \tau) \, \langle \mathbf{v}(0)\mathbf{v}(\tau) \rangle$$

where we have defined $\tau = t'' - t'$ and $\bar{t} = t' + t''$ to perform a change of variables and used the stationarity of the correlation function. In the long time limit, assuming the correlation function decays faster than $1/t$,

$$\lim_{t \to \infty} \frac{\langle \Delta \mathbf{r}^2(t) \rangle}{2t} = \int_0^\infty d\tau \, \langle \mathbf{v}(0)\mathbf{v}(\tau) \rangle = Dd$$

we find a pretty remarkable result. The integral of the velocity autocorrelation function yields the diffusion constant. This is an example of a Green–Kubo relation. It is one of many examples that relate transport coefficients to spontaneous fluctuations around an equilibrium distribution.

Exercise 6.9: Confirm the equality above for time translational invariant time correlation functions

$$\int_0^t dt' \int_0^t dt'' \, \langle \mathbf{v}(t')\mathbf{v}(t'') \rangle = 2 \int_0^t d\tau \, (t - \tau) \, \langle \mathbf{v}(0)\mathbf{v}(\tau) \rangle$$

using the substitution $\tau = t'' - t'$ and $\bar{t} = t' + t''$ and evaluating the associated Jacobian.

Another example of a Green–Kubo result employs the other continuity equation we have derived, the Navier–Stokes equation. Let's take the x component of the momentum flux, for which the momentum balance in the absence of a body force or mean flow is

$$\frac{\partial \bar{p}_x(\mathbf{r}, t)}{\partial t} = \mu \frac{\partial^2}{\partial \mathbf{r}^2} \bar{p}_x(\mathbf{r}, t)$$

in the limit that the fluid is incompressible, $\partial_x \bar{p}_x(\mathbf{r}, t) = 0$. Fourier transforming the balance equation yields

$$\frac{\partial \hat{\bar{p}}_x(\mathbf{k}, t)}{\partial t} = -\mu k^2 \hat{\bar{p}}_x(\mathbf{k}, t)$$

where we denote the Fourier transform of the momentum $\hat{\bar{p}}_x(\mathbf{k}, t)$.

Aside: Laplace transforms. The Laplace transform of a function, $g(t)$ is defined as

$$L[g(t)] \equiv \int_0^\infty dt\, g(t)e^{-st} = \tilde{g}(s)$$

which is an exponential transform, on half of the real axis. It is a generalization of the Fourier transform. Its inverse is given by

$$L^{-1}[\tilde{g}(s)] \equiv \frac{1}{2\pi i} \int_L ds\, \tilde{g}(s)e^{st}$$

where the domain of integration L is along a vertical line in the complex plane such that the real part is greater than the real part of all singularities of $\tilde{g}(s)$. The forward transforms for a couple of typical functions are listed below:

$$g(t) = 1 \quad , \quad \tilde{g}(s) = \frac{1}{s}$$

$$g(t) = e^{-\gamma t} \quad , \quad \tilde{g}(s) = \frac{1}{s+\gamma}$$

$$g(t) = \sin(\omega t) = \frac{1}{2\pi i}\left(e^{i\omega t} - e^{-i\omega t}\right) \quad , \quad \tilde{g}(s) = \frac{\omega}{s^2 + \omega^2}$$

$$g(t) = \cos(\omega t) \quad , \quad \tilde{g}(s) = \frac{s}{s^2 + \omega^2}$$

Note that from integration by parts $L[\ddot{g}(t)] = s^2 L[g(t)] - sg(0) - \dot{g}(0)$, which shows that the Laplace transform, like the Fourier transform, reduces differential equations to algebraic equations. Finally, the convolution theorem,

$$g(t) = \int dt'\, h(t - t')c(t') \quad , \quad \tilde{g}(s) = \tilde{h}(s)\tilde{c}(s)$$

holds exactly as it does for Fourier transforms.

The momentum equation in Fourier space can be used to write an equation for the momentum–momentum correlation function, $C_k(t) = \langle \bar{p}_x(-\mathbf{k}, 0)\bar{p}_x(\mathbf{k}, t)\rangle$, by multiplying both sides by $\hat{\bar{p}}(-\mathbf{k}, 0)$ and taking an average within the steady state

$$\frac{\partial C_k(t)}{\partial t} = -\mu k^2 C_k(t)$$

which could be formally integrated. Laplace transforming both sides,

$$s\tilde{C}_k(s) - C_k(0) = -\mu k^2 \tilde{C}_k(s)$$

$$\implies \tilde{C}_k(s) = \frac{C_k(0)}{s + \mu k^2}$$

where s is the Laplace variable and $C_k(0)$ is the initial value of the correlation function. This correlation function is a Lorenztian with a single purely imaginary pole. We consider fluctuations in the limit that $k \to 0$, for which we have

$$\tilde{C}_k(s) = \frac{C_k(0)}{s + \mu k^2} \approx \frac{C_k(0)}{s}\left(1 - \frac{\mu k^2}{s}\right).$$

Additionally, if we then take a limit where $s \to 0$, equivalent to a long time limit, we can relate the viscosity to the momentum correlation function,

$$\begin{aligned}
\mu &= -\lim_{s \to 0}\lim_{k \to 0}\frac{s^2}{k^2}\frac{\tilde{C}_k(s)}{C_k(0)} \\
&= -\lim_{s \to 0}\lim_{k \to 0}\frac{s^2}{k^2}\frac{1}{C_k(0)}\int_0^\infty dt\, e^{-st}\,\langle \bar{p}_x(-\mathbf{k},0)\bar{p}_x(\mathbf{k},t)\rangle \\
&= -\lim_{s \to 0}\lim_{k \to 0}\frac{1}{k^2}\frac{1}{C_k(0)}\int_0^\infty dt\, e^{-st}\,\langle \dot{\bar{p}}_x(-\mathbf{k},0)\dot{\bar{p}}_x(\mathbf{k},t)\rangle
\end{aligned}$$

where in the second line we have inserted the definition of the Laplace transform, and in the last time we have integrated the argument of the integral by parts to relate the viscosity to the momentum-flux autocorrelation function. Within a thermal distribution, $C_0(0) = k_\mathrm{B}TNm = k_\mathrm{B}TV\rho_m$, which follows from equipartition where ρ_m is the mass density. Relating the momentum flux to the shear stress $\dot{\bar{p}}_x(\mathbf{k},t) = ik\sigma_{x,y}(\mathbf{k},t)$ using the momentum balance, we can rewrite the correlation function

$$\mu = \frac{\beta}{V\rho_m}\int_0^\infty dt\,\langle \sigma_{x,y}(0)\sigma_{x,y}(t)\rangle$$

which is the standard Green–Kubo expression for the shear viscosity. Note that like the expression for the diffusion constant, this follows directly from the form of the balance equation and the linear constituent relation. The only equilibrium assumption is the identification of $C_0(0)$ and the lack of a body force or net drift.

6.8   von Neumann equation

We have previously discussed the equation of motion for the probability distribution on a classical phase space, the so-called Liouville equation. A reasonable question to ask is how these results translate into the quantum realm. To start, we need to first define an operator that encodes the likelihood that a quantum system is found in a particular state.

To do so, we need to remind ourselves that observables in quantum mechanics are operators, so while classically the Hamiltonian of the system, $\mathcal{H}$, is just a number, quantum mechanically it becomes an operator, $\hat{\mathcal{H}}$. The action of this operator is to evolve the system in time,

$$i\hbar\frac{d\,|\psi\rangle}{dt} = \hat{\mathcal{H}}\,|\psi\rangle$$

where $|\psi\rangle$ is the wavefunction, describing the state of the system, and $\hbar$ is Plank's reduced constant. Apart from $i\hbar$, this continuity equation looks just like the Liouville equation we have been using to understand motion in phase space, and it is of course known as the time-dependent Schrodinger equation. Indeed, this equation can be formally integrated,

$$|\psi(t)\rangle = e^{-i\hat{\mathcal{H}}t/\hbar}\,|\psi(0)\rangle\ .$$

An important distinction between this propagation rule and the Liouville equation, is that $|\psi\rangle$ is not a probability density, but rather an amplitude. We will see shortly that the consequence of this is that expectation values have a slightly different form. A corollary to this is that the stationary $|\psi\rangle$ is not a Boltzmann form, but rather satisfies the eigenvalue equation,

$$\hat{\mathcal{H}}\,|\psi_k\rangle = E_k\,|\psi_k\rangle$$

where $|\psi_k\rangle$ are the eigenvectors of $\hat{\mathcal{H}}$ and E_k are the associated eigenvalues, corresponding to the energies of the eigenstates.

Consider the *density operator* or matrix, $\hat{\rho}(t)$, defined as an outer production of the wavefunction at time t,

$$\hat{\rho}(t) = |\psi(t)\rangle\langle\psi(t)| = \sum_{nm} C_n^*(t)C_m(t)\,|\psi_n\rangle\langle\psi_m|$$

where in the second equality we have expanded the wavefunction $|\psi(t)\rangle$ into a complete set of orthonormal states with coefficients $C_n(t)$, such that $|\psi(t)\rangle = \sum_n C_n(t)\,|\psi_n\rangle$. The elements of the density operator are thus

$$\rho_{nm}(t) = \langle\psi_n|\,\hat{\rho}(t)\,|\psi_m\rangle = C_n^*(t)C_m(t)$$

and the normalization of the wavefunction implies $\sum_n|C_n(t)|^2 = 1$. The function of the density operator can be clarified by considering a quantum expectation value of an arbitrary operator $\hat{A}$,

$$\langle A(t)\rangle = \langle\psi(t)|\,\hat{A}\,|\psi(t)\rangle = \sum_{nm} C_n^*(t)C_m(t)\,\langle\psi_m|\,\hat{A}\,|\psi_n\rangle = \sum_{nm}\rho_{nm}(t)A_{mn}$$

which is equivalent to the trace definition of the expectation we have invoked previously

$$\langle A(t)\rangle = \mathrm{Tr}[\hat{\rho}(t)\hat{A}]$$

as follows from the completeness of the basis $\sum_n |\psi_n\rangle\langle\psi_n| = 1$. Thus, the elements of the density operator ρ_{nm} encode the weights associated with a quantum expectation value. In this way $\hat{\rho}$ is the quantum analogue to the phase space distribution $f(\mathbf{x})$ and the integral over phase space is replaced by a trace operation.

The time evolution of the density operator can be found straightforwardly. Like any other operator, it satisfies a Heisenberg equation of motion, derived simply for a system described by the Hamiltonian $\hat{\mathcal{H}}$

$$\frac{d}{dt}\hat{\rho} = \frac{d}{dt}|\psi(t)\rangle\langle\psi(t)| = \left(\frac{d}{dt}|\psi(t)\rangle\right)\langle\psi(t)| + |\psi(t)\rangle\left(\frac{d}{dt}\langle\psi(t)|\right)$$

$$= -\frac{i}{\hbar}\hat{\mathcal{H}}\,|\psi(t)\rangle\langle\psi(t)| + \frac{i}{\hbar}\,|\psi(t)\rangle\langle\psi(t)|\,\hat{\mathcal{H}}$$

where we have introduced the time-dependent Schrodinger equation for the time evolution of the wavefunction. More compactly, we can define a linear operator that propagates the density operator

$$\frac{d}{dt}\hat{\rho} = -\frac{i}{\hbar}\left[\hat{\mathcal{H}}, \hat{\rho}\right] \equiv -i\hat{\mathcal{L}}\hat{\rho}$$

where $\hat{\mathcal{L}}$ is known as the quantum Liouville operator, and the above linear evolution equation of the density operator is known as the von Neumann, or sometime Liouville-von Neumann, equation. As a linear first-order differential equation it can be integrated straightforwardly,

$$\hat{\rho}(t) = e^{-i\mathcal{L}t}\hat{\rho}(0) = e^{-i\hat{\mathcal{H}}t/\hbar}\hat{\rho}(0)e^{i\hat{\mathcal{H}}t/\hbar}$$

provided an initial condition $\hat{\rho}(0)$. Like its classical analogue the Boltzmann density matrix,

$$\hat{\rho}_{\text{eq}} = \frac{e^{-\beta\hat{\mathcal{H}}}}{Q}$$

is stationary under the action of the quantum Liouville operator since $\hat{\mathcal{H}}$ commutes with itself.

6.9 Quantum correlation functions

Suppose we want a thermally averaged quantum operator, $\hat{A}$. Its thermal, time dependent average is defined as

$$\langle A(t)\rangle = \sum_k p_k \langle\psi_k(t)|\,\hat{A}\,|\psi_k(t)\rangle$$

where p_k are the probabilities associated with the densities $\langle\psi_k(t)|\psi_k(t)\rangle$, which are Boltzmann distributed in equilibrium,

$$p_k = \frac{e^{-\beta E_k}}{\sum_k e^{-\beta E_k}}.$$

Just as we did with the Liouvillian, we can pull out the time-dependence from the state of the system,

$$\begin{aligned}
\langle A(t)\rangle &= \sum_k p_k \langle\psi_k|\,e^{i\hat{\mathcal{H}}t/\hbar}\hat{A}e^{-i\hat{\mathcal{H}}t/\hbar}\,|\psi_k\rangle \\
&= \sum_k \langle\psi_k|\,e^{-\beta\hat{\mathcal{H}}}e^{i\hat{\mathcal{H}}t/\hbar}\hat{A}e^{-i\hat{\mathcal{H}}t/\hbar}\,|\psi_k\rangle \Big/ \sum_k \langle\psi_k|\,e^{-\beta\mathcal{H}}\,|\psi_k\rangle \\
&= \text{Tr}\left[e^{-\beta\hat{\mathcal{H}}}\hat{A}(t)\right]\Big/\text{Tr}\left[e^{-\beta\hat{\mathcal{H}}}\right]
\end{aligned}$$

where in the second line we have used the fact that $\langle\psi_k|$ is an eigenvector of $\hat{\mathcal{H}}$, with eigenvalue E_k. We have also identified the so-called Heisenberg operator

$$\hat{A}(t) = e^{i\hat{\mathcal{H}}t/\hbar}\hat{A}e^{-i\hat{\mathcal{H}}t/\hbar}$$

which is time-dependent, and we have identified the trace of the states of the system in the last line. Note the time evolution of observables is the opposite of the density matrix, like we found for expectation values classically. Put in this form, the expectation value looks very similar to what we had before, with the trace taking the place of the integral over phase space.

Using these different operators, there is also a simple way to express correlation functions. Consider the time correlation function between operators $\hat{A}(t)$ and $\hat{B}(t')$. Like our previous quantum expectation value, we can write this as a sum over energy eigenstates

$$\left\langle \hat{A}(t)\hat{B}(t') \right\rangle = \sum_k p_k \left\langle \psi_k \right| \hat{A}(t)\hat{B}(t') \left| \psi_k \right\rangle$$

$$= \sum_{k,m} p_k \left\langle \psi_k \right| \hat{A}(t) \left| \psi_m \right\rangle \left\langle \psi_m \right| \hat{B}(t') \left| \psi_k \right\rangle$$

$$= \sum_{k,m} p_k A_{km} B_{mk} e^{i(E_m - E_k)(t-t')/\hbar}$$

where in the second line we introduce the resolution of the identity as a sum over eigenstates and in the third we evaluate the time dependence of the Heisenberg operators by the action of the Hamiltonian on its eigenstates. There are a few things to note about quantum correlation functions. First, quantum correlation functions are complex. This means they are not directly observable. However, we will see that their real and imaginary parts are related to experimental observables. Second, they oscillate at characteristic frequencies $\omega_{m,k} = (E_m - E_k)/\hbar$ related to energy gaps in the spectrum of the Hamiltonian. Finally, the ordering of the operators in the expectation value matters, since in general they will not commute. However, changing their order can be done with the equation above, resulting in

$$\left\langle \hat{A}(0)\hat{B}(t) \right\rangle^* = \left\langle \hat{B}(0)\hat{A}(-t) \right\rangle = \left\langle \hat{B}(t)\hat{A}(0) \right\rangle$$

where the star denotes complex conjugation, and the last line follows from the stationarity of the correlation function.

We can also consider the frequency components of a quantum correlation function by Fourier transforming them, as is natural considering they generally have oscillatory components. Consider the autocorrelation function

$$C_{AA}(t) = \left\langle \hat{A}(0)\hat{A}(t) \right\rangle$$

its Fourier transform, $\hat{C}_{AA}(\omega)$ is given by

$$\hat{C}_{AA}(\omega) = \int dt\, e^{i\omega t} C_{AA}(t)$$

$$= \sum_{k,m} p_k |A_{km}|^2 \int dt\, e^{i\omega t} e^{-i(E_m - E_k)t/\hbar}$$

$$= 2\pi \sum_{k,m} p_k |A_{km}|^2 \delta[\omega - (E_m - E_k)/\hbar]$$

where we have employed the decomposition of the evolution of the operators by resolving the energy eigenstates of the system. The correlation function is particularly simple in frequency space, given simply by a Boltzmann weighted sum of poles evaluated at the characteristic energy gaps of the system. If we evaluate the correlation function at one of those characteristic frequencies and its negative, assuming that the characteristic frequencies are non-degenerate, we find their ratio is

$$\frac{\hat{C}_{AA}(\omega_{k,m})}{\hat{C}_{AA}(\omega_{m,k})} = \frac{p_m}{p_k} = e^{-\beta(E_m - E_k)}$$

which is a quantum reflection of microscopic reversibility, since these correlation functions are proportional to rates between eigenstates.

Exercise 6.10: Given the limiting values of a time correlation function at short times $\langle \hat{A}\hat{A} \rangle$ and long times $\langle \hat{A} \rangle \langle \hat{A} \rangle$ are equilibrium expectation values, what does this imply about the initial and final imaginary components of a quantum time autocorrelation function?

6.10 Quantum work fluctuations

The Jarzynski equality we derived from classical equations of motion required only that the dynamics were time-reversal symmetric and that the density of phase space was conserved under the evolution of a Hamiltonian. These latter properties are also shared by the quantum mechanical evolution of a closed system, and as such it is reasonable to expect that the Jarzynski equality also holds for quantum systems. Indeed, we will show shortly that it does, but with a caveat— to know the work done on a closed system requires knowing exactly the energy change of that system. To know precisely the value of any observable quantum mechanically requires a measurement of that observable. Thus, the quantum analogue of the Jarzynski equality, and Crooks' work fluctuation theorem, entail making measurements at the start and end of the process.

Let us consider a set up analogous to the classical one considered previously. Namely we envision initially preparing a quantum mechanical system in an equilibrium state at a fixed temperature. At $t = 0$ we will begin changing a parameter in the Hamiltonian, $\Lambda(t)$, from its initial value $\Lambda(0)$ to some final value $\Lambda(\tau)$ at $t = \tau$. The evolution of the closed quantum system is given by the operator $\hat{U}(t)$, deduced from integrating the time-dependent Schrodinger equation

$$\hat{U}(t) = \mathcal{T}_+ e^{-\frac{i}{\hbar} \int_0^t dt' \mathcal{H}[\Lambda(t')]}$$

where $\mathcal{T}_+$ is the time ordering operator, necessary because a time-dependent Hamiltonian does not necessarily commute with itself at different times. The time ordering operator is short for

$$\hat{U}(t) = \sum_{n=0}^{\infty} \frac{(-i/\hbar)^n}{n!} \int_0^t dt_1 \int_0^{t_1} dt_2 \dots \int_0^{t_{n-1}} dt_n \, \mathcal{H}[\Lambda(t_1)] \mathcal{H}[\Lambda(t_2)] \dots \mathcal{H}[\Lambda(t_n)]$$

an infinite series. We can denote the initial thermal density matrix as

$$\hat{\rho}_{\Lambda(0)} = \frac{e^{-\beta \mathcal{H}[\Lambda(0)]}}{Q_{\Lambda(0)}}$$

with associated partition function $Q_{\Lambda(0)}$. Let us define a projection operator

$$\hat{P}_\alpha(t) = |\psi_\alpha(t)\rangle \langle \psi_\alpha(t)|$$

that acts as a means of collapsing the system into instantaneous energy eigenstate α with probability p_α. Because the system is prepared initially at equilibrium, the initial probability of starting in state α

$$p_\alpha(0) = \frac{e^{-\beta E_\alpha[\Lambda(0)]}}{Q_{\Lambda(0)}}$$

is just the Boltzmann distribution. The conditional probability of initially starting in state α and at time τ ending in instantaneous energy eigenstate γ is

$$P(\alpha \to \gamma) = \mathrm{Tr}\left[\hat{U}^\dagger(\tau) \hat{P}_\gamma \hat{U}(\tau) \hat{P}_\alpha \right]$$

just the trace over the evolved projection operator for γ. The corresponding change in energy can be identified as work, $\mathcal{W}_{\gamma,\alpha} = E_\gamma(\tau) - E_\alpha(0)$ as the system is closed. Because there are many potential pairs of energy α and γ that yield the same value of $\mathcal{W}$, there is a distribution of work values computed from

$$P(\mathcal{W}) = \sum_{\alpha,\gamma} \delta(\mathcal{W} - \mathcal{W}_{\gamma,\alpha}) P(\alpha \to \gamma) p_\alpha$$

which is the probability of starting in an initial state α times the probability of transitioning from α to a particular final state γ, summed over all possible initial and final states. The underlying microscopic reversibility of the dynamics allows us to relate the work done in this forward process to the work done in the corresponding reverse process, the Crooks fluctuation theorem,

$$P(\mathcal{W}) = \sum_{\alpha,\gamma} \delta(\mathcal{W} - \mathcal{W}_{\gamma,\alpha}) P(\alpha \to \gamma) \frac{e^{-\beta E_\alpha[\Lambda(0)]}}{Q_{\Lambda(0)}}$$

$$= \sum_{\alpha,\gamma} \delta(\mathcal{W} - \mathcal{W}_{\gamma,\alpha}) P(\alpha \to \gamma) \frac{e^{-\beta E_\gamma[\Lambda(\tau)]}}{Q_{\Lambda(\tau)}} e^{\beta \mathcal{W}_{\gamma,\alpha} - \beta \Delta A}$$

$$= \sum_{\alpha,\gamma} \delta(\mathcal{W} - \mathcal{W}_{\gamma,\alpha}) P^*(\gamma \to \alpha) \frac{e^{-\beta E_\gamma[\Lambda(\tau)]}}{Q_{\Lambda(\tau)}} e^{\beta \mathcal{W}_{\gamma,\alpha} - \beta \Delta A}$$

$$= P(-\mathcal{W}) e^{\beta \mathcal{W} - \beta \Delta A}$$

where we have used the properties of the evenness of the delta function and the cyclic permutation symmetry of the trace. The latter allows us to equate $P(\alpha \to \gamma) = P^*(\gamma \to \alpha)$, where P^* denotes the time-reversed process, and employ $\mathcal{W}_{\gamma,\alpha} = -\mathcal{W}_{\alpha,\gamma}$. The Jarzynski equality follows directly from this quantum Crooks fluctuation theorem.

Further reading

The kinetic theory discussed in this chapter is also covered in Mehran Kardar's *Statistical physics of particles*, and the corresponding liquid state dynamics can be read in Jean-Pierre Hansen and Ian McDonald's *Theory of simple liquids*, as well as Denis Evans and Gary Morriss' *Statistical mechanics of nonequilbrium liquids*. However, the material concerning the fluctuation theorems and Jarzynski equality is only really in the primary literature, for example the seminal work from Chris Jarzynski, *Nonequilibrium equality for free energy differences* in Physical Review Letters 1997.

Additional exercises

Exercise 6.11: In this problem, you will explore models of the self-correlation, the van Hove function, which is the Fourier transform of the single particle density. Consider an isotropic fluid of identical particles, and assume that for such a system $\Delta \mathbf{r}(t) = \mathbf{r}(t) - \mathbf{r}(0)$ for a tagged particle is a Gaussian random variable.

1. Compute the generating function,

$$\hat{F}_s(k, t) = \langle e^{i\mathbf{k}\cdot\Delta\mathbf{r}(t)} \rangle$$

   and show that

$$\hat{F}_s(k, t) = \exp\left[-k^2 R^2(t)/2d\right],$$

   where $R^2(t) = \langle |\mathbf{r}(t) - \mathbf{r}(0)|^2 \rangle$ and d is dimensionality.

2. In the Gaussian approximation, $\langle |\mathbf{r}(t) - \mathbf{r}(0)|^4 \rangle$ can be expressed in terms of $R^2(t)$. Derive such an expression. One way to do so is to note that in general you may generate all moments of a distribution from the generating functional,

$$\left(\partial^n \hat{F}/\partial k_i \partial k_j \ldots\right)\Big|_{\mathbf{k}=0} = i^n \langle r_i r_j \ldots \rangle.$$

   where the subscripts denote vector components.

3. For an isotropic fluid, $F_s(\mathbf{r}, t)$ is dependent only on the length, $r = |\mathbf{r}|$. Show that for an isotropic fluid in $d = 3$,

$$F_s(r, t) = \frac{1}{(2\pi)^3} \int d\mathbf{k}\, e^{-i\mathbf{k}\mathbf{r}}\, \hat{F}_s(k, t) = \frac{1}{2r\pi^2} \int_0^\infty dk\; k\sin(kr)\hat{F}_s(k, t).$$

4. For the ideal gas, $\mathbf{r}(t) = \mathbf{r}(0) + \mathbf{v}t$. As a result,

$$\hat{F}_s(k, t) = \langle e^{i\mathbf{k}\cdot\mathbf{v}t} \rangle$$

   Use this formula together with the fact that the velocity distribution of a classical equilibrated system are Gaussian to show that for $d = 3$

$$\hat{F}_s(k, t) = e^{-k^2 t^2 k_{\mathrm{B}} T/2m}$$

   where m is the particle's mass and the gas is kept at constant T.

5. Invert this Fourier transform and show that the distribution of displacements is

$$F_s(r,t) = \left(\frac{m}{2\pi t^2 k_{\rm B} T}\right)^{3/2} e^{-mr^2/2t^2 k_{\rm B} T}$$

for an ideal gas, and comment on the long-time behavior.

Exercise 6.12: Here you will work through an exactly solvable example of the Jarzynski equality and Crooks work fluctuation theorem. The example we will consider is a piston containing a small number of particles, N, where the volume of the piston, V, can be manipulated in order to do work on the gas. We will neglect interactions between gas particles and treat the gas dynamics as classical. In such an approximation, the energy is just given by the kinetic energy of the gas, $E = \sum_{i=1}^{N} m\mathbf{v}^2/2$.

1. If the piston is held fixed and the gas is brought to thermal equilibrium with an external heat bath at inverse temperature β, the distribution of internal energy is given by

$$P(E,\beta) = \frac{\Omega(E,N,V)}{Q(\beta,N,V)} e^{-\beta E}$$

where $Q(\beta,N,V)$ is the canonical partition function and $\Omega(E,N,V)$ is the density of states. Evaluate both $\Omega(E,N,V)$ and $Q(\beta,N,V)$.

2. Using your results from the previous part, show that $P(E,\beta)$ can be written as

$$P(E,\beta) = \frac{\beta}{(3N/2 - 1)!} (\beta E)^{3N/2-1} e^{-\beta E}$$

where we will assume that N is an even number so that the factorial is well-defined (otherwise, we could introduce the gamma function).

3. Let us consider changing the volume of the piston from V_0 to V_1 in the absence of the bath such that the process is adiabatic. Using properties of an ideal gas, show that the change in energy is equal to

$$\Delta E = E_1 - E_0 = \alpha E_0$$

where $\alpha = (V_0/V_1)^{2/3} - 1$.

4. Since the process is adiabatic, the change in energy is equal to the work, $W = \Delta E$, which is uniquely determined by the initial energy of the gas in the piston. Using $P(W) = \langle \delta(W - \alpha E_0) \rangle$, show that the distribution of work values is

$$P(W) = \frac{\beta}{|\alpha|(3N/2 - 1)!} \left(\frac{\beta W}{\alpha}\right)^{3N/2-1} e^{-\beta W/\alpha} \Theta(\alpha W)$$

5. By computing the work to perform the reverse process V_1 to V_0 adiabatically, show that Crooks' work fluctuation theorem holds as

$$\frac{P(W)}{P(-W)} = \left(\frac{V_1}{V_0}\right)^N e^{\beta W}$$

where you can independently confirm $\beta \Delta A = -N \ln V_1/V_0$ using the ratio of partition functions for an isothermal process.

6. Make a plot of $P(\mathcal{W})$ versus $\beta\mathcal{W}$ for the forward and reverse processes using $N = 2, 4$, and 6, with $V_1/V_0 = 2$, and comment on the form of the distributions.

Exercise 6.13: In this problem, you will determine the average frequency of collisions between spheres in a one-component hard sphere fluid where the particles have diameter σ, and mass m, density is ρ, and temperature is T. Within a geometrical factor $(2/3)$, this collision frequency is given by the Enskog relaxation time defined as

$$\tau_{\mathrm{E}}^{-1} = \lim_{t \to 0^+} \left[-\frac{d}{dt} \langle \mathbf{v}(0) \cdot \mathbf{v}(t) \rangle / \langle v^2 \rangle \right]$$

where $\mathbf{v}(t)$ is the velocity of a tagged hard sphere. For hard spheres, the slope of the velocity autocorrelation function is finite as $t \to 0^+$. In other words, the function is nonanalytic at small times.

1. Provide a physical explanation of the nonanalyticity. That is, explain why the velocity can change instantaneously and also discuss why τ_{E}^{-1}, defined as the initial slope, is related so simply to the average collision frequency.

2. Consider the dynamics of two colliding hard spheres, particles 1 and 2. Let $\mathbf{v}_{12} \equiv \mathbf{v}_2 - \mathbf{v}_1$ denote the relative velocity before the collision, and let $\mathbf{n} = (\mathbf{r}_2 - \mathbf{r}_1)/\sigma$ be the unit vector between the centers of the spheres at the instant they collide. After colliding with particle 2, the velocity of particle 1 is $\mathbf{v}_1 + \Delta\mathbf{v}_1$. Show that the change in velocity is $\Delta\mathbf{v}_1 = \mathbf{n} \cdot \mathbf{v}_{21}$. Hint: This is an exercise in the principles of conservation of momentum, angular momentum, and energy.

3. Any one of $N-1$ particles might collide with particle 1 at a given instant. Assume the fluid is at thermal equilibrium, and show that

$$\lim_{t \to 0^+} \left[-\frac{d}{dt} \langle \mathbf{v}(0) \cdot \mathbf{v}(t) \rangle / \langle v^2 \rangle \right] = \rho g(\sigma^+)\pi \int_0^\sigma db\, b \int d\mathbf{v}_{21}$$

$$\times\, \Phi(\mathbf{v}_{21})\, |\, \mathbf{v}_{21}\, |\, (\mathbf{v}_{21} \cdot \Delta\mathbf{v}_1)$$

where $g(\sigma^+)$ is the contact value of the radial distribution function, b is the impact parameter, and

$$\Phi(\mathbf{v}_{21}) \propto \exp\left(-\beta\mu v_{21}^2/2\right)$$

is the relative velocity distribution (μ denoting the reduced mass). Perform the indicated integrations and show that

$$\tau_{\mathrm{E}}^{-1} = \frac{8}{3}(\pi/\beta\mu)^{-1/2}\rho\sigma^2\, g(\sigma^+)$$

4. The Percus–Yevick equation explored in Chapter 3 provides an expression for $g(\sigma^+)$ as a function of ρ

$$g(\sigma^+) = \frac{1 + \eta/2}{(1 - \eta)^2}$$

where $\eta = \pi\rho\sigma^3/6$ is the packing fraction. Use this expression and the result of part 3 to compute τ_{E}^{-1} for particles of mass 40 amu and a temperature 100 K for a few densities in the range $0.7 \le \rho\sigma^3 \le 0.95$.

5. Assuming the velocity correlations decay exponentially,

$$\langle \mathbf{v}_1(0) \cdot \mathbf{v}_1(t) \rangle = \langle v_1^2 \rangle \exp(-t/\tau_{\mathrm{E}}), \quad t \geq 0$$

compute the self-diffusion constant, D, over the same range considered in part 4. Compare this theory for diffusion, known as Enskog theory, with the empirical results from computer simulation that are well fit to the polynomial

$$D = \frac{3}{8}(\beta\mu\pi)^{-1/2}\sigma \left[3.7043 - 6.3355\rho\sigma^3 + 2.718(\rho\sigma^3)^2 \right].$$

Explain why the simulation gives lower values for D than Enskog theory.

Exercise 6.14: Here we will work out a concrete quantum system where the Jarzynski equality can be demonstrated explicitly. Consider a harmonic oscillator in one dimension

$$\hat{\mathcal{H}} = \frac{1}{2m}\hat{p}^2 + \frac{m\omega^2(t)}{2}\hat{x}^2 \qquad \omega^2(t) = \omega_0^2 + \left(\omega_1^2 - \omega_0^2\right)t/\tau$$

where we will change the frequency between ω_0 and ω_1 linearly in time over an observation time, τ. We will imagine that the system is first brought into contact with a bath at temperature T, so that it is initially Boltzmann distributed. It is then disconnected from the bath, the energy is measured, the frequency is changed, and the energy is measured again.

1. Show that the time evolution operator $\hat{U}(x, x_0, \tau)$ is given by

$$U(x, x_0, \tau) = \sqrt{\frac{m}{2\pi i h X(\tau)}} \exp\left[\frac{im}{2\hbar X(\tau)} \left(\dot{X}(\tau)x^2 - 2xx_0 + Y(\tau)x_0^2 \right) \right]$$

   where $X(t)$ and $Y(t)$ are the solutions to the classical equation of motion for the time-dependent Hamiltonian with boundary conditions $X(0) = 0$, $\dot{X}(0) = 1$, and $Y(0) = 1$, $\dot{Y}(0) = 0$.
2. The transition probability between two instantaneous energy eigenstates of the system can be computed, $P(\alpha \to \gamma) = |\langle \alpha(x_0)| U(x, x_0, \tau) |\gamma(x)\rangle|^2$, where $|\alpha(x_0)\rangle$ is the α'th harmonic oscillator eigenstate with frequency ω_0, while $|\gamma(x)\rangle$ is the γ'th eigenstate with frequency ω_1. Evaluate $P(\alpha \to \gamma)$.
3. The generating function for the work, $\langle \exp[-\beta \mathcal{W}]\rangle$, is expressible as a sum over all α and γ. Perform this sum and show that it returns the expected change in the free energy. You will find the following relation useful,

$$\sum_{\alpha,\gamma} u^\alpha v^\gamma P(\alpha \to \gamma) = \sqrt{h(1 - u^2)(1 - v^2) + (1 + u^2)(1 + v^2) - 4uv}$$

where $h = [\omega_0^2(\omega_1^2 X^2(\tau) + \dot{X}^2(\tau)) + \omega_1^2 Y^2(\tau) + \dot{Y}^2(\tau)]/4\omega_0\omega_1$.

7

Irreversibility and dynamical response

Linear response theory forms the basis for many of the modern experimental approaches like scattering and spectroscopy used to elucidate molecular behavior. The central concept of linear response is the response function, which describes how a system's properties change as a result of an applied perturbation. We have previously developed this framework in the context of static response, allowing us to relate two distinct thermal equilibrium states and devise linear models of materials that would be otherwise difficult to study. Here we will generalize this perspective to dynamic response, and the evolution between distinct thermal equilibrium states. Of particular importance will be the derivation of the fluctuation–dissipation theorem, which relates the response of a system to its time-dependent fluctuations, clarifying an equivalence between how a system responds to a small perturbation and how it regresses back to equilibrium in the absence of a perturbation. With this framework we can develop linear models of relaxation, classically and quantum mechanically, and derive equations of motion suitable for describing the dynamics of thermal ensembles held at fixed temperature rather than fixed energy. These models are Gaussian, and approximate the dynamical correlations of interacting systems. These effective equations of motion are rendered stochastic due to the incomplete description of the microscopic degrees of freedom that evolve the system.

7.1 Dynamically forced harmonic oscillator

With the tools to describe collective time-dependent phenomena developed in Chapter 6, we are prepared to study how a system responds dynamically to a small perturbation. We will first consider a system that always responds linearly, and for which we can arrive at some exact results. Consider a harmonic oscillator perturbed by a force that is time-dependent. The Hamiltonian for such a system can be written as,

$$\mathcal{H} = \frac{m}{2}\dot{x}^2 + \frac{m\omega^2}{2}x^2 - f(t)x$$

where m is its mass and ω is its characteristic frequency. For $f(t) = 0$, and an initial position $x(0)$ and velocity $\dot{x}(0)$, we can solve the differential equation,

$$\ddot{x}(t) = -\omega^2 x(t) \quad \rightarrow \quad x(t) = x(0)\cos(\omega t) + \frac{\dot{x}(0)}{\omega}\sin(\omega t)$$

resulting in a harmonic series. We see that the frequency, ω, indeed sets the time of typical oscillations about an equilibrium position. In order to solve the equation with $f(t) \neq 0$, we need to invoke Laplace's theory for linear dynamical systems, and subsequently invoke Laplace transforms.

Exercise 7.1: Prove that $L[\ddot{g}(t)] = s^2 L[g(t)] - sg(0) - \dot{g}(0)$ by integrating by parts.

Exercise 7.2: Evaluate the Laplace transform for a polynomial $g(t) = t^\alpha$ for $\alpha > 1$.

For $f = 0$, we can Laplace transform the linear second-order differential equation. This yields,

$$s^2 \tilde{x}^\circ(s) - sx^\circ(0) - \dot{x}^\circ(0) = -\omega^2 \tilde{x}^\circ(s)$$

$$\tilde{x}^\circ(s) = x^\circ(0)\frac{s}{s^2 + \omega^2} + \dot{x}^\circ(0)\frac{1}{s^2 + \omega^2}$$

where we have denoted by superscript o the solution $\tilde{x}^\circ$ with no force. We can invert this expression, yielding the form of the equation we found previously. However, for nonzero $f(t)$, the solution takes the form,

$$\tilde{x}(s) = \tilde{x}^\circ(s) + \frac{\tilde{f}(s)}{ms^2 + m\omega^2}$$

where $\tilde{f}(s)$ is the Laplace transform of the time-dependent force. If we define the denominator

$$\tilde{G}(s) = \frac{1}{ms^2 + m\omega^2} \qquad G(t) = \frac{1}{m\omega}\sin(\omega t)$$

which is inverted to the sine function, we can invoke the convolution theorem in reverse, to obtain the complete solution of the equation of motion

$$x(t) - x^\circ(t) = \int_{-\infty}^{t} dt'\, G(t - t')f(t')$$

$$= \int_{-\infty}^{t} dt'\, \frac{\delta x(t)}{\delta f(t')}f(t')$$

where we identify $G(t)$ as the Green's function, or response function, equivalently defined as the change in the trajectory $\delta x(t)$ for a given modification of the applied force $\delta f(t)$. This relationship, exact for a harmonic oscillator, is a statement of linear response in the time domain. It relates the motion of the particle in a harmonic potential, to that with an additional applied force. We can identify the Green's function with our response function by imposing causality,

$$\chi(t) = \Theta(t)G(t) \qquad \Theta(t) = \begin{cases} 0 & t < 0 \\ 1 & t > 0 \end{cases}$$

where $\Theta(t)$ is a step function, where we imagine the force is turned on at $t = 0$.

For this simple case, we can relate the response function to thermal fluctuations by noting

$$x(0)x(t) = x^2(0)\cos(\omega t) + x(0)\frac{\dot{x}(0)}{\omega}\sin(\omega t)$$

$$\langle x(0)x(t)\rangle_{\mathrm{o}} = \langle x^2\rangle_{\mathrm{o}}\cos(\omega t)$$

$$= (1/\beta m\omega^2)\cos(\omega t)$$

where we have multiplied the equation of motion by $x(0)$ and taken its thermal average. The velocity term averages to zero due to the factorization of thermal averages of position and momentum. We have also explicitly evaluated $\langle x^2\rangle$ within the reference system. Taking the time derivative of the position–position correlation function,

$$\frac{d}{dt}\langle x(0)x(t)\rangle_{\mathrm{o}} = -(1/\beta m\omega)\sin(\omega t) = -G(t)/\beta$$

we can identify it with the Green's function, up to a factor of $-\beta$. In terms of the response function, $\chi(t - t') = \delta x(t)/\delta f(t')$,

$$\chi(t) = -\beta\Theta(t)\frac{d}{dt}\langle x(0)x(t)\rangle_{\mathrm{o}}$$

this is known as the fluctuation–dissipation theorem. While it is exact for a harmonic oscillator, it holds for nonlinear systems to first order in f, as will be shown below.

Exercise 7.3: By direct differentiation of

$$x(t) - x^{\mathrm{o}}(t) = \int_{-\infty}^{t} dt'\, G(t - t')f(t')$$

confirm that this solution is consistent with the original second-order differential equation.

7.2 Linear response in the time domain

More generally, we can probe the dynamical response of an arbitrarily complex system to a linear perturbation. Let's decompose the Hamiltonian of such a system as

$$\mathcal{H}(\mathbf{x}) = \mathcal{H}_{\mathrm{o}}(\mathbf{x}) - f(t)A(\mathbf{x})$$

where $\mathcal{H}_{\mathrm{o}}(\mathbf{x})$ is a reference and $f(t)$ a time dependent force acting on A. In the presence of a time-dependent $f(t)$, the system is not stationary, and the observed values of A are nonequilibrium averages, denoted $\bar{A}(t)$. When $f(t) = 0$, however, the system will approach or remain at equilibrium where the observed value will be $\langle A\rangle$. While the meaning of the equilibrium average, $\langle A\rangle$, is general and independent of many details, we will see that $\bar{A}(t)$ depends upon the preparation and other specifics of the system.

In order to understand the time dependence of $\bar{A}(t)$, we can write it down as a functional Taylor series,

$$\bar{A}(t) \approx \bar{A}_\mathrm{o}(t) + \int_{-\infty}^{\infty} dt' \left. \frac{\delta \bar{A}(t)}{\delta f(t')} \right|_\mathrm{o} f(t') + \mathcal{O}(f^2)$$

where we have kept only terms up to linear order in $f(t)$. Again, the response function is identified as

$$\left. \frac{\delta \bar{A}(t)}{\delta f(t')} \right|_\mathrm{o} = \chi(t, t')$$

which relates how A responds at a time t to the perturbation at a time t', evaluated in the reference system. Two important properties of response functions are

$$\chi(t, t') = \chi(t - t') \quad , \quad \chi(t) = 0 \quad t < 0$$

where the first equation results from stationarity and the second from causality.

Since $\chi(t - t')$ is independent of $f(t)$, our choice of $f(t)$ for deriving a formula for $\chi(t - t')$ is simply a matter of convenience. One specifically convenient thought experiment is to imagine that the system is equilibrated from $t = -\infty$ with $f(t) = \bar{f}$, and then at $t = 0$, f is switched off. Mathematically this is written as,

$$f(t) = \begin{cases} \bar{f} & t < 0 \\ 0 & t > 0 \end{cases} = \bar{f}\,[1 - \Theta(t)]$$

where Θ is the step function, illustrated in Figure 7.1.

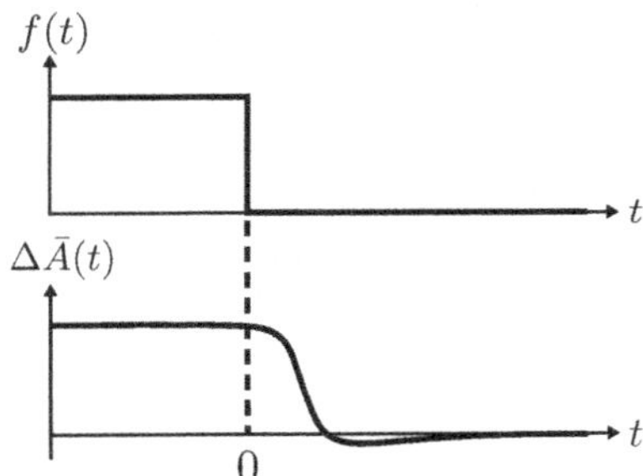

Fig. 7.1 Relaxation from one prepared equilibrium state to another equilibrium state after changing the conditions that stabilized the former, where $\Delta \bar{A}(t) = \bar{A}(t) - \bar{A}_\mathrm{o}(t)$.

For times up to $t = 0^-$, the probability distribution for phase space points is given by a Boltzmann distribution of the form,

$$P_{\bar{f}}(\mathbf{x}) = \frac{e^{-\beta \mathcal{H}_\mathrm{o} + \beta \bar{f} A}}{Q_{\bar{f}}}$$

where $Q_{\bar{f}}$ is the partition function computed in the presence of $\bar{f}$. For $t \to \infty$, the system is assumed to relax to a distribution given by

$$P_{\mathrm{o}}(\mathbf{x}) = \frac{e^{-\beta \mathcal{H}_{\mathrm{o}}}}{Q_{\mathrm{o}}}$$

where Q_{o} is the partition function computed in the absence of $\bar{f}$. For $t > 0$, the average value of A as a function of time can be computed from,

$$\bar{A}(t) = \int d\mathbf{x}\, P_{\bar{f}}(\mathbf{x}) A(t, \mathbf{x})$$

where the initial conditions are drawn from $P_{\bar{f}}$, but the evolution of $A(t)$ is determined solely by $\mathcal{H}_{\mathrm{o}}$. Generically, the two partition functions are related by

$$Q_{\bar{f}} = Q_{\mathrm{o}} \left\langle e^{\beta \bar{f} A} \right\rangle_{\mathrm{o}}$$

which is the usual form of the cumulant generating function. For small $\bar{f}$, the two probability distributions, $P_{\bar{f}}(\mathbf{x})$ and $P_{\mathrm{o}}(\mathbf{x})$, are related by,

$$P_{\bar{f}}(\mathbf{x}) \approx \left(1 - \beta \bar{f} \langle A \rangle_{\mathrm{o}} + \dots \right)\left(1 + \beta \bar{f} A + \dots \right) P_{\mathrm{o}}(\mathbf{x})$$

which follows from expanding by the argument of the exponential, as well as the ratio of partition functions. Substituting this distribution back into the definition of the average, we have

$$\bar{A}(t) = \int d\mathbf{x}\, A(t, \mathbf{x}) \left(1 - \beta \bar{f} \langle A \rangle_{\mathrm{o}} + \dots \right)\left(1 + \beta \bar{f} A + \dots \right) P_{\mathrm{o}}(\mathbf{x})$$

$$= \left\langle A(t) \left(1 - \beta \bar{f} \langle A \rangle_{\mathrm{o}} + \dots \right)\left(1 + \beta \bar{f} A + \dots \right) \right\rangle_{\mathrm{o}}$$

$$= \langle A \rangle_{\mathrm{o}} + \beta \bar{f} \langle \delta A(0) \delta A(t) \rangle_{\mathrm{o}}$$

where we find that the change in the average value of A is determined to first order in f, by the time correlation function of A in the reference ensemble.

From equating this result with our initial functional Taylor series expansion, we have

$$\bar{A}(t) = \langle A \rangle_{\mathrm{o}} + \bar{f} \int_{-\infty}^{\infty} dt'\, \chi(t - t')\left[1 - \Theta(t')\right]$$

where taking the time derivative of both sides, and using $d\Theta(t')/dt' = \delta(t')$,

$$\frac{d}{dt} \beta \bar{f} \langle \delta A(0) \delta A(t) \rangle_{\mathrm{o}} = \frac{d}{dt} \bar{f} \int_{-\infty}^{\infty} dt'\, \chi(t - t')\left[1 - \Theta(t')\right]$$

$$= -\bar{f} \int_{-\infty}^{\infty} dt'\, \frac{d\chi(t - t')}{dt'}\left[1 - \Theta(t')\right]$$

$$= \bar{f} \int_{-\infty}^{\infty} dt'\, \chi(t - t')\frac{d}{dt'}\left[1 - \Theta(t')\right] = -\bar{f}\chi(t)$$

we arrive at a relationship between the response function and a derivative of the time correlation function. Finally, reimposing causality for generic forces,

$$\chi(t) = -\beta \Theta(t) \frac{d}{dt} \langle \delta A(0) \delta A(t) \rangle_{\mathrm{o}}$$

we have the same statement of the fluctuation-dissipation theorem. Historically, this was alluded to by Onsager in his formulation of the *regression hypothesis*, which stated, "the regression of microscopic thermal fluctuations at equilibrium follows the macroscopic law of relaxation of small non-equilibrium disturbances." It was first proved by Callen and Welton in 1951.

> **Exercise 7.4:** Show that the response function encoding the change of a variable $\Delta\bar{B}(t)$ is given by
>
> $$\chi(t) = -\beta\Theta(t)\frac{d}{dt}\langle\delta A(0)\delta B(t)\rangle_{\mathrm{o}}$$
>
> if the Hamiltonian is perturbed by $-f(t)A$.

7.3   Quantum regression hypothesis

In order to prove the regression hypothesis and derive a linear response relation quantum mechanically, we can proceed as before by positing a Hamiltonian of the form

$$\hat{\mathcal{H}} = \hat{\mathcal{H}}_{\mathrm{o}} - f(t)\hat{A}$$

where

$$f(t) = \begin{cases} \bar{f} & t < 0 \\ 0 & t > 0 \end{cases}$$

now we can re-express our time-dependent average as,

$$\bar{A}(t) = \sum_k \langle\psi_k|\, e^{-\beta\hat{\mathcal{H}}}e^{i\hat{\mathcal{H}}_{\mathrm{o}}t/\hbar}\,\hat{A}\,e^{-i\hat{\mathcal{H}}_{\mathrm{o}}t/\hbar}\,|\psi_k\rangle \Big/ \sum_k \langle\psi_k|\, e^{-\beta\hat{\mathcal{H}}}\,|\psi_k\rangle$$

where $|\psi_k\rangle$ are the eigenstates of $\hat{\mathcal{H}}$. If

$$\left[\hat{\mathcal{H}}_{\mathrm{o}}, \hat{A}\right] = 0$$

then A will be a constant of motion and subsequently will not relax. However, if

$$\left[\hat{\mathcal{H}}_{\mathrm{o}}, \hat{A}\right] \neq 0$$

the operators do not commute, and thus A could change with time, but

$$e^{-\beta\hat{\mathcal{H}}} \neq e^{-\beta\hat{\mathcal{H}}_{\mathrm{o}}}e^{\beta\bar{f}\hat{A}} = e^{-\beta\hat{\mathcal{H}}_{\mathrm{o}}}\left(1 + \beta\bar{f}\hat{A} + \dots\right)$$

so we cannot use the expansion that was so useful before. Instead, we can define a *resolvant*, or Laplace transform of the Boltzmann operator,

$$L\left[e^{-\beta\hat{\mathcal{H}}}\right] = \tilde{G}(s) = \int_0^\infty d\beta\, e^{-\beta s}e^{-\beta\hat{\mathcal{H}}}$$

$$= \frac{1}{s + \hat{\mathcal{H}}_{\mathrm{o}} - \bar{f}\hat{A}}$$

$$= \frac{1}{s + \hat{\mathcal{H}}_{\mathrm{o}}}\frac{1}{1 - \bar{f}\hat{A}\left(\frac{1}{s+\hat{\mathcal{H}}_{\mathrm{o}}}\right)}.$$

This is a form that can be easily expanded for small $\bar{f}$,

$$\tilde{G}(s) \approx \tilde{G}_{\mathrm{o}}(s) + \tilde{G}_{\mathrm{o}}(s)\bar{f}\hat{A}\tilde{G}_{\mathrm{o}}(s) + \mathcal{O}(\bar{f}^2)$$

where $\tilde{G}_{\mathrm{o}}(s)$ is the Laplace transform evaluated at $f = 0$. Taking the inverse Laplace transform and invoking the convolution theorem, we get

$$e^{-\beta\hat{\mathcal{H}}} \approx e^{-\beta\hat{\mathcal{H}}_{\mathrm{o}}} + \int_0^\beta d\beta'\, e^{-(\beta-\beta')\hat{\mathcal{H}}_{\mathrm{o}}}\,\bar{f}\hat{A}e^{-\beta'\hat{\mathcal{H}}_{\mathrm{o}}}\,.$$

The kernel in the integral can be identified with a peculiar type of Heisenberg operator,

$$\hat{A}(-i\tau) = e^{\beta'\hat{\mathcal{H}}_{\mathrm{o}}}\hat{A}e^{-\beta'\hat{\mathcal{H}}_{\mathrm{o}}}$$

where $\tau = \beta'\hbar$, and $i\beta'\hbar$ is an imaginary time. Putting this approximation back into the trace operation for the average of $\hat{A}$, we get

$$\bar{A}(t) \approx \frac{\langle A\rangle_{\mathrm{o}} + \frac{\bar{f}}{\hbar}\int_0^{\beta\hbar} d\tau\,\left\langle \hat{A}(-i\tau)\hat{A}(t)\right\rangle_{\mathrm{o}}}{1 + \frac{\bar{f}}{\hbar}\int_0^{\beta\hbar} d\tau\,\left\langle \hat{A}(-i\tau)\right\rangle_{\mathrm{o}}}$$

$$= \langle A\rangle_{\mathrm{o}} + \frac{\bar{f}}{\hbar}\int_0^{\beta\hbar} d\tau\,\left\langle \delta\hat{A}(-i\tau)\delta\hat{A}(t)\right\rangle_{\mathrm{o}}$$

where we have now codified the quantum version of linear response, from which follows a fluctuation–dissipation relation. Apart from the integral and $\hbar$, this linear response relationship looks just like the classical version, i.e., the time-dependent change in the average of a variable being perturbed, is related to the time correlation function of that variable in the unperturbed state. The distinction comes from the integral over imaginary time, which is a direct consequence of the uncertainty principle, just as we saw earlier in Chapter 3. In this case, the uncertainty associated with any thermal state has to be integrated over in order to understand the response of the system.

The definition of the quantum response function can be determined following the same procedure as we followed classically. Namely, defining $\chi(t - t') = \delta\bar{A}(t)/\delta f(t')$, we can determine it using a step function for $f(t)$ so that

$$\chi(t) = -\frac{1}{\hbar}\Theta(t)\frac{d}{dt}\int_0^{\beta\hbar} d\tau\,\left\langle \delta\hat{A}(-i\tau)\delta\hat{A}(t)\right\rangle_{\mathrm{o}}$$

where as before $\Theta(t)$ enforces causality. This form is often referred to as the Kubo-transformed correlation function. An alternative form can be derived noting that from time translational invariance, the argument of the integral is equal to $\left\langle \delta\hat{A}(0)\delta\hat{A}(t + i\tau)\right\rangle_{\mathrm{o}}$, and thus

$$\frac{d}{dt}\int_0^{\beta\hbar} d\tau\,\left\langle \delta\hat{A}(-i\tau)\delta\hat{A}(t)\right\rangle_{\mathrm{o}} = \int_0^{\beta\hbar} d\tau\,\frac{d}{dt}\left\langle \delta\hat{A}(0)\delta\hat{A}(t + i\tau)\right\rangle_{\mathrm{o}}$$

$$= \int_0^{\beta\hbar} d\tau\,\frac{d}{d(i\tau)}\left\langle \delta\hat{A}(0)\delta\hat{A}(t + i\tau)\right\rangle_{\mathrm{o}}$$

$$= -i\left[\left\langle \delta\hat{A}(0)\delta\hat{A}(t + i\beta\hbar)\right\rangle_{\mathrm{o}} - \left\langle \delta\hat{A}(0)\delta\hat{A}(t)\right\rangle_{\mathrm{o}}\right]$$

where we have performed a change of variables by noting that the t dependence of the correlation function is the same as its $i\tau$ dependence, and used the fundamental theorem of calculus to evaluate the integral over imaginary time. Using the cyclic properties of the trace it is also true that

$$\left\langle \delta\hat{A}(0)\delta\hat{A}(t + i\beta\hbar)\right\rangle_{\mathrm{o}} = \left\langle \delta\hat{A}(t)\delta\hat{A}(0)\right\rangle_{\mathrm{o}}$$

so that the response function becomes

$$\chi(t) = \frac{i}{\hbar}\Theta(t)\left[\left\langle \delta\hat{A}(t)\delta\hat{A}(0)\right\rangle_{\mathrm{o}} - \left\langle \delta\hat{A}(0)\delta\hat{A}(t)\right\rangle_{\mathrm{o}}\right] = \frac{i}{\hbar}\Theta(t)\left\langle [\delta\hat{A}(t), \delta\hat{A}(0)]\right\rangle_{\mathrm{o}}$$

or the commutator between the two time orderings of the quantum time correlation function.

> **Exercise 7.5:** Confirm that
> $$\left\langle \delta\hat{A}(0)\delta\hat{A}(t + i\beta\hbar)\right\rangle = \left\langle \delta\hat{A}(t)\delta\hat{A}(0)\right\rangle$$
> using the definition of the thermal average and the cyclic properties of the trace.

7.4 Linear absorption

With linear response we have a relationship between how the system responds to a disturbance, and the intrinsic dynamics of the system in the unperturbed state. With this we can begin to ask questions about how to implement this relation experimentally. To illustrate this point, let's consider a simple absorption measurement. Absorption is the measure of the energy transferred to a system by an external force acting on it. The average absorption for a perturbation of the form $-f(t)A$ is given by,

$$\mathrm{abs} = -\frac{1}{T}\int_0^T dt\, \dot{f}(t)\bar{A}(t)$$

or alternatively,

$$\mathrm{abs} = \frac{1}{T}\int_0^T dt\, f(t)\dot{\bar{A}}(t)$$

where the dot denotes time derivative. For a simple monochromatic force,

$$f(t) = \bar{f}\cos\omega t = \frac{\bar{f}}{2}\left(e^{i\omega t} + e^{-i\omega t}\right)$$

we can use linear response to write out the first-order theory for the change in A. This is

$$\Delta \bar{A}(t) = \bar{A}(t) - \langle A \rangle = \frac{\bar{f}}{2} \int_{-\infty}^{\infty} dt' \, \chi(t - t') \left(e^{i\omega t'} + e^{-i\omega t'} \right)$$

$$= \frac{\bar{f}}{2} \int_{-\infty}^{\infty} d\tau \, \chi(\tau) \left(e^{-i\omega \tau} e^{i\omega t} + e^{i\omega \tau} e^{-i\omega t} \right)$$

$$= \frac{\bar{f}}{2} \left[\hat{\chi}^*(\omega) e^{i\omega t} + \hat{\chi}(\omega) e^{-i\omega t} \right]$$

where in the second line we make a change of variables $\tau = t - t'$, and in the third we identify,

$$\hat{\chi}(\omega) = \int_{-\infty}^{\infty} d\tau \, \chi(\tau) e^{i\omega \tau}$$

as a Fourier transform of the response function. We also note that there is a relationship between the complex conjugate of the Fourier-transformed response function

$$\hat{\chi}(-\omega) = \hat{\chi}^*(\omega)$$

and its value at negative ω. We can further decompose the response function into its real and imaginary parts,

$$\hat{\chi}(\omega) = \hat{\chi}'(\omega) + i\hat{\chi}''(\omega)$$

where $\hat{\chi}'(\omega)$ and $\hat{\chi}''(\omega)$ are purely real. Substituting these back in and expanding the complex exponential,

$$\Delta \bar{A}(t) = \bar{f} \left[\hat{\chi}'(\omega) \cos(\omega t) + \hat{\chi}''(\omega) \sin(\omega t) \right]$$

we find that the response has two parts. The first term on the right is the in-phase component of the response, and the second term is the out-of-phase component.

The response function has a number of symmetries that are useful to note. For example,

$$\hat{\chi}'(\omega) = \hat{\chi}^{*\,'}(\omega) = \hat{\chi}'(-\omega)$$

namely the real part of the response function is even in ω. Analogously $i\hat{\chi}''(\omega)$ is odd in ω. In the time domain,

$$\chi'(t) = \int \frac{d\omega}{2\pi} \hat{\chi}'(\omega) e^{-i\omega t}$$

$$= \int \frac{d\omega}{2\pi} \hat{\chi}'(-\omega) e^{i\omega t} = \int \frac{d\omega}{2\pi} \hat{\chi}'(\omega) e^{i\omega t}$$

thus $\chi'(t)$ is even in time. Similarly, $i\chi''(t)$ is odd in t. Further causality requires that for $t < 0$,

$$\chi(t) = \chi'(t) + i\chi''(t) = 0$$

or

$$\chi'(t) = -i\chi''(t)$$

For $t > 0$,

$$\chi'(t) = i\chi''(t) = \chi(t)/2$$

so $\chi'(t)$ and $i\chi''(t)$ contain redundant information. These connections, and their Fourier analog are known as the *Kramers–Kronig relationships*. An example of the even and odd components of the response function are shown in Figure 7.2.

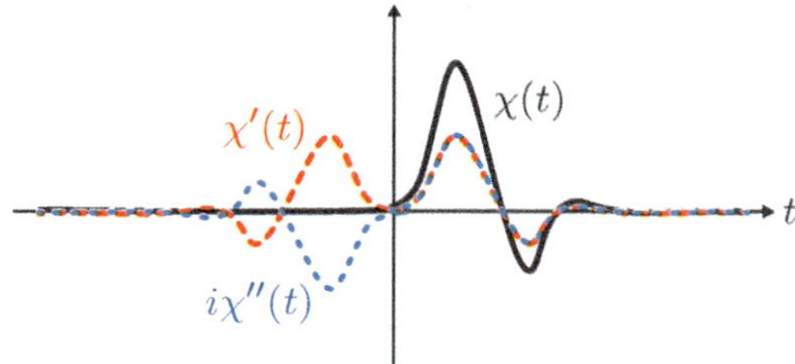

Fig. 7.2 Illustration of the symmetry relationships for the even and odd components of $\chi(t)$.

With our response theory for $\bar{A}(t)$, we can evaluate the expression for absorption,

$$\text{abs}(\omega) = \frac{1}{T} \int_0^T dt\, \bar{f} \cos(\omega t) \frac{d}{dt} \bar{f} \left[\hat{\chi}'(\omega) \cos(\omega t) + \hat{\chi}''(\omega) \sin(\omega t) \right]$$

$$= \frac{\bar{f}^2}{T} \int_0^T dt\, \cos(\omega t) \left[-\omega \hat{\chi}'(\omega) \sin(\omega t) + \omega \hat{\chi}''(\omega) \cos(\omega t) \right]$$

where there are two terms on the right-hand side. One, proportional to $\cos^2(\omega t)$, will accumulate over T, and the other, proportional to $\sin(\omega t) \cos(\omega t)$, will oscillate around zero after each period. In the limit that $T \to \infty$,

$$\text{abs}(\omega) = \omega \bar{f}^2 \hat{\chi}''(\omega)/2$$

we find that only the imaginary part of the response function contributes to absorption, and that it is linear in ω but quadratic in the strength of the force.

From the classical fluctuation–dissipation theorem, we have

$$\chi(t) = -\beta \Theta(t) \frac{d}{dt} C(t)$$

where

$$C(t) = \langle \delta A(0) \delta A(t) \rangle \,.$$

We note that

$$\hat{\chi}(\omega) = -\beta \int_0^\infty dt\, \dot{C}(t) e^{i\omega t}$$

or

$$\hat{\chi}''(\omega) = -\beta \int_0^\infty dt\, \dot{C}(t) \sin(\omega t) = \beta \omega \int_0^\infty dt\, C(t) \cos(\omega t) \,.$$

Give this, we can rewrite out expression for the absorption,

$$\text{abs}(\omega) = \frac{1}{2} \beta \omega^2 \bar{f}^2 \int_0^\infty dt\, \cos(\omega t) \langle \delta A(0) \delta A(t) \rangle$$

to obtain a detailed relationship between measurement, in the form of absorption, and molecular dynamics, in the form of a time correlation function of the unperturbed system! From this we see that absorption occurs at frequencies that coincide with timescales of the system's natural dynamics. We can also note that the above classical expression is even in ω. An example pair of $C(t)$ and its implied absorption spectra, abs(ω), is shown in Figure 7.3.

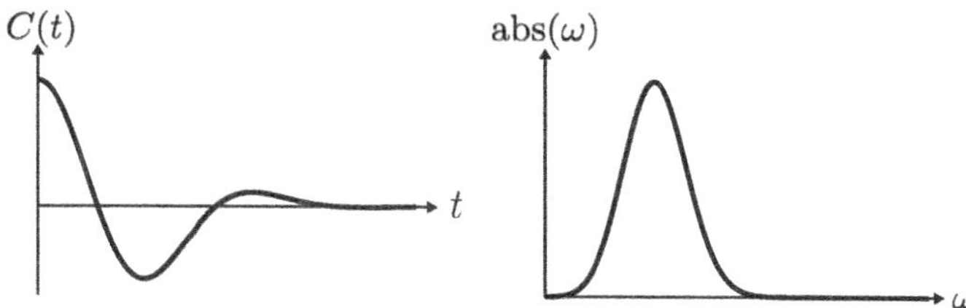

Fig. 7.3 Example of a time correlation function and its the absorption spectra that results from its Fourier transform.

Exercise 7.6: Evaluate the classical absorption as a function of frequency for an oscillatory force of the form $-\bar{f}\cos(\omega t)x$ using a correlation function

$$\langle \delta x(0)\delta x(t)\rangle = \frac{1}{\beta\omega_0^2\eta}\cos(\omega_0 t)e^{-\eta t}$$

which is a model appropriate for a damped harmonic oscillator with decay rate η and characteristic frequency ω_0. You should find that abs(ω) is Lorentzian.

The quantum mechanical absorption is similarly straightforward to deduce. Taking our previously derived response function,

$$\chi(t) = \frac{i}{\hbar}\Theta(t)\left\langle [\delta\hat{A}(t), \delta\hat{A}(0)]\right\rangle_o$$

we can recognize that its Fourier transform is a difference of two Fourier transformed correlation functions

$$\hat{\chi}(\omega) = \mathcal{F}\left\{\frac{i}{\hbar}\Theta(t)\left\langle [\delta\hat{A}(t), \delta\hat{A}(0)]\right\rangle_o\right\}$$

$$= \frac{i}{\hbar}\int_0^\infty dt\, e^{i\omega t}\left(C(t) - C(-t)\right) = \frac{i}{2\hbar}\left(\hat{C}(\omega) - \hat{C}(-\omega)\right)$$

where in the first line we have introduced the definition of the autocorrelation function and used the time translational invariance to rewrite $\left\langle \delta\hat{A}(0)\delta\hat{A}(t)\right\rangle = \left\langle \delta\hat{A}(-t)\delta\hat{A}(0)\right\rangle$. From the relations above its clear that the imaginary part of the quantum response function is $\hat{\chi}''(\omega) = \left(\hat{C}(\omega) - \hat{C}(-\omega)\right)/2\hbar$ which can be rewritten as

$$\hat{\chi}''(\omega) = \frac{1}{2\hbar}\left(1 - e^{-\beta\hbar\omega}\right)\hat{C}(\omega)$$

$$= \frac{1}{2\hbar}\left(1 - e^{-\beta\hbar\omega}\right)\int_{-\infty}^{\infty} dt\, e^{i\omega t}\left\langle \delta\hat{A}(t)\delta\hat{A}(0)\right\rangle$$

using the relation between complex conjugate pairs of Fourier-transformed quantum correlation functions from Chapter 6. This implies that the quantum mechanical absorption coefficient is

$$\text{abs}(\omega) = \omega\frac{\bar{f}^2}{2\hbar}\left(1 - e^{-\beta\hbar\omega}\right)\int_{-\infty}^{\infty} dt\, e^{i\omega t}\left\langle \delta\hat{A}(t)\delta\hat{A}(0)\right\rangle$$

which is proportional to ω and reduces to the classical expression in the high temperature limit, where $k_\mathrm{B}T \gg \hbar\omega$. This quantum mechanical expression for absorption forms the basis for all molecular spectroscopy. Probing vibrations, rotations, or even electronic excitations is just now a matter of identifying the pertinent light-matter coupling operator, and from that, deducing the subsequent correlation function that is responsible for its *lineshape* or the frequency dependence of its absorption.

> **Exercise 7.7:** For vibrational spectroscopy of molecules, the perturbation in the Hamiltonian is well described by a classical electric field interacting with the dipole, $\hat{m}$, of the system, $\hat{\mathcal{H}} = \hat{\mathcal{H}}_\mathrm{o} - \xi\cos(\omega t)\hat{m}$. Using this perturbation, work out an expression for abs(ω) employing a time correlation function.

7.5 Linear model for motion in solution

Having worked out the relationship between fluctuations and the response to a time-dependent perturbation, we are in a position to derive a simple linear model of dynamics in a liquid. The derivation that we will follow will not be the most general, but will get to the basic idea fairly easily.

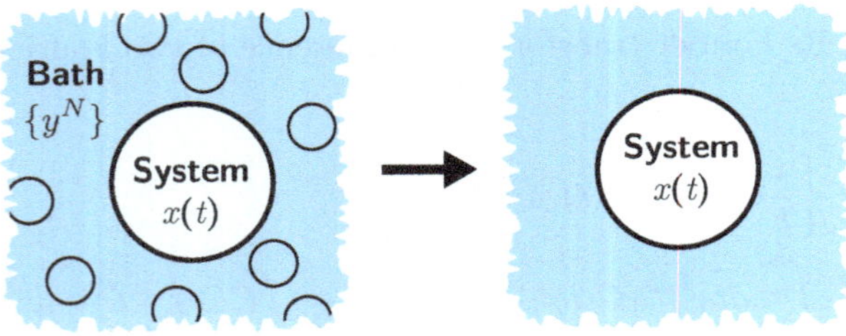

Fig. 7.4 System bath model, upon integrating out the degrees of freedom.

Let's imagine that the world is decomposable into a part we are interested in, called the system, and a part we do not explicitly wish to follow, called the bath. This is sketched in Figure 7.4, where x denotes the state of the system, and $\{y^N\}$ the collection of variables in the bath. Generically, we can imagine the Hamiltonian for such a world is decomposable into,

$$\mathcal{H} = \mathcal{H}_o(x) + \mathcal{H}_B(\{y^N\}) + V(x, \{y^N\})$$

where $\mathcal{H}_o(x)$ is the Hamiltonian of the system, $\mathcal{H}_B(\{y^N\})$, that of the bath, and $V(x, \{y^N\})$ is the interaction between them. For simplicity, let's imagine that the bath is made out of independent harmonic oscillators, and they are each linearly coupled to the system,

$$V(x, \{y^N\}) = -x \sum_i a_i y_i$$

with coupling strength a_i. Note that one could rewrite the coupling in a translationally invariant way if desired, which will just modify the exact form of $\mathcal{H}_o(x)$. It is the bilinear coupling is what makes the derivation manageable.

Let's assume that at time $t = 0$, the system starts at the origin so that $x(0) = 0$. The harmonic degrees of freedom can be integrated formally,

$$y_i(t) = y_i^o(t) + \int_0^t dt' \, \chi(t - t') a_i x(t')$$

where $y_i^o(t)$ is the evolution of y_i in the absence of the coupling to the system, and

$$\chi(t) = -\beta\Theta(t) \frac{d}{dt} C_i(t)$$

is the response function. As usual, $\chi(t)$ is related to a correlation function $C_i(t) = \langle y_i(0) y_i(t) \rangle$. Plugging in the response function and integrating by parts,

$$y_i(t) = y_i^o - \beta \int_0^t dt' \, C_i(t - t') a_i \dot{x}(t') + \beta C_i(0) a_i x(t)$$

where $\beta C_i(0) a_i x(t)$ is the boundary term evaluated at $t' = t$, and the other boundary term $-\beta C_i(t) a_i x(0)$ vanishes because of the assumed boundary condition.

Similarly, the equation of motion for the system can be written down as,

$$m\ddot{x} = -\frac{d\mathcal{H}_o(x)}{dx} + \sum_i^N a_i y_i$$

$$= -\mathcal{H}_o'(x) + \sum_i^N a_i \left(y_i^o - \beta \int_0^t dt' \, C_i(t - t') a_i \dot{x}(t') + \beta C_i(0) a_i x(t) \right)$$

$$= -\mathcal{H}_o'(x) + \sum_i^N \beta C_i(0) a_i^2 x(t) + \sum_i^N a_i y_i^o - \beta \int_0^t dt' \sum_i^N C_i(t - t') a_i^2 \dot{x}(t')$$

where we have inserted our expression for $y_i(t)$. We can simplify this expression by identifying a couple of specific terms, first defining

$$C_{\mathrm{B}}(t) = \sum_i^N C_i(t) a_i^2$$

as the total correlation function from the bath. The first two terms depend on the dynamics of x but not on y, and so can be rewritten together,

$$w(x) \equiv \mathcal{H}_{\mathrm{o}}(x) - \frac{\beta}{2} C_{\mathrm{B}}(0) x^2$$

where we recognize that one part of the action of the bath is to renormalize the effective potential. Generically, the fluctuations of the bath renormalize the potential to a potential of mean force discussed in Chapter 2. The third term depends on y but not x,

$$\eta(t) \equiv \sum_i^N a_i y_i^{\mathrm{o}}(t) \, .$$

This term can be interpreted as the fluctuating total force acting on the system, in a manner that is independent of the system. In the limit that N is large, this uncorrelated sum could be modeled as a Gaussian random variable. Finally, if we identify

$$\gamma(t) \equiv \beta C_{\mathrm{B}}(t)/2$$

the last term can be identified as a time-dependent friction. It exists under the integral, so its influence results in the equation of motion no longer being time-local, and but depends on its history. For this reason, it is identified as a memory kernel. Putting these terms all together,

$$m\ddot{x} = -w'(x) - 2 \int_0^t dt' \, \gamma(t - t') \dot{x}(t') + \eta(t)$$

we find that our initially deterministic set of equations has become a stochastic differential equation generated from our lack of information about the bath!

To understand this more carefully, we need to evaluate the statistics of the fluctuating forces. Its mean is

$$\langle \eta(t) \rangle = \left\langle \sum_i^N a_i y_i^{\mathrm{o}}(t) \right\rangle = 0$$

from translational invariance, and the variance of the uncorrelated bath variables,

$$\langle \eta(t) \eta(t') \rangle = \left\langle \sum_i^N a_i^2 y_i^{\mathrm{o}}(t) y_i^{\mathrm{o}}(t') \right\rangle$$

$$= \sum_i^N a_i^2 C_i(t - t') = C_{\mathrm{B}}(t - t')$$

is given by the bath correlation function. That is to say,

$$\langle \eta(t) \eta(t') \rangle = 2 k_{\mathrm{B}} T \gamma(t - t')$$

the random force and the friction are exactly balanced, with a relative magnitude that is set by the available thermal energy. This relationship between random fluctuations and friction is known as the *second fluctuation-dissipation theorem*, and the equation of motion is the *generalized Langevin equation*. The generalized Langevin equation plays an important role in nonequilibrium statistical mechanics. It was originally written down by Zwanzig in a manner similar to how we derived it, and then later with the help of Mori, was derived under very general conditions.

If we take the generalized Langevin equation,

$$m\ddot{x} = -w'(x) - 2 \int_0^t dt' \, \gamma(t - t')\dot{x}(t') + \eta(t)$$

and assume that the bath relaxes very quickly relative to the system, then the memory kernel becomes

$$\gamma(t) = \gamma\delta(t)$$

This is the so-called *Markovian* approximation, differentiating it from the *non-Markovian* equation when the memory is present. Inserting this into our equation of motion, and noting that the time integral covers only $t > 0$,

$$m\ddot{x} = -w'(x) - \gamma\dot{x}(t) + \eta(t)$$

we are left with an equation that is now time local, though still stochastic due to random realizations of $\eta(t)$. This equation is known as the *Langevin equation*. The random force and friction are still delicately balanced

$$\langle \eta(t)\eta(t') \rangle = 2k_{\mathrm{B}}T\gamma\delta(t - t')$$

to ensure equilibration and thermodynamic consistency.

In the limit that the friction is very large, $\gamma \gg 0$, inertia is instantly quenched, $\ddot{x} \approx 0$, leaving us with

$$\gamma\dot{x}(t) = -w'(x) + \eta(t)$$

which is known as the *overdamped Langevin equation*, or Brownian motion. This equation when $w(x) = 0$ has a long history in statistical mechanics. It is named after Brown, who in 1827, while looking through a microscope at particles trapped in cavities inside pollen grains in water, noted that the particles moved through the water in a seemingly random way. Einstein, having heard of this observation, wrote down this equation, and solved it in one of his famous 1905 papers. His theory gave tremendous support for the existence of atoms and molecules, which at the time were still only hypothesized, and was actually used (in the form of the osmotic pressure produced from this motion) to measure Avogadro's number more accurately than was possible for a long time.

7.6   Velocity correlations in a dense fluid

The Langevin equations provide a Gaussian model for the motion of a system in contact with a bath. As an application of its use, consider an atom moving in a simple fluid, the latter treated as the bath. Writing the generalization of the Langevin equation in terms of the atom's velocity, with no mean force $\omega(x) = 0$,

$$m\frac{d}{dt}v(t) = -2\int_0^t dt'\,\gamma(t-t')v(t') + \eta(t)$$

we can multiply through by the initial velocity, $v(0)$, and average over initial conditions

$$m\frac{d}{dt}\langle v(0)v(t)\rangle = -2\int_0^t dt'\,\gamma(t-t')\,\langle v(0)v(t')\rangle + \langle v(0)\eta(t)\rangle$$

$$\implies m\dot{C}(t) = -2\int_0^t dt'\,\gamma(t-t')C(t')$$

where $C(t) = \langle v(0)v(t)\rangle$. We now have an explicit expression for the velocity autocorrelation function. This equation can be solved by Laplace transforms,

$$s\tilde{C}(s) - C(0) = -2\tilde{\gamma}(s)\tilde{C}(s)/m$$

$$\tilde{C}(s) = \frac{C(0)}{s + 2\tilde{\gamma}(s)/m}$$

which is true for any friction kernel. Taking, $\gamma(t) = \gamma\delta(t)$, in which $\tilde{\gamma}(s) = \gamma$, we can insert this into our expression and invert the Laplace transform, finding,

$$C(t) = C(0)e^{-\gamma t/m}\,.$$

In the Markovian approximation, when the bath relaxes quickly, the velocity correlations are exponential, with a time constant equal to m/γ. Recalling the Green–Kubo equation, which relates the diffusion constant to the integral over the velocity autocorrelation function,

$$D = \int_0^\infty dt\,C(t) = k_{\mathrm{B}}T/\gamma$$

we see that the diffusivity is inversely related to the friction. This is known as the Einstein relation. Inserting the Stokes drag law for a sphere, $\gamma = 6\pi R\mu/\rho$, we find

$$D = k_{\mathrm{B}}T/(6\pi R\mu/\rho)$$

we arrive at the Stokes–Einstein relation. In this relation, R is the particle radius and μ/ρ is the viscosity of the medium divided by its density.

While the Markovian approximation is useful when we are really imagining that the system is a large colloidal particle, if the system is instead made up of the same molecules as the bath, then there is no separation of timescales between the dynamics of the bath and the particle, and the Markovian approximation breaks down. Indeed, in a dense liquid, the velocity autocorrelation function is not a monotonically decreasing function, reflecting the recoil of the momentum from collisions. Examples of $C(t)$ are shown in Figure 7.5. For a dense liquid, a more accurate model must account for the

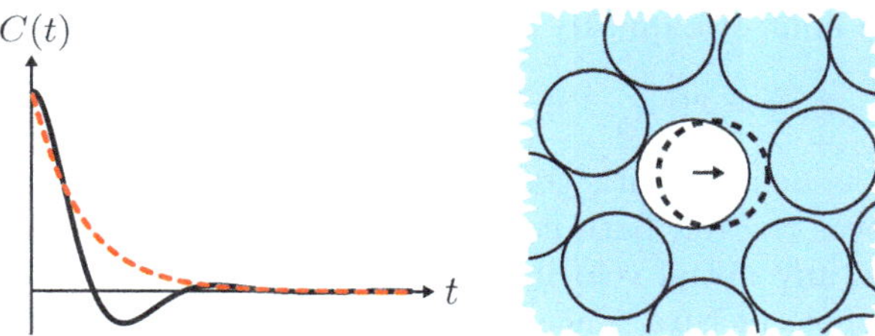

Fig. 7.5 Velocity autocorrelation with the exponential model valid for a dilute solution (dashed line) and from simulations of a dense fluid (solid line). The latter exhibits recoil as illustrated on the right for a tagged particle at high densities.

finite relaxation time of the bath. The simplest such model would assume that the bath relaxes with a characteristic timescale τ_B, so the friction kernel becomes in Laplace space,

$$L\left[\gamma(t) = \gamma e^{-t/\tau_\mathrm{B}}/\tau_\mathrm{B}m\right] = \frac{\gamma/m}{s\tau_\mathrm{B}+1}$$

which upon inserting into our previous expression,

$$\tilde{C}(s) = \frac{C(0)}{s + \frac{2\gamma/m}{s\tau_\mathrm{B}+1}}$$

gives a continued fraction. This can be rewritten and inverse Laplace transformed, which yields,

$$C(t) = C(0)\left[\cos(\bar{\omega}t) + \frac{1}{2\tau_\mathrm{B}\bar{\omega}}\sin(\bar{\omega}t)\right]e^{-t/2\tau_\mathrm{B}}$$

where

$$\bar{\omega} = \sqrt{2\gamma/m\tau_\mathrm{B} - 1/4\tau_\mathrm{B}^2}\,.$$

This expression for the velocity autocorrelation function relates the two parameters, $\bar{\omega}$ and τ_B, to averages of the bath, through which the diffusivity could be evaluated. It also models observed and simulation velocity correlations in dense liquids very well.

Exercise 7.8: Confirm the form of the correlation function $C(t)$ given an exponential memory by performing the explicit inverse Laplace transform.

7.7　Models for diffusive motion

We have shown how by starting with a full set of dynamical equations for a system linearly coupled to a bath of Gaussian variables, we can integrate out the degrees of freedom associated with the bath. The exact result was the generalized Langevin equation, which in the limit of a quickly relaxing bath took the form,

$$\dot{x}(t) = v(t) \qquad m\dot{v}(t) = -\gamma v(t) + \eta(t)$$

where γ is a friction associated with the bath degrees of freedom, and is precisely equal to an integral over the full bath correlation function. If the term proportional to γ was all there was, any initial velocity would decay exponentially to 0, which we know cannot be the whole picture, since at equilibrium, $\langle v^2 \rangle = k_B T/m$. The second term, η, is a random force with zero mean and variance,

$$\langle \eta(t)\eta(t') \rangle = 2k_B T\gamma\delta(t - t')$$

that precisely balances the frictional force, ensuring equilibration with the bath.

The real utility of the Langevin equation is as a model with which to approximate time correlation functions for interacting systems. Here we will consider two correlation functions, those associated with translational and rotational motion. For translational motion, we can take the equation of motion above and integrate it, yielding

$$\Delta x(t) = x(t) - x(0) = \int_0^t dt' \, v(t')$$

where we have defined the displacement $\Delta x(t)$ as we have before. The expectation value of this is zero, so considering the second moment,

$$\langle \Delta x^2(t) \rangle = \int_0^t dt' \int_0^t dt'' \, \langle v(t')v(t'') \rangle$$

$$= 2 \int_0^t d\tau \, (t - \tau) \, \langle v(0)v(\tau) \rangle$$

where we have made the substitutions, $\tau = t'' - t'$, and $\bar{t} = t'' + t'$ and noted the time-translational invariance of the equilibrium correlation function. In order to evaluate this expression, we have to derive an expression for how the velocity changes with time. We can do this by integrating its equation of motion using Laplace transforms,

$$v(t) = v(0)e^{-\gamma t/m} + \frac{1}{m} \int_0^t dt' \, \eta(t')e^{-\gamma(t-t')/m}$$

we find the homogenous initial value part, and an integral over the noise. Multiplying this equation by $v(0)$ and averaging over the noise,

$$\langle v(0)v(t) \rangle = \langle v^2 \rangle \, e^{-\gamma t/m} + \frac{1}{m} \int_0^t dt' \, \langle v(0)\eta(t') \rangle \, e^{-\gamma(t-t')/m}$$

$$= \frac{k_B T}{m} e^{-\gamma t/m}$$

we find that velocity correlations are exponentially decaying with a time constant m/γ, because the noise in the future is uncorrelated with the initial velocity. Inserting this into our expression for the mean squared displacement, and evaluating the integral,

$$\langle \Delta x^2(t) \rangle = \frac{2k_\mathrm{B}T}{m} \int_0^t d\tau \, (t - \tau) e^{-\gamma\tau/m}$$

$$= \frac{2k_\mathrm{B}T}{\gamma} \left(t - m/\gamma + me^{-\gamma t/m}/\gamma \right)$$

we have an expression valid for all times. In Chapter 6 we argued that the limiting behavior of the mean-squared displacement should be simple, that at short times we expect ballistic motions to dominate, and so it should scale as t^2. Evaluating the correlation function for $t \ll m/\gamma$ we find,

$$\langle \Delta x^2(t) \rangle = \frac{k_\mathrm{B}T}{m} t^2$$

that indeed this is the case. At long times we assumed that displacements should become uncorrelated, and should scale linearly in t. Evaluating the correlation function above for $t \gg m/\gamma$, we have

$$\langle \Delta x^2(t) \rangle = \frac{2k_\mathrm{B}T}{\gamma} t = 2Dt$$

which is indeed linear in t with a prefactor that is exactly two times the diffusion constant D.

Another useful calculation is to consider rotational motion in a dense liquid. For example consider the fluctuations of a dipole, $\mu(t)$, as described by the dipole–dipole correlation function. For a molecule with a permanent, fixed dipole,

$$\langle \mu(t) \cdot \mu(t') \rangle = \mu^2 \langle \mathbf{u}(t) \cdot \mathbf{u}(t') \rangle$$

where $\mathbf{u}(t)$ is a unit vector along the dipole. In two dimensions, the unit vector on the circle can be described with polar coordinates

$$\mathbf{u}(t) = (\cos \theta(t), \, \sin \theta(t)) \, .$$

as illustrated in Figure 7.6. The fluctuating part of the correlation function can be rewritten as

$$\langle \mathbf{u}(t) \cdot \mathbf{u}(t') \rangle = \left\langle e^{-i\theta(t)} e^{i\theta(t')} \right\rangle = \left\langle e^{i\Delta\theta(t'-t)} \right\rangle$$

where the dot product is represented by a complex exponential of $\Delta\theta(t - t') \equiv \theta(t') - \theta(t)$.

Now let's turn to a Langevin equation for rotational motion. This equation takes a form identical that in Cartesian coordinates,

$$\dot{\theta}(t) = \omega(t) \qquad I\dot{\omega}(t) = -\gamma_r \omega(t) + \eta(t)$$

where ω is the angular velocity, I is the moment of inertia, and γ_r is the rotational friction. The random force still satisfies a fluctuation–dissipation theorem,

$$\langle \eta(t)\eta(t') \rangle = 2k_\mathrm{B}T\gamma_r\delta(t - t') \, .$$

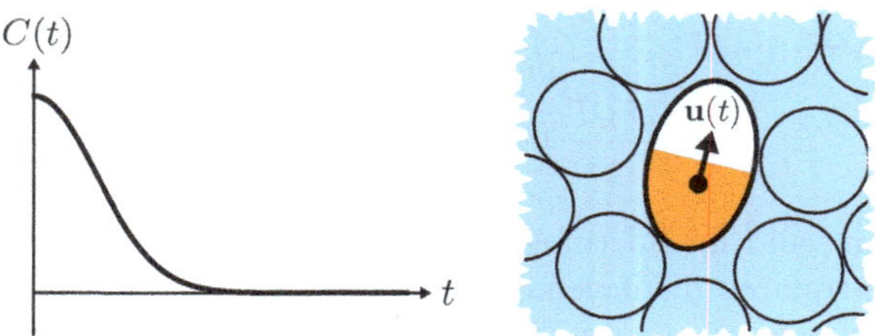

Fig. 7.6 Correlation function for the orientation for rotational diffusive motion within the Langevin model.

Just as before, it is straightforward to show that the average direction is zero,

$$\langle \Delta\theta(t) \rangle = 0$$

and the variance of the direction is

$$\langle \Delta\theta^2(t) \rangle = \frac{2k_{\mathrm{B}}T}{\gamma_r}\left(t - I/\gamma_r + I e^{-\gamma_r t/I}/\gamma_r \right)$$

using the equivalent result from the mean-squared-displacement. Since $\theta(t)$ is linear in the noise, which is a Gaussian random variable by construction, θ is also a Gaussian random variable, and so knowing these two moments are sufficient to compute the correlation function. Specifically,

$$\left\langle e^{i\Delta\theta(t)} \right\rangle = \int d\Delta\theta(t)\, P[\Delta\theta(t)] e^{i\Delta\theta(t)}$$

$$= \frac{1}{\sqrt{2\pi\left\langle \Delta\theta^2(t) \right\rangle}} \int d\Delta\theta(t)\, e^{-\Delta\theta^2(t)/2\left\langle \Delta\theta^2(t) \right\rangle} e^{i\Delta\theta(t)}$$

$$= e^{-\left\langle \Delta\theta^2(t) \right\rangle/2}$$

where we recognize that the correlation function is nothing but a characteristic function. In the limit of $t \ll \gamma_r$

$$\langle \mathbf{u}(0) \cdot \mathbf{u}(t) \rangle = e^{-k_{\mathrm{B}}Tt^2/I}$$

we see that the correlation function decays like a Gaussian, which is a consequence of ballistic motion. The alternative limit, $t \gg \gamma_r$

$$\langle \mathbf{u}(0) \cdot \mathbf{u}(t) \rangle = e^{-2k_{\mathrm{B}}Tt/\gamma_r} = e^{-2D_r t}$$

the decay is like a regular exponential, where we have identified $2k_{\mathrm{B}}T/\gamma_r$ as a rotational diffusion constant.

7.8 Fokker–Planck equation

There exists a complementary description of stochastic processes, analogous to the Liouvillian description for classical mechanics, that constructs the equation of motion

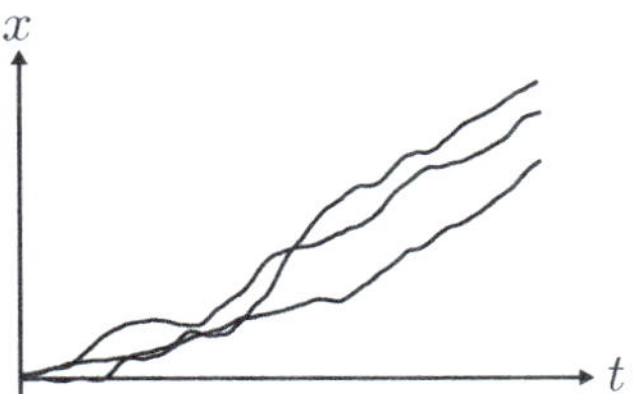

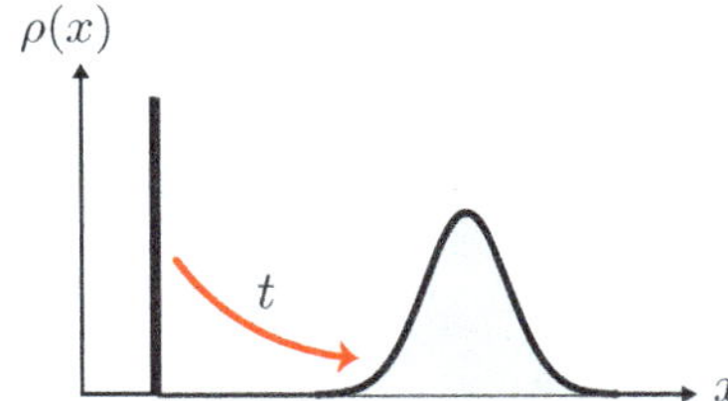

Fig. 7.7 Alternative perspective of stochastic dynamics, as either individual trajectories or their collective ensemble.

for an ensemble of trajectories. These two perspectives are sketched in Figure 7.7. As a concrete case, let's first consider the $1d$ overdamped Langevin equation,

$$\gamma \dot{x}(t) = F(x) + \eta(t) \qquad \langle \eta(t) \rangle = 0 \qquad \langle \eta(t)\eta(t') \rangle = 2k_{\mathrm{B}}T\gamma\delta(t - t')\,.$$

This equation has a deterministic part from the external force, $F(x)$, and a stochastic part from $\eta(t)$. For a fixed time series of η, however, the equation can be viewed analogously to any other classical equation of motion.

As we expect from our discussion of the Liouville equation, the time-dependent probability for phase space points for fixed noise, $p(x, t; \eta(t))$, should obey a continuity equation for the form,

$$\frac{\partial p}{\partial t} = -\frac{\partial}{\partial x}\dot{x}p$$

$$= -\gamma^{-1}\frac{\partial}{\partial x}(F + \eta)p$$

where in the second line we have substituted our equation of motion into the flux of position x. The first piece we can associate with the operator,

$$\mathcal{L} = \gamma^{-1}\frac{\partial}{\partial x}F$$

and, taking this as the homogenous part of the differential equation, formally integrate it to yield,

$$p(x, t; \eta(t)) = e^{-\mathcal{L}t}p(x, 0; \eta(0)) - \int_0^t dt'\, e^{-\mathcal{L}(t-t')}\frac{\partial}{\partial x}\gamma^{-1}\eta(t')p(x, t'; \eta(t'))$$

where we have an integral equation, with an initial value part, and $p(x, t; \eta(t))$ on both sides. It is important to note that the $p(x, t; \eta(t))$ under the integral only depends on the noise up to $t < t'$. Inserting this back into the original expression for $\dot{p}$, we get

$$\frac{\partial p}{\partial t} = -\mathcal{L}\left(e^{-\mathcal{L}t}p(x, 0; \eta(0)) - \int_0^t dt'\, e^{-\mathcal{L}(t-t')}\frac{\partial}{\partial x}\gamma^{-1}\eta(t')p(x, t'; \eta)\right)$$

$$- \frac{\partial}{\partial x}\gamma^{-1}\eta(t)\left(e^{-\mathcal{L}t}p(x, 0; \eta(0)) - \int_0^t dt'\, e^{-\mathcal{L}(t-t')}\frac{\partial}{\partial x}\gamma^{-1}\eta(t')p(x, t'; \eta)\right)$$

a big mess. Really what we care about is $f(x,t) = \langle p(x,t;\eta(t))\rangle$, or the distribution of phase space points averaged over the noise history. Making this substitution and taking the average of both sides of the equation, we get

$$\frac{\partial f}{\partial t} = -\,\mathcal{L}f(x,t) + \mathcal{L}\int_0^t dt'\, e^{-\mathcal{L}(t-t')}\frac{\partial}{\partial x}\gamma^{-1}\,\langle \eta(t')p(x,t';\eta(t'))\rangle$$
$$-\,\frac{\partial}{\partial x}\gamma^{-1}e^{-\mathcal{L}t}\,\langle \eta(t)p(x,0;\eta(0))\rangle$$
$$+\int_0^t dt'\,\frac{\partial}{\partial x}e^{-\mathcal{L}(t-t')}\frac{\partial}{\partial x}\gamma^{-2}\,\langle \eta(t)\eta(t')p(x,t';\eta(t'))\rangle$$

which can be dramatically simplified due to the underlying assumption of Gaussian noise. Specifically, the first term on the right-hand side is independent of the noise, so it stays the same. The second term has an explicit dependence on the noise at time t' and an implicit dependence at time $t' \leq t$. These are uncorrelated and will average to zero. This is the same for the third term, with a dependence on the noise at t and $t = 0$. The last term, however, includes a factor that evaluates to something nonzero. Inserting the delta function and the Einstein relation for the diffusion constant $D = k_{\mathrm{B}}T/\gamma$, we arrive at,

$$\frac{\partial f}{\partial t} = -\mathcal{L}f(x,t) + D\frac{\partial^2}{\partial x^2}f(x,t)$$
$$= \gamma^{-1}\frac{\partial}{\partial x}\left[\frac{\partial U}{\partial x}f(x,t)\right] + D\frac{\partial^2}{\partial x^2}f(x,t)$$

where the last line follows from re-expressing our linear operator. This equation, known as the *Smoluchowski equation*, includes a piece that derives straightforwardly from the deterministic part of the equation of motion, and a piece that results from the noise. The first piece looks like a conventional convective forcing, while the second looks like the spreading associated with diffusion. Noting that the force is the gradient of a potential, this equation becomes,

$$\frac{\partial f}{\partial t} = D\frac{\partial}{\partial x}e^{-\beta U}\frac{\partial}{\partial x}e^{\beta U}f(x,t)$$

assuming that the fluctuation–dissipation theorem holds, $\beta D = 1/\gamma$. In steady state, this expression can be solved, $f_{\mathrm{eq}}(x,t) \propto \exp[-\beta U]$ where we again find that the equilibrium distribution relaxes to a Boltzmann form.

More generally, for any higher-dimensional space obeying a stochastic equation of motion of the form,

$$\dot{\mathbf{a}} = \mathbf{v}(\mathbf{a}) + \mathbf{b} \qquad \langle \mathbf{b}(t) \otimes \mathbf{b}(t')\rangle = \mathbf{B}(\mathbf{a})\delta(t - t')$$

where the $\mathbf{a}$'s are some degrees of freedom, with $\mathbf{v}$ being their velocity, and $\mathbf{b}$ a vector of Gaussian white noise with correlation matrix $\mathbf{B}(\mathbf{a})$, the analogous continuity equation for a given noise history can be written as,

$$\frac{\partial p(\mathbf{a},t;\mathbf{b})}{\partial t} = -\frac{\partial}{\partial \mathbf{a}}\dot{\mathbf{a}}p(\mathbf{a},t;\mathbf{b})\,.$$

By taking an average over different realizations of the noise, and defining the mean distribution as $f(\mathbf{a}, t) = \langle p(\mathbf{a}, t; \mathbf{b}(t)) \rangle$, we can go through the same steps as before. In this case we arrive at the celebrated Fokker–Planck equation,

$$\frac{\partial f(\mathbf{a}, t)}{\partial t} = -\frac{\partial}{\partial \mathbf{a}} \cdot \mathbf{v} f(\mathbf{a}, t) + \frac{1}{2} \frac{\partial}{\partial \mathbf{a}} \frac{\partial}{\partial \mathbf{a}} \mathbf{B} f(\mathbf{a}, t) = -\mathcal{L} f(\mathbf{a}, t)$$

where as previously, there is a part that comes from the deterministic part of the equation of motion, and looks like convection, and a piece that is analogous to diffusion.

Aside: Stochastic calculus and Ito's lemma. The normal rules of calculus need to be generalized to handling the sorts of stochastic differential equations we have derived in this chapter. For concreteness, consider a stochastic differential equation of the form

$$\dot{a} = v(a) + b$$

where $v(a)$ is a deterministic drift and b is a Gaussian random variable with $\langle b \rangle = 0$ and $\langle b(t)b(t') \rangle = B\delta(t - t')$. Since $b(t)$ is a Gaussian variable, its integral is also a Gaussian variable. Let us define an increment db as an integral of b over time δt. The mean of db is 0, and its variance is

$$\int_t^{t+\delta t} dt' \int_t^{t+\delta t} dt'' \, \langle b(t')b(t'') \rangle = B \int_t^{t+\delta t} dt' \int_t^{t+\delta t} dt'' \, \delta(t' - t'')$$

$$= B\delta t$$

implying that db scales naturally like $\delta t^{1/2}$. Since b is singular, it is best interpreted through the differential increment associated db. The significance of this scaling is evident in a distinct chain rule that emerges for stochastic variables. For a function $g(a, t)$, where a obeys the stochastic differential equation, its total derivative is

$$dg = \frac{\partial g}{\partial t} dt + \frac{\partial g}{\partial a} da + \frac{1}{2} \frac{\partial^2 g}{\partial a^2} da^2 + \cdots = \left(v \frac{\partial g}{\partial a} + \frac{B}{2} \frac{\partial^2 g}{\partial a^2} \right) dt + \frac{\partial g}{\partial a} db + \mathcal{O}(dt^{3/2})$$

where due to the sub-linear scaling of db we have to expand the total derivative to second order. Gathering terms up to first order in dt we have

$$\frac{dg}{dt} = \frac{\partial g}{\partial t} + v \frac{\partial g}{\partial a} + \frac{B}{2} \frac{\partial^2 g}{\partial a^2} + \frac{\partial g}{\partial a} b$$

which is known as Ito's lemma.

As a concrete example of this more general equation, consider the Langevin equation of motion. In terms of the variables above it can be written as,

$$\mathbf{a} = \begin{bmatrix} \mathbf{r} \\ \mathbf{p} \end{bmatrix} \qquad \mathbf{v} = \begin{bmatrix} \mathbf{p}/m \\ \mathbf{F}(\mathbf{r}) - \gamma \mathbf{p}/m \end{bmatrix} \qquad \mathbf{B} = \begin{bmatrix} 0 & 0 \\ 0 & 2k_{\mathrm{B}}T\gamma \end{bmatrix}.$$

Substituting these expressions into the Fokker–Planck equations we get,

$$\frac{\partial f(\mathbf{r}, \mathbf{p}, t)}{\partial t} = -\frac{\partial}{\partial \mathbf{r}} \frac{\mathbf{p}}{m} f - \frac{\partial}{\partial \mathbf{p}} \left[\mathbf{F}(\mathbf{r}) - \frac{\gamma}{m} \mathbf{p} \right] f + \gamma k_{\mathrm{B}} T \frac{\partial^2}{\partial \mathbf{p}^2} f$$

$$= -\frac{\mathbf{p}}{m} \frac{\partial}{\partial \mathbf{r}} f - \left[\mathbf{F}(\mathbf{r}) - \frac{\gamma}{m} \mathbf{p} \right] \frac{\partial}{\partial \mathbf{p}} f + \frac{\gamma d}{m} f + \gamma k_{\mathrm{B}} T \frac{\partial^2}{\partial \mathbf{p}^2} f = -\mathcal{L} f(\mathbf{r}, \mathbf{p}, t)$$

which is often referred to as *Kramers' equation*. The steady-state solution of this equation can be solved, yielding $f_{\mathrm{eq}}(\mathbf{r}, \mathbf{p}) \propto \exp[-\beta \mathcal{H}(\mathbf{r}, \mathbf{p})]$, again a Boltzmann distribution with $\mathcal{H}(\mathbf{r}, \mathbf{p})$ the Hamiltonian whose spatial gradient returns the force vector $\mathbf{F}$ and includes a contribution from kinetic energy. In general, Fokker Planck equations are not solvable, and need not support a steady state.

Note that like the Liouville equation, the Fokker–Planck operator, $\mathcal{L}$, can be used to evolve expectation values. However, to do this we need to employ the techniques of stochastic calculus as the stochastic differential equation of motion is not smooth and thus the notion of a derivative requires some care to formulate. In particular, we must employ Ito's lemma. For an observable $g(\mathbf{r}, \mathbf{p})$, this yields

$$\frac{d \langle g \rangle}{dt} = \dot{\mathbf{r}} \frac{\partial \langle g \rangle}{\partial \mathbf{r}} + \dot{\mathbf{p}} \frac{\partial \langle g \rangle}{\partial \mathbf{p}} + \gamma k_{\mathrm{B}} T \frac{\partial^2 \langle g \rangle}{\partial \mathbf{p}^2}$$

$$= \frac{\mathbf{p}}{m} \frac{\partial \langle g \rangle}{\partial \mathbf{r}} + \left[\mathbf{F}(\mathbf{r}) - \frac{\gamma \mathbf{p}}{m} \right] \frac{\partial \langle g \rangle}{\partial \mathbf{p}} + \gamma k_{\mathrm{B}} T \frac{\partial^2 \langle g \rangle}{\partial \mathbf{p}^2} = \mathcal{L}^{\dagger} \langle g \rangle$$

where $\mathcal{L}^{\dagger}$ is identified as the adjoint to the Fokker–Planck operator, with a negative drift on the other side of the first derivative compared to the regular Fokker–Planck operator.

7.9 Ionic conductivity of a dilute electrolyte

We have previously derived the response to a time-dependent perturbation in a somewhat *ad hoc* way. Specifically, we have implicitly assumed that the system regresses to a new equilibrium state after the perturbation, and we have not really approached the response theory from a dynamical perspective. With the discussion of the Fokker–Planck equation, we are now in a position to do so.

Let us consider this concretely in the context of electrical conductivity, or the response of a charged particle to an applied electric field like that shown in Figure 7.8. The equations of motion for an otherwise free particle in a motionless solvent are the Langevin equations

$$\dot{x} = v \qquad m\dot{v} = -\gamma v + q\mathcal{E}(t) + \eta \qquad \langle \eta \rangle = 0 \qquad \langle \eta(t)\eta(t') \rangle = 2\gamma k_{\mathrm{B}} T \delta(t - t')$$

where we work with position, x, and velocity, v rather than momentum) and we will restrict our attention to $1d$. As usual m is the mass, and γ is the friction due to the environment. The charge of the particle is q, and the time-dependent electric field is $\mathcal{E}(t)$. The corresponding Fokker-Planck equation can be written as

$$\frac{\partial f(x, v, t)}{\partial t} = \left(-v \frac{\partial}{\partial x} + \frac{\partial}{\partial v} \frac{\gamma}{m} v + \frac{\gamma k_{\mathrm{B}} T}{m^2} \frac{\partial}{\partial v^2} \right) f + \left(-\frac{1}{m} q\mathcal{E}(t) \frac{\partial}{\partial v} \right) f$$

$$= -\mathcal{L}_0 f - \mathcal{L}_1(t) f$$

where we separate the equilibrium and time-dependent parts of the operator.

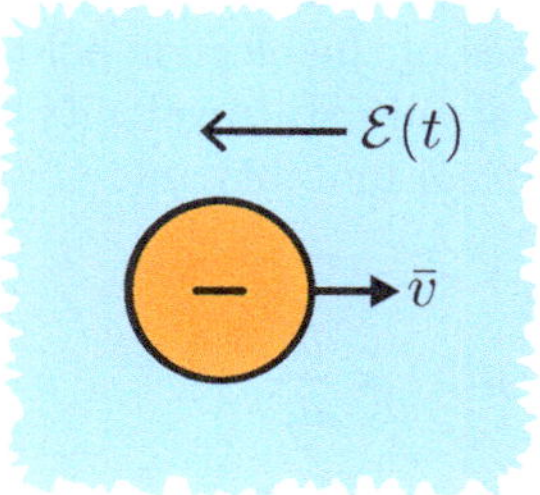

Fig. 7.8 The mobility of an ion in solution relates the steady-state velocity to the applied electric field.

Until now everything is exact. If we were able to solve the Fokker–Planck equation above, we could predict the average motion of the particle under the action of the applied electric field. However, in general this is not possible. Instead we will imagine the usual linear response set up where the particle begins in equilibrium, where there is no applied field. At $t = 0$ the electric field is turned on, but its amplitude is small, so the change in the stationary distribution is small. In such a perturbative limit we can assume that the distribution $f(x, v, t)$ can be written as

$$f(x, v, t) = f_{\text{eq}}(x, v) + \Delta f(x, v, t) \qquad f(x, v, 0) = f_{\text{eq}}(x, v)$$

where $\Delta f(x, v, t)$ is the presumed time-dependent correction to the equilibrium state assumed to be linear in the field, and we note explicitly that the particle begins at equilibrium. Under this assumption, to first-order in perturbation theory we arrive at

$$\frac{\partial f_{\text{eq}}(x, v)}{\partial t} = 0 \qquad \frac{\partial \Delta f(x, v, t)}{\partial t} = -\mathcal{L}_0 \Delta f(x, v, t) - \mathcal{L}_1(t) f_{\text{eq}}(x, v)$$

where the first equation represents the zeroth-order condition of stationarity of the equilibrium distribution. The second equation is a first-order inhomogeneous differential equation which can be integrated, yielding,

$$\Delta f(x, v, t) = e^{-\mathcal{L}_0 t} \Delta f(x, v, 0) - \int_0^t dt' \, e^{-\mathcal{L}_0(t-t')} \mathcal{L}_1(t') f_{\text{eq}}(x, v)$$

$$= -\int_0^t dt' \, e^{-\mathcal{L}_0(t-t')} \mathcal{L}_1(t') f_{\text{eq}}(x, v)$$

where the second line follows from our assumed initial condition. Given this approximate form of the distribution function, we can compute averages of some test function $A(x, v)$ as integrals over the phase space of x and v

$$\bar{A}(t) = \int dx \int dv \, A(x, v) \left[f_{\text{eq}}(x, v) + \Delta f(x, v, t) \right]$$

$$= \langle A \rangle + \int dx \int dv \, A(x, v) \Delta f(x, v, t)$$

where we have the contribution from equilibrium and its correction.

Let's imagine that the applied electric field is monochromatic, i.e., $\mathcal{E}(t) = \mathcal{E}\cos(\omega t)$ and turned on at $t = 0$. In the presence of the electric field, the particle is expected to move. The differential change in the steady state velocity of the particle with electric field amplitude is known as the electrical mobility, $\mu(\omega)$, which in general depends on the frequency of the applied field,

$$\mu(\omega) = \frac{d\bar{v}}{d\mathcal{E}}\,.$$

Putting this together we can compute the average velocity in linear response,

$$\bar{v}(t) = \langle v \rangle + \beta q\mathcal{E}\int dx \int dv \int_0^t dt'\,\cos(\omega t')v(0)e^{-\mathcal{L}_0(t-t')}v(0)f_{\text{eq}}(x,v)$$

$$= \langle v \rangle + \beta q\mathcal{E}\int dx \int dv \int_0^t dt'\,\cos(\omega t')v(0)v(t'-t)f_{\text{eq}}(x,v)$$

$$= \beta q\mathcal{E}\int_0^t dt'\,\cos(\omega t')\,\langle v(0)v(t'-t)\rangle$$

where we acted $\mathcal{L}_1(t')$ on the Boltzmann distribution to bring down a factor of βvm, and find as before that at first-order in $\mathcal{E}$ we are left with a response that depends on an equilibrium correlation function. Taking the derivative and the long time limit

$$\mu(\omega) = \beta q\int_0^\infty dt'\,\cos(\omega t')\,\langle v(0)v(t')\rangle = \beta q C(\omega)$$

we find the frequency-dependent ionic mobility is the real part of the Fourier transform of the velocity autocorrelation function, $C(\omega) = \mathcal{F}[\langle v(0)v(t)\rangle]$, which follows from the evenness of the autocorrelation function and that $\mathcal{E}(t)$ is 0 for $t < 0$.

Exercise 7.9: Evaluate the frequency-dependent mobility, $\mu(\omega)$, using the Markovian approximation to the Langevin equation

$$m\dot{v} = -\gamma v + \eta \qquad \langle \eta(t)\rangle = 0 \qquad \langle \eta(t)\eta(t')\rangle = 2k_{\text{B}}T\gamma\delta(t-t')$$

and show that it has a Lorentzian form, $\mu(\omega) = q/[\gamma(1 + \omega^2\tau^2)]$ where $\tau = m/\gamma$.

7.10 Fluctuating hydrodynamics

The stochastic equations of motion we have derived by integrating out some degrees of freedom, illustrate how particles evolve effectively within weakly interacting surroundings. This evolution is part deterministic and part random. Observables that depend on the particles' position thus similarly must evolve through a combination of a deterministic and random motion. To see this explicitly, we can consider developing an equation of motion for a collective property of the system like its density field, as we did previously within the Hamiltonian mechanical framework in Chapter 6.

For concreteness take the overdamped equation of motion, $\dot{\mathbf{r}}_i = \mu \mathbf{F}_i[\mathbf{r}^N] + \boldsymbol{\eta}_i$ where $\mathbf{F}_i$ is the conservative force, $\mathbf{F}_i(\mathbf{r}^N) = -\nabla_i \sum_j u_2(|\mathbf{r}_i - \mathbf{r}_j|)$, given by the gradient of a sum of pair potentials $u_2(r)$, while $\mu = 1/\gamma$ is the mobility of a particle, and $\boldsymbol{\eta}_i$ is a Gaussian random force with $\langle \boldsymbol{\eta}_i \rangle = 0$ and $\langle \boldsymbol{\eta}_i(t)\boldsymbol{\eta}_j(t')\rangle = 2k_\mathrm{B}T\mu \mathbf{1}\delta_{ij}\delta(t - t')$. Any function of a single point in space can be mapped to the time-dependent particle positions using the delta function relation

$$f[\mathbf{r}_i(t)] = \int d\mathbf{r}\, f(\mathbf{r})\delta[\mathbf{r} - \mathbf{r}_i(t)]$$

where f is an arbitrary test function. Taking the time derivative of this test function we obtain an equation of motion for the density.

In order to evaluate derivatives of the form

$$\frac{df(\mathbf{r}_i)}{dt} = \int d\mathbf{r}\, \delta[\mathbf{r} - \mathbf{r}_i(t)]\frac{d}{dt}f(\mathbf{r})$$

we need to apply Ito's lemma. Using this rule for our test function we have,

$$\frac{df(\mathbf{r}_i)}{dt} = \int d\mathbf{r}\, \delta[\mathbf{r} - \mathbf{r}_i(t)]\left[-\mu\left(\frac{\partial}{\partial\mathbf{r}}\sum_j u_2(|\mathbf{r} - \mathbf{r}_j|)\right)\frac{\partial}{\partial\mathbf{r}} + \mu k_\mathrm{B}T\frac{\partial^2}{\partial\mathbf{r}^2} + \eta_i\frac{\partial}{\partial\mathbf{r}}\right]f(\mathbf{r})$$

$$= \int d\mathbf{r}\, f(\mathbf{r})\left[\mu\frac{\partial}{\partial\mathbf{r}}\left(\frac{\partial}{\partial\mathbf{r}}\sum_j u_2(|\mathbf{r} - \mathbf{r}_j|)\right) - \frac{\partial}{\partial\mathbf{r}}\eta_i + \mu k_\mathrm{B}T\frac{\partial^2}{\partial\mathbf{r}^2}\right]\delta[\mathbf{r} - \mathbf{r}_i(t)]$$

where in the second line we have employed Ito's lemma and in the third we have integrated by parts. Equivalently, the time derivative could be consider as being contained by the time dependence of the delta function,

$$\frac{df(\mathbf{r}_i)}{dt} = \int d\mathbf{r}\, f(\mathbf{r})\frac{\partial}{\partial t}\delta[\mathbf{r} - \mathbf{r}_i(t)]$$

and both of these statements must be true for any test function. Since the fluctuating density field is given by a sum over the time-dependent delta functions,

$$\rho(\mathbf{r}, t) = \sum_i \delta[\mathbf{r} - \mathbf{r}_i(t)]$$

then setting the arguments of the integrals over our test function equal to each other and summing over all particle indices, we arrive at

$$\frac{\partial}{\partial t}\rho(\mathbf{r}, t) = \left[\mu\frac{\partial}{\partial\mathbf{r}}\left(\frac{\partial}{\partial\mathbf{r}}\sum_j u_2(|\mathbf{r} - \mathbf{r}_j|)\right) + \mu k_\mathrm{B}T\frac{\partial^2}{\partial\mathbf{r}^2}\right]\rho(\mathbf{r}, t) - \frac{\partial}{\partial\mathbf{r}}\sum_i \eta_i\delta[\mathbf{r} - \mathbf{r}_i(t)]$$

which is a stochastic equation for the density field. However, it is not closed, as it depends on the individual noises from the particle's stochastic equations of motion. These can be eliminated by introducing a new Gaussian random variable $N(\mathbf{r}, t)$ with zero mean and a variance that depends on $\rho(\mathbf{r}, t)$. Specifically, defining $N(\mathbf{r}, t) = \sum_i \boldsymbol{\eta}_i(t)\delta[\mathbf{r} - \mathbf{r}_i(t)]$, its covariance is

$$
\langle N(\mathbf{r}, t)N(\mathbf{r}', t')\rangle = \sum_{i,j} \langle \boldsymbol{\eta}_i(t)\boldsymbol{\eta}_j(t')\rangle \, \delta[\mathbf{r} - \mathbf{r}_i(t)]\delta[\mathbf{r}' - \mathbf{r}_j(t')]
$$

$$
= 2\mu k_{\mathrm{B}} T \rho(\mathbf{r}, t)\delta(\mathbf{r} - \mathbf{r}')\delta(t - t')
$$

where we have used the properties of the delta functions and the noise. Additionally, we can rewrite the pair potential as

$$
\sum_j u_2(|\mathbf{r} - \mathbf{r}_j|) = \int d\mathbf{r}' u_2(\mathbf{r}' - \mathbf{r})\rho(\mathbf{r}')
$$

an integral of the density at another point in space. Inserting this and the noise into the above equation of motion for the density we find the *Kawasaki–Dean equation*

$$
\frac{\partial}{\partial t}\rho(\mathbf{r}, t) = \mu\nabla\rho(\mathbf{r}, t)\nabla \int d\mathbf{r}' \rho(\mathbf{r}', t)u_2(\mathbf{r}' - \mathbf{r}) + \mu k_{\mathrm{B}} T \nabla^2 \rho(\mathbf{r}, t) + \nabla N(\mathbf{r}, t)
$$

which is now a closed equation of motion for the density field, that fluctuates due to the noise $N(\mathbf{r}, t)$ and is non-local due to the potential that couple two points in space.

The structure of this stochastic differential equation has an important difference from those we have previously seen. In the expression above, the density field is a fluctuating variable, and the noise that drives its random motion depends on the density. This is known as multiplicative noise, as the noise is multiplied by the random variable. This causes some trouble in the interpretation of the stochastic differential equation, as there is no unique means of evaluating integrals over short time intervals. Two standard conventions exist, known as Ito and Stratonovich. The discussion of Ito's lemma in the previous aside uses the Ito convention implicitly. While we will rarely need to be concerned with this distinction, as most physical equations do not have multiplicative noise, when we do we will use the Ito convention.

The fluctuating equation for the density field can be made more compact by defining a functional of the density field $\mathcal{F}[\rho]$,

$$
\beta\mathcal{F}[\rho] = \frac{\beta}{2} \int d\mathbf{r} \int d\mathbf{r}' \rho(\mathbf{r}, t)u_2(\mathbf{r}' - \mathbf{r})\rho(\mathbf{r}', t) + \int d\mathbf{r}\rho(\mathbf{r}, t)\left[\ln \rho(\mathbf{r}, t) - 1\right]
$$

which is a sum of the interaction term dependent on densities at two points in space, while the second term is a self-term dependent on only one point in space. This functional is often equated to the free energy, as the first term is clearly identified as the potential energy, while the second term is equal to the contribution to the entropy from translational degrees of freedom. This identification together with Einstein's relation $D = \mu k_{\mathrm{B}} T$ allows us to write the fluctuating density equation as

$$
\frac{\partial}{\partial t}\rho(\mathbf{r}, t) = D\nabla \left[\rho(\mathbf{r}, t)\nabla\frac{\delta \beta\mathcal{F}[\rho]}{\delta \rho(\mathbf{r})}\right] + \nabla N(\mathbf{r}, t)
$$

where $\delta\beta\mathcal{F}[\rho]/\delta\rho(\mathbf{r})$ is equivalent to a chemical potential. On average in the limit that particles do not interact, this expression is identical to Fick's law of diffusion.

Aside: Multiplicative noise. The mathematical difficulty in interpreting stochastic differential equations with multiplicative noise can be understood by considering an equation of the form

$$\dot{x} = g(x)\eta(t) \qquad \langle\eta(t)\rangle = 0 \qquad \langle\eta(t)\eta(t')\rangle = \delta(t-t')$$

which is equivalent to $\dot{x} = \tilde{\eta}(t)$, with $\langle\tilde{\eta}(t)\rangle = 0$ and $\langle\tilde{\eta}(t)\tilde{\eta}(t')\rangle = g(x)^2\delta(t-t')$ by a change of Gaussian variables. Integrating the equation over δt,

$$x(t+\delta t) - x(t) = \int_t^{t+\delta t} dt'\, g[x(t')]\eta(t')$$

we have an integral of the form

$$I = \int_t^{t+\delta t} dt'\, g[x(t')]\eta(t') = g[x(t)] + \alpha(x(t+\delta t) - x(t)) \int_t^{t+\delta t} dt'\eta(t')$$

for which there is a continuum of potential ways to evaluate it, here parameterized by $\alpha = [0,1]$. This is due to the fact that η is not smooth, so the mean-value theorem does not hold. In other words, even as $\delta t \to 0$, I does not go to $g[x(t)]\eta(t)\delta t$. If $g(x)$ is a smooth function then,

$$g[x(t)] + \alpha(x(t+\delta t) - x(t)) \approx g[x(t)] + \alpha[x(t+\delta t) - x(t)]\frac{dg[x(t)]}{dx} + \dots$$

and inserting this into the integrated equation of motion

$$x(t+\delta t) - x(t) = \left(g[x(t)] + \alpha[x(t+\delta t) - x(t)]\frac{dg[x(t)]}{dx}\right) \int_t^{t+\delta t} dt'\eta(t')$$

$$= g[x(t)] \int_t^{t+\delta t} dt'\eta(t') + \alpha g[x(t)]\frac{dg[x(t)]}{dx} \int_t^{t+\delta t} dt' \int_t^{t+\delta t} dt''\eta(t')\eta(t'')$$

and averaging over the noise

$$\langle x(t+\delta t) - x(t)\rangle = \alpha g[x(t)]\frac{dg[x(t)]}{dx}\delta t$$

we find an average drift of the particle even in the absence of a force. If $\alpha = 0$, the Ito convention, this drift vanishes. If $\alpha = 1/2$, the Stratonovich convention, it persists and for thermodynamic consistency the drift would need to be subtracted from an equation of motion or its associated Fokker–Planck equation.

Exercise 7.10: Demonstrate that in the limit that the particles do not interact with each other, the fluctuating equation of motion for the density field reduces to

$$\frac{\partial}{\partial t}\rho(\mathbf{r}, t) = D\nabla^2 \rho(\mathbf{r}, t) + \nabla N(\mathbf{r}, t)$$

which is the stochastic version of Fick's second law.

The non-local density dependence entering through the pair potential term in the fluctuating density field equation makes it difficult to study this equation analytically. Strategies to simplify it to make the equation tractable either ignore this term, or assume a relatively spatially local functional form for $\beta\mathcal{F}[\rho]$. For example, to study the behavior of a fluid near phase-coexistence, one could employ the Landau theory we developed in Chapter 4, and postulate

$$\beta\mathcal{F}[\rho] = \frac{1}{2}\int d\mathbf{r}\, a\rho(\mathbf{r})^2(\rho(\mathbf{r}) - \bar{\rho})^2 + \gamma|\nabla\rho(\mathbf{r})|^2$$

which supports stable phases for $\rho \approx 0$ and $\bar{\rho}$, meant to approximate the stability of a liquid and vapor. The parameters a and γ are phenomenological constants. Inserting this density functional into the fluctuating density equation results in the so-called *Cahn–Hilliard equation*

$$\frac{\partial}{\partial t}\rho = D\nabla^2\left[a\rho(2\rho - \bar{\rho})(\rho - \bar{\rho}) - \gamma\nabla^2\rho\right] + \nabla N(\mathbf{r}, t)$$

which due to the cubic dependence on ρ is still not solvable analytically, but is often studied numerically. However, imagine starting in an equilibrium state of the fluid above the critical point where the density is uniform everywhere. At $t = 0$ consider quenching the temperature to below T_c, domains of the stable phase will emerge in a process known as *spinodal decomposition*. The equation of motion for the density field can tell us how order will evolve at short times.

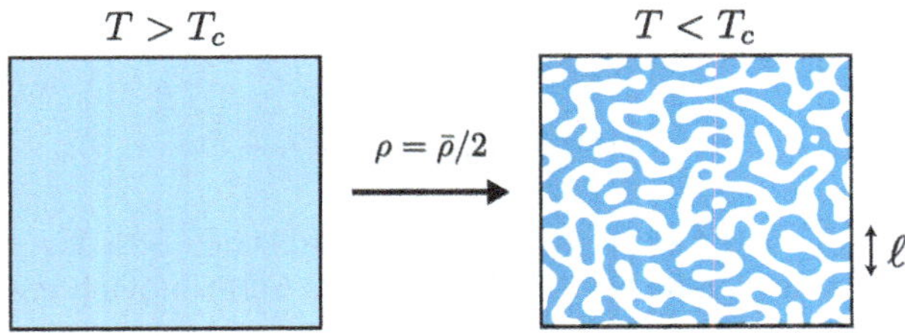

Fig. 7.9 Illustration of spinodal decomposition predicted from the Cahn–Hilliard equation.

Introducing a Fourier representation of the density field of the form

$$\rho(\mathbf{r}, t) = \int d\mathbf{r}\, e^{i\mathbf{k}\cdot\mathbf{r}} \hat{\rho}(\mathbf{k}, t)$$

we find on average, after linearizing ρ about $\bar{\rho}/2$,

$$\partial_t \langle \hat{\rho}(\mathbf{k}, t) \rangle = k^2 a D \bar{\rho}^2 \left(1 - 2k^2\ell^2\right) \langle \hat{\rho}(\mathbf{k}, t) \rangle / 2$$

where $\ell^2 = \gamma/a\bar{\rho}^2$ is a characteristic lengthscale related to the correlation length in Landau theory. This expression predicts that fluctuations grow fastest for modes with wavelengths equal to $1/2\ell$. A cartoon of this phase transition mechanism is shown in Figure 7.9. The growth of density fluctuations is the essence of spinodal decomposition or the process by which phase separation occurs in the absence of metastability.

7.11   Linear model for quantum relaxation

The von Neumann equation specifies the time evolution of a closed quantum system. Like its classical analog, it admits a Boltzmann distribution as a stationary state, but also conserves energy. As such, it cannot describe real, irreversible relaxation. To describe relaxation quantum mechanically, we follow the same path we did classically and ask about the dynamics of a subset of degrees of freedom embedded in an infinite environment. Such a reduced description, under sensible approximations, should admit a description of energy transfer and subsequently relaxation. An illustration of what we will do is shown in Figure 7.10.

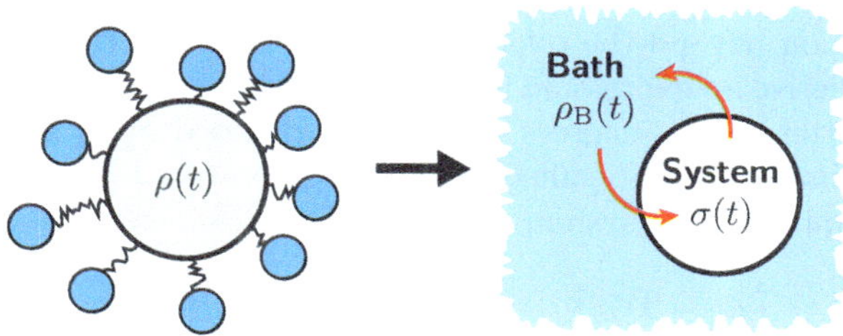

Fig. 7.10 Schematic of tracing out a system coupled to a set of harmonic bath modes.

Consider a system defined by the Hamiltonian $\hat{\mathcal{H}}$ that is separable into three terms

$$\hat{\mathcal{H}} = \hat{\mathcal{H}}_S + \hat{\mathcal{H}}_B + \hat{V}$$

where $\hat{\mathcal{H}}_S$ acts on the Hilbert space of the degrees of freedom we wish to follow—the system, $\hat{\mathcal{H}}_B$ acts on the Hilbert space of the environment that will be integrated out—the bath, and $\hat{V}$ couples the two. We will assume that the bath is made out of harmonic oscillators, and that the coupling takes a bilinear form

$$\hat{V} = \hat{S} \otimes \sum_{n}^{N} a_n \hat{x}_n$$

where $\hat{S}$ operates in the system, $\hat{x}_n$ are displacements in the bath degrees of freedom, weighted by a_n, and $\otimes$ denotes a Kroniker product.

The corresponding von Neumann equation for the full density matrix of the system plus bath can be written as

$$\frac{\partial}{\partial t}\hat{\rho} = -i\hat{\mathcal{L}}_0\hat{\rho} - \frac{i}{\hbar}[\hat{V}(t), \hat{\rho}(t)]$$

where we separate a reference operator $\hat{\mathcal{L}}_0\cdot = [\hat{\mathcal{H}}_S + \hat{\mathcal{H}}_B, \cdot]/\hbar$ that determines the evaluation of the density matrix in the absence of the coupling $\hat{V}$ and a part that reflects the coupling. In principle we could try to solve for the evolution of the full density matrix, but for example if the number of bath degrees of freedom was large, or infinite, this would not be tractable. What we would like is to understand the impact of the bath on the dynamics of the system. To proceed, it is useful to move into the so-called *interaction picture* in which we follow the evolution of the density matrix referenced to the phase generated in the absence of the system bath coupling. Denoting operators under this similarity transform with subscripts I, we define

$$\hat{\rho}_I(t) = e^{i\hat{\mathcal{H}}_0 t/\hbar}\hat{\rho}(t)e^{-i\hat{\mathcal{H}}_0 t/\hbar} \qquad \hat{V}_I(t) = e^{i\hat{\mathcal{H}}_0 t/\hbar}\hat{V}e^{-i\hat{\mathcal{H}}_0 t/\hbar}$$

where $\hat{\mathcal{H}}_0 = \hat{\mathcal{H}}_S + \hat{\mathcal{H}}_B$. In this gauge, the evolution equation is particularly simple,

$$\frac{\partial}{\partial t}\hat{\rho}_I = -i\hat{\mathcal{L}}_V\hat{\rho}_I = -\frac{i}{\hbar}[\hat{V}_I(t), \hat{\rho}_I(t)]$$

where additional motion beyond the reference is given by the action of the coupling.

We would like to derive an expression for the *reduced density matrix* that only spans the Hilbert space of the system degrees of freedom, denoted $\hat{\sigma}_I(t)$ in the interaction picture. To do so we introduce a so-called projection operator, $\hat{\mathcal{P}}$, whose action on the density matrix is to pull out the system part,

$$\hat{\mathcal{P}}\hat{\rho}_I = \mathrm{Tr}_B[\hat{\rho}_I(t)] \otimes \hat{\rho}_B = \hat{\sigma}_I(t) \otimes \hat{\rho}_B$$

where $\hat{\rho}_B$ is the equilibrium state of the bath. We can also define the complement of $\hat{\mathcal{P}} = 1 - \hat{\mathcal{Q}}$, where $\hat{\mathcal{P}}^2 = \hat{\mathcal{P}}$ and $\hat{\mathcal{P}}\hat{\mathcal{Q}} = \hat{\mathcal{Q}}\hat{\mathcal{P}} = 0$. Introducing these operators into the evolution equation for the density matrix, we arrive at two coupled equations

$$\frac{\partial}{\partial t}\hat{\mathcal{P}}\hat{\rho}_I = -i\hat{\mathcal{P}}\hat{\mathcal{L}}_V\hat{\mathcal{P}}\hat{\rho}_I - i\hat{\mathcal{P}}\hat{\mathcal{L}}_V\hat{\mathcal{Q}}\hat{\rho}_I \qquad \frac{\partial}{\partial t}\hat{\mathcal{Q}}\hat{\rho}_I = -i\hat{\mathcal{Q}}\hat{\mathcal{L}}_V\hat{\mathcal{P}}\hat{\rho}_I - i\hat{\mathcal{Q}}\hat{\mathcal{L}}_V\hat{\mathcal{Q}}\hat{\rho}_I$$

which encode the evolution of the reduced density matrix of the system, and the time dependence of an irrelevant part $\hat{\mathcal{Q}}\hat{\rho}_I$. We can eliminate the expression for the irrelevant part by integrating it

$$\hat{\mathcal{Q}}\hat{\rho}_I = \hat{U}(t,0)\hat{\mathcal{Q}}\hat{\rho}_I(0) - i\int_0^t dt' \hat{U}(t,t')\hat{\mathcal{Q}}\hat{\mathcal{L}}_V(t')\hat{\mathcal{P}}\hat{\rho}_I(t')$$

noting the propagator

$$\hat{U}(t, t_0) = \mathcal{T}_+ e^{-i \int_{t_0}^t dt' \hat{\mathcal{Q}} \hat{\mathcal{L}}_V(t')}$$

as a time-ordered exponential as we saw in Chapter 6. Introducing the equation for $\hat{\mathcal{Q}} \hat{\rho}_I$ into the equation for $\hat{\mathcal{P}} \hat{\rho}_I$, we get

$$\frac{\partial}{\partial t} \hat{\mathcal{P}} \hat{\rho}_I = -i \hat{\mathcal{P}} \hat{\mathcal{L}}_V \hat{\mathcal{P}} \hat{\rho}_I - i \hat{\mathcal{P}} \hat{\mathcal{L}}_V \left(\hat{U}(t, 0) \hat{\mathcal{Q}} \hat{\rho}_I(0) - i \int_0^t dt' \hat{U}(t, t') \hat{\mathcal{Q}} \hat{\mathcal{L}}_V(t') \hat{\mathcal{P}} \hat{\rho}_I(t') \right)$$

which can be rewritten compactly as a closed expression for the reduced density matrix

$$\frac{\partial}{\partial t} \hat{\sigma}_I(t) = -i \hat{\mathcal{P}} \hat{\mathcal{L}}_V \sigma_I(t) - i \mathcal{I}(t) - \int_0^t dt' \hat{\mathcal{K}}(t, t') \hat{\sigma}_I(t')$$

a formally exact equation of motion known as the *Nakajima–Zwanzig equation*. This equation of motion is non-Markovian, with a memory kernel operator

$$\hat{\mathcal{K}}(t, t') = \hat{\mathcal{P}} \hat{\mathcal{L}}_V(t) \hat{U}(t, t') \hat{\mathcal{Q}} \hat{\mathcal{L}}_V(t') \hat{\mathcal{P}}$$

and an additional inhomogeneous term, $\mathcal{I}(t) = \hat{\mathcal{P}} \hat{\mathcal{L}}_V \hat{U}(t, 0) \hat{\mathcal{Q}} \hat{\rho}_I(0)$ that depends on the initial condition, in addition to a mean contribution from the bath encoded in $\hat{\mathcal{P}} \hat{\mathcal{L}}_V \sigma_I(t)$. For the factorized initial condition in the definition of $\hat{\mathcal{P}}$, $\mathcal{I}(t) = 0$.

The Nakajima–Zwanzig equation is formally exact, analogous to the generalized Langevin equation, but requires the evaluation of a memory kernel to use, which is often just as cumbersome as evaluating the full dynamics of the full system plus bath. If we assume that the system and bath interact weakly, we can simplify its expression using the *Born approximation*. Specifically, we can ignore the additional propagation of the bath, $\hat{U}(t, t') \hat{\mathcal{Q}} \approx 1$,

$$\hat{\mathcal{K}}(t, t') \approx \hat{\mathcal{P}} \hat{\mathcal{L}}_V(t) \hat{\mathcal{L}}_V(t')$$

which has the effect of treating the system bath coupling to second order in perturbation theory. This reduces the equation of motion to

$$\frac{\partial}{\partial t} \hat{\sigma}_I(t) = -i \mathrm{Tr}_B \left\{ \hat{\mathcal{L}}_V \sigma_I(t) \otimes \rho_B \right\} - \int_0^t dt' \mathrm{Tr}_B \left\{ \hat{\mathcal{L}}_V(t) \hat{\mathcal{L}}_V(t') \hat{\sigma}_I(t') \otimes \rho_B \right\}$$

a coherent, unitary term and an additional dissipative, non-Markovian term. Within the Born approximation, the terms in the integrated evolution equation for $\hat{\sigma}_I(t)$ can be worked out explicitly. For example, the first term on the right-hand side of the equation can be evaluated as

$$\mathrm{Tr}_B \left\{ \hat{\mathcal{L}}_V \sigma_I(t) \otimes \rho_B \right\} = \mathrm{Tr}_B \{ [V_I(t'), \sigma_I(t) \otimes \rho_B] \} / \hbar = \lambda [\hat{S}_I(t'), \hat{\sigma}_I(t')] / \hbar$$

where $\lambda = \sum_n \mathrm{Tr}_B \{ a_n \hat{x}_n(t) \rho_B \}$ is a constant energy shift due to the coupling from the bath. This term is not especially consequential, as we could have defined our initial oscillators in the bath to have a mean position of 0. The later term on the right-hand side, however, is nontrivial. It can be evaluated with the following bit of algebra

$$
\mathrm{Tr}_B[\hat{\mathcal{L}}_V(t)\hat{\mathcal{L}}_V(t')\sigma_I(t')\otimes\rho_B] = C_B(t'-t)\rangle[\hat{S}_I(t),\hat{S}_I(t')\hat{\sigma}_I(t')]/\hbar^2
$$
$$
- C_B^*(t'-t)\rangle[\hat{S}_I(t),\hat{\sigma}_I(t')\hat{S}_I(t')]/\hbar^2
$$

where

$$
C_B(t'-t) = \sum_n a_n^2 \langle \hat{x}_n(t)\hat{x}_n(t')\rangle
$$

is the equilibrium weighted bath correlation function, and $C_B(t'-t) = C_B^*(t-t')$. Note there is a sum over only n, as we will assume that the oscillators in the bath are uncorrelated.

Exercise 7.11: Show that for a classical harmonic bath, $C_B(t)$ can be written as

$$
C_B(t) = \frac{1}{\beta}\int_0^\infty d\omega \frac{1}{\omega^2} J(\omega)\cos(\omega t)
$$

where $J(\omega) = \sum_n a_n^2 \delta(\omega - \omega_n)$ is the so-called spectral density.

Plugging these terms back into the integrated evaluation equation, we find

$$
\frac{\partial}{\partial t}\hat{\sigma}_I(t) = -\frac{i}{\hbar}\lambda[\hat{S}_I,\hat{\sigma}_I(t)] - \frac{1}{\hbar^2}\int_0^t dt' \left\{ C_B(t-t')[\hat{S}_I(t),\hat{S}_I(t')\hat{\sigma}_I(t')] \right.
$$
$$
\left. - C_B^*(t-t')[\hat{S}_I(t),\hat{\sigma}_I(t')\hat{S}_I(t')] \right\}
$$

which is now an intergro-differential equation for $\sigma_I(t)$. Finally, putting it back into Schrodinger representation, and changing variables in the integral from $t-t' \to t'$,

$$
\frac{\partial}{\partial t}\hat{\sigma}(t) = -\frac{i}{\hbar}[\hat{\mathcal{H}}_S + \lambda\hat{S},\hat{\sigma}(t)] - \frac{1}{\hbar^2}\int_0^t dt' C_B(t')[\hat{S}, e^{-i\mathcal{H}_S t'/\hbar}\hat{S}\hat{\sigma}(t-t')e^{i\mathcal{H}_S t'/\hbar}]
$$
$$
- C_B^*(t')[\hat{S}, e^{-i\mathcal{H}_S t'/\hbar}\hat{\sigma}(t-t')\hat{S}e^{i\mathcal{H}_S t'/\hbar}]
$$

we find an equation that is not time local, and is hence non-Markovian. This equation is a result of second-order time-dependent perturbation theory in the system-bath coupling. Its connection to linear response is clarified by the observation of the bath correlation functions entering into the equations, and thus is valid when the interaction is weak or the bath is Gaussian.

To make the solution of the reduced density dynamics more tractable, in addition to assuming the bath is weakly coupled, we can assume it relaxes quickly relative to the characteristic timescales of the system. Applying such a Markovian approximation in this case is most straightforwardly done element-wise in the density matrix. Evaluating the density matrix $\sigma_{ab}(t) = \langle a|\hat{\sigma}(t)|b\rangle$ with $\langle a|[\hat{\mathcal{H}}_S,\hat{\sigma}(t)]|b\rangle/\hbar = \omega_{ab}\sigma_{ab}$, we find

$$\frac{\partial}{\partial t}\sigma_{ab}(t) = -i\omega_{ab}\sigma_{ab} - \frac{i}{\hbar}\lambda\sum_c S_{ac}\sigma_{cb} - S_{cb}\sigma_{ac}$$

$$-\frac{1}{\hbar^2}\sum_{cd}\int_0^t dt' C_B(t')\left[S_{ac}S_{cd}\sigma_{db}(t-t')e^{-i\omega_{cb}t'} - S_{ac}S_{db}\sigma_{cd}(t-t')e^{-i\omega_{ad}t'}\right]$$

$$-C_B^*(t')\left[S_{ac}S_{db}\sigma_{cd}(t-t')e^{-i\omega_{cb}t'} - S_{cd}S_{db}\sigma_{ac}(t-t')e^{-i\omega_{ad}t'}\right]$$

where we have inserted two resolutions of the identity. Defining the tensor elements,

$$R_{ac,cd}(t') = \frac{1}{\hbar^2}C_B(t')S_{ac}S_{cd}$$

we can simplify our notation,

$$\frac{\partial}{\partial t}\sigma_{ab}(t) = -i\omega_{ab}\sigma_{ab} - \frac{i}{\hbar}\lambda\sum_c S_{ac}\sigma_{cb} - S_{cb}\sigma_{ac}$$

$$-\sum_{cd}\int_0^t dt'\left[R_{ac,cd}(t')\sigma_{db}(t-t')e^{-i\omega_{cb}t'} - R_{ac,db}(t')\sigma_{cd}(t-t')e^{-i\omega_{ad}t'}\right.$$

$$\left. -R_{ac,db}(-t')\sigma_{cd}(t-t')e^{-i\omega_{cb}t'} + R_{cd,db}(-t')\sigma_{ac}(t-t')e^{-i\omega_{ad}t'}\right].$$

In order to apply the Markovian approximation, we would be tempted to pull the density matrix out of the integral directly. However this would not be justified, as non-diagonal elements of the density matrix can have phase factors that oscillate quickly. To understand these, consider the first integral term in the above equation

$$\int_0^t dt' R_{ac,cd}(t')\sigma_{db}(t-t')e^{-i\omega_{cb}t'} = e^{-i\omega_{db}t}\int_0^t dt' R_{ac,cd}(t')\sigma_{db}^I(t-t')e^{i(\omega_{db}-\omega_{cb})t'}$$

$$\approx e^{-i\omega_{db}t}\sigma_{db}^I(t)\int_0^\infty dt' R_{ac,cd}(t')e^{i\omega_{db}t'}e^{-i\omega_{cb}t'}$$

$$= \sigma_{db}(t)\int_0^\infty dt' R_{ac,cd}(t')e^{i\omega_{dc}t'}$$

$$= \sigma_{db}(t)R_{ac,cd}(\omega_{dc})$$

in the first line we have reintroduced the interaction picture, and in the second we have performed the Markovian approximation assuming that the reduced density matrix in the interaction picture is slowly varying relative to the bath correlation function, $\sigma_{db}^I(t-t') \approx \sigma_{db}^I(t)$. In the third line, we have moved back to the Schrodinger picture and taken the integral to infinity, and used this to define the one-sided Fourier transform of $R_{ac,cd}(t')$.

If we follow the above procedure for each of the terms in our non-Markovian equation, we arrive at

$$\frac{\partial}{\partial t}\sigma_{ab}(t) = -\,i\omega_{ab}\sigma_{ab} - \frac{i}{\hbar}\lambda\sum_c S_{ac}\sigma_{cb} - S_{cb}\sigma_{ac}$$

$$-\sum_{cd} R_{ac,cd}(\omega_{dc})\sigma_{db}(t) + R^*_{bd,dc}(\omega_{cd})\sigma_{ac}(t)$$

$$+\sum_{cd}\left[R_{ac,db}(\omega_{ca}) + R^*_{ca,bd}(\omega_{db})\right]\sigma_{cd}(t)$$

which is known as the *Redfield equation*. It is the standard means of describing a quantum system weakly coupled to a bath. It was originally introduced to understand nuclear spin relaxation in the context of magnetic resonance. Reflecting back on how we got here, we assumed there was no entanglement between the system and bath—Born approximation, that the bath was weakly coupled to the system—perturbation theory, and that the bath relaxes more quickly than the system—Markovian approximation. A consequence of these approximations is the system's natural frequencies, the first term on the right-hand side, are altered by the static coupling by the second term on the right-hand side. In addition, the third set of terms on the right-hand side alters the dynamics. In particular, the different kinetic coefficients $R_{ab,cd}$ are elements of the so-called Redfield tensor, and describe relaxation, dissipation, and dephasing. They are given by integrals over bath correlations, evaluated at resonant frequencies in the system.

7.12 Spin dephasing and dissipation

In order to gain some physical insight into the evolution of the reduced density matrix due to the Redfield tensor, let us consider a few simplified cases. First we introduce a decomposition of a tensor element into its pure real and pure imaginary parts

$$R_{ab,cd}(\omega) = \Gamma_{ab,cd}(\omega) + iD_{ab,cd}(\omega)\,.$$

For simplicity let us imagine that $\lambda = 0$ and consider the equation of motion just between diagonal elements of the density matrix. From the Markovian equation in the last section we get,

$$\frac{\partial}{\partial t}\sigma_{aa}(t) = \sum_c \left[R_{ca,ac}(\omega_{ca}) + R^*_{ca,ac}(\omega_{ca})\right]\sigma_{cc}(t)$$

$$- \left[R_{ac,ca}(\omega_{ac}) + R^*_{ac,ca}(\omega_{ca})\right]\sigma_{aa}(t)$$

$$= \sum_c k_{c\to a}\sigma_{cc}(t) - \sum_c k_{a\to c}\sigma_{aa}(t)$$

this is a set of equations for the transfer of population between different energy levels. As the dynamics are determined by the purely real components of the tensor, we can define the transition rates

$$k_{a\to c} = 2\Gamma_{ac,ca}(\omega_{ca}) \quad k_{c\to a} = 2\Gamma_{ca,ac}(\omega_{ac})$$

resulting in an effective classical master equation. If we trace back our definitions, we find the rate of population transfer is given by

$$k_{a\to c} = \frac{1}{\hbar^2}|S_{a,c}|^2 \int_{-\infty}^{\infty} dt\, C_B(t) e^{i\omega_{ca}t}$$

the Fourier transform of the bath correlation function, evaluated at the energy gap between the two eigenstates multiplied by a factor of the coupling matrix element squared. One way of interpreting this is to say that the Redfield equation includes only transitions that are resonant with a single bath excitation.

Exercise 7.12: Show that the rates $k_{a\to c}$ and $k_{c\to a}$ obey detailed balance,

$$\frac{k_{a\to c}}{k_{c\to a}} = e^{\beta\hbar\omega_{ac}}$$

and thus will evolve a Boltzmann distribution as its unique steady state.

Similarly, we can ask what happens to an off-diagonal element of the density matrix, a *coherence*, due to its own dynamics. The equation of motion for $\sigma_{ab}(t)$ for $a \neq b$, considering only those self-terms is

$$\frac{\partial}{\partial t}\sigma_{ab} = -i\omega_{ab}\sigma_{ab} - \left[\sum_c R_{ac,ca}(\omega_{ac}) + R^*_{bc,cb}(\omega_{bc})\right]\sigma_{ab}$$
$$+ \left[R_{bb,aa}(0) + R^*_{aa,bb}(0) - R_{aa,aa}(0) - R^*_{bb,bb}(0)\right]\sigma_{ab}\,.$$

Inserting the decomposition of the Redfield tensor into real and imaginary parts, we can rewrite the above as

$$\frac{\partial}{\partial t}\sigma_{ab} = -i\tilde{\omega}_{ab}\sigma_{ab} - \frac{1}{2}\left[\sum_{c\neq a} k_{a\to c} + \sum_{c\neq b} k_{b\to c}\right]\sigma_{ab} - \gamma_{ab}\sigma_{ab}$$

where we have a slightly renormalized frequency, a *Lamb shift*

$$\tilde{\omega}_{ab} = \omega_{ab} + \sum_{c\neq a} D_{ac,ca}(\omega_{ac}) + \sum_{c\neq b} D_{bc,cb}(\omega_{bc})$$

and two processes that act to decay the coherence. The first represents decoherence due to population transfer, $t_1^{-1} = \frac{1}{2}\left[\sum_{c\neq a} k_{a\to c} + \sum_{c\neq b} k_{b\to c}\right]$ and is often called the "t_1" time. The second is more subtle, occurring with rate γ_{ab}. It is given by

$$\gamma_{ab} = \frac{1}{\hbar^2}(|S_{a,a}|^2 - |S_{b,b}|^2)\int_0^{\infty} dt\, C_B(t)$$

and represent decoherence through what is called *pure dephasing*, or the "t_2" process. It manifests decoherence due to fluctuations in the energy gap ω_{ab} resulting from the coupling to the bath. An example of these relaxation dynamics for a simple two-level system is illustrated in Figure 7.11.

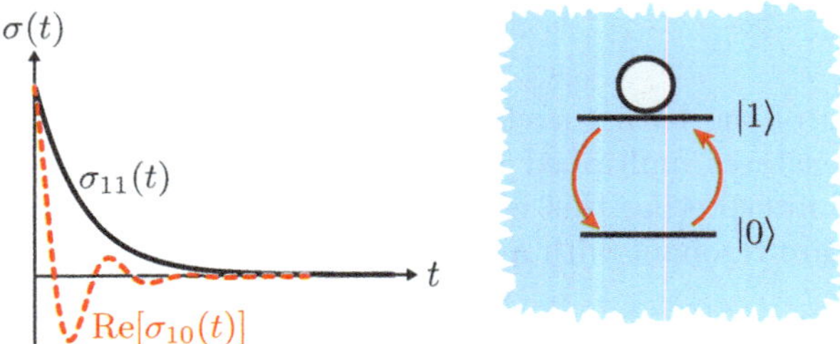

Fig. 7.11 (Right) A two-level system coupled to a harmonic bath, or spin-boson model. (left) Relaxation dynamics predicted from the Redfield equation of motion for the population of the excited state and the coherence.

Further reading

The perspective of the time dependent response theory and the Langevin equation developed in the chapter is largely influenced by Robert Zwanzig's *Nonequilibrium statistical mechanics* and David Chandler's *Introduction to modern statistical mechanics*. More information on open quantum system dynamics can be found in Abraham Nitzan's *Chemical dynamics in condensed phases* and Heinz-Peter Breuer and Francesco Petruccione's *The theory of open quantum systems*. A canonical reference on stochastic calculus in physical science is Andy Lau and Tom Lubensky's *State-dependent diffusion: Thermodynamic consistency and its path integral formulation* in *Physical Review E*, 2007.

Additional exercises

Exercise 7.13: In this exercise, you will work through a theory for the friction that a fluid feels when it flows over a wall. To start, consider a large planar wall with mass M in contact with a fluid, whose motion is constrained within the plane. Integrating out the fluctuations of the fluid and invoking its lateral translational invariance, the resulting generalized Langevin equation for the fluctuating wall velocity is

$$M\frac{dv}{dt} = -\lambda A v_s(t) + \eta(t)$$

where A is the lateral surface area, λ is the wall-fluid friction coefficient, and v_s is the relative velocity of the fluid to the surface of the wall, defined as $v_s = v - v_f$ where v_f is the fluid velocity. From the generalized Langevin equation in the absence of the Markovian assumption, v_s and v are related by

$$v_s(t) = \int_0^t dt'\, \gamma(t - t') v(t')$$

where $\gamma(t)$ is the memory kernel generated from integrating out the fluid.

1. For the generalized Langevin equation above, determine the mean and variance of the random noise η using the fluctuation-dissipation theorem.

2. Show that the velocity–velocity correlation function, $C_{vv} = \langle v(0)v(t) \rangle$, satisfies

$$M \frac{d}{dt} C_{vv}(t) = -\lambda A \int_0^t dt' \, \gamma(t - t') C_{vv}(t')$$

using the generalized Langevin equation.

3. Show that as $t \to 0^+$,

$$C_{vv}(0^+) = \frac{k_{\mathrm{B}} T}{M} \qquad \text{and} \qquad \frac{d}{dt} C_{vv}(0^+) = \frac{k_{\mathrm{B}} T A \lambda}{M^2}$$

using equipartition and noting that the wall cannot respond instantaneously to the fluid, so $\gamma(t) \to \delta(t)$ as $t \to 0^+$.

4. The force acting on the wall along the direction of its motion, F, is related to its velocity through Newton's equation of motion $F = M \, dv/dt$. Using this relationship and time translational invariance, show that the force–force correlation function $C_{FF}(t) = \langle F(0)F(t) \rangle$ is given by

$$C_{FF}(t) = -M^2 \frac{d^2}{dt^2} C_{vv}(t)$$

5. Using the method of Laplace transforms show that

$$\hat{C}_{FF}(s) = k_{\mathrm{B}} T \lambda A \left(1 + \frac{s \hat{\gamma}(s)}{s + \lambda A \hat{\gamma}(s)/M} \right)$$

where $\hat{C}_{FF}(s)$ and $\hat{\gamma}(s)$ are the Laplace transforms of $C_{FF}(t)$ and $\gamma(t)$.

6. In the long time limit, or equivalently $s \to 0$, show that

$$\lambda = \frac{1}{k_{\mathrm{B}} T A} \int_0^\infty dt \, \langle F(0)F(t) \rangle$$

the fluid-wall interfacial friction obeys a Green–Kubo relation given by the time integral of the fluid-wall force autocorrelation function.

Exercise 7.14: Dephasing of vibrating molecules in liquid phases is probed through measurement of lineshapes of infrared vibrational spectra. In this exercise, you will work through the theory for it originally due to Kubo and Anderson. To begin, recall that quantum transitions in molecules are the result of the perturbation

$$\Delta \mathcal{H} = -\mathcal{E}(t) M,$$

where M is the net dipole moment of the sample in the direction of the time-dependent electric field $\mathcal{E}(t)$. We shall assume that the system is composed of many vibrating molecules, each with a dipole moment that changes as its chemical bond length vibrates. Accordingly, at time t we write

$$M(t) = \sum_{i=1}^{N} m_i(t), \qquad m_i(t) \approx \bar{m}_i \left[1 + a q_i(t) \right],$$

where $\bar{m}_i(t)$ is the vibrationally averaged dipole of the ith molecule, $q_i(t)$ is the vibrational coordinate of that molecule, and a, the rate of change of that molecule's dipole due to changes in $q_i(t)$, is assumed to be a constant. $\bar{m}(t)$ varies on time scales over which molecules rotate—times of the order of picoseconds—so it is effectively a constant $\bar{m}_i(t) \approx \bar{m}$. In contrast, $q(t)$ varies on times scales of vibrational motions—femtoseconds. Infrared spectra are dominated by the latter motions. We will assume that vibrations of one molecule are weakly correlated to those of another, and as such, the infrared vibrational spectrum can be approximated as

$$\text{abs}(\omega)_{\text{IR}} \approx \omega N |\mathcal{E}_\omega|^2 F(\beta \hbar \omega) \bar{m}^2 a^2 \int_{-\infty}^{\infty} dt \; e^{i\omega t} \text{Re}[\langle q(0) q(t) \rangle]$$

where $F(\beta \hbar \omega)$ is a factor depending upon $\beta \hbar \omega$ in a way that ensures detailed balance between emission and absorption. This equation is the starting point of our analysis. We will first derive it.

1. For a monochromatic field, $\mathcal{E}(t) = \text{Re}[\mathcal{E}_\omega e^{-i\omega t}]$, show that the absorption spectrum is related to the Fourier transform of the response function by

$$\text{abs}(\omega) = -\frac{i\omega}{4} |\mathcal{E}_\omega|^2 (\hat{\chi}(\omega) - \hat{\chi}(-\omega))$$

2. Starting from the quantum version of the fluctuation-dissipation theorem,

$$\bar{A}(t) = \langle A \rangle + \frac{\bar{f}}{\hbar} \int_0^{\beta \hbar} d\tau \left\langle \delta \hat{A}(-i\tau) \delta \hat{A}(t) \right\rangle$$

derive a relationship between the response function $\chi(t)$ and the quantum correlation functions $C(t) = \langle \delta M(t) \delta M(0) \rangle$ and $C(-t)$,

$$\chi(t) = \frac{i}{\hbar} \Theta(t) (C(t) - C(-t))$$

3. Provide an explicit expression for the function $F(\beta \hbar \omega)$. It will be useful to note an identity between Fourier transforms of quantum correlation functions.

4. Intramolecular vibrations are largely controlled by the Born–Oppenheimer potential as a function of the intermolecular bond length, $U_0(R)$. Near the minimum energy, such potential contains small anharmonic corrections described by the Taylor series,

$$U_0(R) \approx \frac{m\omega_0^2}{2} (R - R_0)^2 - \frac{b}{6} (R - R_0)^3,$$

where R_0 refers to the equilibrium bond length in an isolated molecule, m is the reduced mass for the vibrations of that bond, ω_0 is the harmonic oscillator frequency for that bond, and b is a positive constant, indicative of the fact that the anharmonic term causes curvature of the potential to decrease as R increases. In a liquid, this potential is modified,

$$U(R) \approx -(R - R_0)f + U_0(R)$$

where f is the force exerted by the surrounding liquid on the tagged molecule's bond. This force depends upon the configuration of the entire system—the coordinates of the tagged molecule and the coordinates of all the other molecules in the liquid. Use these relationships and the assumption that forces associated with the anharmonic term and the liquid forces are relatively small to show that the frequency for vibrational transitions is altered by the liquid as

$$\omega_0 \to \omega(t) \approx \omega_0 \left(1 - \frac{b}{2m^2\omega_0^4} f(t) \right) .$$

Hence, compressive forces cause a blue shift to the vibrational frequency, while expansive forces cause a red shift. This frequency changes with time due to molecular motions that cause the force $f(t)$ to change with time.

5. The modulation of frequency by the time-varying force, $f(t)$, causes the dephasing of vibrational dynamics. The vibrational coordinate, q, is then $R - R_0$. The resulting correlation function is often approximated by

$$\langle q(0)\, q(t)\rangle \approx \bar{q}^2 \, \left\langle e^{i \int_0^t dt'\, \omega(t')} \right\rangle ,$$

where $\bar{q}^2$ is a constant. Derive this result and identify $\bar{q}^2$ in terms of matrix elements for q between adjacent vibrational energy levels. You should note that the relevant propagator when the Hamiltonian is time dependent takes the form

$$|\psi(t)\rangle = e^{-i/\hbar \int_0^t dt'\, \hat{\mathcal{H}}(t')} |\psi(0)\rangle$$

and that for a complete set of states, a resolution of the identity means that $\sum_n |n\rangle\langle n| = 1$. Also, remember we are considering absorption into a vibrational transition, whose matrix elements must obey specific selection rules.

6. With the result of Part (5), assume the statistics of the modulating force $f(t)$ is Gaussian to derive

$$\langle q(0)\, q(t)\rangle \approx \bar{q}^2 \, e^{i(\omega_0 + \langle \Delta\omega\rangle)t} \, e^{-\Gamma(t)} ,$$

where

$$\langle \Delta\omega\rangle = -(b/2m^2\omega_0^3)\,\langle f\rangle, \qquad \Gamma(t) = \frac{b^2}{4m^4\omega_0^6} \int_0^t dt'\,(t - t')\langle \delta f(0)\, \delta f(t')\rangle .$$

7. Assume the correlation function for the modulating force decays exponentially with time-scale τ, i.e.,

$$\langle \delta f(0)\, \delta f(t)\rangle \approx \langle (\delta f)^2\rangle\, e^{-|t|/\tau}$$

Derive analytical expressions for the line shape, $\mathrm{abs}(\omega)_{\mathrm{IR}}$, for the two limiting cases: τ small and τ large. For the first, the so-called *homogeneous broadening* case, the line shape is a Lorentzian, with a width controlled by τ. For the second, the so-called *inhomogeneous broadening* case, the line shape is Gaussian and independent of τ.

Exercise 7.15: In this chapter, we encountered the general Fokker–Planck equation. Here we will construct and use the Fokker–Planck equation for rotational diffusion.

1. Adapt the result to the case of single-particle rotational Langevin dynamics

$$\frac{d\theta}{dt} = \Omega \qquad I\frac{d\Omega}{dt} = -\gamma\Omega + \eta$$

where I and γ are positive constants. Specifically, determine an equation for the phase space distribution $f(\theta, \Omega; t)$, where Ω is the particle's angular velocity, and θ its direction. Write your result as

$$\frac{\partial f(\theta, \Omega; t)}{\partial t} = -\mathcal{D}f(\theta, \Omega; t)$$

and specifically identify $\mathcal{D}$ as a linear operator.

2. Verify that the Boltzmann distribution, $f_{\text{eq}}(\Omega)$, is stationary under $\mathcal{D}$.

3. Time correlation functions, such as

$$C_l(t) = \left\langle e^{-il\theta(0)} e^{il\theta(t)} \right\rangle$$

where l is an integer $\{1, 2, \dots\}$, can be computed from the Fokker–Planck equation by rewriting the average of the phase space variables and the propagator,

$$C_l(t) = \int d\theta \int d\Omega\, e^{-il\theta} e^{il\theta(t)} f_{\text{eq}}(\Omega)$$

$$= \int d\theta \int d\Omega\, e^{-il\theta} e^{-\mathcal{D}t} e^{il\theta} f_{\text{eq}}(\Omega).$$

In order to compute this correlation function we need to solve the Fokker–Planck equation with an initial condition,

$$f(0) = e^{il\theta} f_{\text{eq}}(\Omega).$$

Show that the time-dependent solution of the Fokker–Planck equation will have the same θ dependence as the initial condition, and that subsequently

$$C_l(t) = \int d\Omega\, f_l(\Omega, t)$$

where $f_l(\Omega, t)$ is some yet to be determined function of time, t and angular momentum, Ω, which satisfies a new evolution equation, $\mathcal{D}_l$.

4. Now you will compute $f_l(\Omega, t)$. We will do this in a few steps. First, write down the Fokker–Planck equation for $f_l(\Omega, t)$ with the new differential operator $\mathcal{D}_l$.

5. Assuming the solution has the form

$$f_l(\Omega, t) = e^{\Psi(\Omega, t)} \qquad \Psi(\Omega, t) = a(t) + b(t)\Omega - \frac{I}{2kT}\Omega^2$$

Determine the ordinary differential equations that $a(t)$ and $b(t)$ satisfy.

6. Solve the initial value to determine $a(t)$ and $b(t)$, using $a(0) = a_0$ and $b(0) = 0$.

7. Using $a(t)$ and $b(t)$, compute $C_l(t)$ by evaluating the integral over $f_l(\Omega, t)$.

8. Using your expression for the correlation function from part 6, determine the behavior in the limit of i) $t \ll I/\gamma$ and ii) $t \gg I/\gamma$.

8

Chemical kinetics and rare events

Some dynamical processes happen very infrequently, yet when they occur qualitatively alter the subsequent properties of the system. Examples of such rare but important events are chemical reactions, where long-lived reactants and products are interconverted through molecular rearrangements that are improbable. Other examples include the dynamics that accompany phase transitions, in which a metastable arrangement of particles is interconverted to a more stable arrangement through an unlikely fluctuation of the system. The perspective we will adopt to describe these processes and the accompanying theoretical formalism to predict the rates at which they occur will leverage the developments in the past chapters, including dynamical linear response, and effective equations of motion. Those historical developments are supplemented with contemporary ideas using path ensembles and dynamical reweighting. By identifying metastable states and the rates at which they interconvert, we also effectively deduce a reduced description of the system's dynamics, in the space of a lower-dimensional set of discrete states.

8.1 Rare but important events

Imagine you are interested in motion on a free energy surface where there are multiple stable states. Here we will imagine there are two, which we will denote as basin A and basin B. If those states are distinguishable, or long-lived, then in between them must be a saddle point. If that saddle point is high in energy, say relative to the available thermal energy, $k_B T$, the probability, f, of seeing a transition in some fixed time is $f \ll 1$. The stable states could correspond to different chemical species or isomers, but they could also correspond to different phases of a condensed material or conformations of a polymer. This setup is shown in Figure 8.1, where q is some order parameter capable of distinguishing between states A and B, while q_A and q_B denote the locations of their corresponding minima, and q^* is the location of the saddle point.

Let's use τ_{mol} as the basic unit of time, which should correspond to a characteristic relaxation time in one of the basins. It is also the typical time for a transition to occur because the saddle point is an unstable position on the potential, and a trajectory initialized there will move quickly to a new stable state. On average, in order to see one barrier crossing, we need to wait $n\tau_{\mathrm{mol}}$, where $nf \sim 1$. If we imagine that the system starts in basin A, the probability that it is still in basin A, n steps later, which we define as the conditional probability $\bar{h}_A(t)$, is

$$\bar{h}_A(t) = (1 - f)^n$$

$$\approx e^{-nf} = e^{-tf/\tau_{\mathrm{mol}}} = e^{-tk_{AB}}$$

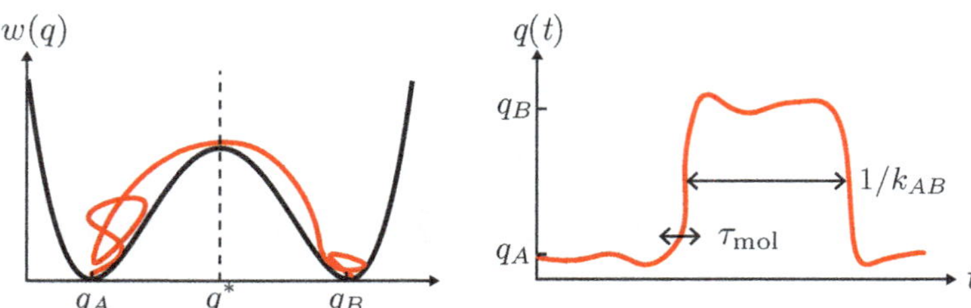

Fig. 8.1 Illustration of a typical barrier crossing event, on a surface defined by the order parameter q and potential al of mean force $w(q)$.

where we assume that each n is an uncorrelated attempt to hop over the barrier. If we assume that n is large, we can replace the product with an exponential. We find that the rate constant, k_{AB}, which determines the characteristic timescale to leave basin A, is just the probability to leave per unit of time, f/τ_{mol}. This relationship can be made explicit by taking the time derivative of $\bar{h}_A(t)$,

$$\frac{d\bar{h}_A(t)}{dt} = -k_{AB}e^{-tk_{AB}}$$
$$= -k_{AB}\bar{h}_A(t)$$

and observe that indeed, we recover a macroscopic law of first-order kinetics. Now though, we have a probabilistic interpretation of a rate constant.

Exercise 8.1: Demonstrate that

$$(1 - f)^n \approx e^{-nf}$$

for $f \ll 1$. Hint: you might note that $1 - f = \exp[\ln(1 - f)]$.

We have shown in previous chapters that relaxation from a prepared nonequilibrium state is related to spontaneous fluctuations about equilibrium. If we want to understand how the system relaxes from an initially prepared ensemble in basin A into an equilibrium amount of A and B, there should exist an equilibrium time correlation function that will characterize the fluctuations that result in that barrier crossing. Indeed, such a correlation function can be written as

$$C_{AB}(t) = \frac{\langle h_A(0)h_B(t)\rangle}{\langle h_A \rangle}$$
$$= \bar{h}_B(t)$$

where h_X is an indicator function that reports whether the system is in state X, and $\bar{h}_B(t)$ denotes h_B given the system is initially in A. Here we will take

$$h_A = \begin{cases} 1 & \text{if } q < q^* \\ 0 & \text{else} \end{cases} \qquad h_B = \begin{cases} 1 & \text{if } q > q^* \\ 0 & \text{else} \end{cases}$$

where q^* is a dividing surface, separating states A and B. Note that $h_A + h_B = 1$ for all t. The correlation function $C_{AB}(t)$ has a simple interpretation. It is the conditional probability that the system is in state B at a time t given the system starts in state A at $t = 0$. In this way, its average is the complement to $\bar{h}_A = 1 - \bar{h}_B$.

The limits of $C_{AB}(t)$, the *side–side correlation function*, are simple to evaluate. At $t = 0$,

$$C_{AB}(0) = 0$$

by construction, as the system cannot be in both A and B. In the long time limit,

$$\lim_{t \to \infty} C_{AB}(t) = \langle h_B \rangle$$

the correlation function reports on the equilibrium probability of being in state B, assuming the indicator functions become uncorrelated. From first-order kinetics, we expect that this limiting value is approached exponentially. Taking its time derivative, evaluating it at 0^+, where 0^+ is some time $\tau_{\mathrm{mol}} < t \ll 1/k_{AB}$, where the flux into state B is constant from the linearization of the exponential,

$$\left. \frac{dC_{AB}(t)}{dt} \right|_{0^+} = \left. \frac{d\bar{h}_B(t)}{dt} \right|_{0^+} = \left. \frac{d(1 - \bar{h}_A(t))}{dt} \right|_{0^+} = k_{AB}$$

we indeed find that this correlation function is related to the rate constant.

8.2 Transition state theory

The expression relating the correlation function and the rate constant is exact. Indeed, it does not even rely on an equilibrium state, though it does assume the system is in a steady state. However, to use this expression requires a correlation function to be computed that is dominated by rare fluctuations. It is both instructive and potentially useful to try to work out an alternative expression.

To start, we note that the time derivative of $C_{AB}(t)$ is

$$\dot{C}_{AB}(t) = \frac{\left\langle h_A(0) \dot{h}_B(t) \right\rangle}{\langle h_A \rangle} = -\frac{\left\langle \dot{h}_A(0) h_B(t) \right\rangle}{\langle h_A \rangle}$$

where we used the time-translational invariance of the equilibrium correlation function. Evaluating the time derivative of the indicator function we observe,

$$\dot{h}_A(0) = \frac{dh_A}{dq} \frac{dq}{dt} = -\delta\left[q(0) - q^*\right] \dot{q}(0)$$

where $\dot{q}$ is the time derivative of q and the delta function is the derivative of a Heaviside function. Inserting this back in and multiplying top and bottom by $\langle \delta[q(0) - q^*] \rangle$

$$\dot{C}_{AB}(t) = \frac{\langle \delta[q(0) - q^*] \dot{q}(0) h_B(t) \rangle}{\langle \delta[q(0) - q^*] \rangle} \frac{\langle \delta[q(0) - q^*] \rangle}{\langle h_A \rangle}.$$

Both of the terms on the right-hand side have a simple interpretation. The first is a time correlation function of the initial flux, $\dot{q}(0)$, and the likelihood of being in B at t,

$$\frac{\langle \delta(q(0) - q^*)\dot{q}(0)h_B(t)\rangle}{\langle \delta(q(0) - q^*)\rangle} = \langle \dot{q}(0)h_B(t)\rangle_{q(0)=q^*}$$

conditioned on starting at q^* at time 0. The second term can be evaluated,

$$\frac{\langle \delta(q(0) - q^*)\rangle}{\langle h_A\rangle} = \frac{e^{-\beta w(q^*)}}{\int_{-\infty}^{q^*} dq \, e^{-\beta w(q)}} = \frac{e^{-\beta \Delta w(q^*)}}{\ell}$$

where we find that it is just the probability of being at q^* relative to being in basin A. Assuming an equilibrium state with free energy $w(q)$, it is given by a Boltzmann factor with the change in the free energy $\Delta w(q^*)$, and a lengthscale ℓ characterizing the number of thermally accessible states in basin A. This exponential sensitivity to the barrier height is known phenomenologically as the *Arrhenius' law*.

Putting this together, we find that the rate constant is given by the probability of being at the transition state, times the flux over the barrier conditioned on ending in state B, or

$$\dot{C}_{AB}(t) = \langle \dot{q}(0)h_B(t)\rangle_{q^*} \, e^{-\beta \Delta w(q^*)}/\ell \, .$$

This is known as the reactive flux correlation function, originally due to Chandler. Up to this point it is still exact, and we have made our life easier by evaluating the expression at the top of the barrier. In fact, the notion of the steady state reactive flux implies that the relationship is invariant to our choice of q^*. Different q^* will change the relative likelihood of observing that fluctuation, but that is exactly offset by an accompanying change in flux. However, to evaluate it we still need to propagate dynamics in order to see if the system ends in state B at time $t > \tau_{\text{mol}}$, and thus in principle need to model those potentially complicated trajectories. If we take the short time limit, by setting $t = 0^+$, the free energy does not change but the flux term does. In fact, at $t = 0^+$, only $\dot{q} > 0$ will result in $h_B = 1$, so this term becomes

$$\langle \dot{q}(0)h_B(0^+)\rangle_{q^*} = \frac{1}{2}\langle |\dot{q}|\rangle$$

where in equilibrium fluxes are symmetric and uncorrelated with positions. Inserting this approximation back into the correlation function, we find

$$\dot{C}_{AB}(0) = \frac{1}{2\ell}\langle |\dot{q}|\rangle \, e^{-\beta \Delta w(q^*)} = k_{AB}^{\text{TST}}$$

that the rate is given by the average flux times the likelihood of reaching the dividing surface. This short time approximation illustrated in Figure 8.2 is called *transition state theory*. It has a long history in chemical physics, beginning with Wigner and Eyring, placed into the language of correlation functions by Chandler. It is remarkable, as it is an estimate of the rate of a dynamical event in which only thermodynamic averages enter—the interpretation being that reactive events are mediated by improbable fluctuations. Transition state theory refers to the location of the dividing surface q^*, which in chemical physics is known as the transition state, and it typically coincides with a saddle point, but in high dimensional systems does not have to.

Transition state theory is rigorously an upper bound on the rate. In order to understand this, Figure 8.3 illustrates paths that are neglected in this approximation.

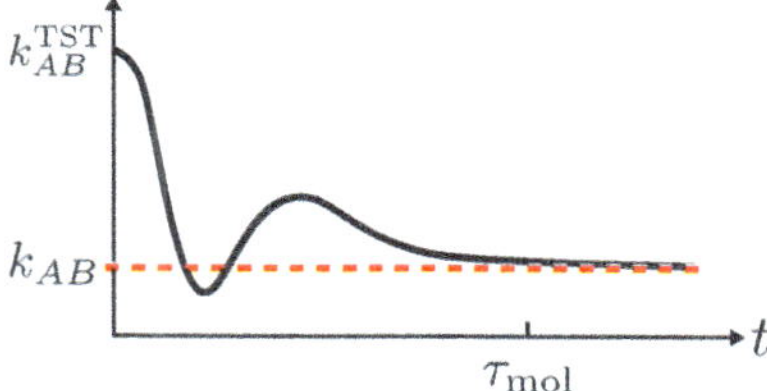

Fig. 8.2 The $t = 0$ value of the flux-side correlation function is the transition state theory approximation, while its asymptotic value is the observable rate.

Specifically, instances of $\dot{q}(0) > 0$ that wind up in state A by turning around, or instances of $\dot{q}(0) < 0$ that end up in B are ignored. In both cases, the overall rate decreases. In the first, a reactive event was counted that should not have been, over-counting the flux. In the second, a reactive event was not counted that should have been, though with a negative contribution to the flux. Because of this variational structure, it is useful to define the ratio of the transition state theory rate to the full rate,

$$\kappa(t) = \frac{\dot{C}_{AB}(t)}{\dot{C}_{AB}(0^+)} \leq 1$$

which is known as the transmission coefficient. This is a quantity that is intrinsically dynamical and requires a model of the barrier crossing to estimate.

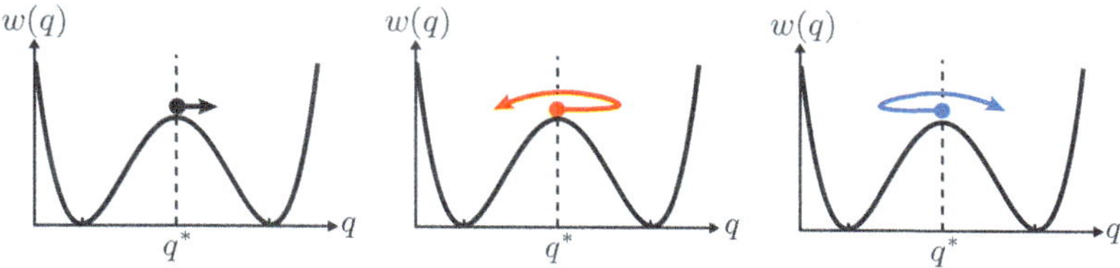

Fig. 8.3 Illustration of motion on the potential. Transition state theory only includes terms like the left panel, where the initial velocity is in the positive direction. This overcounts the reactive events by including trajectories that turn around and end up in A like in the middle panel, or those that start with negative velocity but end up in B.

As an example of a specific form for $\langle |\dot{q}| \rangle$ and ℓ, let's assume that the potential is locally harmonic. Around basin A and the transition state we have,

$$w(q) \approx \frac{1}{2} m \omega_{\mathrm{o}}^2 (q - q_A)^2 \qquad w(q) \approx -\frac{1}{2} m \omega_{\mathrm{b}}^2 (q - q^*)^2 + \Delta w(q^*)$$

where ω_{o} is the frequency of the basin and ω_{b} the frequency of the saddle point. The constant term, $\Delta w(q^*)$, is the offset relating the location of the minimum of basin A to the maximum of the barrier. With these forms we can evaluate,

$$\frac{1}{2}\langle|\dot{q}|\rangle = \int_0^\infty d\dot{q}\,\dot{q}\,\frac{e^{-\beta m\dot{q}^2/2}}{\sqrt{2\pi/\beta m}} = \frac{1}{\sqrt{2\pi\beta m}}$$

which is just given by equipartition. Similarly,

$$\ell = \int_{-\infty}^{q^*} dq\, e^{-\beta m\omega_o^2(q-q_A)^2/2} \approx \sqrt{\frac{2\pi}{\beta m\omega_o^2}}$$

where we assume that the integral is dominated by points around $q = 0$. Putting these together,

$$k_{AB}^{\text{TST}} = \frac{\omega_o}{2\pi}e^{-\beta\Delta w(q^*)}$$

we find that the transition state theory estimate is given by the frequency of the stable basin times a Boltzmann factor.

Exercise 8.2: If q is identified as a Cartesian coordinate, and $\dot{q}$ a canonical velocity, show that transition state theory can be written as

$$k_{AB}^{\text{TST}} = \frac{k_{\text{B}}T}{h}\frac{Q(q^*)}{Q(A)}$$

where $Q(q^*)$ is the partition function for the transition state and $Q(A)$ is the partition function in the A basin

$$Q(A) = \frac{m}{h}\int_{-\infty}^{q^*} dq \int_{-\infty}^{\infty} d\dot{q}\, e^{-\beta w(q)}e^{-\beta m\dot{q}^2/2}$$

where m is the mass associated with q.

8.3 Kramers' theory

Transition state theory is an approximate way to compute the rate of a rare event like a barrier crossing. The remarkable thing about transition state theory is that it relies on only static thermal expectation values as input, and as such abstracts away the potentially complicated dynamics associated with such a transition. While we evaluated an example where the expectation values were taken over a canonical ensemble, we could have just as well done this in a microcanonical ensemble yielding what is known as RRKM theory, which is more appropriate in the gas phase where reactants and products do not thermalize very quickly.

If we want to move beyond transition state theory, we need to put back in some information about the dynamics at the top of the barrier. One way to formulate such corrections is to make approximations to the transmission coefficient,

$$\kappa(t) = \frac{k_{AB}(t)}{k_{AB}(0^+)}$$

which is just the ratio of the exact rate to the transition state theory estimate. In particular, what we really care about is the plateau value of κ at times long relative to τ_{mol} but short relative to the average time between transitions. This timescale would correspond to a steady-state flux over the barrier.

Kramers was one of the first to consider such dynamical corrections, and we will begin with his theory in the high friction limit, which is particularly straightforward. A particle traversing a barrier in the presence of other degrees of freedom obeys a Langevin equation. In the high friction limit, this equation of motion is overdamped,

$$\gamma \dot{q} = -w'(q) + \eta(t) \qquad \langle \eta(t)\eta(t') \rangle = 2k_{\mathrm{B}}T\gamma\delta(t - t')$$

where $-w'(q)$ is the mean force acting on the particle, and γ is the friction due to the degrees of freedom orthogonal to q. The distribution function for q at time t follows from the Smoluchowski equation. A particularly revealing way to write the Smoluchowski equation is,

$$\frac{\partial p(q,t)}{\partial t} = \frac{k_{\mathrm{B}}T}{\gamma} \frac{\partial}{\partial q} e^{-\beta w(q)} \frac{\partial}{\partial q} \left(e^{\beta w(q)} p(q,t) \right)$$

$$= -\frac{\partial}{\partial q} J(q,t)$$

where $J(q,t)$ is identified as the probability flux. One way to compute the rate of barrier crossing is by imposing boundary conditions that act as a source in basin A and a sink in basin B. With those boundary conditions, at long times the system will evolve a steady state with constant, nonzero, flux. That flux will be the reactive flux

$$J_{AB} = p_A k_{AB}$$

where p_A is the probability of being in state A and k_{AB} is the rate.

By manipulating the definition of the flux,

$$J_{AB} = -\frac{k_{\mathrm{B}}T}{\gamma} e^{-\beta w(q)} \frac{\partial}{\partial q} \left(e^{\beta w(q)} p(q) \right)$$

we can solve for the steady state distribution $p(q)$ consistent with the boundary conditions,

$$p(q) = \frac{J_{AB}}{k_{\mathrm{B}}T/\gamma} e^{-\beta w(q)} \int_q^\infty dq' e^{\beta w(q')}$$

where we chose to evaluate the integral at large q. Substituting the steady state distribution into the definition of p_A as the integral up to some dividing surface parameterized by q^*,

$$p_A = \int_{-\infty}^{q*} dq\, p(q)$$

$$= \frac{J_{AB}}{k_{\mathrm{B}}T/\gamma} \int_{-\infty}^{q*} dq\, e^{-\beta w(q)} \int_q^\infty dq' e^{\beta w(q')}$$

$$= J_{AB}/k_{AB}$$

where in the third line we rewrite p_A in terms of the steady state flux and the rate constant. Solving for the rate constant,

$$k_{AB} = \frac{k_{\mathrm{B}}T}{\gamma} \left[\int_{-\infty}^{q*} dq\, e^{-\beta w(q)} \int_{q}^{\infty} dq'\, e^{\beta w(q')} \right]^{-1}$$

we find that it is given by a double integral over the Boltzmann distribution. Even without an explicit equation for the rate, we have found that in the overdamped limit, the rate goes down with increasing friction. This result is probably not surprising, as the friction is inversely proportional to the diffusion constant, and in the overdamped limit, the rate is given by the probability of getting to the barrier times the flux, where the flux is modeled by a random walk.

We can get an explicit expression for the rate by assuming that the potential is locally harmonic about the barrier top and basin A. Proceeding as we did before, we assume that we can decouple the two integrals and evaluate them as

$$\int_{-\infty}^{q*} dq\, e^{-\beta w(q)} \approx \int_{-\infty}^{\infty} dq\, e^{-\beta m \omega_{\mathrm{o}}^2 (q - q_A)^2/2} = \sqrt{2\pi/\beta m \omega_{\mathrm{o}}^2}$$

where ω_{o} is the frequency of basin A. The second integral is evaluated as,

$$\int_{q}^{\infty} dq'\, e^{\beta w(q')} \approx \int_{-\infty}^{\infty} dq'\, e^{\beta \Delta w(q^*) - \beta m \omega_{\mathrm{b}}^2 (q - q^*)^2/2} = \sqrt{2\pi/\beta m \omega_{\mathrm{b}}^2}\, e^{\beta \Delta w(q^*)}$$

where ω_{b} is the frequency of the barrier and $\Delta w(q^*)$ is the height of the barrier. Putting these two terms together we find the total rate is

$$k_{AB} = \frac{m \omega_{\mathrm{o}} \omega_{\mathrm{b}}}{2\pi\gamma} e^{-\beta \Delta w(q^*)}$$

or comparing this result to our previous transition state theory calculations, the transmission coefficient is

$$\kappa = m \omega_{\mathrm{b}}/\gamma$$

which is inversely proportional to the friction and now depends on the shape of the barrier. This limit is known as the *spatial diffusion limit* where the transmission coefficient is dominated by diffusive motion over the barrier.

The alternative limit, when motion is underdamped, is very different. To overcome the barrier, the reaction coordinate requires a boost in energy. After that boost, however, low friction means that the energy is dissipated slowly. In the limit of no friction, energy would be conserved and the system would never relocalize in a single basin. The rate of energy being removed is proportional to the friction. As a consequence in the underdamped limit, the rate determining step is given by the dissipation of energy, which is known as the *energy diffusion limit*. To derive the transmission coefficient in this limit, we first go back to the original underdamped Langevin equation,

$$\dot{q} = v \qquad m\dot{v} = -\gamma v - w'(q) + \eta$$

and instead of manipulating the Fokker–Planck equation in real space, we write it down in energy space. Specifically, the equations of motion in terms of the position and energy, $E = mv^2/2 + w(q)$, are

$$\dot{q} = \sqrt{2\left[E - w(q)\right]/m} \qquad \dot{E} = -2\gamma\left[E - w(q)\right]/m + \tilde{\eta}$$

which follows from the definition of the total energy in terms of kinetic and potential energy. Here $\tilde{\eta}$ is the noise associated with the energy. From these equations for the drift, we can compute the associated Fokker–Planck equation in q and E,

$$\frac{\partial p(E, q, t)}{\partial t} = 2\frac{\gamma}{m}\frac{\partial}{\partial E}\left[E - w\right]p + \frac{\partial}{\partial E}D(E)\frac{\partial}{\partial E}p - \frac{\partial}{\partial q}\sqrt{2\left[E - w\right]/m}\,p$$

where $D(E)$ is the diffusion constant in energy space. This can be simplified to an equation for just $p(E, t)$ by integrating over q,

$$\frac{\partial p(E, t)}{\partial t} = \frac{\partial}{\partial E}\left(\frac{\gamma}{m}A(E)p(E, t) + D(E)\frac{\partial}{\partial E}p(E, t)\right)$$

where $A(E) = 2\left[E - w(q)\right]$ is the action and from the fluctuation–dissipation theorem, $D(E) = k_{\mathrm{B}}T\gamma A(E)/m$.

Exercise 8.3: Verify the Fokker Planck equation for the energy by demonstrating the its stationary distribution is given by

$$p(E) = e^{-\beta E}/Q$$

where Q is the partition function.

The analysis now follows directly as in the previous spatial diffusion case where we take the absorbing boundary condition to be at $E = E_b$ the energy at the top of the barrier. With this boundary condition, we arrive at

$$k_{AB} = \frac{\gamma k_{\mathrm{B}}T}{m}\left(\int_0^{E_b} dE\,e^{-\beta E}\int_E^{E_b} dE'\,\frac{e^{\beta E'}}{A(E')}\right)^{-1}.$$

In this case we find that the rate is proportional to the friction coefficient as anticipated and distinct from the overdamped limit. In the large barrier limit, this expression simplifies to

$$k_{AB} = \frac{\gamma}{mk_{\mathrm{B}}T}A(E_b)e^{-\beta E_b}.$$

The different dependence of the transmission coefficient in the low and high friction limits means that in between these limits is a maximum. This maximum is known as Kramers' turnover. It and a summary of the limiting regimes are illustrated in Figure 8.4.

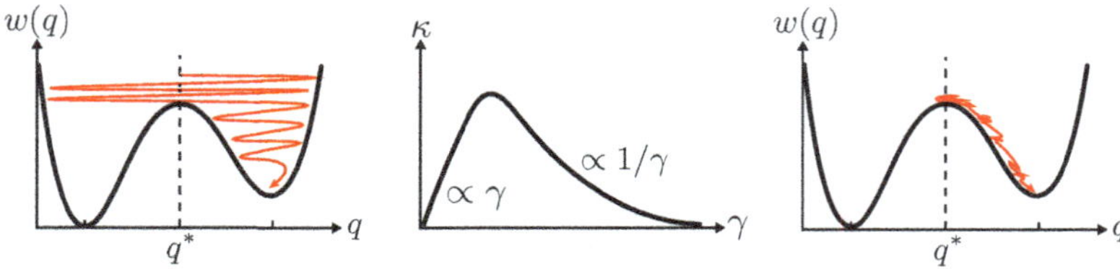

Fig. 8.4 Variation of the transmission coefficient, κ, as a function of the coupling between a reaction coordinate and a bath, characterized by friction γ. Example trajectories in the low and high friction limit are shown on the left and right, respectively.

Exercise 8.4: Verify the limiting behavior of the energy diffusion regime for a large barrier using a saddle point approximation to the integrals appearing in the expression for k_{AB}.

8.4   Nucleation of stable phases

An important application of transition state theory and its corrections is in the estimation of the rate of conversion of a metastable phase into a stable one. As discussed in Chapter 4, the thermodynamic state of a system can favor one phase over another through the relationships between the chemical potentials of the two phases. If a system can support distinct liquid ℓ and vapor v phases, for conditions below the critical point it is possible to prepare a vapor at some high temperature, and then quench the temperature just below the condensation point and have a stable liquid phase emerge. The transition does not happen instantaneously, but with some associated rate. The process by which the liquid emerges from its vapor by crossing the line of first order transitions is known as nucleation, and the associated prediction of the rate of transformation between the two phases is given by *classical nucleation theory.*

The basic idea behind classical nucleation theory is to model the free energetics of a droplet of the stable phase immersed within the metastable phase. In the context of condensation, if the chemical potential of the liquid is μ_l and that of the vapor is μ_v, then provided the system is held at constant temperature and pressure with $\Delta\mu = \mu_l - \mu_v < 0$, the liquid phase is stable and thermodynamically a system prepared in the vapor phase should spontaneously transition into a liquid. The corresponding decrease in the free energy of the system upon converting the system from vapor to liquid will be proportional to the volume of the material transitioned. However, we know from our discussion in Chapter 4, that the appearance of an interface between two phases increases the free energy of the system by an amount proportional to the area of the interface with a constant of proportionality equal to the surface tension, σ. This surface tension will thus oppose the formation of the liquid phase. The balancing of these two terms, one that decreases the free energy and scales like the volume, and one that increases the free energy but scales like an area, results in a free energy barrier provided $\Delta\mu$ is not too negative. These scaling behaviors and the resultant Gibbs free energy change, ΔG as a function of the droplet size n are shown in Figure 8.5.

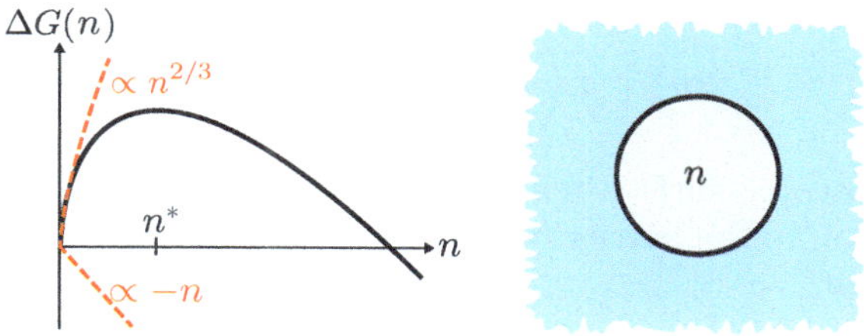

Fig. 8.5 Model for the nucleation barrier within classical nucleation theory.

The simplest model for the Gibbs free energy change, ΔG associated with growing in a droplet of liquid is to imagine that the droplet is made of n liquid particles,

$$\Delta G(n) = n\Delta\mu + a\sigma n^{2/3}$$

where the first term is the favorable change in chemical potential $\Delta\mu < 0$ times the number of particles converted into the liquid droplet. The second term is the surface tension σ times an area per particle, a, times $n^{2/3}$ which is how the number of particles at the surface changes with n. The location of the barrier is easily computed by extremizing ΔG,

$$\left.\frac{d\Delta G}{dn}\right|_{n*} = 0 \qquad \rightarrow \qquad n^* = \left(\frac{2a\sigma}{3|\Delta\mu|}\right)^3$$

where n^* is referred to as the critical nucleus. Droplets of size n^* within this model are equally likely to grow into macroscopic domains or to dissolve back into the vapor, and as such can be identified with the transition state of the nucleation event. Phase transformation, like a chemical reaction, requires a rare fluctuation to proceed. The associated barrier height is

$$\Delta G(n^*) = \frac{(2a\sigma/3)^3}{2|\Delta\mu|^2}$$

which we see decreases with increasing driving force as $(\Delta\mu)^{-2}$. Since the metastable phase is a vapor, it is often a good approximation to assume that the chemical potential difference is given by an ideal gas form, $\Delta\mu = k_{\mathrm{B}}T \ln p/p^*$ where the pressure p is relative to the coexistence pressure, p^*.

Exercise 8.5: One important correction to classical nucleation theory is due to Tolman, who noted that the effective surface tension changes with droplet size as

$$\sigma(n) = \sigma \left(1 - \frac{\delta}{\sqrt{a}n^{1/3}} \right)$$

where σ is the macroscopic surface tension and δ known as the Tolman length. Work out the correction to the barrier height $\Delta G(n^*)$ to first order in δ, assuming the nuclei are spherical.

To compute the rate of formation of the stable phase, we can adapt Kramers' theory for a system in which the reaction coordinate is discrete, like the number of particles in our critical nucleus. Specifically, we can imagine that the way clusters form is by sequential addition, a cluster of some size n can grow to a cluster of size $n + 1$ by adding a particle with some attachment rate k_n^+, or shrink to a cluster of size $n - 1$ with a loss rate k_n^-. In a solution, the concentration of a cluster of size n, denoted ρ_n would obey a simple kinetic equation,

$$\frac{d\rho_n}{dt} = -k_n^+ \rho_n - k_n^- \rho_n + k_{n-1}^+ \rho_{n-1} + k_{n+1}^- \rho_{n+1}$$

where in addition to the two loss processes, we note that the concentration of n-sized clusters can increase by growing from an $(n - 1)$-sized cluster or by an $(n + 1)$-sized cluster shrinking. The principle of detailed balance first explored in Chapter 5 implies that these rates are not independent. In particular,

$$\frac{k_n^+}{k_{n+1}^-} = \frac{\bar{\rho}_{n+1}}{\bar{\rho}_n} = e^{-\beta[\Delta G(n+1) - \Delta G(n)]}$$

the ratio of any two rates is equal to their corresponding equilibrium concentrations, denoted here by the overbar, which in turn depend on their free energies.

With the set of coupled equations for the mass flow between different cluster sizes, we can proceed as we did with Kramers' theory and relate the rate of spontaneous nucleation to a steady-state flux generated by a judicious choice of boundary conditions. Specifically, we will compute the net flux given an absorbing state at large n, imposed by $\rho_n = 0$ as $n \to \infty$, and an equilibrium state for clusters of size $n = 1$, or $\rho_1 = \bar{\rho}_1$. From the kinetic equations, net flux J for clusters of size n to $n + 1$ is

$$J = k_n^+ \rho_n - k_{n+1}^- \rho_{n-1}$$

which is valid for all n. This can be rearranged as

$$J = k_n^+ \left(\rho_n - \frac{k_{n+1}^-}{k_n^+} \rho_{n+1} \right)$$

$$= k_n^+ \left(\rho_n - \frac{\bar{\rho}_n}{\bar{\rho}_{n+1}} \rho_{n+1} \right) = k_n^+ \bar{\rho}_n \left(\frac{\rho_n}{\bar{\rho}_n} - \frac{\rho_{n+1}}{\bar{\rho}_{n+1}} \right)$$

where in the second line we have used detailed balance to relate the ratio of rates to the ratio of equilibrium concentrations. Moving the factor of $k_n^+ \bar{\rho}_n$ over to the left-hand side, and summing over n we find

$$J \sum_{n=1}^{\infty} \frac{1}{k_n^+ \bar{\rho}_n} = \sum_{n=1}^{\infty} \left(\frac{\rho_n}{\bar{\rho}_n} - \frac{\rho_{n+1}}{\bar{\rho}_{n+1}} \right) = 1$$

where we use the steady state condition to pull J out of the sum and in the second equality employ our boundary conditions. This results in the Becker–Doring equation for the net flux, which can be manipulated using detailed balance,

$$J = \left(\sum_{n=1}^{\infty} \frac{1}{k_n^+ \bar{\rho}_n} \right)^{-1} \approx k_{n^*}^+ \bar{\rho}_{n^*} = k_1^+ \bar{\rho}_1 e^{-\beta(2a\sigma/3)^3/2|\Delta\mu|^2}$$

where the sum is expected to be dominated by the cluster with the smallest $\bar{\rho}_{n^*}$, or the critical nucleus whose concentration is determined by the nucleation barrier. For a condensation transition, the attachment rate $k_{1^*}^+$ is well described by kinetic theory as the flux of particles from the gas on the area given by size of the critical nucleus.

8.5 Reactions in non-Markovian environments

The preceding sections illustrated some important effects that the bath can have on the flux over the barrier. If the system is weakly coupled, the rate-limiting step is the dissipation of energy. If the system is strongly coupled, then diffusion over the barrier is rate-determining. In both limits, however, we have assumed that the bath is well characterized by a single number, γ. This is equivalent to assuming that the *dynamics* of the bath do not play an important role in the barrier crossing. This assumption breaks down if the bath relaxes on a similar timescale as τ_{mol}. In such a case, a non-Markovian theory is needed. The canonical way to do this was first introduced by Grote and Hynes. Below we outline their basic calculation.

Consider a parabolic barrier and a Gaussian, non-Markovian bath. The generalized Langevin equation in mass-weighted coordinates is,

$$\ddot{q} = \omega_{\mathrm{b}}^2 q - \int_0^t dt'\, \xi(t-t')\dot{q}(t') + \eta(t) \qquad \langle \eta(t)\eta(t') \rangle = k_{\mathrm{B}} T \xi(t-t')$$

with the random force η, a frequency atop the barrier ω_{b} and a memory kernel $\xi(t-t')$. Since the equation is linear, it can be easily solved with Laplace transforms. Transforming the equation we get,

$$s^2 \tilde{q}(s) - s q_0 - \dot{q}_0 = \omega_{\mathrm{b}}^2 \tilde{q}(s) - \tilde{\xi}(s)(s\tilde{q}(s) - q_0) + \tilde{\eta}(s)$$

where $\tilde{q}(s)$ is the Laplace transform of $q(t)$ and q_0 and $\dot{q}_0$ are the initial position and velocity. We assume that the Laplace transform of the memory kernel $\tilde{\xi}(s)$ and noise

$\tilde{\eta}(s)$ exist. Since the barrier is centered around 0, within the reactive flux calculation $q_0 = 0$. Solving for $\tilde{q}$,

$$\tilde{q}(s) = \frac{\dot{q}_0 + \tilde{\eta}(s)}{s^2 + s\tilde{\xi}(s) - \omega_b^2}.$$

Laplace transforms conserve sign, which means that at long times $q > 0$ if

$$\dot{q}_0 + \tilde{\eta}(\lambda) > 0$$

where $\lambda > 0$ is the frequency associated with the fastest-growing mode. Because the particle is atop a parabolic barrier, its dynamics are unstable, and we need only to concern ourselves with this timescale. The limit of $t \to \infty$ is an absorbing state at $q = \pm\infty$. The frequency determined by the inverse Laplace transform is referred to as the Grote–Hynes frequency and given by the largest positive solution to

$$\lambda^2 + \lambda\tilde{\xi}(\lambda) - \omega_b^2 = 0.$$

It is bounded by ω_b from above and $\omega_b^2/\xi(0)$ from below.

In view of the inequality above, the transmission coefficient for this model can be computed from,

$$\kappa = \langle \dot{q}_0\Theta[\dot{q}_0 + \tilde{\eta}(\lambda)]\rangle / \langle \dot{q}_0\Theta(\dot{q}_0)\rangle$$

where Θ is the Heaviside step function which has been substituted for our more general indicator function $h_B(t)$. It has a convenient Fourier representation,

$$\Theta(x) = \int_{-\infty}^{\infty} \frac{dk}{2\pi i k} e^{ikx}$$

which follows from the differentiation and identification of the delta function.

Substituting the Fourier representation of the step function into the expression for κ,

$$\kappa = \frac{\int_{-\infty}^{\infty} \frac{dk}{k} \langle \dot{q}_0 e^{ik(\dot{q}_0 + \tilde{\eta}(\lambda))}\rangle}{\int_{-\infty}^{\infty} \frac{dk}{k} \langle \dot{q}_0 e^{ik\dot{q}_0}\rangle}$$

The averages can be computed noting that $\dot{q}_0$ and η are uncorrelated Gaussian random variables. This yields,

$$\kappa = \frac{\int_{-\infty}^{\infty} dk \exp\left[-\frac{1}{2\beta}k^2 - \frac{1}{2}k^2 \langle \delta^2\tilde{\eta}(\lambda)\rangle\right]}{\int_{-\infty}^{\infty} dk \exp\left[-\frac{1}{2\beta}k^2\right]}$$

where,

$$\beta \left\langle \delta^2 \tilde{\eta}(\lambda) \right\rangle = \int_0^\infty dt \int_0^\infty dt'\, \xi(t-t')e^{-\lambda(t+t')}$$
$$= \frac{1}{\lambda}\tilde{\xi}(\lambda)\,.$$

Finally putting this all together we get that the transmission coefficient is simply the ratio of the frequency of the barrier to the Grote–Hynes frequency,

$$\kappa = \frac{\lambda}{\omega_{\mathrm{b}}}$$

which characterizes how the friction kernel affects the periodic motion of the particle. In the large friction limit, $\xi(t) = \xi_0\delta(t)$, and $\lambda_0 = \omega_{\mathrm{b}}^2/\xi_0$ or

$$\kappa \approx \frac{\omega_{\mathrm{b}}}{\xi_0}$$

which is the same as Kramers' theory. In general, however, the Grote–Hynes estimate is larger than Kramers. It is particularly useful in cases of small ω_{b}.

8.6 Bimolecular reactions

While phenomenologically rate laws can be complex, they typically result from sequences of elementary steps comprised of monomolecular and bimolecular reactions. The transition state theory we have worked out thus far is suited for monomolecular processes. Here we will generalize our results to a generic bimolecular process,

$$A + B \to C$$

in which two species A and B diffuse in solution and react to form species C. The basic macroscopic chemical kinetics expression for the rate of change of the concentration of species B, $\bar{\rho}_B$, due to this reaction is

$$\frac{d\bar{\rho}_B}{dt} = -k\bar{\rho}_B\bar{\rho}_A$$

which defines the rate constant k. This is the standard expression for second-order kinetics. The probabilistic interpretation of $k\bar{\rho}_B$ is the probability to react per unit of time, as we had previously, times the probability to find a B particle close enough to an A so that they can react. Just as we previously had to introduce a microscopic time scale, τ_{mol}, to define the probability to react per time, we will introduce a microscopic unit of length that A and B need to be closer than in order to react, R^*. This geometry is illustrated in Figure 8.6.

If the species are diffusing in solution, then we can assume that their dynamics are well described by an overdamped Langevin equation. The corresponding Smoluchowski equation for the density, $\rho_B(\mathbf{r},t) = N_B p(\mathbf{r},t)$, is

$$\frac{\partial \rho_B(\mathbf{r},t)}{\partial t} = -\nabla \cdot \mathbf{J}(\mathbf{r},t) \qquad \mathbf{J}(\mathbf{r},t) = -De^{-\beta w(r)}\nabla e^{\beta w(r)}\rho_B(\mathbf{r},t)$$

where D is the diffusion constant of particle B, and $w(r)$ is the spherically symmetric potential of mean force that A and B interact through. For simplicity we will assume

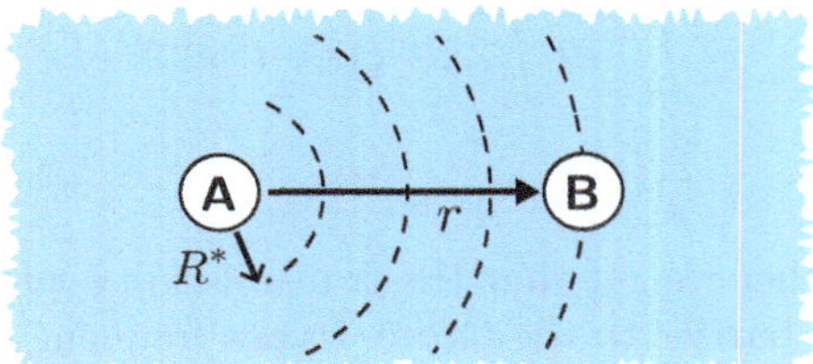

Fig. 8.6 Geometry of bimolecular reaction in which diffusion and barrier crossing must simultaneously occur. Dashed lines denote surfaces of constant separation distance r.

A and B are spherically symmetric. We can proceed as we did before and compute the rate from the steady state flux provided boundary conditions that allow us to evolve an appropriate steady state. Here we will consider pinning particle A at the origin, $r = 0$. Further we will assume that,

$$\rho_B(\infty, t) = \bar{\rho}_B$$

the bulk concentration of B is fixed. The short-ranged boundary condition is a little more complicated. Here we will assume a constant flux,

$$\partial_t \rho_B(R^*, t) = -k^* \bar{\rho}_B^*$$

across the distance R^* that is proportional to the concentration of B at R^* times the reaction probability k^*. While the former is something we will need to evaluate, the latter is taken as a parameter.

Given this setup, the problem has radial symmetry and in the steady state the flux is a function of r only,

$$J(r) = -D \left[e^{-\beta w(r)} \frac{d}{dr} e^{\beta w(r)} \right] \rho_B(r) \, .$$

At steady state, the total flux across each shell of constant r must be constant. We will denote that constant J_o and note

$$J(r) = -\frac{J_o}{4\pi r^2} \qquad J_o = k \bar{\rho}_B$$

where the first equality comes from the changing surface area of each constant r shell, and the latter follows from the definition of the reactive flux. By defining, $b(r) = e^{\beta w} \rho_B(r)$, we can rearrange the expression for the flux, now in terms of J_o, and obtain

$$\frac{db(r)}{dr} = \frac{J_o}{4\pi D} \frac{e^{\beta w}}{r^2}$$

which can be integrated from $r = R^*$ to $r = \infty$, yielding

$$b(\infty) - b(R^*) = \frac{J_o}{4\pi D \ell} \qquad \ell^{-1} = \int_{R^*}^{\infty} dr \, \frac{e^{\beta w}}{r^2}$$

where we have introduced a lengthscale ℓ that characterizes the scale over which the potential of mean force $w(r)$ changes. Assuming $w(r) \to 0$ as $r \to \infty$ and employing our large r boundary conditions we find,

$$\bar{\rho}_B^* = \left[\bar{\rho}_B - \frac{J_{\mathrm{o}}}{4\pi D\ell} \right] e^{-\beta w^*}$$

or eliminating J_{o},

$$\bar{\rho}_B^* = \left[1 - \frac{k}{4\pi D\ell} \right] \bar{\rho}_B e^{-\beta w^*}$$

we arrive at an expression for the concentration of B at R^*. In order to eliminate it we note that steady state implies,

$$J_{\mathrm{o}} = k\bar{\rho}_B = k^* \bar{\rho}_B^*$$

and substituting $\bar{\rho}_B^* = k\bar{\rho}_B/k^*$, we arrive at

$$k = \frac{4\pi D\ell}{1 + 4\pi D\ell e^{\beta w^*}/k^*}$$

which is the macroscopic rate of a bimolecular reaction. This expression depends on the diffusion constant, D, the reaction at contact, k^*, as well as the potential of mean force directly and through ℓ. This expression has two simple limiting forms. In the limit that $D\ell/k^* \to 0$, where diffusion cannot keep up with the reaction,

$$k \approx 4\pi D\ell$$

we find the reaction depends only on the diffusion constant. This is the *diffusion-limited rate*. In the alternative limit, $D\ell/k^* \gg 1$,

$$k \approx k^* e^{-\beta w^*}$$

we find that the dependence on the diffusion constant goes away and we are left with just the rate at contact R^* times the equilibrium probability of observing contact. This is the *reaction-limited regime*. The interpretation of these two contributions is that for a bimolecular reaction, in principle two separate rare events need to occur. The reactants need to find each other in space, which may require a rare concentration fluctuation. In addition, provided the reactants are close, a rare configurational fluctuation must occur for the products to form.

> **Exercise 8.7:** Water auto-ionization, in which a tagged water in its liquid dissociates into a proton and a hydroxide
>
> $$H_2O \underset{k_a}{\overset{k_d}{\rightleftharpoons}} H^+ + OH^-$$
>
> is a rare event with a small dissociation rate k_d and a high associated barrier. Assuming that the backward association rate, k_a, is diffusion limited, $k_a = 4\pi D\ell$, with a characteristic lengthscale set by the Bjerrum length $\ell = 1/\varepsilon k_B T$, use the known equilibrium constant of water $K = k_d/k_a = 10^{-14}$ M to infer k_d. You can take $D \approx 10^{-5}$ cm^2/s.

8.7 The golden rule

We have considered rare reactive events occurring classically, whose dynamics are well described by classical equations of motion. While such a description is a good approximation for the motion of heavy atoms at high temperature, if light particles like electrons or protons are moving between long-lived metastable states or external conditions like the temperature or pressure are such that quantum mechanical effects cannot be ignored, such a classical perspective will be inappropriate. In order to go beyond a classical approach we need to generalize our formalism.

One way to proceed is to consider a separable Hamiltonian of the form

$$\hat{\mathcal{H}}_0 = \begin{pmatrix} \hat{\mathcal{H}}_A & 0 \\ 0 & \hat{\mathcal{H}}_B \end{pmatrix}$$

where $\hat{\mathcal{H}}_A$ and $\hat{\mathcal{H}}_B$ represent Hamiltonians for species that are not interacting. Concretely one can imagine A and B being an excess electron or proton localized on molecules A or B, each described by a *diabatic* state. For each $X \in \{A, B\}$ there exists a set of eigenstates and eigenvalues

$$\hat{\mathcal{H}}_X |\psi_{X,n}\rangle = E_{X,n} |\psi_{X,n}\rangle$$

where $|\psi_{X,n}\rangle$ are the eigenvectors associated with $\hat{\mathcal{H}}_X$ and $E_{X,n}$ are energies, or their associated eigenvalues. Within each subspace, the eigenvectors are orthogonal, $\langle \psi_{X,n}|\psi_{X,m}\rangle = \delta_{nm}$. The full system obeys the time-dependent Schrodinger equation with wavefunction $|\Psi\rangle$. At $t = 0^+$ a coupling between species A and B is turned on that adds to the Hamiltonian, $\hat{\mathcal{H}} = \hat{\mathcal{H}}_0 + \Delta\hat{\mathcal{H}}$, with

$$\Delta\hat{\mathcal{H}} = \begin{pmatrix} 0 & \hat{V} \\ \hat{V} & 0 \end{pmatrix}$$

where for simplicity we can imagine that $\hat{V}$ is constant. Concretely this could be because the fragments on which an electron or proton are localized are initially far apart, but are brought into proximity with each other.

After the coupling between the two states is turned on, the time-dependent Schrodinger equation becomes

$$i\hbar \frac{d\,|\Psi\rangle}{dt} = \left(\hat{\mathcal{H}}_0 + \Delta\hat{\mathcal{H}}\right)|\Psi\rangle$$

which in the interaction picture of the initial Hamiltonian is

$$i\hbar \frac{d\,|\Psi_I\rangle}{dt} = \Delta\hat{\mathcal{H}}_I\,|\Psi_I\rangle$$

with

$$|\Psi_I(t)\rangle = e^{i\hat{\mathcal{H}}_0 t/\hbar}\,|\Psi(t)\rangle \qquad \Delta\hat{\mathcal{H}}_I = e^{i\hat{\mathcal{H}}_0 t/\hbar}\,\Delta\hat{\mathcal{H}}\,e^{-i\hat{\mathcal{H}}_0 t/\hbar}$$

where we have defined $|\Psi_I(t)\rangle$ and $\Delta\hat{\mathcal{H}}_I$ as the wavefunction and perturbation in the interaction picture. Within the interaction picture, the time-dependent Schrodinger equation can be integrated, using the time ordered exponential

$$|\Psi_I(t)\rangle = \mathcal{T}_+ e^{-i\int_0^t dt'\,\Delta\hat{\mathcal{H}}_I(t')/\hbar}\,|\Psi_I(0)\rangle$$

$$\approx \left(1 - \frac{i}{\hbar}\int_0^t dt'\,\Delta\hat{\mathcal{H}}_I(t')\right)|\Psi_I(0)\rangle$$

which if we assume $\Delta\hat{\mathcal{H}}$ is small, we can expand to first-order. This is an assumption that is consistent with our notion that A and B are well-defined, long-lived states of the system. If V were large, these states would be mixed quantum mechanically. Moving back out of the interaction picture, we have

$$|\Psi(t)\rangle = e^{-i\hat{\mathcal{H}}_0 t/\hbar}\,|\Psi(0)\rangle - \frac{i}{\hbar}\int_0^t dt'\,e^{i\hat{\mathcal{H}}_0(t'-t)/\hbar}\,\Delta\hat{\mathcal{H}}\,e^{-i\hat{\mathcal{H}}_0 t'/\hbar}\,|\Psi(0)\rangle$$

which is a sum of the unperturbed evolution from some initial condition under $\hat{\mathcal{H}}_0$, and a first-order correction due to the perturbation, which is a convolution with the linear action of V on the initial condition and the reference Green's function.

Imagine the system is prepared in state A, in a particular eigenstate state n such that $|\Psi(0)\rangle = |\psi_{A,n}\rangle$. The probability at time t that the system is found state in B, in particular eigenstate m is determined by the amplitude,

$$\langle\psi_{B,m}|\Psi(t)\rangle = -\frac{i}{\hbar}\int_0^t dt'\,\langle\psi_{B,m}|\,e^{i\hat{\mathcal{H}}_0(t'-t)/\hbar}\,\Delta\hat{\mathcal{H}}\,e^{-i\hat{\mathcal{H}}_0 t'/\hbar}\,|\psi_{A,n}\rangle$$

$$= -\frac{i}{\hbar}\int_0^t dt'\,e^{iE_{B,m}(t'-t)/\hbar - iE_{A,n}t'/\hbar}V\,\langle\psi_{B,m}|\psi_{A,n}\rangle$$

where in the second line we have used the fact that $|\psi_{A,n}\rangle$ and $|\psi_{B,m}\rangle$ are eigenstates of $\hat{\mathcal{H}}_0$ to resolve their associated energies $E_{A,n}$ and $E_{B,m}$. Taking the modulus of that probability amplitude, we find that the probability of transitioning between $|\psi_{A,n}\rangle$ and $|\psi_{B,m}\rangle$ a time t after $\Delta\hat{\mathcal{H}}$ was turned on is

$$P_{A,n \to B,m}(t) = \frac{4}{\hbar^2}|V|^2|\langle\psi_{B,m}|\psi_{A,k}\rangle|^2 \frac{\sin^2\left(\omega_{B_m,A_n}t/2\right)}{\omega^2_{B_m,A_n}}$$

where we have explicitly evaluated the integral over the phase factors and introduced the frequency $\omega_{B_m,A_n} = (E_{B,m} - E_{A,n})/\hbar$. The probability of a transition is a complicated function of time. However, in the long time limit

$$\lim_{t \to \infty} \frac{\sin^2(\omega t/2)}{\omega^2} = \pi t \delta(\omega)/2$$

as motivated by Figure 8.7. This implies that in the long time limit, the probability to transition grows linearly with time,

$$P_{A,n \to B,m}(t) \approx \frac{2\pi}{\hbar} |V|^2 |\langle \psi_{B,m} | \psi_{A,n} \rangle|^2 t \delta\left(E_{B,m} - E_{A,n}\right)$$

where we have used the fact that $\delta(ax) = \delta(x)/a$. In other words, the rate to transfer between $|\psi_{A,n}\rangle$ and $|\psi_{B,m}\rangle$ is

$$k_{A,n \to B,m} = \frac{2\pi}{\hbar} |V|^2 |\langle \psi_{B,m} | \psi_{A,n} \rangle|^2 \delta\left(E_{B,m} - E_{A,n}\right)$$

which is the rate of change of $P_{A,n \to B,m}(t)$.

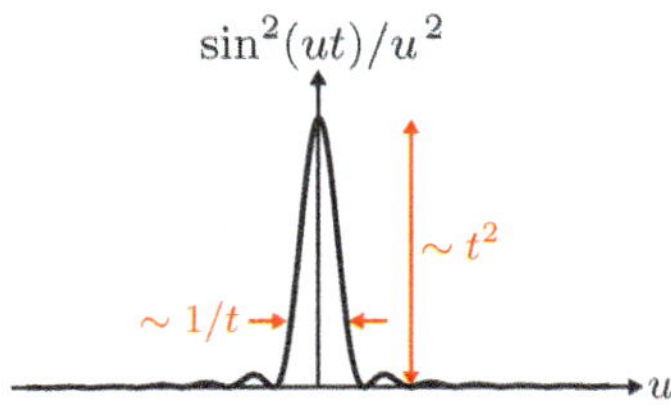

Fig. 8.7 The function $\sin^2(ut)/u^2$ approaches a delta function as $t \to \infty$.

Note that this expression, colloquially referred to as *Fermi's golden rule*, or the golden rule, has the structure of a prefactor $2\pi|V|^2|\langle \psi_{B,m}|\psi_{A,n}\rangle|^2/\hbar$ times a delta function. Again, there are no explicit dynamics in this expression, only static quantities enter. Interpreting this equation within our classical transition state theory, the prefactor is analogous to the flux, while the delta function is the probability of reaching a transition state, which in this case is just given by a resonance condition, that $E_{B,m} = E_{A,n}$.

Exercise 8.8: Generalized our result for the transition probability $k_{A,n \to B,m}$ to the case where the coupling between the states is time dependent, $V(t) = v \cos(\omega t)$ and show that

$$k_{A,n \to B,m} = \frac{\pi}{2\hbar^2} |v|^2 |\langle \psi_{B,m}|\psi_{A,n}\rangle|^2 [\delta\left(\omega_{B_m,A_n} + \omega\right) + \delta\left(\omega_{B_m,A_n} - \omega\right)]$$

where v and ω are the amplitude and frequency of the coupling.

The golden rule rate is between two specific quantum states. If we rather wanted an aggregate rate between any state initially in A to any state in B, we need only sum over all initial and final states,

$$k_{AB} = \frac{2\pi}{\hbar}|V|^2 \sum_{n,m} p_n |\langle \psi_{B,m}|\psi_{A,n}\rangle|^2 \delta\left(E_{B,m} - E_{A,n}\right)$$

where we have weighted the rate by the likelihood of being in a particular initial state with its equilibrium initial likelihood p_n. This expression can be rewritten by first using the Fourier representation of the delta function,

$$k_{AB} = \frac{1}{\hbar^2}|V|^2 \sum_{n,m} p_n |\langle \psi_{B,m}|\psi_{A,n}\rangle|^2 \int_{-\infty}^{\infty} dt\, e^{i(E_{B,m} - E_{A,n})t/\hbar}$$

and then noting that the sum can be re-expressed as

$$\sum_{n,m} p_n |\langle \psi_{B,m}|\psi_{A,n}\rangle|^2 e^{i(E_{B,m} - E_{A,n})t/\hbar} = \sum_n p_n \langle \psi_{A,n}| e^{i\hat{\mathcal{H}}_B t/\hbar} e^{-i\hat{\mathcal{H}}_A t/\hbar} |\psi_{A,n}\rangle$$

using the resolution of the identity and the action of the reference Hamiltonian on its eigenstate basis. Inserting this expression into the full rate from A to B we obtain

$$k_{AB} = \frac{1}{\hbar^2}|V|^2 \int_{-\infty}^{\infty} dt\, \left\langle e^{i\hat{\mathcal{H}}_B t/\hbar} e^{-i\hat{\mathcal{H}}_A t/\hbar} \right\rangle_A$$

which relates the rate as a prefactor times an integral over an equilibrium average restricted to the initial state. In general $\hat{\mathcal{H}}_A$ and $\hat{\mathcal{H}}_B$ do not commute. As a consequence, the rate depends on the complex dynamics that evolve the averages over the phase factors. Note we found a similar sort of result when we treated the interaction with the bath within perturbation theory in the context of the Redfield equation. There the rate between eigenstates was given by the coupling to the bath squared times an integral of a bath correlation function.

If we assume the evolution of $\hat{\mathcal{H}}_A$ and $\hat{\mathcal{H}}_B$ are well described classically and ignore terms resulting from their non-communtation, we could combine the two complex exponentials. Doing so we would recognize another Dirac delta function, yielding

$$k_{AB} = \frac{2\pi}{\hbar}|V|^2 \langle \delta(\mathcal{H}_B - \mathcal{H}_A)\rangle_A$$

which is an ensemble-averaged version of the golden rule result, neglecting potential interference effects. Outside of this classical limit, the time correlation function entering into the rate can be evaluated for harmonic systems.

Exercise 8.9: Show that for harmonic oscillator Hamiltonians of the form

$$
\mathcal{H}_X = E_X + \sum_j \frac{p_j^2}{2m_j} + \frac{m_j \omega_j^2}{2} \left(q_j - \gamma_j \frac{q_X}{m_j \omega_j^2} \right)^2
$$

for $X = \{A, B\}$, that the rate within the classical limit of the golden rule is

$$
k_{AB} = \frac{|V|^2}{\hbar} \sqrt{\frac{\pi}{\lambda k_{\mathrm{B}} T}} e^{-\beta (E_B - E_A + \lambda)^2 / 4\lambda}
$$

where $\lambda = \sum_j \gamma_j^2 (q_A - q_B)^2 / 2m_j \omega_j^2$, by averaging $\delta(\mathcal{H}_B - \mathcal{H}_A)$ within a classical equilibrium ensemble at fixed temperature and Hamiltonian $\mathcal{H}_A$.

8.8 Marcus' theory of electron transfer

Electron transfer reactions, or oxidation-reduction reactions, are among the most significant processes in chemistry and biology. Perhaps the simplest example comes from inorganic aqueous electrochemistry,

$$
\mathrm{Fe}^{2+} + \mathrm{Fe}^{3+} \rightleftharpoons \mathrm{Fe}^{3+} + \mathrm{Fe}^{2+}
$$

or the so-called ferrous-ferric exchange. In such a reaction, a donor species, in this case the Fe^{2+}, transfers an electron to an acceptor, Fe^{3+}. No bonds are made or broken, but charge is significantly reorganized. For context, the 3rd ionization energy for iron is around $E_{\mathrm{ion}} = 10eV$. If we naively make a transition state estimate of the rate to climb out of the Coulomb potential set by that ionization energy, and note that a typical timescale for electronic motion is on the order of attoseconds, we would expect spontaneous ionization to never occur. On the contrary, in a 1M aqueous solution, the ferrous-ferric exchange occurs with a characteristic timescale, $1/k_e \sim 0.1s$. While still clearly a rare event, the solvent fundamentally changes the nature of the reaction. Recognizing the active role of the solvent, Marcus was able to accurately predict the rate of electron transfer.

Exercise 8.10: Use transition state theory to estimate the electron transfer rate using a barrier height equal to a typical ionization energy, $10eV$ and a characteristic frequency of the reactant as $1/10^{-18}s$.

Figure 8.8 illustrates the general pathway for electron transfer. There are two localized electronic states, or redox states, pictured. In one, a transferable electron is located at site A, the donor, in the other, the transferable charge is located on site B, the acceptor. When the electron is localized on one of the two species, the asymmetric charge distribution polarizes the surrounding solvent. The polarization is

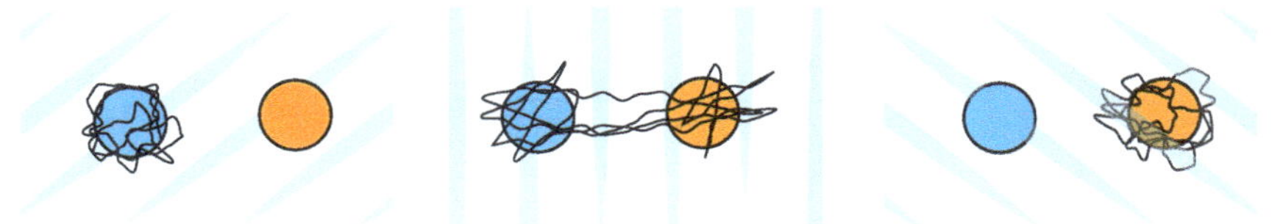

Fig. 8.8 Pathway for electron transfer. An electron is initially localized on the donor site (blue), polarizing the environment. Through a spontaneous solvent fluctuation that creates two energetically degenerate locations, the electron becomes delocalized. Finally, with the rearrangement of the solvent the electron relocalizes on the acceptor (orange).

associated with a reorientation of the dipolar field. Accompanying that polarization is a favorable solvation energy as discussed in Chapter 3. The loss of asymmetry in the charge distribution will therefore be associated with a loss of that favorable energy. Passing from a state in which electronic charge is localized on one redox center to a state where the charge is distributed over two redox centers will be associated with an increase in energy. The charge-distributed state is the transition state for electron transfer, and the increase in energy to reach that state is the activation energy.

The nuclear motion that results in polarization fluctuations occurs on a timescale much slower than the timescale of electronic motion. Typical dipole–dipole correlation times are on the order of picoseconds, 10^{-12}s, nearly 6 orders of magnitude slower than electronic motion. A key insight that Marcus had was to acknowledge that in such systems with dynamics occurring on vastly different timescales, the fast motion can only take place in such configurations of the slow degrees of freedom for which the fast transition conserves energy. In the case of electron transfer, the slow solvent configuration must be the same before and after the electron changes its location. This is known as the Franck–Condon principle. In order for an electronic transition to conserve energy, the energy of the electron residing on either the donor or acceptor must be the same. The energy gap between the two states can thus serve as a useful reaction coordinate.

In order to make a quantitative statement about electron transfer, we need to construct a Hamiltonian. A generic form for an electron interacting with a solvent is,

$$\mathcal{H} = \mathcal{H}_e + \mathcal{H}_b + \mathcal{H}_{e,b}$$

where $\mathcal{H}_e$ is the Hamiltonian for the electron, $\mathcal{H}_b$ is the Hamiltonian for the solvent or bath, and $\mathcal{H}_{e,b}$ is the coupling between the two. From the perspective of the electron, the simplest model is a two-state system

$$\mathcal{H}_e = \begin{pmatrix} 0 & K \\ K & \Delta\epsilon \end{pmatrix}$$

where we have expressed the matrix in a basis of diabatic states associated with the electron localized on state A, $|A\rangle$, or localized on state B, $|B\rangle$. Taking the electron on state A as the zero of energy, $\Delta\epsilon$ is the chemical potential difference for localizing an

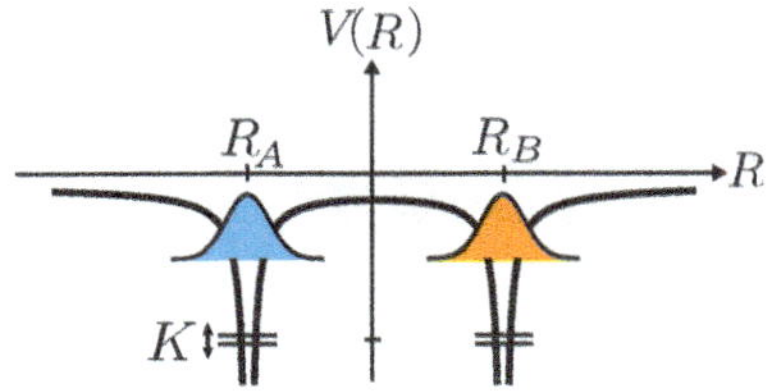

Fig. 8.9 Tunneling potential for an electron transferring between a sum of Coulomb potentials, centered at the two nuclei.

electron on site B, minus that on site A, and $K = \langle A| \mathcal{H}_e |B\rangle$ is the electronic coupling. Typically it is of the form

$$K \sim K_0 e^{-|R_A - R_B - \sigma|/\xi}$$

which captures the exponential decay of the wavefunction as it tunnels out of the Coulomb potential well. For typical inorganic species, K_0 is around 100 cm^{-1}, $\xi = $ 1Å, and σ is the distance of the closest approach, around 1 nm. With these typical parameters, the size of this coupling is a fraction of a k_BT. Indeed, Marcus' theory of electron transfer is valid only in the limit of weak coupling as it will use the golden rule result derived in the previous section. Figure 8.9 shows the potential the electron feels, as a sum of the Coulomb potentials for the donor and acceptor centered at R_A and R_B. The electronic coupling results in a tunnelling splitting of the size K.

From the perspective of the solvent, Marcus employed a continuum description to model its fluctuations. The appropriate continuum theory is the dielectric continuum theory developed in Chapter 3, which is an effective linear model, or Gaussian field theory. Within the context of continuum dielectric fluctuations, the effect of the solvent is the generation of polarization. Putting this together, the total Hamiltonian can be written

$$\mathcal{H} = \begin{pmatrix} 0 & K \\ K & \Delta\epsilon \end{pmatrix} - \mathcal{E}(\mathbf{r}^N) \begin{pmatrix} 1 & 0 \\ 0 & -1 \end{pmatrix} + \mathcal{H}_b(\mathbf{r}^N) \begin{pmatrix} 1 & 0 \\ 0 & 1 \end{pmatrix}$$
$$= \begin{pmatrix} \mathcal{H}_A(\mathbf{r}^N) & K \\ K & \mathcal{H}_B(\mathbf{r}^N) \end{pmatrix}$$

where the coupling of the system to the polarization field, $\mathcal{E}$, is through the dipole generated by the asymmetric charge distribution. The second equality is a more compact notation that clarifies that the effect of the bath is to generate fluctuations in the energy gap between states A and B. Indeed the energy gap, ΔE, is given by

$$\Delta E(\mathbf{r}^N) = \mathcal{H}_A(\mathbf{r}^N) - \mathcal{H}_B(\mathbf{r}^N) = -2\mathcal{E}(\mathbf{r}^N) - \Delta\epsilon.$$

Consistent with our development so far, we anticipate that the transition state is that for which the energy gap between the two electronic states vanishes. Since $\Delta\epsilon$ is independent of the bath, the probability of being at the transition state is the probability of a polarization of magnitude, $\mathcal{E}(\mathbf{r}^N) = -\Delta\epsilon/2$.

For a system at thermal equilibrium, which underpins transition state theory estimates, the likelihood of any fluctuation is given by

$$P(\mathcal{E}) = e^{-\beta F(\mathcal{E})}$$

where $F(\mathcal{E})$ is the free energy to generate a polarization of magnitude $\mathcal{E}$. Within
dielectric continuum theory, polarization fluctuations are assumed to be Gaussian.
Therefore the free energy function takes the form

$$F(\mathcal{E}) = \frac{1}{\lambda}\mathcal{E}^2$$

where λ sets the scale of fluctuations. For the charge localized on state A, the free
energy is

$$F_A(\mathcal{E}) = \frac{1}{\lambda}\mathcal{E}^2 - \mathcal{E}$$

whereas for the charge localized on state B, the free energy is

$$F_B(\mathcal{E}) = \frac{1}{\lambda}\mathcal{E}^2 + \mathcal{E} + \Delta\epsilon$$

where we have taken into account the offset and asymmetric response for charges on
A or B. Our criterion for the transition state is $F_A(\mathcal{E}) = F_B(\mathcal{E})$. Solving for the value
of $\mathcal{E}$ that satisfies this, $\mathcal{E}^*$ and putting it into the free energy for state A less the value
at its minimum, we find that the barrier height is given by

$$\Delta F_A(\mathcal{E}^*) = F_A(\Delta\epsilon/2) - F_A(\lambda/2) = \frac{(\Delta\epsilon + \lambda)^2}{4\lambda}$$

which we note is quadratic in the thermodynamic driving force, $\Delta\epsilon$, and minimized at
$\Delta\epsilon = -\lambda$. Specific expressions for λ , known as the reorganization energy, depend on
the context of electron transfer. For homogeneous polar fluids, dielectric continuum
theory gives an estimate of

$$\lambda = \frac{q^2}{\sigma}\left(1/\epsilon_\infty - 1/\epsilon_s\right)$$

where ϵ_∞ is the optical dielectric constant. This is the energy difference of an ion of
charge q and diameter σ solvated in a fluid with static dielectric constant ϵ_s and that
solvated with dielectric fluctuations only from the optical modes of the solvent. In the
vertically excited state, only the other electrons have time to polarize in response to
the charge, the nuclear components do not change, and so their contribution to the
polarization is not present.

To compute the rate, we use the golden rule assuming the bath is classical,

$$k_e = \frac{2\pi}{\hbar}|\langle A| H_e |B\rangle|^2 \langle\delta(\mathcal{H}_A - \mathcal{H}_B)\rangle_A$$

which takes the usual form of the coupling squared times the density of states to ensure
conservation of energy, which for a process at finite temperature must be thermally
averaged. Plugging in terms,

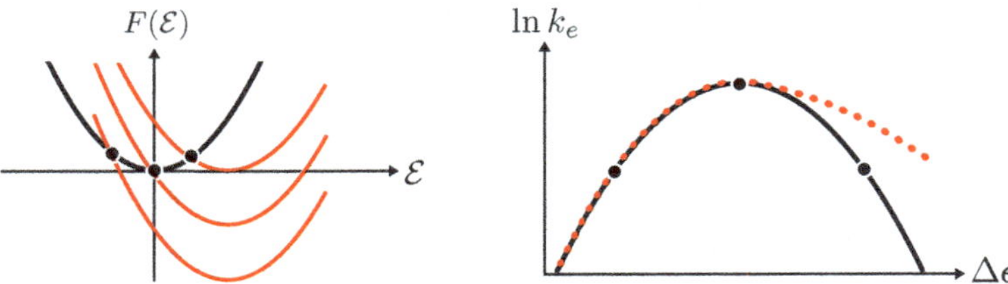

Fig. 8.10 Variation of the barrier height and associated electron transfer rate with thermodynamic driving force. The inverted and normal regimes correspond to the thermodynamic driving force, $\Delta\epsilon$, being larger and smaller, respectively, than the reorganization energy, λ. While in the normal regime, Marcus theory predicts well the experimentally observed rate (sketched in red), in the inverted regime, experiments differ substantially from the classical prediction of an inverted parabolic functional.

$$k_e = \frac{2\pi}{\hbar} K^2 P(\mathcal{E}^*)$$

$$= \frac{\pi}{\hbar} \frac{K^2}{\sqrt{\pi\lambda/\beta}} e^{-\beta \frac{(\Delta\epsilon+\lambda)^2}{4\lambda}}$$

we finally arrive at our transition state theory estimate for the rate of electron transfer. It has the usual components, of a flux over a barrier, which in this case is quantum mechanical, $4\pi K^2/\hbar$ times the probability of being at the barrier, $\exp\{-\beta(\Delta\epsilon + \lambda)^2/4\lambda\}/\sqrt{4\pi\lambda/\beta}$.

Qualitative free energy surfaces for increasing $\Delta\epsilon$ and their associated rates, as predicted from Marcus theory, are shown in Figure 8.10. A nontrivial prediction of Marcus theory is the so-called *inverted region*, whereby the electron transfer rate decreases with increasing exothermicity. The existence of the inverted region was considered a most striking and counter-intuitive result. Its verification by Miller and Closs is often cited as the pivotal experiments validating the theory. While its existence is correctly predicted by the classical theory described above, the quantitative behavior of the rate as a function of $\Delta\epsilon$ is not correctly captured. In particular, pathways involving nuclear tunneling become especially prevalent in the inverted region. These pathways augment those that are allowed classically, resulting in a rate that is larger than that predicted by Marcus theory.

8.9 Splitting probabilities and committors

All of our discussion on rates has assumed that we can identify a dynamically relevant reaction coordinate, q. However, in most applications of modern interest identifying this coordinate is the most difficult part of the problem. Consider, for example, embedding our usual free energy $w(q)$ in the two-dimensional space shown in Figure 8.11. Here we have introduced a plausible additional coordinate, q', such that the potential the system evolves in is described by the surface $w(q, q')$. We have done this in such a way that integrating q' out returns a bimodal potential $w(q)$. However, $w(q)$ clearly hides the dynamically relevant q'. Given the geometry of $w(q, q')$, to make a transition between the two basins, a particle has to move along *both* q and q'.

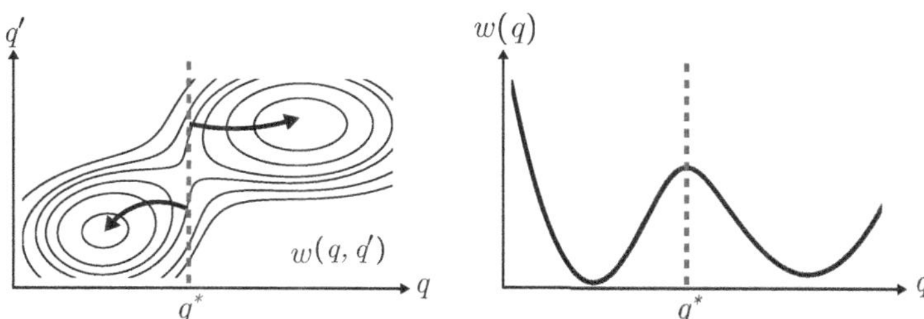

Fig. 8.11 Illustration of the potential difficulty of correctly identifying the transition state in a high-dimensional system. The left plot is a 2d free energy surface, which when q' is integrated out results in the right plot. Trajectories initiated at the putative q^* transition state are likely to recross.

While in principle our reactive flux correlation function makes no assumptions as to the form of the underlying potential, the ease of its evaluation certainly depends on choosing a coordinate that admits as few recrossings as possible. Otherwise, the transmission coefficient will obtain a very small value at its plateau. Solving this problem is in general complicated, as it requires finding a way to *discover* variables like q'. To imagine how we might do this, consider having access to many realizations of a reaction, i.e. many trajectories that start in basin A and proceed to basin B. A sensible question could be where along the trajectory is the transition state located? If in general we do not know about the features of the underlying potential, we have to be a little more precise about what we mean by a transition state.

To develop a generalization of a a transition state, it is useful to introduce a function that reports on the probability of a given configuration to react. Such a function, known as a *committor* or splitting probability, could be defined for a specific configuration $\mathbf{r}^N$, as the probability, averaged over all possible velocities, of evolving to state B

$$\phi_B \left(\mathbf{r}^N \right) = \int d\mathbf{v}^N \, p \left(\mathbf{v}^N \right) h_B \left[\tau_{\text{mol}}; \mathbf{v}^N, \mathbf{r}^N \right]$$

over a time τ_{mol}, where $p(\mathbf{v}^N)$ is just the Maxwell-Boltzmann distribution. This function returns the probability that the configuration $\mathbf{r}^N$ reacts, and as such the ensemble of configurations where $\phi_B(\mathbf{r}^N) = 1/2$ are those configurations equally likely to react as to proceed to the initial state. They are the most undecided configurations, which is exactly what we mean by a transition state. Here however, we no longer have a specific point along a collective coordinate, rather we have a transition state ensemble, a collection of configurations that are equally likely to react as to not. A picture of how this function evolves over the configurations of a reactive trajectory is shown in Figure 8.12.

Let us examine this function first in a simple example of an overdamped particle in a potential $w(q)$. The evolution of such a particle's position is given by $\gamma \dot{q} = -w'(q) + \eta$,

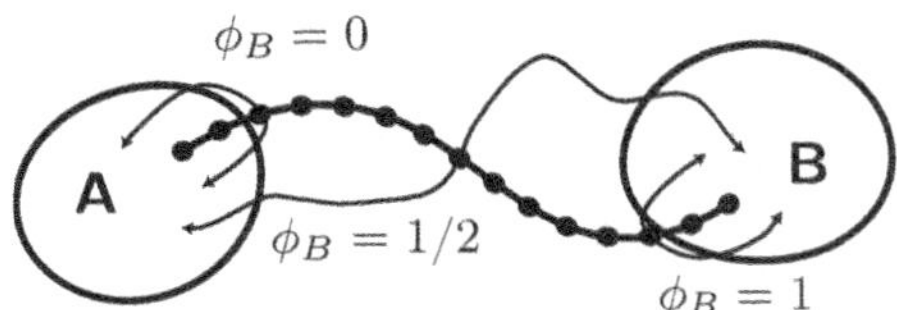

Fig. 8.12 Example of the characteristic evolution of the committor along a reactive trajectory, in which configurations are arranged sequentially in time.

where as usual, γ denotes the friction due to the solvent and η the random noise. The distribution of particle positions at any time is given by the Schmolowski equation

$$\frac{\partial p(q,t)}{\partial t} = -\frac{\partial}{\partial q} J(q,t)$$

$$= D\frac{\partial}{\partial q} e^{-\beta w(q)} \frac{\partial}{\partial q} e^{\beta w(q)} p(q,t)$$

and if we solve this equation with the boundary conditions,

$$p(q,0) = \delta(q - q_0) \qquad p(q_A,t) = 0 \qquad p(q_B,t) = 0$$

which input an initial source at q_0 and sinks at q_A and q_B, we can compute the committor as

$$\phi_B(q_0) = \int_0^\infty dt\, J(q_B,t)$$

which is just the integrated flux into state B, or fraction of trajectories that end in B having started at q_0. The result is,

$$\phi_B(q_0) = \frac{\int_{q_A}^{q_0} dq\, e^{\beta w(q)}}{\int_{q_A}^{q_B} dq\, e^{\beta w(q)}}$$

which, for the sorts of bistable potentials we have previously considered, is sigmoidal, increasing from 0 near q_A to 1 near q_B with its largest slope found at q^* and a characteristic width, $w(q^* \pm \Delta q) \approx w(q^*) \pm k_{\mathrm{B}}T$.

This simple closed-form expression is only possible in one dimension. More generally, for an overdamped dynamics of the form

$$\gamma \dot{\mathbf{r}}_i = \mathbf{F}_i(\mathbf{r}^N) + \boldsymbol{\eta}_i \qquad \langle \boldsymbol{\eta}_i(t) \otimes \boldsymbol{\eta}_j(t') \rangle = 2k_{\mathrm{B}}T\gamma \mathbf{1}\delta_{ij}\delta(t - t')$$

where the force $\mathbf{F}_i(\mathbf{r}^N) = -\nabla_i U(\mathbf{r}^N)$ is the gradient of a potential, the committor is the solution to the steady-state adjoint Fokker–Planck equation in the full configuration space of the system,

$$\sum_i^N \gamma^{-1}\mathbf{F}_i(\mathbf{r}^N) \cdot \nabla_i \phi_B(\mathbf{r}^N) + D\nabla_i^2 \phi_B(\mathbf{r}^N) = 0$$

with boundary conditions $\phi_B(A) = 0$ and $\phi_B(B) = 1$. For an equilibrium system, the principle of microscopic reversibility dictates that the complimentary committor for starting in B and ending in A, is $\phi_A = 1 - \phi_B$. The associated probability current, $\mathbf{J}_{AB}$, between A and B is

$$\mathbf{J}_{AB} = \sum_i^N \gamma^{-1} \mathbf{F}_i(\mathbf{r}^N) p(\mathbf{r}^N) \phi_A(\mathbf{r}^N) - D\nabla_i p(\mathbf{r}^N) \phi_A(\mathbf{r}^N)$$

which for an equilibrium system, with $\rho_{\mathrm{eq}}(\mathbf{r}^N) \propto \exp[-\beta U(\mathbf{r}^N)]$, reduces to

$$\mathbf{J}_{AB} = \sum_i^N D p(\mathbf{r}^N) \nabla_i \phi_B(\mathbf{r}^N)$$

which is the diffusive flux in the direction of variations in the committor. While typically intractable, if the committor could be evaluated it would provide an estimate of the rate

$$k_{AB} = D \frac{\langle |\nabla \phi_B|^2 \rangle}{\langle h_A \rangle}$$

where the average is a usual equilibrium average. In a complex system we can use the committor as a means to uncover the dynamically relevant coordinates. But how should we gather sufficient information to perform that analysis?

8.10 Reactive path ensembles

Let's begin addressing the question of discovering reaction coordinates by writing down exactly what we mean by the collection of trajectories that react. Such a collection could be considered an ensemble of reactive trajectories. The probability of observing a particular trajectory can be represented following some basic notions of conditional probabilities. The probability to observe a trajectory or sequence of points in phase space, $\mathbf{X}(t) = \{\mathbf{x}_0, \mathbf{x}_{\Delta t} \ldots, \mathbf{x}_t\}$, of duration t, is given by

$$P_0[\mathbf{X}(t)] = p_{\mathrm{eq}}(\mathbf{x}_0) P[\mathbf{X}(t)|\mathbf{x}_0]$$

where $p_{\mathrm{eq}}(\mathbf{x}_0)$ is the probability to observe the initial conditions $\mathbf{x}_0$, and $P[\mathbf{X}(t)|\mathbf{x}_0]$ is the conditional probability to generate trajectory $\mathbf{X}(t)$ from that initial condition. If the dynamics are Markovian, the probability of a trajectory is

$$P[\mathbf{X}(t)|\mathbf{x}_0] = \prod_{i=0}^{t/\Delta t - 1} P(\mathbf{x}_{(i+1)\Delta t}|\mathbf{x}_{i\Delta t})$$

where $P(\mathbf{x}_{\Delta t}|\mathbf{x}_0)$ is the transition probability, or the probability of reaching $\mathbf{x}_{\Delta t}$ in a time Δt, provided the system is at $\mathbf{x}_0$ at time 0. The fact that the dynamics are assumed Markovian means that the transition probability is independent of how the system arrived at its current state. We have considered Markovian dynamics in the context of the Langevin equation, but indeed even Newton's equations are Markovian,

$$P(\mathbf{x}_{\Delta t}|\mathbf{x}_0) = \delta(\mathbf{x}_{\Delta t} - e^{\mathcal{L}\Delta t}\mathbf{x}_0)$$

where $\mathcal{L}$ is the Liouvillian operator. The transition probability thus explicitly depends on only $\mathbf{x}_{\Delta t}$ and $\mathbf{x}_0$, but is singular, which is a signature of Newton's equations being deterministic. For stochastic equations of motion like our Langevin equation, this measure has support over the full phase space.

The probability, $P_0[\mathbf{X}(t)]$ is for an arbitrary trajectory, which we can condition into a reactive trajectory by multiplying by indicator functions for basins A and B,

$$P_{\mathrm{AB}}[\mathbf{X}(t)] = P_0[\mathbf{X}(t)]h_A[\mathbf{x}_0]h_B[\mathbf{x}_t]/Z_{AB}(t)$$

where, $Z_{AB}(t)$

$$Z_{AB}(t) = \int \mathcal{D}[\mathbf{X}(t)]P_0[\mathbf{X}(t)]h_A[\mathbf{x}_0]h_B[\mathbf{x}_t] = \langle h_A(0)h_B(t)\rangle$$

is the normalization that counts all reactive trajectories. The second equality notes that this *path partition function* is equivalent to the numerator of our side–side correlation function. Analogously, we can define the denominator of the side–side correlation function,

$$Z_A = \int \mathcal{D}[\mathbf{X}(t)]P_0[\mathbf{X}(t)]h_A[\mathbf{x}_0] = \langle h_A\rangle$$

which counts the number of trajectories that start in basin A. The side–side correlation function then follows as

$$C_{AB} = \frac{\langle h_A(0)h_B(t)\rangle}{\langle h_A(0)\rangle} = \frac{Z_{AB}(t)}{Z_A} \sim k_{AB}t$$

which is simply a ratio of path partition functions, or that the rate constant is the ratio of partition functions times the observation time. This is an interesting result, but intuitive. It simply states that the number of trajectories that reactive in a fixed time, divided by the number of trajectories that start in A, is proportional to the rate of reaction. More importantly, with a well-defined distribution, $P_{\mathrm{AB}}[\mathbf{X}(t)]$ can be sampled with a suitably defined Monte Carlo procedure. Transition path sampling is a Monte Carlo procedure to do just that. The algorithm is discussed later in Chapter 10 and provides access to sufficient reactive trajectories to generate an ensemble of transition states with which committors can be evaluated and reaction coordinates discovered.

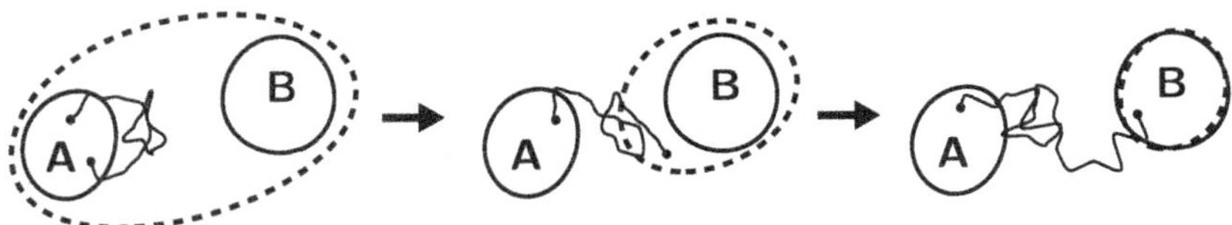

Fig. 8.13 Illustration of a reversible work path that takes an initially unreactive trajectory and transforms it into a reactive trajectory.

However, even without a reaction coordinate, this ratio of partition functions admits a way to evaluate the rate through a thermodynamic analogy. In equilibrium, we know that ratios of partition functions are related to changes in free energy and changes in free energy are nothing but the reversible work to perform some transformation. If we follow this logic, the ratio of path partition functions could be thought of as a reversible work albeit in path space,

$$k_{AB}t = \frac{Z_{AB}(t)}{Z_A} = e^{-\beta \mathcal{W}_B(t)}$$

where $\beta \mathcal{W}_B(t)$ is the "reversible work" to transform a trajectory that starts in A, into a trajectory that transitions from A to B. It helps to visualize how this transformation might actually work. Consider Figure 8.13. In it we have denoted regions A and B with a set of increasingly restrictive constraints on the end point of the trajectory. It is the constraint on this point, and only this point, that differentiates $Z_{AB}(t)$ and Z_A. One can think of the reversible work as that required to stretch a trajectory over to the B state. Like any reversible process, the work required should be independent of the specific path you take. This means that in principle evaluating the rate constant from the ratio of path partition functions offers a way to calculate the rate without any preconceived mechanism or identification of a relevant coordinate.

8.11   Brownian bridges

The dynamics that transition systems between two long-lived states afford a simplicity. Their rates are largely determined by the likelihood of the rare fluctuation that takes the system to its transition state. That the transition state is unstable, means that the transition paths are short, fleeting excursions over the top of potential barriers. There are other rare events that either do not connect two long-lived states but rather many short-lived ones, or that proceed through long-lived intermediates en route between the reactants and products. In the limit that the dynamics are diffusive, then so-called Brownian bridges afford a means of characterizing them.

An overdamped collection of many particles obeys a Fokker–Planck equation,

$$\frac{\partial p(\mathbf{r}^N, t)}{\partial t} = \sum_i^N -\gamma^{-1} \nabla_i \mathbf{F}_i(\mathbf{r}^N) p(\mathbf{r}^N, t) + D\nabla_i^2 p(\mathbf{r}^N, t)$$

where $\mathbf{F}_i(\mathbf{r}^N)$ is the force on particle i, and $p(\mathbf{r}^N, t)$ is the probability of observing the system at position $\mathbf{r}^N$ at time t provided some initial condition, say $\mathbf{r}^N(0) = \mathbf{r}_0^N$. In principle, we could ask a similar question concerning the probability of the system being at $\mathbf{r}_f^N$ at time t_f given it was at position $\mathbf{r}^N$ at time $t < t_f$. Formally this conditional probability, denoted ϕ is given by

$$\phi(\mathbf{r}^N, t) = \frac{\left\langle \delta[\mathbf{r}^N(t) - \mathbf{r}^N]\delta[\mathbf{r}^N(t_f) - \mathbf{r}_f^N] \right\rangle}{\langle \delta[\mathbf{r}^N(t) - \mathbf{r}^N] \rangle}$$

an expectation value integrated over the noise history. As an expectation value, ϕ satisfies an adjoint Fokker–Planck equation, but one going backward in time due to the condition imposed on the final time

$$\frac{\partial \phi}{\partial t} = -\sum_i^N \left(\gamma^{-1} \mathbf{F}_i(\mathbf{r}^N) \nabla_i \phi + D \nabla_i^2 \phi \right)$$

which is the time-dependent version of the equation that the committor function satisfies. In fact, ϕ could be understood as the time-dependent commitment probability for ending in state $\mathbf{r}_f^N(t_f)$. Together with the Fokker–Planck equation we can ask about the probability $\mathcal{P}(\mathbf{r}^N, t)$ of starting at $\mathbf{r}_0^N$ at $t = 0$ and conditioned to end at a given point $\mathbf{r}_f^N(t_f)$ at time t_f, to find the particle at point $\mathbf{r}^N$ at time $t \in \{0, t_f\}$,

$$\mathcal{P}(\mathbf{r}^N, t) = \frac{1}{\mathcal{N}} \phi(\mathbf{r}^N, t) p(\mathbf{r}^N, t)$$

which is the product of ϕ and the regular Fokker–Planck probability evolved from an initial condition $\mathbf{r}^N(0) = \mathbf{r}_0^N$ and $\mathcal{N}$ is a normalization constant. This follows from the Markov property of the transition probabilities. By taking its time derivative and using the two previous equations, it is straightforward to show that $\mathcal{P}(\mathbf{r}^N, t)$ satisfies

$$\frac{\partial \mathcal{P}}{\partial t} = \sum_i^N -\gamma^{-1} \nabla_i \left[\mathbf{F}_i(\mathbf{r}^N) + 2k_{\mathrm{B}} T \nabla_i \ln \phi \right] \mathcal{P}(\mathbf{r}^N, t) + D \nabla_i^2 \mathcal{P}(\mathbf{r}^N, t)$$

which is identical to the original Fokker–Planck equation, but with an added drift, $2D \nabla_i \ln \phi$ with $D = k_{\mathrm{B}} T / \gamma$. This implies the existence of a Langevin equation

$$\gamma \dot{\mathbf{r}}_i = \mathbf{F}_i(\mathbf{r}^N) + 2k_{\mathrm{B}} T \nabla_i \ln \phi(\mathbf{r}^N, t) + \boldsymbol{\eta}_i$$

where due to the added drift, all realizations of the stochastic process will fulfill the conditioning of ending at $\mathbf{r}_f^N(t_f)$ at time t_f. This transformation is known as the *finite time Doob transform*, and produces an equation of motion known as a Brownian bridge. By construction the dynamics generated by this Langevin equation will produce transition paths that are statistically identical to the original Langevin equation, conditioned on the end point. This result is like the fluctuation–dissipation theorem, in that the force that affects a dynamical condition is the gradient of the log of the probability of the system achieving that condition spontaneously.

Like the calculation of the time-independent commitment probability, evaluating $\phi(\mathbf{r}^N, t)$ directly is difficult. However, there are a few systems for which it is feasible. Consider for example free diffusion in one dimension,

$$\gamma \dot{x}(t) = \eta(t) \qquad \langle \eta(0) \eta(t) \rangle = 2k_{\mathrm{B}} T \gamma \delta(t)$$

where the time-dependent commitment probability satisfies

$$\frac{\partial \phi}{\partial t} = -D \frac{\partial^2 \phi}{\partial x^2}$$

with boundary condition $x(t_f) = x_f$ and $D = k_{\mathrm{B}} T / \gamma$ as usual. We solved this free diffusion equation back in Chapter 6, and found

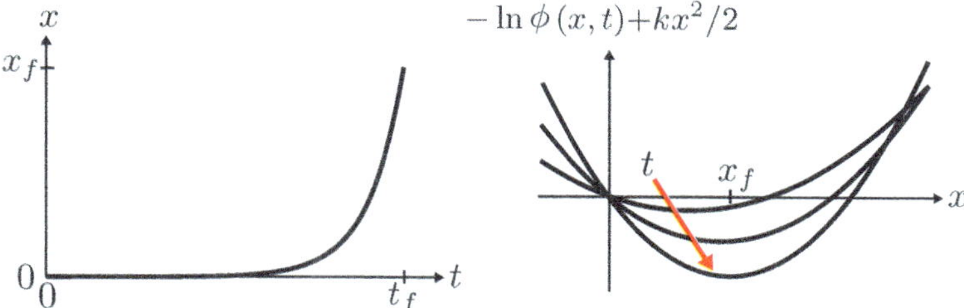

Fig. 8.14 Mean trajectory and effective time dependent potential for the Ornstein–Uhlenbeck process conditioned on ending at x_f at t_f.

$$\phi(x,t) = \frac{1}{\sqrt{4\pi D(t_f - t)}} e^{-(x_f - x)^2/4D(t_f - t)}$$

which is a Gaussian distribution with a variance that disappears as $t \to t_f$. The resultant Langevin equation is

$$\gamma\dot{x} = \gamma\frac{x_f - x}{t_f - t} + \eta(t) \qquad \langle\eta(0)\eta(t)\rangle = 2k_{\mathrm{B}}T\delta(t)$$

where the added drift pushes the particle toward x_f with a force whose magnitude increases as $t \to t_f$. The resultant mean trajectory with initial condition $x(0) = x_0$ is just a straight line, $\langle x(t)\rangle = x_0 + t(x_f - x_0)/t_f$.

Another system for which $\phi(\mathbf{r}^N, t)$ can be evaluated is the so-called *Ornstein–Uhlenbeck process*, or the overdamped diffusion in a harmonic potential. The equation of motion is

$$\gamma\dot{x} = -kx + \eta \qquad \langle\eta(0)\eta(t)\rangle = 2k_{\mathrm{B}}T\gamma\delta(t)$$

where k is the stiffness of the harmonic potential. The conditional probability of ending at x_f at time t_f is

$$\phi(x,t) \propto \exp\left\{\frac{-\beta k}{4\sinh[(t_f - t)/\tau]}\left[(x_f^2 + x^2)\cosh((t_f - t)/\tau) - 2xx_f\right] + \frac{\beta kx^2}{4}\right\}$$

with $\tau = \gamma/k$ the characteristic relaxation time of the particle. The corresponding stochastic trajectory driven to satisfy the conditioning is

$$\gamma\dot{x} = k\frac{x_f - x\cosh[(t_f - t)/\tau]}{\sinh[(t_f - t)/\tau]} + \eta \qquad \langle\eta(0)\eta(t)\rangle = 2k_{\mathrm{B}}T\gamma\delta(t)$$

whose average path and associated time-dependent potential is shown in Figure 8.14. Note that while a new drift has been introduced to affect the transition, the noise is left invariant.

Exercise 8.11: Confirm that the conditional probability $\phi(x,t)$ written for the Ornstein–Uhlenbeck process satisfies

$$\frac{\partial \phi}{\partial t} = \frac{x}{\tau}\frac{\partial}{\partial x}\phi - D\frac{\partial^2 \phi}{\partial x^2}$$

for any $t_f > 0$ and x_f. To demonstrate this, you will need to normalize ϕ by averaging it against the Boltzmann distribution.

8.12 Reweighting reactive path ensembles

The relationship between the rate constant and a ratio of path partition functions offers the intuition that the sort of ensemble reweighting principles used in equilibrium can be brought to bear on dynamical processes and reaction rates. Indeed, if one considers two processes both of which transition the system between states A and B, but do so under slightly different forces, the rates in these two systems can be related. For concreteness, let us consider a system driven by a time-dependent force $\Lambda(t)$. If we imagine that the external potential is bistable, so that there are two long-lived states, then provided metastability persists, the rate in the presence of the time-dependent force Λ is still given by a ratio of path partition functions

$$k_\Lambda = \frac{1}{t}\frac{Z_{AB,\Lambda}(t)}{Z_{A,\Lambda}}$$

where the subscript Λ refers to the presence of the driving force. The dependence of the reactive path partition function on Λ,

$$Z_{AB,\Lambda}(t) = \int \mathcal{D}[\mathbf{X}(t)]P_\Lambda[\mathbf{X}(t)]h_A[\mathbf{x}_0]h_B[\mathbf{x}_t]$$

enters through the probability of observing a particular trajectory $P_\Lambda[\mathbf{X}(t)]$. The above relation is true for any Λ including $\Lambda = 0$,

$$Z_{AB,0}(t) = \int \mathcal{D}[\mathbf{X}(t)]P_0[\mathbf{X}(t)]h_A[\mathbf{x}_0]h_B[\mathbf{x}_t]$$

which means that,

$$Z_{AB,0}(t) = \frac{Z_{AB,\Lambda}(t)}{Z_{AB,\Lambda}(t)}\int \mathcal{D}[\mathbf{X}(t)]P_0[\mathbf{X}(t)]h_A[\mathbf{x}_0]h_B[\mathbf{x}_t]$$

$$= \frac{Z_{AB,\Lambda}(t)}{Z_{AB,\Lambda}(t)}\int \mathcal{D}[\mathbf{X}(t)]P_0[\mathbf{X}(t)]\frac{P_\Lambda[\mathbf{X}(t)]}{P_\Lambda[\mathbf{X}(t)]}h_A[\mathbf{x}_0]h_B[\mathbf{x}_t]$$

$$= Z_{AB,\Lambda}(t)\left\langle \frac{P_0[\mathbf{X}(t)]}{P_\Lambda[\mathbf{X}(t)]} \right\rangle_{AB,\Lambda}$$

where in the first and second lines we have multiplied by 1, and in the third identified the conditional average of the ratio of path probabilities in a reactive ensemble in the presence of the driving force. This change of probability measure of a Markovian process is known as a *Girsanov transform*. An analogous relationship would hold between $Z_{A,\Lambda}$ and $Z_{A,0}$. However, for simplicity, we can consider $\Lambda(t) = 0$ for $t \leq 0$, and take $Z_{A,\Lambda} = Z_{A,0}$.

For the overdamped dynamics considered, the form of the path probability is known. This is typically written as, $P_\Lambda[\mathbf{X}(t)] \propto \exp[-\beta\Gamma_\Lambda]$, where $\beta\Gamma_\Lambda$ is called the *Onsager-Machlup stochastic action* and depends explicitly on $\Lambda(t)$. Therefore we have an expression for the ratio of path probabilities

$$\ln \frac{P_0[\mathbf{X}(t)]}{P_\Lambda[\mathbf{X}(t)]} = -\beta\Delta\Gamma_\Lambda$$

where $\Delta\Gamma_\Lambda = \Gamma_0 - \Gamma_\Lambda$. Combining these relationships with the partition function expression for the rate, we find

$$k_0 = k_\Lambda \left\langle e^{-\beta\Delta\Gamma_\Lambda} \right\rangle_{AB,\Lambda}$$

or through an equivalent set of manipulations

$$k_\Lambda = k_0 \left\langle e^{\beta\Delta\Gamma_\Lambda} \right\rangle_{AB,0} .$$

In either case, we find that the rates in the presence or absence of an extra driving force are related by the change in the probability of observing a reactive trajectory. This result is similar in spirit to those of the fluctuation theorems that relate work or heat statistics to their reversible limits, in that there is no assumption concerning the size of Λ. The force could take the system arbitrarily away from equilibrium, but the equilibrium rate can still be deduced from the statistics of the change in stochastic action. Alternatively, the statistics of reactive trajectories in equilibrium encode how the rate of a reaction is changed when driven arbitrarily far from it.

A number of results follow from these relationships. For example, the convexity of the exponential means that there are variational statements that can be made to relate the rates with and without the force,

$$\ln \frac{k_\Lambda}{k_0} = \ln \left\langle e^{\beta\Delta\Gamma_\Lambda} \right\rangle_{AB,0}$$
$$\geq \beta \left\langle \Delta\Gamma_\Lambda \right\rangle_{AB,0}$$

where we have employed Jensen's inequality. Consider an example of a force that constrains the reaction to occur in time t with probability 1. We can pose the variational question, what is the smallest possible force required to affect that transition? This is equivalent to trying to optimize the relationship from *variational path sampling*

$$\ln k_0 t \leq -\beta \left\langle \Delta\Gamma_\Lambda \right\rangle_{AB,0}$$

where if the reaction occurs almost surely in the driven system, $k_\Lambda t = 1$. For an overdamped Langevin equation in one dimension of the form

$$\gamma\dot{x} = F + \Lambda + \eta \qquad \langle \eta(0)\eta(t) \rangle = 2k_\mathrm{B}T\gamma\delta(t)$$

the conditioned transition probability is computable as an average over the noise

$$P[x(t)|x(0)] = \langle \delta \left[\gamma \dot{x}(t) - F(x) - \Lambda - \eta \right] \rangle$$

$$= \int \mathcal{D}[\eta(t)] e^{- \int_0^t dt' \eta^2(t')/4k_{\mathrm{B}}T\gamma} \delta \left[\gamma \dot{x}(t) - F(x) - \Lambda - \eta \right]$$

$$\propto \exp\left[- \int_0^t dt' \left(\gamma \dot{x} - F - \Lambda \right)^2 / 4k_{\mathrm{B}}T\gamma \right]$$

or $\ln P[x(t)|x(0)] = -\beta \Gamma[\mathbf{X}(t)] + C$, where C is a constant and $\Gamma[\mathbf{X}(t)]$ is the Onsager-Machlup action,

$$\beta \Gamma = \frac{\beta}{4\gamma} \int_0^t dt' \left\{ \gamma \dot{x}(t') - F[x(t')] - \Lambda \right\}^2$$

for an ensemble with a path of length t. The corresponding average change in path action with and without the added force Λ, is

$$\beta \langle \Delta \Gamma_\Lambda \rangle_{AB,0} = \frac{\beta}{4\gamma} \int_0^t dt' \left\langle \left(\gamma \dot{x} - F \right)^2 - \left(\gamma \dot{x} - F - \Lambda \right)^2 \right\rangle_{AB,0}$$

$$= \frac{\beta}{2\gamma} \int_0^t dt' \left\langle \Lambda \left(\gamma \dot{x} - F - \Lambda/2 \right) \right\rangle_{AB,0}$$

$$= \frac{\beta}{2\gamma} \int_0^t dt' \left\langle \Lambda \eta - \Lambda^2/2 \right\rangle_{AB,0}$$

where in the third line we have used the equation of motion in the undriven system, $\eta = \gamma \dot{x} + \partial_x U$. We can find the minimum force by taking the functional derivative of $\delta \beta \langle \Delta \Gamma_\Lambda \rangle_{AB,0} / \delta \Lambda$, and setting it equal to zero which results in,

$$\Lambda^*(t) = \langle \eta(t) \rangle_{AB,0}$$

or an equivalence between the optimal force and the average of the noise conditioned on the reaction occurring, a manifestation of a fluctuation-response relationship or Girsanov theorem. Plugging this optimal force $\Lambda^*(t)$ back into the relative action,

$$\beta \langle \Delta \Gamma_\Lambda \rangle_{AB,0} = \frac{\beta}{4\gamma} \int_0^t dt' \left(\Lambda^* \right)^2$$

which indicates that the rate in the undriven case is

$$\ln k_0 t = - \frac{\beta}{4\gamma} \int_0^t dt' \left(\Lambda^* \right)^2 .$$

We therefore learn that the smallest force that ensures the transition occurs is the one that determines the rate spontaneously in its absence.

We have actually encountered the optimal force already in the previous section on Brownian bridges. In fact, the force that saturates the bound is

$$\Lambda^*(t) = 2k_{\mathrm{B}}T\partial_x \ln \phi$$

where ϕ is the time-dependent commitment probability. This can be proven by noting first that $\ln \phi$ satisfies

$$\frac{\partial \ln \phi}{\partial t} = -\gamma^{-1} F \frac{\partial \ln \phi}{\partial x} - D \frac{\partial^2 \ln \phi}{\partial x^2} + D \left(\frac{\partial \ln \phi}{\partial x} \right)^2$$

which follows directly from multiplying the adjoint Fokker-Planck equation for ϕ by $1/\phi$ on both sides. Secondly, from Ito's lemma we know that the total time derivative of $\ln \phi$ is

$$\frac{d}{dt} \ln \phi = \frac{\partial \ln \phi}{\partial t} + \gamma^{-1} F \frac{\partial \ln \phi}{\partial x} + D \frac{\partial^2 \ln \phi}{\partial x^2}$$
$$= D \left(\frac{\partial \ln \phi}{\partial x} \right)^2$$

where the second line follows from the adjoint Fokker-Planck equation. Putting this together with the expression of the rate,

$$\ln k_0 t = - \int_0^t dt' \frac{d}{dt} \ln \phi = \ln \phi(x_0, 0) - \ln \phi(x_f, t_f)$$
$$\longrightarrow k_0 = \frac{1}{t} \phi(x_0, 0)$$

which is nothing more than the probability of ending in the target state over a time t given an initial state $x_0(0)$ as $\ln \phi(x_f, t_f) = 0$ by construction. Thus $\Lambda^*(t) = 2 k_B T \partial_x \ln \phi$ saturates the variational inequality. While the direct calculation of $\Lambda^*(t)$ is not typically possible, the variational formulation admits a variety of approximations.

Further reading

The perspective on rate theory and time correlation functions are drawn from Robert Zwanzig's *Nonequilibrium statistical mechanics* and Abraham Nitzan's *Chemical dynamics in condensed phases*. David Chandler's notes on rare events found in *Classical and quantum dynamics in condensed phase simulations* was particularly inspiring in the discussion of Marcus theory. Baron Peter's book, *Reaction rate theory and rare events*, is an excellent summary of most of the material in this chapter including a discussion of committors and classical nucleation theory.

Additional exercises

Exercise 8.12: In this exercise we will consider the calculation of splitting probabilities. A coordinate q evolves stochastically on a bistable potential $w(q)$. The stable states A and B, with characteristic values q_A and q_B of the order parameter, might correspond to reactants and products of a chemical reaction. Here we will focus on the splitting probability $\phi_B(q_0)$, i.e., the fraction of trajectories that are initiated from

q_0 and reach q_B before q_A. We sketched the calculation of ϕ_B in the chapter, which begins with the Smoluchowski equation,

$$\frac{\partial p(q,t)}{\partial t} = -\frac{\partial}{\partial q} J(q,t)$$

where,

$$J(q,t) = -De^{-\beta w(q)} \frac{\partial}{\partial q} \left(e^{\beta w(q)} p(q,t) \right)$$

1. The splitting probability $\phi_B(q_0)$ can be written in terms of the solution to the Smoluchowski equation,

$$\phi_B = \int_0^\infty dt\, J(q_B,t)$$

   with boundary conditions

$$p(q,0) = \delta(q - q_0), \qquad p(q_A,t) = 0, \qquad p(q_B,t) = 0$$

   Provide a physical explanation for the expression for ϕ_B and each of these boundary conditions.
2. As a first step toward calculating $\phi_B(q_0)$, integrate the Smoluchowski equation over all times (from $t = 0$ to $t = \infty$). Simplify your result using the specified initial condition $p(q,0)$ and the fact that, over time, our boundary conditions steadily remove probability from the interval $q_A < q < q_B$.
3. Integrate your result from part (2) over the coordinate q. Take q_B as the upper limit of this integration; for the lower limit use an arbitrary value q. In your result, you should be able to identify the splitting probability $\phi_B(q_0)$.
4. Multiply your result from part (3) by the inverse Boltzmann factor $\exp(\beta w)$, and integrate once again over q. This time, take q_A and q_B as the limits of integration. A simple rearrangement should then yield Onsager's result

$$\phi_B(q_0) = \int_{q_A}^{q_0} dq\, e^{\beta w(q)} \Big/ \int_{q_A}^{q_B} dq\, e^{\beta w(q)} .$$

5. Consider the idealized case of a harmonic barrier,

$$w(q) = -\frac{1}{2} m\omega^2 q^2 .$$

   Setting $q_A = -\infty$ and $q_B = \infty$, evaluate the splitting probability using the equation above. Your answer should involve the error function

$$\mathrm{erf}(x) = \frac{2}{\sqrt{\pi}} \int_0^x dz\, e^{-z^2} .$$

6. For the harmonic barrier model, plot $\phi_B(q_0)$ as a function of q_0, over a range that encompasses most of the crossover between its limiting values. Identify the width of this crossover, say, the range of q_0 over which $\phi_B(q_0)$ lies between 0.1 and 0.9. There is no need to be precise with these values. The goal is to determine the rough scale of the crossover. Give a physical interpretation for the crossover.

Exercise 8.13: Here we will work through Grote–Hynes theory for the transmission coefficient, κ, for a classical particle with unit mass crossing a parabolic barrier while coupled to a Gaussian bath. The generalized Langevin equation for this model is

$$\dot{v}(t) = \omega_{\mathrm{B}}^2 x(t) - \int_0^t dt\, \gamma(t-t')v(t') + \eta(t)$$

where $v(t) = \dot{x}(t)$ is the particle's velocity, $\eta(t)$ is the random force, and the auto-correlation function for this random force is proportional to the friction kernel $\gamma(t-t')$. To ensure detailed balance, the proportionality constant is related to the temperature of the bath. For this problem, assume

$$\gamma(t) = \gamma_0 e^{-|t|/\tau}$$

where γ_0 and τ are constants.

1. Compute the reactive flux correlation function relative to that of transition state theory, $\kappa(t) = k(t)/k(0^+)$, in the long time limit. In this exercise, you will go through the calculation presented in the chapter in detail and express your answer in terms of the parameters of the model: the barrier frequency ω_{B}, the friction constant γ_0 and bath relaxation time τ. Plot κ in the two-dimensional space of γ_0/ω_B^2 and $(\tau\omega_B)^{-1}$.

2. Consider the splitting probability, $\phi_B(x)$, that a trajectory initiated at x will end up after a transient time in the region $x > 0$. In transition state theory, this probability is $1/2$ for $x = 0$. Coupling to a bath will produce a distribution of values for this probability, implying that there is an ensemble of transition states. To study the nature of this ensemble, we need to determine the distribution $P[\phi_B(x)] = \langle \delta(\phi_B - \phi_B(x)) \rangle$ for $x = 0$. First, compute the probability that a trajectory starting at $x = 0$ and a given noise history will end up with $x > 0$, by averaging over the initial velocities. Your answer should involve the complementary error function

$$\mathrm{erfc}(x) = \frac{2}{\sqrt{\pi}} \int_x^\infty dz\, e^{-z^2} = 1 - \mathrm{erf}(x)$$

3. Given that splitting probability, average over the noise history to compute the distribution of splitting probabilities, $P[\phi_B(0)]$. Express your answer in terms of the parameters ω_{B}, γ_0 and τ. *Hint*: You may find it convenient to invoke an identity of the δ-function

$$\delta[f(x)] = \sum_{x_i} \frac{\delta(x - x_i)}{|f'(x_i)|}$$

   where x_i's are simple roots of $f(x)$.

4. Plot $P[\phi_B(0)]$ for various combinations of the parameters. Show that $P[\phi_B(0)]$ can be bimodal for a sufficiently sluggish bath with a high friction. What is the physical significance of this bimodality?

Exercise 8.13: In this exercise we will work through a theory of vibrational relaxation using the golden rule result for the quantum mechanical transition rate. We will consider a vibrational system $\hat{\mathcal{H}}_S$ that relaxes through the coupling operator $\hat{V}$ to a continuum of bath states $\hat{\mathcal{H}}_B$ using perturbation theory. In the absence of a coupling term in the Hamiltonian, the Schrodinger equation for the system and bath is

$$\left(\hat{\mathcal{H}}_S + \hat{\mathcal{H}}_B\right)|a,\alpha\rangle = (E_a + E_\alpha)|a,\alpha\rangle$$

where

$$\hat{\mathcal{H}}_S|a\rangle = E_a|a\rangle \qquad \hat{\mathcal{H}}_B|\alpha\rangle = E_\alpha|\alpha\rangle$$

are an independent set of system and bath energies and eigenstates. We will assume the system represents a single harmonic mode and bath is a set of independent harmonic oscillators such that

$$\hat{\mathcal{H}}_S = \frac{1}{2m}\hat{p}^2 + \frac{1}{2}m\omega_s^2\hat{q}^2 \qquad \hat{\mathcal{H}}_B = \sum_n \frac{1}{2m_n}\hat{p}_n^2 + \frac{1}{2}m_n\omega_n^2\hat{q}_n^2$$

1. Show that with the golden rule, the rate for transitions in the system a to b induced by V is

$$k = \frac{2\pi}{\hbar}\sum_{\alpha,\beta} p_{a,\alpha}|\langle a,\alpha|V|b,\beta\rangle|^2\delta\left(E_b - E_a + E_\beta - E_\alpha\right)$$

2. Rewrite the previous expression as an average bath correlation function of the coupling operator

$$k = \frac{1}{\hbar^2}\int_{-\infty}^{\infty} dt \left\langle \hat{V}_{ab}(t)\hat{V}_{ab}(0)\right\rangle_B e^{i\omega_{ba}t}$$

where $\omega_{ba} = (E_b - E_a)/\hbar$ and

$$\hat{V}_{ab}(t) = e^{i\hat{\mathcal{H}}_B t/\hbar}\langle a|\hat{V}|b\rangle e^{-i\hat{\mathcal{H}}_B t/\hbar}$$

is the time-dependent bath coupling operator.
3. In the low-temperature limit, the coupling is well approximated as linear in the bath modes

$$\hat{V} = (|b\rangle\langle a| + |a\rangle\langle b|)\sum_\alpha \xi_\alpha q_\alpha$$

with $q_\alpha = c_\alpha(a_\alpha^\dagger + a_\alpha)$ and $c_\alpha = \sqrt{\hbar/2m\omega_\alpha}$, then show that

$$\left\langle \hat{V}_{ab}(t)\hat{V}_{ab}(0)\right\rangle_B = \sum_\alpha \xi_\alpha^2 c_\alpha^2\left[(\langle n_\alpha\rangle + 1)e^{-i\omega_\alpha t} + \langle n_\alpha\rangle e^{i\omega_\alpha t}\right]$$

where $\langle n_\alpha\rangle$ is the Bose-Einstein distribution for the α'th bath mode.
4. Put your results from 2 and 3 together to show that the rate of vibrational relaxation is given by

$$k = \frac{1}{\hbar^2}\sum_\alpha \xi_\alpha^2 c_\alpha^2\left[(\langle n_\alpha\rangle + 1)\delta\left(\omega_{ba} + \omega_\alpha\right) + \langle n_\alpha\rangle\delta\left(\omega_{ba} - \omega_\alpha\right)\right]$$

5. In the high-temperature limit, we expect the system will be well approximated by classical mechanics. However, with the above expressions this limit is difficult to take as the quantum time correlation function has both real and imaginary parts, while the classical correlation function is purely real. Show that the rate in 2 can be rewritten as

$$k = \frac{4}{\hbar^2} F(\omega_{ba}) \int_0^\infty dt \; \mathrm{Re}\left[\left\langle \hat{V}_{ab}(t) \hat{V}_{ab}(0) \right\rangle_B \right] e^{i\omega_{ba} t}$$

where

$$F(\omega) = \frac{1}{1 + e^{-\beta\hbar\omega}}$$

is a so-called quantum correction factor. Hint: you can use the relation between Fourier-transformed correlation functions $\hat{C}(-\omega) = \hat{C}(\omega)e^{-\beta\hbar\omega}$.

Exercise 8.15: In this problem, we will study the impact of a non-gradient force on the most likely reactive path. Consider the stochastic action for an upside-down parabolic potential with an additional oscillatory force

$$\beta\Gamma = \frac{\beta}{4\gamma} \int_0^t dt' \; \{\gamma\dot{x}(t') - kx(t') - \Lambda(t)\}^2$$

where the force $\Lambda(t)$ has a time dependence but no x dependence.

1. By taking the functional derivative $\delta\beta\Gamma[x(t)]/\delta x(t')$, show that the most likely reactive trajectory satisfies

$$\gamma^2 \ddot{x}(t) = k^2 x(t) + k\Lambda(t) + \gamma\dot{\Lambda}(t)$$

2. Use the method of Laplace transforms to solve the second order differential equation with boundary conditions $x(0) = -x_0$ and $x(t^*) = x_0$. You should find,

$$x(t) = - x_0 \cosh(t/\tau) + \frac{x_0 + x_0 \cosh(t^*/\tau)}{\sinh(t^*/\tau)} \sinh(t/\tau)$$

$$+ f(t)/\gamma - \frac{f(t^*)/\gamma}{\sinh(t^*/\tau)} \sinh(t/\tau)$$

where $f(t)$ is the convolution of the external force with the Green's function,

$$f(t) = \int_0^t dt' \Lambda(t') e^{(t-t')/\tau}$$

and $\tau = \gamma/\tau$.

3. Taking the time-dependent force to be oscillatory $\Lambda(t) = \Lambda_0 \cos(\omega t)$ with amplitude Λ_0 and frequency ω, compute the average stochastic action $\beta\Gamma$ over this most likely transition path.

4. By considering only the first-order term in Λ_0 and taking the $\omega \to 0$ limit of $\beta\Gamma$, show that the change in the rate is given by

$$\ln \frac{k_\Lambda}{k_0} = \beta\Lambda_0 x_0 \geq 0$$

and provide a physical interpretation of this enhancement.

9
Large deviations from equilibrium

We have thus far largely considered systems in equilibrium or close to it. However, there are many cases of general interest where near-equilibrium approximations break down. For example, when matter is manipulated directly on molecular scales, the reversible limits of equilibrium thermodynamics are difficult to approach. Further, when energy is continuously supplied to a system through reservoirs kept out of equilibrium with each other, or by injecting it directly into molecular degrees of freedom, stationary distributions can be established that are non-Boltzmann, and whose form is not generally known. In both cases, in order to evolve a steady state despite the constant injection of energy, or to consider heat flows in addition to work done on the system, we require a description of the dynamics of a system that is open, and can exchange energy with its surroundings. The developments in the previous chapters provide such an open system description in the limit that the system and bath are weakly interacting such that the bath remains in equilibrium. Working within this assumption, we will show in this chapter that it is possible to make quantitative statements about systems far from equilibrium by constructing equations of motion for observables like heat and work. Further, using principles from probability like large deviation theory, we can develop a means of quantifying fluctuations of dynamical quantities like currents of mass, energy, or charge that are outside the scope of traditional statistical mechanics, even if the system is at equilibrium. The combination of large deviation theory, dynamic definitions of thermodynamic quantities, and the structure they encode into ensembles of trajectories form the core of stochastic thermodynamics.

9.1 Dynamics of heat and work

To start, let us employ our framework of stochastic dynamics to write equations of motion for thermodynamic quantities like heat and work. As the underlying equations of motion are stochastic, so too do we expect that thermodynamic quantities will fluctuate. For concreteness, let us consider a system described by a Hamiltonian

$$\mathcal{H} = \sum_i \frac{m_i}{2} \mathbf{v}_i^2 + U[\mathbf{r}^N, \Lambda(t)]$$

where we have the usual sum of kinetic and potential energy, but we also have an explicit time dependence through the potential energy, where $\Lambda(t)$ will be a control parameter we envision we can manipulate deterministically. The rate of change of the energy of the system can be computed as

$$\frac{d\mathcal{H}}{dt} = \frac{\partial \Lambda}{\partial t}\frac{\partial \mathcal{H}}{\partial \Lambda} + \frac{\partial \mathbf{x}}{\partial t} \cdot \frac{\partial \mathcal{H}}{\partial \mathbf{x}}$$
$$= \dot{\mathcal{W}} + \dot{\mathcal{Q}}$$

which through the chain rule naturally decomposes into an explicit time dependence and an implicit time dependence that is inherited through the fluctuating microscopic degrees of freedom—the positions $\mathbf{r}^N$ and velocities $\mathbf{v}^N$ that form the phase space $\mathbf{x}$. The explicit time dependence is identified as the rate of work injected into the system,

$$\dot{\mathcal{W}} = \dot{\Lambda}(t)\frac{\partial U[\mathbf{r}^N, \Lambda(t)]}{\partial \Lambda}$$

which, consistent with our development way back in Chapter 1, reflects changes in energy due to processes we control. The second term, due to changes in the microscopic degrees of freedom, we identify as heat,

$$\dot{\mathcal{Q}} = \sum_i m_i \dot{\mathbf{v}}_i \cdot \mathbf{v}_i - \dot{\mathbf{r}}_i \cdot \mathbf{F}_i[\mathbf{r}^N, \Lambda(t)]$$

which reflect the processes, molecular in origin, that we do not control directly. Here $\mathbf{F}_i[\mathbf{r}^N, \Lambda(t)]$ is the total force acting on particle i, including both the conservative part, $-\partial U/\partial \mathbf{r}_i$, and could be generalized to any non-gradient force from the control parameter Λ. These flows of energy illustrated in Figure 9.1 resolve a stochastic realization of the first law of thermodynamics.

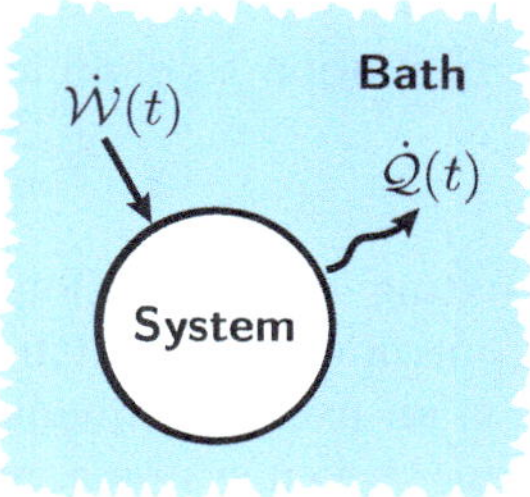

Fig. 9.1 Stochastic version of the first law of thermodynamics in which work is done on the system $\mathcal{W} > 0$ and heat is dissipated into the environment $\mathcal{Q} < 0$.

To render these expressions concrete, we need an equation of motion for the microscopic degrees of freedom. As alluded to already, for heat to flow we need a reservoir for it to flow into, and thus the equations of motion need to be consistent with an incomplete knowledge of the dynamics. This necessarily results in a stochastic equation of motion, and we will work in the limit that the bath, those degrees of freedom we are not following, relaxes quickly and is weakly coupled to the system. For the degrees of freedom we are following, this results in an underdamped Langevin equation

$$m_i \dot{\mathbf{v}}_i = -\gamma_i \mathbf{v}_i + \mathbf{F}_i[\mathbf{r}^N, \Lambda(t)] + \boldsymbol{\eta}_i \qquad \dot{\mathbf{r}}_i = \mathbf{v}_i$$

where $\boldsymbol{\eta}_i$ is the usual uncorrelated Gaussian noise with $\langle \boldsymbol{\eta}_i \rangle = 0$ and $\langle \boldsymbol{\eta}_i(0) \otimes \boldsymbol{\eta}_j(t) \rangle = 2k_{\mathrm{B}}T\gamma_i\delta_{ij}\mathbf{1}\delta(t)$. Manipulating this equation and using the definition above for the rate of heat generation, one can arrive at an explicit expression for the heat production

$$\dot{\mathcal{Q}} = -\sum_i \gamma_i \left(\mathbf{v}_i^2 - \frac{dk_{\mathrm{B}}T}{m_i} \right)$$

which has two notable properties. First, the rate of heat generation is proportional to γ_i, reflecting the fact that heat is exchanged with the bath. In the deterministic limit $\gamma_i \to 0$, there is no heat transfer as there is no bath to dump it into. Second, the rate of heat transfer depends on the temperature of the bath, which by virtue of it being an ideal reservoir maintains a constant temperature. The heat transfer rate depends on the temperature less a measure of the local temperature of the system given by $\mathbf{v}_i^2$, and thus is a microscopic manifestation of the law of heat conduction. On average, in the absence of a time dependent force, $\langle \mathbf{v}_i^2 \rangle = dk_{\mathrm{B}}T/m_i$ from equipartition in d dimensions, and thus there would be no net heat flow. In the presence of a time-dependent force, or upon relaxation from an initially nonequilibrium state, $\langle \mathbf{v}_i^2 \rangle \neq dk_{\mathrm{B}}T/m_i$ and heat would flow. The stochastic formulation of heat and work was introduced by Sekimoto.

> **Exercise 9.1:** Using Ito's lemma and the definitions given earlier, confirm the expression for the heat. Specifically, evaluate the time dependence of the Hamiltonian
> $$\frac{d\mathcal{H}}{dt} = \frac{\partial \mathcal{H}}{\partial t} + \sum_i \dot{\mathbf{r}}_i \cdot \frac{\partial \mathcal{H}}{\partial \mathbf{r}_i} + \dot{\mathbf{v}}_i \cdot \frac{\partial \mathcal{H}}{\partial \mathbf{v}_i} + \frac{k_{\mathrm{B}}T\gamma_i}{m_i^2} \frac{\partial^2 \mathcal{H}}{\partial \mathbf{v}_i^2}$$
> using the underdamped equations of motion to eliminate $\dot{\mathbf{r}}_i$ and $\dot{\mathbf{v}}_i$ and identify the heat.

To arrive at an analogous fluctuating representation of the entropy is perhaps more subtle. The canonical Gibbs' entropy of a system that we defined as a sum over microstates back in Chapter 1 is an averaged quantity. To admit a fluctuating description, appropriate for microscopic driven systems, we can simply not average it over the distribution, and define a fluctuating entropy $S/k_{\mathrm{B}} = -\ln p(\mathbf{r}^N, \mathbf{v}^N, t)$. In statistics, this quantity is often referred to as the *surprisal*, a quantification of the uncertainty in a random variable taking a certain value. Its time dependence can be evaluated for the underdamped Langevin dynamics through Ito's lemma,

$$\frac{1}{k_{\mathrm{B}}}\frac{dS}{dt} = -\frac{\partial \ln p}{\partial t} - \sum_i \frac{\partial \ln p}{\partial \mathbf{r}_i} \cdot \dot{\mathbf{r}}_i + \frac{\partial \ln p}{\partial \mathbf{v}_i} \cdot \dot{\mathbf{v}}_i + \frac{k_{\mathrm{B}}T\gamma_i}{m_i^2} \frac{\partial^2 \ln p}{\partial \mathbf{v}_i^2}$$
$$= -\frac{1}{p}\left[\frac{\partial p}{\partial t} + \sum_i \frac{\partial p}{\partial \mathbf{r}_i} \cdot \dot{\mathbf{r}}_i + \frac{\partial p}{\partial \mathbf{v}_i} \cdot \dot{\mathbf{v}}_i + \frac{k_{\mathrm{B}}T\gamma_i}{m_i^2} \left(\frac{\partial^2 p}{\partial \mathbf{v}_i^2} - \frac{1}{p}\left(\frac{\partial p}{\partial \mathbf{v}_i} \right)^2 \right) \right]$$

where in the second line we have applied the chain rule to simplify the derivative of $\ln p$. The equation of motion for p is given by a Fokker–Planck equation,

$$\frac{\partial p}{\partial t} = \sum_i -\mathbf{v}_i \cdot \frac{\partial}{\partial \mathbf{r}_i} p - \frac{\partial}{\partial \mathbf{v}_i} \frac{-\gamma_i \mathbf{v}_i + \mathbf{F}_i[\mathbf{r}^N, \Lambda(t)]}{m_i} p + \frac{k_\mathrm{B} T \gamma_i}{m_i^2} \frac{\partial^2}{\partial \mathbf{v}_i^2} p$$

where the diffusive term acts on the molecular velocities. This rate of change of the entropy reflects changes just to the system, with a phase space spanned by the set of positions and velocities. It is related to the total entropy change of the system plus the bath by,

$$\frac{dS_\mathrm{tot}}{dt} = \frac{dS}{dt} - \frac{\dot{Q}}{T}$$

where we have used the fact that the bath is ideal and always evolves reversibly to equate $T dS_\mathrm{bath} = dQ_\mathrm{bath} = -dQ_\mathrm{system} \equiv -dQ$, the latter equality following from conservation of energy. Taking the average of the total entropy change by multiplying by p and integrating over all of the positions and velocities we arrive at,

$$\frac{d \langle S_\mathrm{tot} \rangle}{dt} = \sum_i \int d\mathbf{r}_i \int d\mathbf{v}_i \frac{m_i^2}{T \gamma_i p(\mathbf{r}^N, \mathbf{v}^N, t)} j_i^2(\mathbf{r}^N, \mathbf{v}^N, t) \geq 0$$

which is a non-negative quantity, given by constants times a probability flux squared. The probability flux

$$j_i(\mathbf{r}^N, \mathbf{v}^N, t) = \frac{\gamma_i \mathbf{v}_i}{m_i} p(\mathbf{r}^N, \mathbf{v}^N, t) + \frac{k_\mathrm{B} T \gamma_i}{m_i^2} \frac{\partial p(\mathbf{r}^N, \mathbf{v}^N, t)}{\partial \mathbf{v}_i}$$

consists of only the irreversible parts of the Fokker–Planck equation, both proportional to γ_i, reflecting the fact that a deterministic system with $\gamma_i = 0$ would not produce entropy on average. We have thus arrived at another expression of the second law, like that from Clausius. The total rate of change of the entropy on average is nonnegative, or equivalently that the average rate of change of the system entropy is bounded from above by T times the rate at which heat is released to the bath.

Exercise 9.2: Verify the form of the average total entropy production. Hint: Make judicious use of integration by parts.

9.2 Discrete dynamical systems

Much of our discussion on stochastic dynamics has focused on how probability flows between points in a continuous phase. However, quantum mechanically when energy levels are naturally discretized or classically when metastability within a continuous space affords a partitioning into a set of long-lived collections of configurations, it makes sense to consider an analogous equation for dynamics in a discrete state space. As before, the continuity equation that will result from the assumption of conservation of probability will be worked out assuming that the underlying dynamics are Markovian. In the quantum mechanical case, this is assured by assuming that the coupling mediating transitions between states is weak, as in our Redfield theory of Chapter 7 or the golden rule of Chapter 8. In the classical case, this is assured by assuming a separation of timescales between local relaxation within a metastable state and the typical transition times between those states.

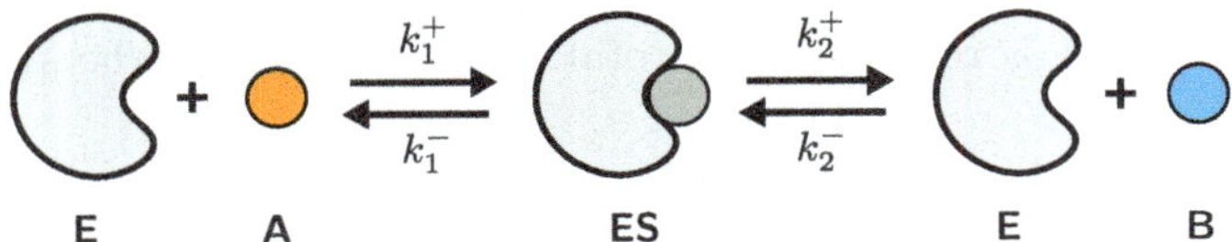

Fig. 9.2 Simple two state model based on Michaelis–Menten kinetics, where an enzyme can reversibly bind a substrate, and chemically convert it into another.

Continuity equations for probability in a discrete state space are known as *master equations*. To start, we can consider a simple example of a two-state system. Imagine the scenario encountered in *Michaelis–Menten* kinetics, where an enzyme can bind a substrate, A, chemically convert it to B, and release the new substrate. We can label the unbound state E and the bound ES, as pictured in Figure 9.2. Each binding is reversible with rates for A, k_1^+, and k_1^-, and those for binding B, k_2^+, and k_2^-. The rate of change of the probability for the enzyme being bound is,

$$\frac{dp_{\mathrm{ES}}}{dt} = k_1^+ p_{\mathrm{E}} + k_2^- p_{\mathrm{E}} - k_1^- p_{\mathrm{ES}} - k_2^+ p_{\mathrm{ES}}$$

where the first two terms are sources of probability dependent on the probability of the enzyme being unbound p_{E}, and the second two terms are drains, losses from unbinding either A or B.

We could write an analogous expression for the rate of change of the probability of being in the unbound state. However, there is a more compact notation we can adopt. If we denote the probability vector for the enzyme as $\mathbf{p} = \{p_{\mathrm{E}}, p_{\mathrm{ES}}\}$, then

$$\frac{d\mathbf{p}}{dt} = \mathcal{L}\mathbf{p}$$

where $\mathcal{L}$ is a matrix of the transition rates. For this two-state model, the transition matrix is

$$\mathcal{L} = \begin{bmatrix} -k_1^+ - k_2^- & k_1^- + k_2^+ \\ k_1^+ + k_2^- & -k_1^- - k_2^+ \end{bmatrix}$$

which can be read off from the individual rate equations. Note that rate matrices have a particular property that the sum in each column is identically 0, which comes about from conservation of probability. That sum rule ensures that the $\mathcal{L}$ has a zero eigenvalue, whose associated eigenvector is the steady-state distribution. For this two-state system it is straightforward to solve for this dominant eigenvector, $\bar{\mathbf{p}}$,

$$\mathcal{L}\bar{\mathbf{p}} = 0 \qquad \bar{\mathbf{p}} = \left\{ \frac{k_1^- + k_2^+}{k_1^- + k_2^+ + k_1^+ + k_2^-}, \frac{k_1^+ + k_2^-}{k_1^- + k_2^+ + k_1^+ + k_2^-} \right\}$$

which is manifestly normalized with $p_{\mathrm{E}} + p_{\mathrm{ES}} = 1$. In the limit that $k_2^- = 0$, and k_2^+ is small, the resultant average rate of formation of product B, is $k_2^+/(k_1^+/k_1^- + 1)$.

> **Exercise 9.3:** Through explicit matrix vector multiplication, confirm that for the two-state model, $\bar{\mathbf{p}}$ is an eigenvector of the matrix $\mathcal{L}$ with eigenvalue 0. Using $\bar{\mathbf{p}}$, compute the average production of B particles within the steady-state.

For a generalization, consider the master equation for an N state system, where we will label the states by m. These states could be physical positions of a single particle, or represent a many bodied configuration of an interacting system. The master equation can be written

$$\dot{p}_m = \sum_{m'} k_{mm'} p_{m'} - k_{m'm} p_m$$

$$= \sum_{m'} k_{mm'} p_{m'}$$

where $k_{mm'}$ are the transition rates from state m' to state m. The second equality follows from the constraint

$$\sum_{m'} k_{m'm} = 0 \qquad k_{mm} = -\sum_{m' \neq m} k_{m'm}$$

that each column sums to zero. This is equivalent to observing that the total rate to exit state m is given by the sum of the rates to transition to another state m'. An alternative way to write down a master equation is to introduce a probability current between states

$$\dot{p}_m = \sum_{m'} k_{mm'} p_{m'} - k_{m'm} p_m = \sum_{m'} J_{mm'}$$

where $J_{mm'} = k_{mm'} p_{m'} - k_{m'm} p_m$ is the flux from state m' to state m. Finally, if we denote $\bar{p}_m$ as the stationary distribution associated with the rate matrix $\mathcal{L}$, then

$$\dot{\bar{p}}_m = \sum_{m'} k_{mm'} \bar{p}_{m'} = 0$$

or the action of that rate matrix on this distribution is identically 0.

With those properties of a general master equation in hand, we can try to prove a statement about the information contained in the distribution of states. In particular, let us take as a definition, Gibbs' form of the entropy,

$$\langle S(t) \rangle /k_{\mathrm{B}} = -\sum_m p_m(t) \ln p_m(t)$$

where the sum is over all states. The time dependence of the entropy results from the time dependence of the probability distribution. In a stationary state, the entropy would be a constant. With this definition, we can derive a balance law for the entropy by taking its total time derivative and rearranging terms

$$
\begin{aligned}
\left\langle \dot{S}(t) \right\rangle / k_{\mathrm{B}} &= -\sum_{m} \dot{p}_m(t) \ln p_m(t) \\
&= -\sum_{mm'} k_{mm'} p_{m'}(t) \ln p_m(t) \\
&= \frac{1}{2} \sum_{m,m'} [k_{mm'} p_{m'}(t) - k_{m'm} p_m(t)] \ln \frac{p_{m'}(t)}{p_m(t)} \\
&= \frac{1}{2} \sum_{m,m'} [k_{mm'} p_{m'}(t) - k_{m'm} p_m(t)] \ln \frac{k_{mm'} p_{m'}(t)}{k_{m'm} p_m(t)} \\
&\quad + \frac{1}{2} \sum_{m,m'} [k_{mm'} p_{m'}(t) - k_{m'm} p_m(t)] \ln \frac{k_{m'm}}{k_{mm'}}
\end{aligned}
$$

we find that the rate of change of the entropy can be decomposed into two terms. The first term on the right-hand side of the last equality is called the total entropy change, or the entropy production, $\dot{S}_{\mathrm{tot}}$, and the second is the entropy change in the environment, or the entropy flux, $\dot{S}_{\mathrm{env}}(t)$, or

$$
\dot{S}(t) = \dot{S}_{\mathrm{tot}}(t) - \dot{S}_{\mathrm{env}}(t)
$$

and within a steady state of the system, $\dot{S}_{\mathrm{tot}}(t) = \dot{S}_{\mathrm{env}}(t)$. Looking at the entropy production we find that it can be written

$$
\begin{aligned}
\left\langle \dot{S}_{\mathrm{tot}} \right\rangle / k_{\mathrm{B}} &= \frac{1}{2} \sum_{m,m'} [k_{mm'} p_{m'}(t) - k_{m'm} p_m(t)] \ln \frac{k_{mm'} p_{m'}(t)}{k_{m'm} p_m(t)} \\
&= \frac{1}{2} \sum_{m,m'} J_{mm'} F_{mm'}
\end{aligned}
$$

where $J_{mm'}$ is the flux and

$$
F_{mm'} = \ln \frac{k_{mm'} p_{m'}(t)}{k_{m'm} p_m(t)}
$$

is known as a conjugate force or sometimes as an *affinity*. As both terms $J_{mm'}$ and $F_{mm'}$ are functions of the same products of rates and probabilities, it is straightforward to use the so-called *log sum inequality* to prove that

$$
\left\langle \dot{S}_{\mathrm{tot}} \right\rangle \geq 0
$$

where equality is only satisfied when

$$
\left\langle \dot{S}_{\mathrm{tot}} \right\rangle = 0 \qquad \text{if} \qquad k_{mm'} p_{m'}(t) = k_{m'm} p_m(t)
$$

which is nothing less than a statement of detailed balance. While we have not made any connections yet to a physical system, or to a Boltzmann distribution, we have nevertheless found that Markov processes satisfying a detailed balance relation produce no entropy on average. Indeed, the extent to which a dynamics, as codified in the rate matrix, breaks detailed balance, it will produce entropy. The second part of the total change in the entropy, the entropy flux,

$$\left\langle \dot{S}_{\text{env}} \right\rangle / k_{\text{B}} = \frac{1}{2} \sum_{m,m'} J_{mm'} \ln \frac{k_{mm'}}{k_{m'm}}$$

is identified as the change of the entropy due to the environment that mediates the transitions. This term exists even when the dynamics respect detailed balance if the system is prepared in a non-steady state, and is equal to the ratio of the forward transition probability to a backward transition probability times the probability flux.

9.3   Local detailed balance

The previous statements were true for arbitrary master equations, provided only that they conserve probability and are Markovian. As such, the above statements contain no real physical information. To understand the physical implications of the above statements, we need a physical model of the rates entering into the master equation. Let us imagine that for all pairs of states m and m', the relative rates between these two states satisfy a so-called local detailed balance relationship,

$$\frac{k_{m'm}}{k_{mm'}} = e^{-\beta(E_{m'} - E_m)}$$

where β is the inverse temperature times Boltzmann's constant of the bath mediating the transition and $E_{m'} - E_m$ is the heat released to that bath. The heat released depends on both conservative time dependent and non-conservative forces acting on the system. We can identify the entropy flux as

$$\left\langle \dot{S}_{\text{env}} \right\rangle / k_{\text{B}} = \frac{\beta}{2} \sum_{m,m'} J_{mm'} \left(E_{m'} - E_m \right)$$

$$= \beta \sum_{m,m'} J_{mm'} E_{m'} = -\beta \left\langle \dot{Q} \right\rangle$$

where $-\dot{Q}$ is the heat flow into the bath. Thus provided the Gibbs definition of entropy and rates that satisfy local detailed balance, we find the usual relationship between the heat and the entropy. Moreover, given a reasonable definition of the average energy,

$$\langle E \rangle = \sum_{m} p_m E_m$$

we can compute the change in the energy as

$$\left\langle \dot{E} \right\rangle = \sum_{m} \dot{p}_m E_m + p_m \dot{E}_m$$

$$= \sum_{m,m'} J_{mm'} E_m + \sum_{m} p_m \dot{E}_m = \left\langle \dot{Q} \right\rangle + \left\langle \dot{W} \right\rangle$$

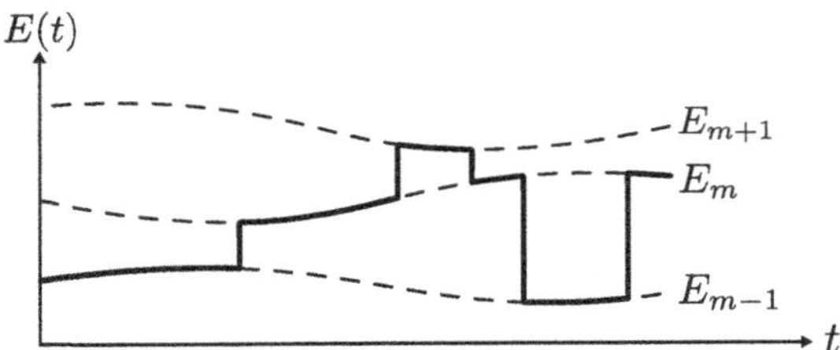

Fig. 9.3 In a discrete state system, work is the change in energy due to changes to the energy level at a fixed state of the system, while heat is exchanged with the surroundings during changes to the state of the system.

where $\dot{E}_m$ is the time derivative of the energy level of the m'th state. Taking the second term as the definition of the change in the work done on the system, we find the first law of thermodynamics. These two processes are illustrated in Figure 9.3, where heat refers to changes of the state $m \to m'$ with the energy level fixed, while work is the change in the energy level with the state m fixed.

Exercise 9.4: The notion of local detailed balance extends beyond transitions mediated by a bath at fixed temperature to those with fixed chemical potential as well. For a system obeying a local detailed balance of the form

$$\frac{k_{m'm}}{k_{mm'}} = e^{-\beta(E_{m'}-E_m)-\beta\mu(N_{m'}-N_m)}$$

determined an explicit form for the entropy produced in the environment in terms of a flux of energy $J_{mm'}E_{m'}$ and particles $J_{mm'}N_{m'}$.

As an example, let us reconsider the Michaelis–Menten system in Figure 9.2. Our previous discussion left the rates k_1^+, k_1^-, k_2^+, and k_2^- unspecified. Within a local detailed balance assumption, envisioning that the reactive events are mediated by a bath at a fixed temperature T and concentrations of species A and B are at constant chemical potentials μ_A and μ_B, then the ratio of rates should be chosen such that

$$k_1^+ = k_1^- e^{-\beta\Delta G+\beta\mu_A} \qquad k_2^+ = k_2^- e^{-\beta\Delta G'-\beta\mu_B}$$

where ΔG is the activation free energy for forming ES, and $\Delta G'$ is the difference in Gibbs free energies of A and B less the activation free energy.

These expressions for the discrete stochastic equation of motion yield analogous expressions for changes in heat, work, and entropy as the equations we employed in the continuum using the Fokker–Planck equation. However, where exactly the local detailed balance assumption enters was not clear. To clarify this, let us reconsider the time dependence for the average entropy of the system, this time in a simplified case of an overdamped equation of motion. Its corresponding Fokker–Planck equation is

$$\frac{\partial p(\mathbf{r},t)}{\partial t} = -\frac{\partial}{\partial \mathbf{r}} \frac{\mathbf{F}[\mathbf{r},\Lambda(t)]}{\gamma} p(\mathbf{r},t) + D\frac{\partial^2}{\partial \mathbf{r}^2} p(\mathbf{r},t) = -\frac{\partial}{\partial \mathbf{r}} j(\mathbf{r},t)$$

where F is the total force acting on the particle at position r and γ and D are the friction from the bath and the particle's diffusion constant. Notice we have defined the probability flux vector j. The change of the entropy of the system is

$$\frac{d\langle S\rangle}{dt} = -k_{\mathrm{B}} \int d\mathbf{r}\,\frac{\partial p(\mathbf{r},t)}{\partial t} \ln p$$

$$= k_{\mathrm{B}} \int d\mathbf{r}\,\frac{\partial j}{\partial \mathbf{r}} \ln p$$

$$= -k_{\mathrm{B}} \int d\mathbf{r}\,\frac{j}{p}\frac{\partial p}{\partial \mathbf{r}} = k_{\mathrm{B}} \int d\mathbf{r}\,\frac{j^2}{Dp} - j \cdot \frac{\mathbf{F}[\mathbf{r},\Lambda(t)]}{D\gamma}$$

where we have used the Fokker–Planck equation to eliminate the time derivative of p, integrated by parts, and finally introduced the expression for the probability flux, $j = \mathbf{F}p/\gamma - D\partial_\mathbf{r}p$. The first term in the last equation is strictly positive, and equivalent to the total entropy production analogously to what we found in the underdamped case. This would suggest that the second term is the heat released into the bath. However, this interpretation requires a local detailed balance assumption.

To see where the local detailed balance assumption comes in, let's compute the heat explicitly. Using Ito's lemma, the time rate of change of the energy of the system provides a form of the first law

$$\frac{d\mathcal{H}}{dt} = \frac{\partial \mathcal{H}}{\partial t} + \dot{\mathbf{r}} \cdot \frac{\partial \mathcal{H}}{\partial \mathbf{r}} + D\frac{\partial^2 \mathcal{H}}{\partial \mathbf{r}^2}$$

$$= \frac{\partial \mathcal{H}}{\partial t} - \frac{1}{\gamma}\mathbf{F}^2 - D\frac{\partial \mathbf{F}}{\partial \mathbf{r}}$$

where we have introduced the force as the derivative of the Hamiltonian, $\mathbf{F} = -\partial_\mathbf{r}\mathcal{H}$. The first term is the usual work rate, and so the second two terms are the heat. On average, the heat released is

$$\beta\langle\dot{Q}\rangle = -\beta \int d\mathbf{r}\,p\left(\frac{1}{\gamma}\mathbf{F}^2 + D\frac{\partial \mathbf{F}}{\partial \mathbf{r}}\right)$$

$$= -\beta \int d\mathbf{r}\,\mathbf{F}\cdot\left(\frac{\mathbf{F}}{\gamma}p - D\frac{\partial p}{\partial \mathbf{r}}\right) = -\beta \int d\mathbf{r}\,j\cdot\mathbf{F}$$

where the second line follows from integration by parts and the last equality requires that $D\gamma = k_{\mathrm{B}}T$. This relationship between the friction of the bath and the diffusivity, where their product is the temperature, is the manifestation of local detailed balance. If the dynamics were not mediated by an ideal heat bath, this relationship would not hold, as it is a consequence of the fluctuation-dissipation theorem.

9.4 Optimal finite-time processes

The ability to express heat and work as time-dependent, fluctuating quantities allows us to interrogate transformations between states in finite time, away from the reversible limits of thermodynamics. For example, while the reversible work theorem tells us that a quasi-static transformation will require the least amount of work, equilibrium thermodynamics is not equipped to answer what is the minimum amount of work required to transition a system in a fixed amount of time. Such a question requires a dynamical theory.

To explore this particular question, let's start with a simple example. Consider a colloidal particle moving in solution under the influence of an optical trap. The corresponding equation of motion is

$$\gamma \dot{r} = -\frac{\partial U(r, \Lambda)}{\partial r} + \eta$$

where γ is the friction from the solution, and $U(r, \Lambda)$ is the potential due to the optical trap that can be changed in time. The noise η obeys the usual equilibrium fluctuation–dissipation relation, $\langle \eta \rangle = 0$ and $\langle \eta(0)\eta(t) \rangle = 2k_B T \gamma \delta(t)$. The equilibrium statistics of the noise reflect an assumption of local detailed balance, or that despite manipulating the colloidal particle in finite time, we can assume that the surrounding solvent stays in equilibrium. For concreteness, assume the optical trap is harmonic,

$$U(r, \Lambda) = \frac{\Lambda(t)}{2} r^2$$

where $\Lambda(t)$ is a tunable stiffness. If we were to manipulate the stiffness, the corresponding average work would be given by

$$\langle \mathcal{W}(\tau) \rangle = \frac{1}{2} \int_0^\tau dt\, \dot{\Lambda}(t) \left\langle r^2(t) \right\rangle$$

where we denote the length of the protocol over which we change Λ as τ. This protocol is illustrated in Figure 9.4.

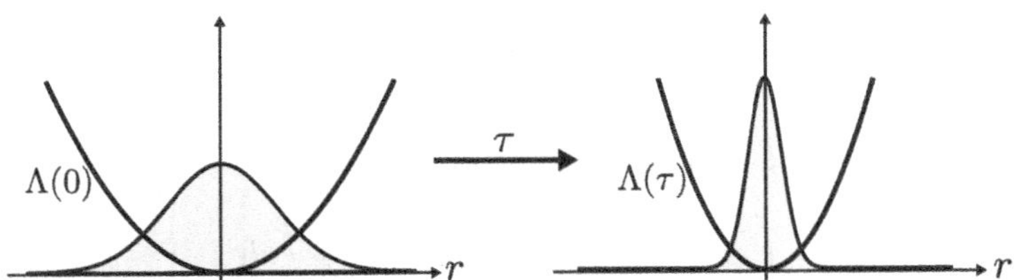

Fig. 9.4 Squeezing of a colloidal particle in a harmonic trap.

The average work is determined by the prescribed protocol—how we change Λ in time—but also how that evolution changes $\left\langle r^2(t) \right\rangle$. While the former is imposed, the latter must be computed. In general this is a difficult task, but for the harmonic system this can be done exactly. The Fokker–Planck equation for the particle is

$$\frac{\partial p(r, t)}{\partial t} = \gamma^{-1} \Lambda(t) \frac{\partial}{\partial r} rp(r, t) + D \frac{\partial^2 p(r, t)}{\partial r^2}$$

where $D = k_{\mathrm{B}}T/\gamma$. Multiplying both sides by r^2 and integrating over all r,

$$\frac{\partial}{\partial t}\left\langle r^2(t)\right\rangle = \gamma^{-1}\Lambda(t)\int dr\, r^2 \frac{\partial}{\partial r} rp(r,t) + D\int dr\, r^2 \frac{\partial^2 p(r,t)}{\partial r^2}$$

$$= -2\gamma^{-1}\Lambda(t)\int dr\, r^2 p(r,t) + 2D\int dr\, p(r,t)$$

$$= -2\gamma^{-1}\Lambda(t)\left\langle r^2(t)\right\rangle + 2D$$

we arrive at an equation of motion for $\left\langle r^2(t)\right\rangle$. In the second line we have integrated by parts assuming the probability distribution vanishes at $r = \pm\infty$, and in the third line we have employed the definitions of integrals of the time-dependent probability distribution. For ease of notation, let's define $\xi(t) = \left\langle r^2(t)\right\rangle$.

Taking the expression of the average work and integrating it by parts to move the time derivative off of the stiffness and onto ξ,

$$\left\langle \mathcal{W}(\tau)\right\rangle = \frac{1}{2}\,\xi(t)\Lambda(t)\big|_0^\tau - \frac{1}{2}\int_0^\tau dt\,\left(\frac{-\gamma\dot{\xi}(t) + 2D\gamma}{2\xi(t)}\right)\dot{\xi}(t)$$

$$= \frac{1}{2}\,[\xi(t)\Lambda(t) - D\gamma\ln\xi(t)]\big|_0^\tau + \frac{\gamma}{4}\int_0^\tau dt\,\frac{\dot{\xi}^2(t)}{\xi(t)}$$

we find that up to boundary terms, the mean work is given uniquely by a function of the time-dependent average of r^2. This closure of the problem would not happen if the potential was not harmonic. Because the mean work is determined solely by the time dependence of ξ, we can minimize the average work by optimizing $\xi(t)$. To do this, we can take the functional derivative of the mean work with respect to $\xi(t)$,

$$\frac{\delta}{\delta\xi(t')}\int_0^\tau dt\,\frac{\dot{\xi}^2(t)}{\xi(t)} = \frac{1}{\xi^2}\left(\dot{\xi}^2 - 2\xi\ddot{\xi}\right) = 0$$

which yields an Euler-Lagrange equation. The general solution is

$$\xi(t) = c_1\left(1 + c_2 t\right)^2$$

where c_1 and c_2 are determined by the protocol. Let us consider protocols that start at one value of $\Lambda(0) = \Lambda_0$ and end at $\Lambda(\tau) = \Lambda_f$. If $\Lambda = \Lambda_0$ from $t < 0$ until the start of the protocol, then $c_1 = k_{\mathrm{B}}T/\Lambda_0$ as determined by the corresponding equilibrium distribution. The other coefficient c_2 is determined by the condition that we wish to minimize the average work.

Substituting ξ into the expression of the mean work,

$$\beta\left\langle \mathcal{W}(\tau)\right\rangle = \frac{c_2^2\gamma\tau}{\Lambda_0} + \frac{\Lambda_f(1 + c_2\tau)^2}{2\Lambda_0} - \ln\left[1 + c_2\tau\right] - 1/2$$

we find a function of c_2. Taking the derivative of $\beta\left\langle \mathcal{W}(\tau)\right\rangle$ with respect to c_2 and setting it equal to zero, we can solve for the coefficient that minimizes the average work. This optimal coefficient, c_2^*, is given by

$$c_2^* \tau = \frac{-1 - \Lambda_f \tau / \gamma + \sqrt{1 + \Lambda_0 \left(2 + \Lambda_f \tau / \gamma\right) \tau / \gamma}}{2 + \Lambda_f \tau / \gamma}$$

for which we find a complex function, but one which has natural timescales $\gamma / \Lambda_{0/f}$ that set the units of τ. The corresponding optimal protocol can be determined using the previous equation of motion for the mean squared position. In terms of the optimal coefficient, the optimal protocol is

$$\Lambda^*(t) = \gamma \frac{2D - \dot{\xi}}{2\xi} = \frac{\Lambda_0 - c_2^*(1 + c_2^* t)\gamma}{(1 + c_2^* t)^2}$$

which is peculiar in that it is not a smooth function. We have solved the expression so that $\Lambda(0) = \Lambda_0$ and $\Lambda(\tau) = \Lambda_f$. However, taking the limit of the optimal protocol toward $t = 0$ from above

$$\lim_{t \to 0^+} \Lambda^*(t) = \Lambda_0 + \left(1 - \sqrt{\frac{\Lambda_0}{\Lambda_f}}\right) \frac{\gamma}{\tau} + \mathcal{O}(1/\tau^2)$$

or toward τ from below

$$\lim_{t \to \tau^-} \Lambda^*(t) = \Lambda_f - \left(1 - \sqrt{\frac{\Lambda_f}{\Lambda_0}}\right) \frac{\gamma}{\tau} + \mathcal{O}(1/\tau^2)$$

we find that to the lowest order in $1/\tau$, the inverse protocol duration, there is a correction to these boundary conditions indicative of a jump discontinuity in the protocol. For any finite time process, the protocol that minimizes the work will include these sharp changes to the control parameter. Note these jumps do not contribute to the mean work. Discontinuous changes occur often in optimal control theory, and are typically referred to as bang-bang solutions.

To the lowest order in the inverse of the protocol duration, the mean work is given by

$$\beta \left\langle \mathcal{W}(\tau) \right\rangle = \frac{1}{2} \ln \left(\frac{\Lambda_f}{\Lambda_0}\right) + \frac{(1 - \sqrt{\Lambda_0/\Lambda_f})^2}{\tau \Lambda_0 / \gamma} + \mathcal{O}(1/\tau^2) \geq \beta \Delta A$$

where the first term, independent of τ, is the reversible work, ΔA, and the leading order correction is positive and scales like $1/\tau$. Counter-intuitively, the correction scales like $1/\sqrt{\Lambda_f}$, so that in the limit that the change in the stiffness is large $\Lambda_f \gg \Lambda_0$, the quasi-static limit is recovered faster. This result can be reconciled by acknowledging that the relaxation time of the Ornstein–Uhlenbeck process is inversely proportional to the spring constant.

The optimal control of the colloidal particle explored above can also serve as a simple model to explore nanoscale engines. Consider, for example, a realization of the Carnot cycle in which isothermal expansion, adiabatic expansion, isothermal compression, and adiabatic compression, are implemented by manipulating the colloidal particle in the harmonic optical trap. The Carnot cycle is designed in order to extract work from the spontaneous transfer of heat from a hot to a cold reservoir. The efficiency of the cycle, ζ, is defined as the work extracted divided by the heat from the hot reservoir that flows into the system.

Exercise 9.5: Consider a similar process to that given earlier, where rather than the spring stiffness being controlled, the spring center is. Specifically, imagine a time-dependent potential of the form

$$U(r, \Lambda) = \frac{k}{2} \left[r - \Lambda(t) \right]^2$$

where k is the spring constant and $\Lambda(t)$ is the controllable center of the spring. Show that the minimum average work for changing Λ from 0 to Λ_f in a time τ is

$$\beta \langle \mathcal{W}(\tau) \rangle = \beta k \Lambda_f^2 / (\tau k / \gamma + 2)$$

using an analogous Euler–Lagrange approach as before. Compare this to the reversible limit.

Figure 9.5 renders the specific steps in which 1) the spring stiffness is first reduced in a bath at temperature T_h over a time $0 < t \le t_1$, 2) the temperature is changed adiabatically and instantly $T_h \to T_c$, 3) the spring stiffness is increased at constant temperature T_c over a time $t_1 < t \le t_3$, and then finally 4) the temperature is changed adiabatically and instantly $T_c \to T_h$.

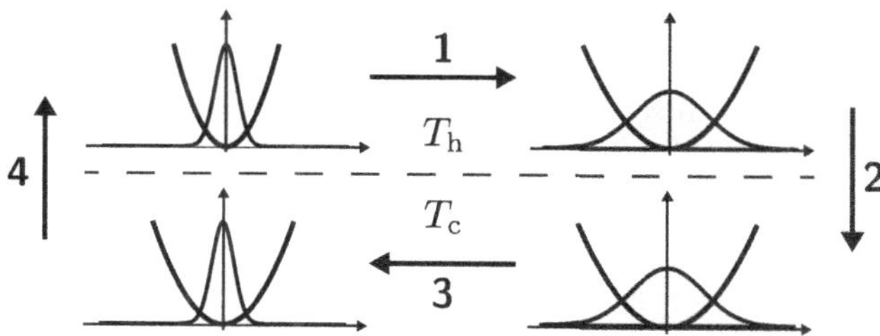

Fig. 9.5 Single particle Carnot cycle with isothermal (1,3) and adiabatic steps (2,4).

Since the protocol is a cycle, the total average change in any state function is zero. This means the total change in the energy and entropy are both 0. As the second and fourth steps are adiabatic and instant, there is no work or heat flow. The total work, $\mathcal{W} = \mathcal{W}_1 + \mathcal{W}_3$ is a sum of the work from the first and third steps, as is the total heat, $\mathcal{Q} = \mathcal{Q}_1 + \mathcal{Q}_3$, while their sum $\mathcal{W}_1 + \mathcal{W}_3 + \mathcal{Q}_1 + \mathcal{Q}_3 = 0$ on average. For each isothermal step, 1 and 3, we can solve the same Euler–Lagrange equations we had previously. However, now we have two protocols, one for each step, and the boundary conditions are altered. Specifically, we require that $\xi(t_1^-)$ at the end of step 1 be equal to $\xi(t_1^+)$ at the start of step 3, as step 2 is envisioned as instantaneous. Analogously, we require that $\xi(t_3)$ at the end of step 3 be equal to $\xi(0)$ at the start of step 1.

With these boundary conditions and minimizing the total work, which is equivalent to maximizing the work extracted from the cycle, the time-dependent mean squared position can be computed. For $\xi_0 = \xi(0)$ and $\xi_2 = \xi(t_1)$, the optimal protocol is

$$
\xi^*(t) = \begin{cases}
\xi_0 \left(1 + \left(\sqrt{\xi_2/\xi_0} - 1\right) \frac{t}{t_1}\right)^2 & 0 < t \le t_1 \\[2mm]
\xi_2 \left(1 + \left(\sqrt{\xi_0/\xi_2} - 1\right) \frac{t-t_1}{t_3-t_1}\right)^2 & t_1 < t \le t_3
\end{cases}
$$

which is a piecewise function, reflecting the two isothermal processes. Using these expressions, we can compute the work. We find the total work is negative, work that is extractable, and is the sum of two terms,

$$
\langle \mathcal{W} \rangle = -\frac{k_B}{2} \left(T_h - T_c\right) \ln \frac{\xi_2}{\xi_0} + \frac{\gamma}{2} \left(t_1^{-1} + (t_3 - t_1)^{-1}\right) \left(\sqrt{\xi_2} - \sqrt{\xi_0}\right)^2
$$

where the first term is proportional to the difference in temperatures of the bath is the reversible contribution. The second term depends on the duration of the cycle. The heat in the first step, $\langle \mathcal{Q}_1 \rangle$ can be calculated by the first law, $\langle \mathcal{Q}_1 \rangle = \langle \Delta U \rangle - \langle \mathcal{W}_1 \rangle$, where the change in energy is computable from $\langle \Delta U \rangle = \xi_2 \Lambda(t_1)/2 - \xi_0 \Lambda(0)/2$. Using the optimal protocol, the heat entering the system is

$$
\langle \mathcal{Q}_1 \rangle = \frac{k_B T_h}{2} \ln \frac{\xi_2}{\xi_0} - \gamma t_1^{-1} \left(\sqrt{\xi_2} - \sqrt{\xi_0}\right)^2
$$

which, like the total work, has a reversible part and an irreversible part depending on the step duration. Defining the efficiency of the cycle as done originally by Carnot, $\zeta = -\langle \mathcal{W} \rangle / \langle \mathcal{Q}_1 \rangle$, we find for this finite time process

$$
\zeta = \frac{k_B \left(T_h - T_c\right) \ln(\xi_2/\xi_0) - \gamma \left(t_1^{-1} + (t_3 - t_1)^{-1}\right) \left(\sqrt{\xi_2} - \sqrt{\xi_0}\right)^2}{k_B T_h \ln(\xi_2/\xi_0) - 2\gamma t_1^{-1} \left(\sqrt{\xi_2} - \sqrt{\xi_0}\right)^2}
$$

which is not strictly positive. At short times, work is actually consumed by the cycle, rather than being extracted. In the opposite limit, where the cycle time is long and the transformations quasi-static,

$$
\lim_{t_1, t_3 \to \infty} \zeta = 1 - T_c/T_h
$$

we recover the famous result of Carnot that the efficiency of an ideal heat engine is determined universally by the difference in temperatures between the two baths.

Exercise 9.6: Determine the optimal cycle time to maximize the power output from the Carnot cycle, $\mathcal{P} = -\mathcal{W}/t_3$, and show that the efficiency at maximum power is

$$
\zeta^* = \frac{2(T_h - T_c)}{3T_h + T_c}
$$

that saturates at $2/3$ for large temperature differences.

9.5   Stochastic path integrals

Thus far, we have considered averages of nonequilibrium processes. To understand fluctuations about the average, it is convenient to introduce some additional formalism. Back in Chapter 8, we introduced an alternative perspective to stochastic dynamics using the notion of ensembles of trajectories, and their associated path probabilities. While then we used it to describe the collection of reactive trajectories, it also provides a powerful means for describing stochastic thermodynamics as we will show over the next few sections. For a Markovian dynamics the probability of observing a trajectory, or time-ordered sequence of configurations, $\mathbf{X}(\tau) = \{\mathbf{x}_0, \mathbf{x}_{\Delta t} \ldots, \mathbf{x}_\tau\}$, is given by

$$P[\mathbf{X}(\tau)] = p(\mathbf{x}_0) \prod_{i=0}^{\tau/\Delta t - 1} P(\mathbf{x}_{(i+1)\Delta t}|\mathbf{x}_{i\Delta t})$$

where $p(\mathbf{x}_0)$ is the probability of observing the initial condition $\mathbf{x}_0$, and $P(\mathbf{x}_{\Delta t}|\mathbf{x}_0)$ is the transition probability, or the probability of reaching $\mathbf{x}_{\Delta t}$ in a time Δt, provided the system is at $\mathbf{x}_0$ at time 0. For either the master equation or the Fokker–Planck equation, the transition probability can be computed

$$P(\mathbf{x}_{\Delta t}|\mathbf{x}_0) = \left\langle \delta(\mathbf{x}_{\Delta t} - e^{\mathcal{L}^\dagger \Delta t}\mathbf{x}_0) \right\rangle$$

where the average is over the appropriate random variables, like the noise history in the Langevin equation and $\mathcal{L}^\dagger$ the adjoint operator.

For the discrete state master equation dynamics, the likelihood of a trajectory is related to the rate of transitioning between two states, as well as the likelihood of a jump happening at a specific time. Unlike in continuous space, where something happens at every time, in discrete space there is a finite probability that nothing occurs over some time interval. This probability is a function of the so-called exit rate, or the total rate out of a particular state. For a system in state m at time t, the total rate of leaving that state, denoted the exit rate $k_m^{\mathrm{ex}}(t)$, is

$$k_m^{\mathrm{ex}}(t) = -k_{mm}(t) = \sum_{m' \neq m} k_{m'm}(t)$$

which is a sum over all potential states m can transition into. Here we have associated a time dependence to the rates, in acknowledgment that these transitions may be controlled by some time-varying parameter. The probability that the system stays in state m over some interval of time is given by a product of 1 minus the probability to exit, which is in turn the exit rate times the time interval

$$\prod_{i=1} [1 - k_m^{\mathrm{ex}}(t_i)\,(t_{i+1} - t_i)] \approx e^{-\int dt\, k_m^{\mathrm{ex}}(t)}$$

which is well approximated by an exponential for sufficiently small times. The resultant trajectory that the system traces out over some observation time τ is broken up into a series of lag times during which the system stays in a particular state and instantaneous jumps into new states. An illustration of such a stochastic trajectory is shown in

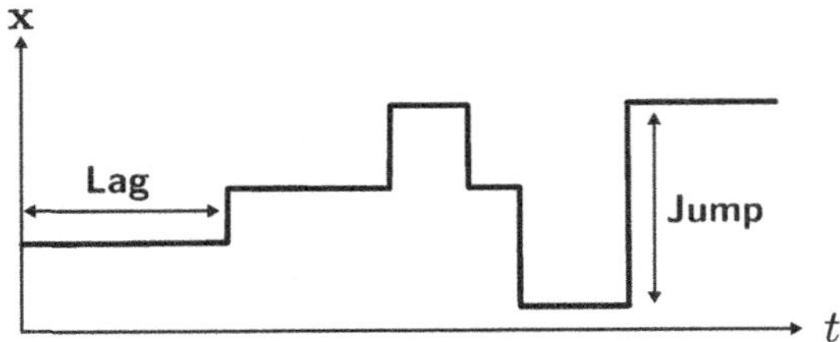

Fig. 9.6 Example stochastic trajectory for a discrete state, continuous time Markov process.

Figure 9.6. The trajectory probability conditioned on starting at some state $\mathbf{x}_0$ is therefore,

$$P[\mathbf{X}(\tau)|\mathbf{x}_0] = e^{-\int_{t_n}^{\tau} dt\, k_{\mathbf{x}_n}^{\mathrm{ex}}(t)} \prod_{i=0}^{n-1} k_{\mathbf{x}_{i+1},\mathbf{x}_i}(t_{i+1}) e^{-\int_{t_i}^{t_{i+1}} dt\, k_{\mathbf{x}_i}^{\mathrm{ex}}(t)}$$

which is a product of the probabilities of staying in a given state until a transition occurs at some time t_i, times the probability of that specific transition given by its associated rate at that time. The product is over all jumps, n, that occur over τ, and the final waiting time outside of the product comes from the probability that another transition does not occur before τ.

Integrals over these path measures require some care in defining. The fluctuating quantities include the states that the system can transition between, the times at which those transitions occur, as well as how many transitions occur over the specified observation time. Thus the associated path measure needs to include a summation over all of these,

$$\int \mathcal{D}[\mathbf{X}(\tau)] = \sum_{n=0}^{\infty} \sum_{\mathbf{x}_0,\dots,\mathbf{x}_n} \int_0^{t_2} dt_1 \dots \int_{t_{n-1}}^{\tau} dt_n$$

where as before we use the notion $\mathcal{D}[\mathbf{X}(\tau)]$ to denote the space of possible stochastic trajectories. Note that with this definition of the continuous measure, we have the following normalization condition dependent on the initial condition $p(\mathbf{x}_0)$,

$$\int \mathcal{D}[\mathbf{X}(\tau)] P[\mathbf{X}(\tau)|\mathbf{x}_0] p(\mathbf{x}_0) = 1$$

with which averages over path space can be taken straightforwardly.

As we have come to understand, thermodynamics is fundamentally connected to time-reversal symmetry and its breaking. As a consequence it is useful to define a time-reversed trajectory of this discrete space, continuous time system as $\mathbf{X}^*(\tau) = \{(\mathbf{x}_n, t_n), (\mathbf{x}_{n-1}, t_{n-1}), \dots, (\mathbf{x}_1, t_1), (\mathbf{x}_0, t_0)\}$ which is the sequence of configurations visited in the forward trajectory inverted in order, with the jumps occurring at the same time. In principle, the reverse trajectory should include time reversed configurations, $\mathbf{x} \to \mathbf{x}^*$. However, we will assume that $\mathbf{x}$ contains only variables that are even in time. The associated probability of observing the time-reversed trajectory is

$$P[\mathbf{X}^*(\tau)|\mathbf{x}_\tau] = e^{-\int_{t_n}^{\tau} dt\, k_{\mathbf{x}_n}^{\mathrm{ex}}(t)} \prod_{i=0}^{n-1} k_{\mathbf{x}_i,\mathbf{x}_{i+1}}(t_{i+1}) e^{-\int_{t_i}^{t_{i+1}} dt\, k_{\mathbf{x}_i}^{\mathrm{ex}}(t)}$$

where the only change from the forward time trajectory is the inversion of the rates, $k_{\mathbf{x}_{i+1},\mathbf{x}_i} \to k_{\mathbf{x}_i,\mathbf{x}_{i+1}}$. Correspondingly, the ratio of forward to reverse paths is

$$\frac{P[\mathbf{X}(\tau)|\mathbf{x}_0]}{P[\mathbf{X}^*(\tau)|\mathbf{x}_\tau]} = \prod_{i=0}^{n-1} \frac{k_{\mathbf{x}_{i+1},\mathbf{x}_i}(t_{i+1})}{k_{\mathbf{x}_i,\mathbf{x}_{i+1}}(t_{i+1})}$$

$$= \prod_{i=0}^{n-1} e^{-\beta\left[E_{\mathbf{x}_{i+1}}(t_{i+1}) - E_{\mathbf{x}_i}(t_{i+1})\right]} = e^{-\beta Q}$$

which assuming the rates obey a local detailed balance with a single bath at inverse temperature β is just given by the change in the energy of the system along the jumps. As these changes in energy occur instantaneously in time and correspond to changes in the state of the system, the log ratio can be interpreted as the heat released into a bath along the forward trajectory. This is known as the *Crooks fluctuation theorem*, and can be viewed as a statement of microscopic reversibility.

For a stochastic equation of motion like the overdamped Langevin equation,

$$\gamma\dot{x} = F(x) + \eta \qquad \langle\eta\rangle = 0 \qquad \langle\eta(0)\eta(t)\rangle = 2k_{\mathrm{B}}T\gamma\delta(t - t')$$

where γ is the friction, $F(x)$ is some force, and η a Gaussian random variable, we can also determine its transition probability, and associated path integral. To start, we will construct this by explicitly discretizing the equation. The singular nature of the noise requires some care to evaluate, but when this is done as in Chapter 7, to first order in Δt, the overdamped Langevin equation reads,

$$x_{\Delta t} - x_0 = F(x_0)\Delta t/\gamma + \sqrt{\Delta t}\,\eta/\gamma$$

which is a linear function of the noise, η, which has a new variance dependent on Δt. The transition probability can be written as an average over the noise, conditioned on starting at $x(0)$ and ending in $x(\Delta t)$, or

$$P(x_{\Delta t}|x_0) = \left\langle \delta\left(x_{\Delta t} - x_0 - F(x_0)\Delta t/\gamma - \sqrt{\Delta t}\,\eta/\gamma\right)\right\rangle$$

$$= \int d\eta\, p(\eta)\delta(x_{\Delta t} - x_0 - F(x_0)\Delta t/\gamma - \sqrt{\Delta t}\,\eta/\gamma)$$

$$= \exp\left[-\left(\gamma\Delta x - F(x_0)\Delta t\right)^2/4k_{\mathrm{B}}T\gamma\Delta t\right]/\sqrt{4\pi\Delta t k_{\mathrm{B}}T/\gamma}$$

where $\Delta x = x_{\Delta t} - x_0$. In the second line we have expressed the average over the noise, with distribution $p(\eta)$, and in the third we exploited the fact that the noise is a Gaussian random variable. Rearranging this we see

$$P(\Delta x) = \exp\left[-\frac{\beta\gamma}{4\Delta t}\left(\Delta x - F(x_0)\Delta t/\gamma\right)^2\right]/\sqrt{4\pi\Delta t/(\beta\gamma)}$$

that the probability of making a displacement of size Δx over a time Δt is Gaussian with mean $F(x)\Delta t/\gamma$ and variance $2\Delta t/(\beta\gamma)$.

We have already encountered the continuous limit of this transition probability in Chapter 8. The probability of observing any given trajectory is just the probability of drawing a sequence of noises, resulting in

$$P[\mathbf{X}(\tau)|x_0] = \exp\left[-\int_0^\tau dt\, (\gamma\dot{x} - F[x(t)])^2/4k_\mathrm{B}T\gamma\right]$$

where the likelihood of tracing out a specific trajectory takes the form of a Gaussian. This formulation of overdamped dynamics is known as the Onsager–Machlup stochastic path action. In the limit that the temperature T goes to zero, this probability is maximized by the minimum path action as would be true for Hamiltonian dynamics.

Exercise 9.7: For an underdamped equation of motion of the form

$$m\ddot{\mathbf{r}}(t) = -\gamma\dot{\mathbf{r}}(t) + \mathbf{F}[\mathbf{r}(t)] + \boldsymbol{\eta}(t)$$

where $\langle\boldsymbol{\eta}(t)\rangle = 0$ and $\langle\boldsymbol{\eta}(t)\boldsymbol{\eta}(t')\rangle = 2k_\mathrm{B}T\gamma\delta(t - t')$, show that the probability of observing a trajectory is

$$P[\mathbf{X}(\tau)|\mathbf{x}_0] = \exp\left\{-\int_0^\tau dt\, [m\ddot{\mathbf{r}}(t) + \gamma\dot{\mathbf{r}}(t) - \mathbf{F}(t)]^2/4k_\mathrm{B}T\gamma\right\}$$

using the same procedure as above. Show that the ratio of this probability and its time reversed also satisfies a Crooks fluctuation theorem.

9.6 Work fluctuation relations

With the path integral representation, we can explore some relationships for dynamical systems far from equilibrium. To start, let's revisit the relative probability of seeing a trajectory in the forward direction of time, $\mathbf{X}(\tau)$, relative to the time-reversed trajectory trajectory, denoted $\mathbf{X}^*(\tau)$. These conjugate paths are illustrated in the discrete limit in Figure 9.7. The ratio of likelihoods in the forward and backward direction of time is

$$\frac{P[\mathbf{X}(\tau)|x_0]}{P[\mathbf{X}^*(\tau)|x_\tau]} = \frac{\exp\left[-\int_0^\tau dt\, \gamma\dot{x}^2/4k_\mathrm{B}T + F^2/4k_\mathrm{B}T\gamma - \dot{x}F/2k_\mathrm{B}T\right]}{\exp\left[-\int_0^\tau dt\, \gamma\dot{x}^2/4k_\mathrm{B}T + F^2/4k_\mathrm{B}T\gamma + \dot{x}F/2k_\mathrm{B}T\right]}$$

$$= \exp\left[\int_0^\tau dt\, \dot{x}F/k_\mathrm{B}T\right]$$

whereas before, we find significant cancellation. After simplifying the ratio, we are left with an integral over the velocity times the force. To understand this, let's take some specific examples of the force on the system. For a standard gradient force,

$$F = -\frac{dU(x)}{dx(t)}$$

we can substitute it in yielding

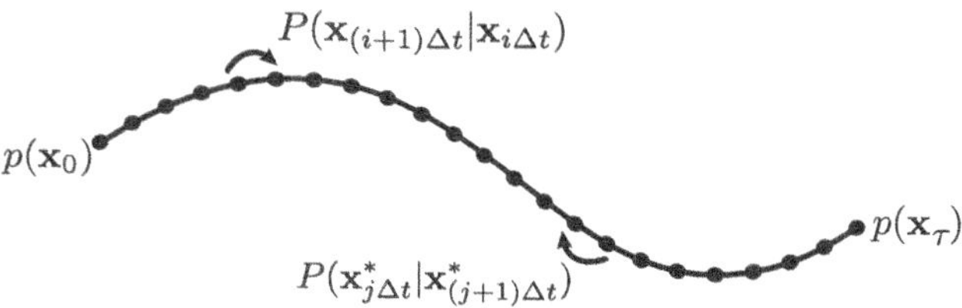

Fig. 9.7 Forward and reverse trajectories determined by transition probabilities $P(\mathbf{x}_{(i+1)\Delta t}|\mathbf{x}_{i\Delta t})$ between states $\mathbf{x}_{(i+1)\Delta t}$ and $\mathbf{x}_{i\Delta t}$ and their time-reversed $P(\mathbf{x}^*_{i\Delta t}|\mathbf{x}^*_{(i+1)\Delta t})$.

$$\frac{P[\mathbf{X}(\tau)|x_0]}{P[\mathbf{X}^*(\tau)|x_\tau]} = \exp\left[-\beta \int_0^\tau dt\, \frac{dx}{dt}\frac{dU(x)}{dx}\right]$$

$$= \exp\left[-\beta \int_0^\tau dt\, \frac{dU}{dt}\right] = \exp\left[-\beta(U(x_\tau) - U(x_0))\right]$$

where the second line follows from the fundamental theorem of calculus. With a gradient force, we again arrive at a statement of detailed balance. This change in the potential energy comes from the bath, held at fixed T. It has an interpretation of heat, though one in this case that depends only on the temporal boundaries.

More generally, the force on the system could be some combination of a gradient force and a time-dependent force,

$$F = -\frac{dU(x)}{dx} + \Lambda(t)f(x)$$

where $\Lambda(t)$ is some time-dependent protocol. For a potential with an explicit time dependence, the total change in the potential with time has contributions from both heat and work. For fixed initial and final conditions, the ratio of forward to backward probabilities

$$\frac{P[\mathbf{X}(\tau)|\mathbf{x}_0]}{P[\mathbf{X}^*(\tau)|\mathbf{x}_\tau]} = \exp\left[-\beta \int_0^\tau dt\, \dot{Q}\right] = e^{-\beta Q[\mathbf{X}(\tau)]}$$

is just given by the heat dissipated to the environment. This is another formulation of the Crooks fluctuation theorem, now for dynamics in a continuous space. As a consequence of it being a fundamentally thermodynamic result, its form is independent of the underlying dynamics. It can be viewed as a more general statement of the second law of thermodynamics. This is clear by writing down the so-called *integral fluctuation theorem*,

$$\int \mathcal{D}[\mathbf{X}(\tau)]\, P[\mathbf{X}(\tau)|\mathbf{x}_0]e^{\beta Q[\mathbf{X}(\tau)]} = \int \mathcal{D}[\mathbf{X}^*(\tau)]\, P[\mathbf{X}^*(\tau)|\mathbf{x}_\tau]$$

$$\left\langle e^{\beta Q[\mathbf{X}(\tau)]}\right\rangle = 1$$

where the second line follows from the normalization of probability, and the definition of the average. From Jensen's inequality, the traditional statement of the second law is recovered

$$-\langle \mathcal{Q} \rangle \geq 0 \qquad \text{or} \qquad \langle \Delta S_{\text{tot}} \rangle \geq 0$$

where we assert the standard definition of the change in the entropy of the bath, $-\mathcal{Q}[\mathbf{X}(\tau)]/T = \Delta S_{\text{env}}$, and because the initial and final points are fixed, there is no associated change in the entropy of the system.

We can use the Crooks fluctuation theorem to prove another important identity. Consider the case for the time-dependent force,

$$\Lambda(t)f(\mathbf{x}) = \begin{cases} \Lambda_0 f(\mathbf{x}) & t < 0 \\ \Lambda(t) f(\mathbf{x}) & 0 < t < \tau \\ \Lambda_\tau f(\mathbf{x}) & t > \tau \end{cases}$$

which means that the system is prepared in an equilibrium state with Λ_0 and will eventually evolve to an equilibrium with Λ_τ. Previously, we could only tease useful information out of this sort of setup if the change was small, which led to our theory of linear response, or when the dynamics could be solved exactly. However, using the Crooks fluctuation theorem and integrating over the initial equilibrium states

$$\left\langle e^{-\beta \mathcal{W}[\mathbf{X}(\tau)]} \right\rangle = \int d\mathbf{x}_0 \int \mathcal{D}[\mathbf{X}(\tau)] e^{-\beta \mathcal{W}[\mathbf{X}(\tau)]} P[\mathbf{X}(\tau)|\mathbf{x}_0] p(\mathbf{x}_0)$$

$$= \int d\mathbf{x}_0 \int \mathcal{D}[\mathbf{X}(\tau)] e^{-\beta \mathcal{W}[\mathbf{X}(\tau)]} \frac{P[\mathbf{X}(\tau)|\mathbf{x}_0]}{P[\mathbf{X}^*(\tau)|\mathbf{x}_\tau]} P[\mathbf{X}^*(\tau)|\mathbf{x}_\tau] \frac{p(\mathbf{x}_0)}{p(\mathbf{x}_\tau)} p(\mathbf{x}_\tau)$$

$$= \int d\mathbf{x}_\tau \int \mathcal{D}[\mathbf{X}^*(\tau)] P[\mathbf{X}^*(\tau)|\mathbf{x}_\tau] p(\mathbf{x}_\tau) e^{-\beta \Delta A} = e^{-\beta(A[\Lambda_\tau] - A[\Lambda_0])}$$

we find the Jarzynski equality first encountered in Chapter 6, which then was derived under deterministic dynamics. The recovery of this relation in the case of stochastic equations of motion is a general consequence of its origin in thermodynamics, which is again independent of kinetic details.

9.7 Single molecule pulling experiments

The same manipulations we used to prove the Jarzynski equality in instances where the system can exchange heat with a bath can be used to show the Crooks work fluctuation theorem

$$P(\mathcal{W}) = P(-\mathcal{W}) e^{\beta \mathcal{W} - \beta \Delta A}$$

which relates the distribution of work done in the forward process to that done in the reverse. While in the reversible limit $\langle \mathcal{W} \rangle = \Delta A$, the Crooks work fluctuation theorem says that $P(\mathcal{W}) = P(-\mathcal{W})$ for $\mathcal{W} = \Delta A$, or that the distributions of work fluctuations intersect at its reversible value, so the statistics of $P(\mathcal{W})$ encode the reversible limit even far from equilibrium. This equality has been exceedingly useful in molecular experiments where approaching the reversible limit is difficult, yet the free energetics of the system are desired. Single molecule pulling experiments using optical tweezers performed on RNA hairpins by Bustamante have been used to measure work fluctuations and provided early experimental verification of the Crooks work fluctuation theorem and Jarzynski equality.

One of the most difficult aspects of working with nonequilibrium systems is that the likelihood of configurations is only known implicitly. An explicit form is not generally available even in a nonequilibrium steady-state. While the distribution of configurations satisfies a Liouville equation or Fokker–Planck equation depending on the ensemble, solving that equation is not generally easy, and its functional dependence on microscopic parameters is thus obscured. However, the fact that there is structure to the probabilities associated with trajectory spaces provides some useful relationships for distributions of configurations under nonequilibrium driving conditions.

Consider, for example, the Crooks fluctuation theorem associated with starting and ending in states able to equilibrate,

$$\frac{p(\mathbf{x}_0)}{p(\mathbf{x}_\tau)}\frac{P[\mathbf{X}(\tau)|\mathbf{x}_0]}{P[\mathbf{X}^*(\tau)|\mathbf{x}_\tau]} = e^{\beta\mathcal{W}[\mathbf{X}(\tau)]-\beta\Delta A}$$

where as usual, the first law connects the work to the heat and the change in internal energy, $\mathcal{W} = \Delta U - \mathcal{Q}$, which we have used to eliminate the heat. This expression implies that the initial and final states are not independent of each other, even within stochastic dynamics. Rearranging, we have

$$p(\mathbf{x}_0)P[\mathbf{X}(\tau)|\mathbf{x}_0] = p(\mathbf{x}_\tau)P[\mathbf{X}^*(\tau)|\mathbf{x}_\tau]e^{\beta\mathcal{W}[\mathbf{X}(\tau)]-\beta\Delta A}$$

where we find a generalized balance expression that in the limit that the work $\mathcal{W}[\mathbf{X}(t)]$ done on the system over the observation time τ is equal to the reversible limit, given by the free energy ΔA, we would recover the statement of detailed balance. More generally, this expression provides a link between averages in and out of equilibrium.

Take, for example, an arbitrary observable $O(\mathbf{x}_\tau)$ depending on the end point of the trajectory. Multiplying the generalized balance expression on both sides by this observable

$$O(\mathbf{x}_\tau)p(\mathbf{x}_\tau)P[\mathbf{X}^*(\tau)|\mathbf{x}_\tau] = O(\mathbf{x}_\tau)p(\mathbf{x}_0)P[\mathbf{X}(\tau)|\mathbf{x}_0]e^{-\beta\mathcal{W}[\mathbf{X}(\tau)]+\beta\Delta A}$$

and then summing over all paths, we would find a relationship between averages in the forward ensemble and those in the reverse ensemble. However, the left-hand side of the equality refers only to the initial state in the time-reversed dynamics. As we have assumed that the initial and final distributions are equilibrium, then since $O(\mathbf{x}_\tau)$ only depends on the final time not the subsequent evolution, its average is indistinguishable from an equilibrium average under conditions equivalent to the end of the trajectory

$$\langle O(\mathbf{x})\rangle_{\mathrm{eq}} = \left\langle O(\mathbf{x}_\tau)e^{-\beta\mathcal{W}[\mathbf{X}(\tau)]+\beta\Delta A}\right\rangle_{\mathrm{neq}}$$

and we arrive at an equality between an equilibrium average and an average over the nonequilibrium trajectories driven under some control protocol. The two are related just by a reweighting factor that depends on the work done relative to the equilibrium free energy difference, or the dissipated work.

Take for example the case illustrated in Figure 9.8, where the nonequilibrium protocol consists of pulling a polymer by applying a time-dependent force in order to change its end-to-end distance R, as can be done with optical tweezers. If we take

$O(\mathbf{x}) = \delta(R - R[\mathbf{x}])$, then the average of $\langle \delta(R - R[\mathbf{x}]) \rangle_{\mathrm{eq}}$ in equilibrium is the probability of finding the system at a given end-to-end distance $p_{\mathrm{eq}}(R)$, and it is related to a conditioned expectation value,

$$p_{\mathrm{eq}}(R) = p_{\mathrm{neq}}(R) \left\langle e^{\beta W[\mathbf{X}(\tau)] - \beta \Delta A} \right\rangle_{R,\mathrm{neq}}$$

the probability of finding the end-to-end distance at the end of the pulling protocol, $p_{\mathrm{neq}}(R)$, times an average of the extra work, the work less its reversible limit, required to affect that transformation. This relationship, originally due to Hummer and Szabo, allows one to reconstruct free energy surfaces from single molecule pulling experiments.

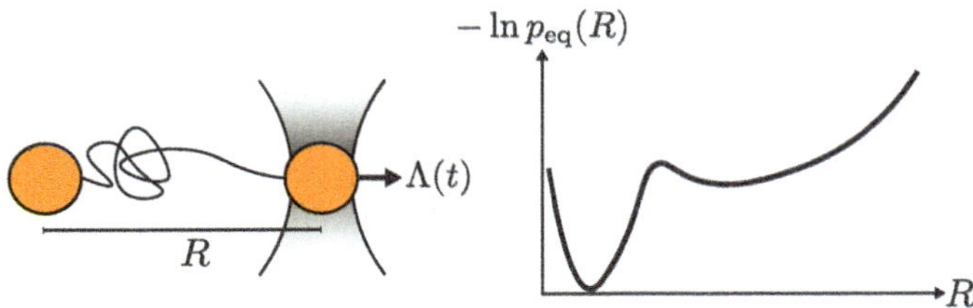

Fig. 9.8 Illustration of the Hummer–Szabo relation relating a nonequilibrium measurement to an equilibrium probability distribution.

The Hummer–Szabo relation affords a means of recovering detailed equilibrium information from nonequilibrium processes. Manipulating the expression in reverse provides a means of decoding fluctuations away from equilibrium. Consider for example the generalized balance relationship, only with the work term left on the time-reversed side of the equality. Averaging an observable dependent on the final time results in

$$\langle O(\mathbf{x}_\tau) \rangle_\Lambda = \left\langle O(\mathbf{x}_\tau) e^{-\beta W[\mathbf{X}^*(\tau)] + \beta \Delta A} \right\rangle_{\Lambda^*} = \frac{\left\langle O(\mathbf{x}_\tau) e^{-\beta W[\mathbf{X}^*(\tau)]} \right\rangle_{\Lambda^*}}{\left\langle e^{-\beta W[\mathbf{X}^*(\tau)]} \right\rangle_{\Lambda^*}}$$

where the left-hand side is averaged at the end of some nonequilibrium protocol, which we find is equal to measuring the observable at the start of the time-reversed protocol weighted by the extra work accumulated over the subsequent trajectory. Using Jarzynski's equality, we can write the free energy as an exponential average of the work, eliminating any trace of the equilibrium process. Indeed, the free energy is introduced initially as bookkeeping for the entropy change of the system, which need not refer to an equilibrium system. Note we have adopted the notation of a subscript Λ on averages over the forward time dynamics under some external protocol, and subscript Λ^* as an average taken under the time-reversed protocol.

Consider a case where the nonequilibrium protocol is capable of establishing a new steady-state that is not described by an equilibrium distribution. This could be for example by introducing time-independent non-gradient forces into the system, or by connecting the system to multiple reservoirs that are not in equilibrium with each other. Under such conditions in the long time limit, $\tau \to \infty$, the system will evolve a stationary distribution computed as $p_{\mathrm{neq}}(\mathbf{r}^N) = \langle \delta(\mathbf{r}^N - \mathbf{r}_\tau^N) \rangle_\Lambda$. Taking our observable as this Dirac delta function, we find

$$
\begin{aligned}
p_{\mathrm{neq}}(\mathbf{r}^N) &= \frac{\langle \delta(\mathbf{r}^N - \mathbf{r}_\tau^N) e^{-\beta \mathcal{W}[\mathbf{X}^*(\tau)]} \rangle_{\Lambda^*}}{\langle e^{-\beta \mathcal{W}[\mathbf{X}^*(\tau)]} \rangle_{\Lambda^*}} \\
&= \langle \delta(\mathbf{r}^N - \mathbf{r}_\tau^N) \rangle_{\Lambda^*} \frac{\langle \delta(\mathbf{r}^N - \mathbf{r}_\tau^N) e^{-\beta \mathcal{W}[\mathbf{X}^*(\tau)]} \rangle_{\Lambda^*}}{\langle \delta(\mathbf{r}^N - \mathbf{r}_\tau^N) \rangle_{\Lambda^*} \langle e^{-\beta \mathcal{W}[\mathbf{X}^*(\tau)]} \rangle_{\Lambda^*}} \\
&= p_{\mathrm{eq}}(\mathbf{r}^N) \frac{\langle e^{-\beta \mathcal{W}[\mathbf{X}^*(\tau)]} \rangle_{\mathbf{r}^N, \Lambda^*}}{\langle e^{-\beta \mathcal{W}[\mathbf{X}^*(\tau)]} \rangle_{\Lambda^*}}
\end{aligned}
$$

where in the second line we have multiplied by 1, and in the third identified the average of the delta function at the final time as an equilibrium average. We have also introduced a conditional average with the subscript $\mathbf{r}^N$ denoting the starting configuration of the reverse dynamics. We refer to this compact representation of the nonequilibrium distribution function as the Kawasaki–Crooks relation, as this specific form was first arrived at by Crooks but similar expressions go back to Kawasaki. In the limit that the system is close to equilibrium, the work done on the system by the control parameter is small, and we can Taylor expand the exponential averages

$$
\ln p_{\mathrm{neq}}(\mathbf{r}^N)/p_{\mathrm{eq}}(\mathbf{r}^N) \approx -\beta \left(\langle \mathcal{W} \rangle_{\mathbf{r}^N, \Lambda^*} - \langle \mathcal{W} \rangle_{\Lambda^*} \right)
$$

and find that the first-order correction to the equilibrium configuration of the system is given by the work done conditioned on starting in a particular state. The second, non-conditioned average, serves to provide a well-defined long time limit, as provided a finite correlation time, the initial conditioning will be lost for protocols of sufficient duration and $\langle \mathcal{W} \rangle_{\mathbf{r}^N, \Lambda^*} \to \langle \mathcal{W} \rangle_{\Lambda^*}$.

9.8 Heat conduction and fluctuation theorems

Over the last few sections, we have explored how nonequilibrium systems produce entropy, whether by their relaxation to equilibrium or through the explicit injection of energy. This production of entropy is intimately related to a notion of time's arrow, and the emergence of time reversal symmetry breaking. Specifically, the log ratio of the probability of observing a trajectory $\mathbf{X}$, relative to observing its time-reversed, $\mathbf{X}^*$

$$
\begin{aligned}
\ln \frac{P(\mathbf{X})}{P(\mathbf{X}^*)} &= \ln \frac{p(\mathbf{x}_0)}{p(\mathbf{x}_\tau^*)} + \sum_i^{\tau/\Delta t} \ln \frac{P(\mathbf{x}_{(i+1)\Delta t}|\mathbf{x}_{i\Delta t})}{P(\mathbf{x}_{i\Delta t}^*|\mathbf{x}_{(i+1)\Delta t}^*)} \\
&= \Delta S/k_{\mathrm{B}} + \Delta S_{\mathrm{env}}/k_{\mathrm{B}} = \Delta S_{\mathrm{tot}}/k_{\mathrm{B}}
\end{aligned}
$$

is the entropy production. Provided the ratio of transition probabilities is mediated by a bath that obeys a local detailed balance, the second term in the first line can be identified with the heat released into the environment. The first term is the information entropy change of the system. So very sharply, entropy encodes our ability to distinguish the direction of time, the difference between systems evolving forward in

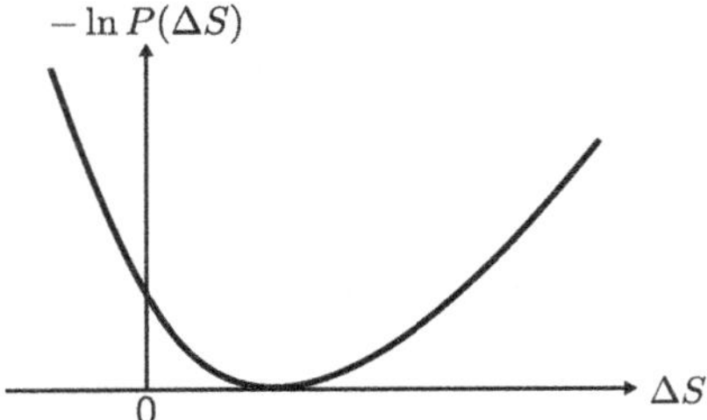

Fig. 9.9 Distribution of total entropy production with exponentially suppressed negative entropy fluctuations, indicative of fluctuation theorem symmetry.

time or being shown to us in reverse. For a system evolving between two equilibrium states with a dynamics that preserves detailed balance, our ability to distinguish the direction of time is localized to the boundaries of the trajectories. For a system driven out of equilibrium, say in contact with baths at different chemical potentials, this direction is apparent throughout the trajectory—we can tell the difference between mass flowing up or down a chemical potential gradient. The longer that trajectory, the more entropy produced, the more certain we would be of time's arrow.

With this definition of the entropy generated along a trajectory, it is straightforward to prove a relationship between the likelihood of producing a given amount of entropy to that of destroying the same amount. Specifically, if $P(\Delta S_{\text{tot}})$ is the probability of producing a total amount of entropy ΔS_{tot}, then integrating over all possible

$$
\begin{aligned}
P(\Delta S_{\text{tot}}) &= \int \mathcal{D}[\mathbf{X}(\tau)]\, P[\mathbf{X}(\tau)]\, \delta\left(\Delta S_{\text{tot}}/k_{\mathrm{B}} - \ln \frac{P[\mathbf{X}(\tau)]}{P[\mathbf{X}^*(\tau)]}\right) \\
&= \int \mathcal{D}[\mathbf{X}(\tau)]\, P[\mathbf{X}^*(\tau)] e^{\Delta S_{\text{tot}}[\mathbf{X}(\tau)]/k_{\mathrm{B}}}\, \delta\left(\Delta S_{\text{tot}}/k_{\mathrm{B}} - \ln \frac{P[\mathbf{X}(\tau)]}{P[\mathbf{X}^*(\tau)]}\right) \\
&= \int \mathcal{D}[\mathbf{X}(\tau)]\, P[\mathbf{X}(\tau)] e^{-\Delta S_{\text{tot}}[\mathbf{X}(\tau)]/k_{\mathrm{B}}}\, \delta\left(\Delta S_{\text{tot}}/k_{\mathrm{B}} + \ln \frac{P[\mathbf{X}(\tau)]}{P[\mathbf{X}^*(\tau)]}\right) \\
&= P(-\Delta S_{\text{tot}}) e^{\Delta S_{\text{tot}}/k_{\mathrm{B}}}
\end{aligned}
$$

we find the *fluctuation theorem*, which says that it is exponentially less likely to destroy entropy than it is to produce it. This was first proposed by Evans, Cohen, and Morriss under specific conditions, and proved by Seifert, Kurchan, Lebowitz, Spohn, and Gallavotti among others in specific contexts. Indeed, it follows from the normalization of probability that the above equality implies

$$
\left\langle e^{-\Delta S_{\text{tot}}/k_{\mathrm{B}}} \right\rangle = 1 \longrightarrow \langle \Delta S_{\text{tot}} \rangle \geq 0
$$

where the equality is historically known as an integral fluctuation theorem, and the inequality is the second law of thermodynamics. An example distribution consistent with this symmetry is shown in Figure 9.9. Note the exponential tail for $\Delta S_{\text{tot}} < 0$ makes it increasingly unlikely to observe a system destroy large amounts of entropy.

An alternative way to characterize entropy production fluctuations is to compute its associated generating function. Let $Z(\lambda)$ be a moment-generating function, which can be computed from an average over paths,

$$Z(\lambda) = \left\langle e^{-\lambda \Delta S_{\text{tot}}/k_{\text{B}}} \right\rangle$$

$$= \int d\Delta S_{\text{tot}} \, P(\Delta S_{\text{tot}}) e^{-\lambda \Delta S_{\text{tot}}/k_{\text{B}}} .$$

The generating function, $Z(\lambda)$, has a symmetry that owes its origin to microscopic reversibility. This can be shown by

$$Z(\lambda) = \int d\Delta S_{\text{tot}} \, P(\Delta S_{\text{tot}}) e^{-\lambda \Delta S_{\text{tot}}/k_{\text{B}}}$$

$$= \int d\Delta S_{\text{tot}} \, P(-\Delta S_{\text{tot}}) e^{\Delta S_{\text{tot}}/k_{\text{B}}} e^{-\lambda \Delta S_{\text{tot}}/k_{\text{B}}} = Z(1-\lambda)$$

where we have replaced $P(\Delta S_{\text{tot}})$ using the fluctuation theorem by $P(-\Delta S_{\text{tot}})$ times $\exp(\Delta S_{\text{tot}}/k_{\text{B}})$. This reflection symmetry about $\lambda = 1/2$ is known as Lebowitz–Spohn symmetry.

Exercise 9.8: The cumulant generating function, $\ln Z(\lambda)$, has a number of useful properties. Prove that

$$\ln Z(0) = 0 \qquad \frac{d^2 \ln Z(\lambda)}{d\lambda^2} \geq 0$$

namely, that $\ln Z(\lambda)$ is a convex function that passes through the origin.

To explore the implications of these results, let's imagine that the system is exposed to two different heat baths, at distinct temperatures. Assuming that each bath mediates transitions that obey a local detailed balance relation, then for each process

$$\frac{k^v_{m'm}}{k^v_{mm'}} = e^{-\beta^v(E_{m'}-E_m)}$$

where $v = 1, 2$ labels rates associated with baths at temperatures β^v. Within the steady state produced with these two heat baths, the contribution to the total entropy production is given by the heat dissipated to the bath. The corresponding rate of change of the entropy of the bath is

$$\dot{S}_{\text{env}}/k_{\text{B}} = -\sum_v \beta^v \dot{Q}^v = -\Delta\beta\dot{Q}$$

where stationarity requires that $\dot{Q}^1 = -\dot{Q}^2$, so $\Delta\beta = \beta^1 - \beta^2$. The total entropy change is given by the time integral,

$$\Delta S_{\text{tot}}/k_{\text{B}} = \int_0^\tau dt' \, \dot{S}_{\text{env}} = -\Delta\beta Q$$

which in the steady state has no contribution from the entropy change in the system. We have proved previously that the total entropy production obeys a fluctuation symmetry. Plugging in our definition of entropy production in terms of heat and the difference in inverse temperatures

$$\frac{P(\Delta S_{\text{tot}} = -\Delta\beta\mathcal{Q})}{P(\Delta S_{\text{tot}} = \Delta\beta\mathcal{Q})} = \frac{P(-\mathcal{Q})}{P(\mathcal{Q})} = e^{\Delta\beta\mathcal{Q}}$$

where the second equality follows from the Jacobians of the numerator and denominator being identical. Such an identity for the time-integrated current is a consequence of the linear relationship between the entropy production and the current, where $\Delta\beta$ is identifiable as the force or affinity.

Analogously, the generating function for the current,

$$Z(\lambda) = \langle e^{-\lambda\mathcal{Q}} \rangle = \int d\mathcal{Q}\, P(\mathcal{Q}) e^{-\lambda\mathcal{Q}}$$
$$= \int d\mathcal{Q}\, P(-\mathcal{Q}) e^{-\Delta\beta\mathcal{Q}} e^{-\lambda\mathcal{Q}} = Z(-\Delta\beta - \lambda)$$

is symmetric about $\lambda = -\Delta\beta/2$. We can use this symmetry for the current generating function to prove an important relationship between the cumulants of the current. We remind ourselves that the cumulants can be computed by derivatives of $\ln Z$,

$$-\frac{d\ln Z(\lambda)}{d\lambda}\bigg|_{\lambda=0} = \langle \mathcal{Q}\rangle \qquad \frac{d^2\ln Z(\lambda)}{d\lambda^2}\bigg|_{\lambda=0} = \langle \delta\mathcal{Q}^2\rangle$$

where the averages are taken in the steady state. Using the Lebowitz–Spohn symmetry, we can replace a derivative with respect to λ, with a derivative with respect to the force, $\Delta\beta$, in the limit that $\Delta\beta \to 0$,

$$\frac{d\ln Z}{d\lambda}\bigg|_{\lambda=0} = \frac{d\ln Z}{d\Delta\beta}\bigg|_{\lambda=0}$$

and plugging in our definitions for the cumulants we arrive at a relationship between the change in the average current and a temperature difference,

$$\kappa \equiv -\frac{d\langle\mathcal{Q}\rangle/\tau}{d\Delta\beta}\bigg|_{\Delta\beta=0} = \langle \delta\mathcal{Q}(\tau)^2\rangle/2\tau$$

and the fluctuations of that current about a steady state with fixed temperature difference. This is a linear response relationship, like a Green–Kubo relation, in this case for the thermal conductivity κ. We saw previously that such results for the transport of mass or momentum could be derived from underlying hydrodynamics. However, here, all that we assumed to arrive at this important generalization is microscopic reversibility, or local detailed balance. Apart from this, the current of interest needed to be linearly related to the total entropy production. With those two conditions, we have arrived at the fluctuation–dissipation theorem. Environments able to mediate the transfer of mass or charge allow for analogous results.

> **Exercise 9.9:** For a particle moving under a constant applied force in steady state, the total entropy production is given by $\Delta S/k_B = \beta f J$ where f is the constant applied force and $J = \int dt\, \dot{x}(t)$ is integral of the particle current over an observation time τ. Demonstrate a Lebowitz–Spohn symmetry $Z(\lambda) = \langle \exp[-\lambda J]\rangle = Z(\beta f - \lambda)$, and derive a corresponding Green–Kubo expression in the long time limit
>
> $$\frac{d}{df}\langle \dot{x}\rangle\bigg|_{f=0} = \beta \int_0^\infty dt\, \langle \dot{x}(0)\dot{x}(t)\rangle$$
>
> which equates the particle mobility to the velocity autocorrelation function.

9.9   Fluctuations of time extensive quantities

The total entropy is a sum that is extensive in an observation time, τ. If we imagine that we can block time into $\mathcal{T} = \tau/\Delta t$ independent segments of length Δt, then the entropy produced is a sum over the entropies produced at each discrete time, s_t,

$$S/k_B = \sum_{t=1}^{\mathcal{T}} s_t$$

where each member of the sum is statistically independent. If that assumption is valid, then the moment-generating function acquires a simple structure,

$$\left\langle e^{-\lambda S/k_B}\right\rangle = \left\langle e^{-\lambda s_1} e^{-\lambda s_2} e^{-\lambda s_3} \cdots e^{-\lambda s_{\mathcal{T}}}\right\rangle$$

$$= \left\langle e^{-\lambda s_1}\right\rangle \left\langle e^{-\lambda s_2}\right\rangle \left\langle e^{-\lambda s_3}\right\rangle \langle\cdots\rangle \left\langle e^{-\lambda s_{\mathcal{T}}}\right\rangle$$

$$= \left[\left\langle e^{-\lambda s_1}\right\rangle\right]^{\mathcal{T}} = \left[\left\langle e^{-\lambda s_1}\right\rangle^{1/\Delta t}\right]^{\tau}$$

which is a product of $\mathcal{T}$ terms. We can identify the term in the brackets as dependent on s, the entropy over Δt,

$$\left\langle e^{-\lambda s}\right\rangle^{\tau/\Delta t} \sim e^{-\tau \psi(\lambda)}$$

where $\psi(\lambda)$ is defined as a large deviation function, or precisely a scaled, cumulant generating function for fluctuations of the total entropy.

The generating function is a Laplace transform of the distribution function,

$$Z(\lambda) = \int dS\, P(S) e^{-\lambda S/k_B}$$

and as such, it contains the same information as the original distribution function. Indeed, for smooth functions, Laplace transforms are invertible, so the distribution function can be written as an inverse Laplace transform

$$P(S) = \frac{1}{2\pi i} \int_C d\lambda \, Z(\lambda) e^{\lambda S / k_{\mathrm{B}}}$$
$$= \frac{1}{2\pi i} \int_C d\lambda \, e^{-\tau[\psi(\lambda) - \lambda s]}$$

where in the second line we have replaced $Z(\lambda)$ with its associated large deviation function, and identified $S/k_{\mathrm{B}} = \tau s$. In the long time limit, the exponential will be dominated by its maximum, so it can be evaluated by a saddle point approximation,

$$\frac{d}{d\lambda} \left[\psi(\lambda) - \lambda s \right]\big|_{\lambda^*} = 0$$

or,

$$s = \frac{d\psi}{d\lambda}\bigg|_{\lambda^*(s)} .$$

Inserting this back in we find the Gartner–Ellis theorem, ignoring sub-exponential terms from the prefactor,

$$P(S) = \exp\left[-\tau \psi(\lambda^*) + \tau \lambda^* s \right]$$
$$= \exp\left[-\tau I(s) \right]$$

that the exponentially small fluctuations are given by a function, $I(s)$ known as a rate function. This function and it complement, $\psi(\lambda)$, are related to each other by a Legendre transform, just as the entropy and free energy are in equilibrium. Putting together the definition of the rate function and the fluctuation theorem

$$\frac{e^{-\tau I(s)}}{e^{-\tau I(-s)}} = e^{\tau s} \implies I(s) - I(-s) = -s$$

which is a property of the rate function known as Gallavotti–Cohen symmetry.

This exponential structure within distribution functions of large sums of arbitrary observables is generic. These results follow from an area of probability theory known as large deviation theory, which is considered to be a generalization of the results of the central limit theorem, or the theory of typical fluctuations. We have employed large deviation theory already earlier in this textbook, as a means of understanding the connection between equilibrium statistical mechanics and the emergent structure of traditional thermodynamics by taking a thermodynamic limit of many particles. However, much of the structure we uncovered in that context has analogies where rather than the large system limit, we consider the long time limit.

Because these results provide some additional structure for understanding fluctuations of nonequilibrium states, it is worth considering some examples of dynamical large deviations. Let us consider a sum of uncorrelated random variables

$$Y = \sum_{i=1}^{\mathcal{T}} y_i$$

where there are $\mathcal{T}$ members of the sum, and we will be interested in cases where $\mathcal{T} \gg 1$. The distribution of the summed variable is given by

$$P(Y) = \int dy_1 \int dy_2 \cdots \int dy_{t_N}\, p(y_1)\, p(y_2) \cdots p(y_{\mathcal{T}}) \delta \left(Y - \sum_i^{\mathcal{T}} y_i \right)$$

which is just a $\mathcal{T}$ dimensional integral with a delta function constraint. Because the y_i's are uncorrelated, we can rewrite the joint distribution as,

$$P(Y) = \left[\prod_i^{\mathcal{T}} \int dy_i\, p(y_i) \right] \delta \left(Y - \sum_i^{\mathcal{T}} y_i \right)$$

$$= \int \frac{dk}{2\pi} \left[\prod_i^{\mathcal{T}} \int dy_i\, p(y_i) \right] e^{ik(Y - \sum_i^{\mathcal{T}} y_i)} = \int dk\, e^{ikY} \left[\int dy_1\, p(y_1) e^{-iky_1} \right]^{\mathcal{T}}$$

where we have introduced an auxiliary variable to replace the delta function and identified the product structure under the integral. These manipulations are identical to what we did in Chapter 3 when we discussed the central limit theorem. In that case, we took the argument of the exponential, Taylor expanded it, and argued that independent of the distribution, $p(y_1)$, the leading order terms yielded a Gaussian distribution. Here we will see how different underlying distributions change the structure of rare fluctuations.

For concreteness let us first consider that each variable is drawn from an identical set of Gaussian distributions,

$$p(y) = \frac{1}{\sqrt{2\pi\sigma^2}} e^{-y^2/2\sigma^2}$$

characterized by a mean of 0 and a variance of σ^2. The distribution for the sum can be easily computed,

$$P(Y) = \int \frac{dk}{2\pi} e^{ikY} \left[\int dy\, \frac{1}{\sqrt{2\pi\sigma^2}} e^{-y^2/2\sigma^2} e^{-iky} \right]^{\mathcal{T}}$$

$$= \int \frac{dk}{2\pi} e^{ikY} e^{-k^2\sigma^2\mathcal{T}/2}$$

$$= \frac{1}{\sqrt{2\pi\mathcal{T}\sigma^2}} e^{-Y^2/2\mathcal{T}\sigma^2}$$

which yields an expected Gaussian result, with the same mean and a variance that is $\mathcal{T}\sigma^2$. If we rewrite this in terms of the intensive variables $y = Y/\mathcal{T}$, we arrive at

$$P(Y) \sim e^{-\mathcal{T}I(y)} \qquad I(y) = y^2/2\sigma^2$$

where the rate function is just the original Gaussian distribution. An analogous calculation can be done for an exponential distribution

$$p(y) = \frac{1}{\mu} e^{-y/\mu}$$

characterized by a mean μ. However, the result is a rate function

$$P(Y) \sim e^{-\mathcal{T}I(y)} \qquad I(y) = y/\mu - 1 - \ln y/\mu$$

which is clearly distinct from the rate function for the Gaussian random variables, and distinct from the original exponential distribution.

> **Exercise 9.10:** Confirm that for a set of exponentially distributed random numbers that their large deviation rate function is
>
> $$I(y) = y/\mu - 1 - \ln y/\mu$$
>
> where μ is the mean of y.

Consistent with the central limit theorem, if we examine the rate function for the exponential distribution around its maximum,

$$I(y) \approx I(\mu) + I'(\mu)(y - \mu) + I''(\mu)(y - \mu)^2/2$$
$$= (y - \mu)^2/2\mu^2$$

we find that the fluctuations are Gaussian. However, the rate function, shown in Figure 9.10 in comparison to a Gaussian rate function, exhibits a substantially increased likelihood of fluctuations to large values of the mean. This is a reflection of the breadth of the underlying exponential distribution. Fluctuations to values smaller than the mean are suppressed relative to Gaussian, as the domain of the exponential distribution is $y \geq 0$ while the Gaussian is $-\infty \leq y \leq \infty$. Thus, the tails of a distribution reflect more strongly the underlying process that generates it.

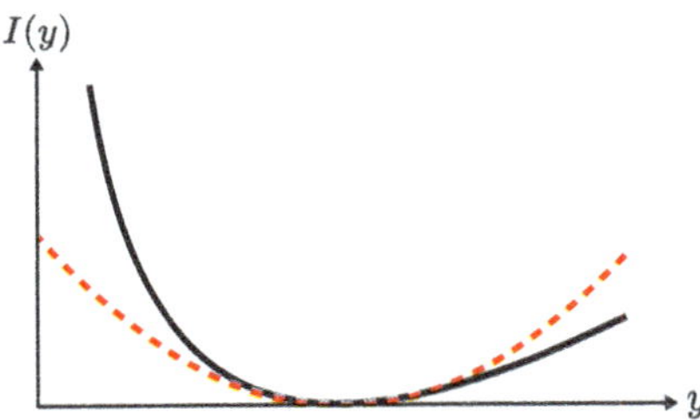

Fig. 9.10 Rate function for a sum of exponentially distributed variables in black compared to its Gaussian approximation in red.

9.10 Tilting and Doob transform

For a time-extensive sum of random variables, provided finite correlations, a rate function and its scaled cumulant generating function will exist. Provided one, the other can be computed from a Legendre transform. The theory of dynamical large deviations provides a means of computing rare fluctuations, and also characterizing them in terms of auxiliary processes that render rare fluctuations typical. We have seen this working a barrier-crossing event in Chapter 8. There we found that a trajectory conditioned on reacting could be driven to do so in a manner that was indistinguishable from the original conditioned process. The driving force resulted from a so-called *Doob transform*. This perspective can be generalized to fluctuations of time extensive quantities in the long time limit and to trajectory ensembles that are statistically biased, or *tilted* to make a rare fluctuation more probable.

Let us again consider the generating function for entropy fluctuations,

$$e^{-\lambda S/k_{\mathrm{B}}} = \left(e^{S/k_{\mathrm{B}}}\right)^{-\lambda} = [P(\mathbf{X})/P(\mathbf{X}^*)]^{-\lambda}$$

where we remind ourselves of the definition of the entropy as a ratio of path probabilities. The generating function can then be written for discrete, fixed waiting time,

$$Z(\lambda) = \int \mathcal{D}[\mathbf{X}]\, P(\mathbf{X}) \left[\frac{P(\mathbf{X})}{P(\mathbf{X}^*)}\right]^{-\lambda}$$

$$= \int dx_0 \int dx_1 \cdots p_{x_0} k_{x_1 x_0} k_{x_2 x_1} \cdots \left[\frac{p_{x_0} k_{x_1 x_0} k_{x_2 x_1} \cdots}{p_{x_\tau} k_{x_0 x_1} k_{x_1 x_2} \cdots}\right]^{-\lambda}$$

where in the second line we have written out explicitly our definition of the path probabilities as products of transition probabilities, $k_{x_1 x_2}$. We can identify a different rate matrix, known as a tilted operator or Lebowitz–Spohn operator,

$$k_{ji}(\lambda) = k_{ji}^{1-\lambda} k_{ij}^{\lambda}$$

which when replaced into the generating function equation reduces to

$$Z(\lambda) = \int dx_0 \int dx_\tau\, p_{x_0}^{1-\lambda} p_{x_\tau}^{\lambda} \left[k_{x_\tau x_0}(\lambda)\right]^{\tau}$$

which includes a boundary term, $p_{x_0}^{1-\lambda} p_{x_\tau}^{\lambda}$, and the tilted operator raised to the power τ. In the long time limit, this operator is determined solely by its largest eigenvalue. If we denote the tilted operator $\mathcal{L}_\lambda = \mathcal{L}^{1-\lambda}\mathcal{L}^{\lambda}$, at long times

$$Z(\lambda) \sim e^{-\tau \psi(\lambda)}$$

the large deviation function is given by the largest eigenvalue.

Exercise 9.11: The tilted operator for the entropy production associated with a particle hopping around a three site ring with rates a and b is given by

$$\mathcal{L}_\lambda = \begin{pmatrix} -a-b & a^{1-\lambda}b^{\lambda} & a^{\lambda}b^{1-\lambda} \\ a^{\lambda}b^{1-\lambda} & -a-b & a^{1-\lambda}b^{\lambda} \\ a^{1-\lambda}b^{\lambda} & a^{\lambda}b^{1-\lambda} & -a-b \end{pmatrix}$$

Show that the dominant eigenvalue and associated large deviation function is given by $\psi(\lambda) = -a\left((b/a)^{1-\lambda} + (b/a)^{\lambda} - (b/a) - 1\right)$.

An analogous calculation can be brought to bear on large deviations associated with integrals over continuous states. Consider for example, an observable O that is an integral over a function f that depends on the instantaneous state of the system

$$O = \int_t^\tau dt'\, f[\mathbf{r}(t')]$$

where the integral is over an observation time τ that we will take to be large. The associated generating function providing a position of the system can be defined as

$$Z(\lambda, \mathbf{r}) = \left\langle \delta[\mathbf{r} - \mathbf{r}(t)]e^{-\lambda O} \right\rangle$$

where λ now has units of $1/O$, but acts in the same way as the other counting parameters we have introduced. By taking the time derivative of Z,

$$\frac{d}{dt}Z(\lambda, \mathbf{r}) = \frac{\partial}{\partial t}Z(\lambda, \mathbf{r}) + \frac{\mathbf{F}}{\gamma} \cdot \frac{\partial}{\partial \mathbf{r}}Z(\lambda, \mathbf{r}) + D\frac{\partial^2}{\partial \mathbf{r}^2}Z(\lambda, \mathbf{r})$$

where we have used Ito's lemma assuming an overdamped equation of motion with drift $\mathbf{F}$, diffusion constant D, and friction γ. We could have alternatively taken the time derivative of the definition of Z, which would yield

$$\frac{d}{dt}Z(\lambda, \mathbf{r}) = \lambda f[\mathbf{r}(t)]Z(\lambda, \mathbf{r})$$

where the factor of $\lambda f[\mathbf{r}(t)]$ comes from the argument of the exponential. Putting these two results together,

$$\frac{\partial}{\partial t}Z(\lambda, \mathbf{r}) = -\left\{\frac{\mathbf{F}}{\gamma} \cdot \frac{\partial}{\partial \mathbf{r}}Z(\lambda, \mathbf{r}) + D\frac{\partial^2}{\partial \mathbf{r}^2}Z(\lambda, \mathbf{r})\right\} + \lambda f Z(\lambda, \mathbf{r})$$
$$= -\left(\mathcal{L}^\dagger - \lambda f\right)Z(\lambda, \mathbf{r})$$

we find that the generating function satisfies a differential equation given by the adjoint of the Fokker–Planck operator, $\mathcal{L}^\dagger$, with an added term $-\lambda f Z(\lambda, \mathbf{r})$. This is known as the Feynman–Kac theorem. By solving this parabolic partial differential equation, we could in principle obtain the generating function. Indeed, just as we found with the Lebowitz–Spohn operator for the discrete system, the maximum eigenvalue of $\mathcal{L}_\lambda = \left(\mathcal{L}^\dagger - \lambda f\right)$ returns the large deviation function.

The Feynman–Kac theorem applies directly to observables that depend on the state of the system. However, observables of the form

$$O = \int dt\, \mathbf{g}(\mathbf{r}) \cdot \dot{\mathbf{r}} + f(\mathbf{r})$$

which depend on the state of the system as well as its stochastic increments, $\dot{\mathbf{r}}$, require special attention. These observables are typically referred to as *current type* observables, as they depend on time-displaced coordinates. Consider the definition of the generating function for such an observable as an average over trajectories. For overdamped dynamics,

$$\langle e^{-\lambda O}\rangle = \int \mathcal{D}[\mathbf{X}(t)]P[\mathbf{X}]e^{-\lambda \int dt\, \mathbf{g}(\mathbf{r})\cdot\dot{\mathbf{r}}+f(\mathbf{r})}$$

$$= \int \mathcal{D}[\mathbf{X}(t)]e^{-\frac{1}{4k_{\mathrm{B}}T\gamma}\int dt(\gamma\dot{\mathbf{r}}-\mathbf{F}(\mathbf{r}))^2-\lambda\int dt\,\mathbf{g}(\mathbf{r})\cdot\dot{\mathbf{r}}+f(\mathbf{r})}$$

$$= \int \mathcal{D}[\mathbf{X}(t)]e^{-\frac{1}{4k_{\mathrm{B}}T\gamma}\int dt(\gamma\dot{\mathbf{r}}-\mathbf{F}(\mathbf{r})+2k_{\mathrm{B}}T\mathbf{g}(\mathbf{r})\lambda)^2-\lambda\int dt\,\mathbf{g}(\mathbf{r})\cdot\mathbf{F}(\mathbf{r})/\gamma-D\lambda\mathbf{g}^2(\mathbf{r})+f(\mathbf{r})}$$

we have found a way to write the generating function as an average over a dynamics with a λ dependent drift and a new observable $\mathbf{g}(\mathbf{r})\cdot\mathbf{F}(\mathbf{r})/\gamma - D\lambda\mathbf{g}^2(\mathbf{r}) + f(\mathbf{r})$ that depends only on the state of the system. This manipulation follows from inserting the Onsager–Machlup stochastic action and then rewriting the sum of the action and the current type part of the observable,

$$(\gamma\dot{\mathbf{r}}-\mathbf{F})^2 + 4k_{\mathrm{B}}T\gamma\mathbf{g}\cdot\dot{\mathbf{r}}\lambda = (\gamma\dot{\mathbf{r}}-\mathbf{F}+2k_{\mathrm{B}}T\mathbf{g}(\mathbf{r})\lambda)^2 + 4k_{\mathrm{B}}T\mathbf{F}\cdot\mathbf{g}\lambda - (2k_{\mathrm{B}}T\mathbf{g}(\mathbf{r})\lambda)^2$$

where we have completed the square by adding the two additional terms on the right-hand side of the equation. These tilted dynamics, with added drift $-2k_{\mathrm{B}}T\mathbf{g}(\mathbf{r})\lambda$ put the generating function into a form where we can use the Feynman–Kac theorem. The resultant equation of motion for Z becomes

$$\partial_t Z(\lambda,\mathbf{r},t) = -\mathcal{L}_\lambda Z(\lambda,\mathbf{r},t)$$

where the tilted operator is

$$\mathcal{L}_\lambda = \left(\frac{\mathbf{F}}{\gamma}-2D\lambda\mathbf{g}\right)\cdot\frac{\partial}{\partial\mathbf{r}} + D\frac{\partial^2}{\partial\mathbf{r}^2} - \frac{\mathbf{F}}{\gamma}\cdot\mathbf{g}\lambda + D\mathbf{g}^2\lambda^2 - f\lambda$$

$$= \frac{\mathbf{F}}{\gamma}\cdot\left(\frac{\partial}{\partial\mathbf{r}}-\lambda\mathbf{g}\right) + D\left(\frac{\partial}{\partial\mathbf{r}}-\lambda\mathbf{g}\right)^2 - f\lambda$$

which is rewritten compactly for $\mathbf{g}$ independent of $\mathbf{r}$. Integrating the expression,

$$Z(\lambda,\mathbf{r},\tau) = e^{-\mathcal{L}_\lambda\tau}Z(\lambda,\mathbf{r},0)$$

which in the long time limit will be dominated by the leading eigenvalue associated with $\mathcal{L}_\lambda$, allowing us to extract $\psi(\lambda) = -\ln Z/\tau$ in the limit $\tau\to\infty$. The eigenvalue problem can be cast in a time-independent manner with the linear equation

$$\mathcal{L}_\lambda\phi_\lambda = \psi(\lambda)\phi_\lambda$$

where ϕ_λ is the dominant eigenvector associated with $\mathcal{L}_\lambda$ whose eigenvalue is $\psi(\lambda)$. One must be a little more careful with a nonequilibrium system, as $\mathcal{L}_\lambda$ is not Hermitian and so is diagonalized by distinct left and right eigenvectors. Strictly, ϕ_λ is the dominate right eigenvector.

The Legendre transform relation between the rate function and the scaled cumulant generating function is a manifestation of ensemble equivalence. The rate function represents a collection of trajectories conditioned on a given time integrated observable, while the cumulant generating function represents an ensemble that obtains the same typical value of the observable under a statistical bias. This is like the relation between the microcanonical ensemble which collects configurations conditioned on the energy, and the canonical one where β biases the likelihood of configurations to affect a specific average value of the energy. Unlike in equilibrium, the parameter λ that tilts the likelihood of trajectories is not in general related to a physical quantity. However, there is a dynamic process that realizes a λ or tilted ensemble of trajectories. This can be understood through a trajectory reweighting principle like in Chapter 8.

The scaled cumulant generating function can be written as an average over an ensemble of trajectories in the presence of a control force Λ. This follows by expressing the generating function as a path integral,

$$
\begin{aligned}
e^{-\tau\psi(\lambda)} &= \int \mathcal{D}[\mathbf{X}(\tau)]P[\mathbf{X}(\tau)]e^{-\lambda O[\mathbf{X}(\tau)]} \\
&= \int \mathcal{D}[\mathbf{X}(\tau)]P_\Lambda[\mathbf{X}(\tau)]e^{-\beta\Delta\Gamma_\Lambda[\mathbf{X}]}e^{-\lambda O[\mathbf{X}(\tau)]} = \left\langle e^{-\beta\Delta\Gamma-\lambda O}\right\rangle_\Lambda
\end{aligned}
$$

where $\beta\Delta\Gamma_\Lambda = \ln P_\Lambda[\mathbf{X}(\tau)]/P[\mathbf{X}(\tau)]$ is the relative stochastic action for trajectories in the absence and presence of Λ. The subscript Λ on the average denotes that it is generated with an equation of motion that includes this additional force. While this equality is true for any force, applying Jensen's inequality

$$
\begin{aligned}
\psi(\lambda) &\leq \lambda \langle O\rangle_\Lambda /\tau + \beta \langle\Delta\Gamma_\Lambda\rangle_\Lambda /\tau \\
&= \min_\Lambda \left[\lambda \langle O\rangle_\Lambda /\tau + \beta \langle\Delta\Gamma_\Lambda\rangle_\Lambda /\tau\right]
\end{aligned}
$$

we find that $\psi(\lambda)$ is bounded by an average of the observable and the average relative stochastic action, or Kullback–Leibler divergence. This relationship can be considered a variational principle for the large deviation function over possible forces that realize rare fluctuations of O. There is an optimal force that saturates this inequality given by $\Lambda^* = 2k_BT\left(\nabla\ln\phi_\lambda - \lambda g\right)$, the dominant eigenvector of the tilted generator. The change to the dynamical generator accompanying the addition of this force is known as the time-independent Doob transform, the analog to the finite time version discussed previously. The addition of the force can be represented compactly as a similarity transform on the generator, $\hat{\mathcal{L}}_\lambda = \phi_\lambda^{-1}\mathcal{L}_\lambda\phi_\lambda - \psi(\lambda)$, restoring $\hat{\mathcal{L}}_\lambda$ as a normalized process. In the context of physical dynamical systems, this Doob transform has been deduced by Spohn, Evans, Jack, Sollich, Chetrite and Touchette.

Exercise 9.12: For freely diffusing overdamped particle,

$$
\gamma\dot{r} = \eta \qquad \langle\eta\rangle = 0 \qquad \langle\eta(0)\eta(t)\rangle = 2k_BT\gamma\delta(t)
$$

show that for tilting on $O = \int dt\,\dot{r}(t)$, the optimal control force is a constant $\Lambda^* = -2k_BT\lambda$ with $\psi(\lambda) = -k_BT\lambda^2/\gamma$ using the variation theorem from earlier.

9.11   Current fluctuations and response

A fundamental distinction between equilibrium and nonequilibrium systems is the presence of persistent currents. There are a number of insights that large deviation theory provides on the structure of current distribution functions, as its integral is a stochastic variable whose extent can be taken arbitrarily large. Consider a time-integrated current, J, or generalized displacement,

$$J = \int_0^\tau dt\, j(t)$$

where $j(t)$ is an instantaneous flux at time t, and τ is the observation time, taken to be larger than any characteristic correlation time of $j(t)$. Explicit currents of interest might include a single particle velocity, or collections of them, or might be the rate of heat transfer to the bath. The fluctuations of J can be characterized by a probability distribution, or alternatively by its characteristic function. The fundamental principle of large deviation theory is that currents that are correlated over a finite amount of time admit an asymptotic, time-intensive form of the logarithm of their distribution function

$$I_E(j) = - \lim_{\tau \to \infty} \frac{1}{\tau} \ln \langle \delta(j\tau - J[\mathbf{X}]) \rangle_E$$

where $I_E(j)$ is a natural function of the time-intensive current, $j = J/\tau$. We will adopt a notation that distinguishes averages in the presence of an external field E that drives the current, where $\langle .. \rangle_E$ denotes an average in the steady state generated by field E, and $\delta(j\tau - J[\mathbf{X}])$ is Dirac's delta function evaluated using a fluctuating current $J[\mathbf{X}]$ that depends on a trajectory $\mathbf{X}$. This asymptotic form implies that deviations away from the mean are exponentially unlikely, with a rate set by $I_E(j)$. In cases where the entropy production is linear in the current, $S/k_\mathrm{B} = \beta E J$, which occurs if the only source of dissipation is from the applied field conjugate to J, then from Gallavotti–Cohen symmetry $I_E(j) - I_E(-j) = -\beta E j$. From large deviation theory, the characteristic function associated with fluctuations of J can be computed from the Laplace transform of $I_E(j)$,

$$\psi_E(\lambda) = - \lim_{\tau \to \infty} \frac{1}{\tau} \ln \left\langle e^{-\lambda J} \right\rangle_E$$

where $\psi_E(\lambda)$ is the scaled cumulant generating function, and depends on the Laplace parameter λ but not τ. Derivatives of $\psi_E(\lambda)$ with respect to λ evaluated at $\lambda = 0$ yield the time intensive cumulants of J. For those cases where Gallavotti–Cohen symmetry holds, Lebowitz–Spohn symmetry follows, $\psi_E(\lambda) = \psi_E(\beta E - \lambda)$.

In equilibrium, where $E = 0$, microscopic reversibility requires that $I_0(j)$ is an even function of j. This means for small fluctuations, the rate function is Gaussian

$$I_0(j) \approx \beta j^2 / 4\chi$$

where χ is the twice the variance of the current. Assuming that the departure from equilibrium is small, Onsager originally conjectured that the log probability of a current fluctuation was given by the entropy production to create it. In what he called a dissipation function, now identifiable as a rate function, the log probability of a current fluctuation was given by $I_E(j) \approx I_0(j) - \beta E j/2$. The corresponding typical current will be $j = \chi E$ which is consistent with phenomenological linear laws relating current to the applied field with a constant of proportionality χ that is observed to be a conductivity. The rate functions in and near equilibrium are illustrated in Figure 9.11.

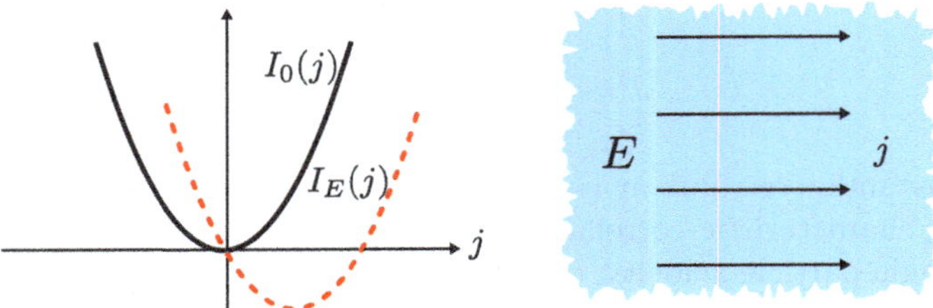

Fig. 9.11 Rate function for current fluctuations in equilibrium or driven out of equilibrium through an applied field E which couples to current j.

In the Gaussian approximation for current fluctuations, the scaled cumulant generating function is obtained as a Legendre transform of the rate function and given by

$$\psi_0(\lambda) = -\lambda^2 \chi/\beta$$

where $\psi_0(\lambda)$ is quadratic in λ. As a scaled cumulant generating function, derivatives of $\psi_0(\lambda)$ with respect to its argument provide the intensive cumulants of J. For example,

$$-\left.\frac{d^2\psi_0(\lambda)}{d\lambda^2}\right|_{\lambda=0} = \frac{1}{\tau}\langle \delta J^2 \rangle_0 = 2\chi/\beta$$

which relates the conductivity with the variance of the current, $\chi = \beta \langle \delta J^2 \rangle_0/2\tau$. In the long time limit this is equivalent to

$$\chi = \beta \int_0^\infty dt \, \langle j(0)j(t) \rangle_0$$

assuming current correlations decay faster than $1/t$. The first form of the fluctuation dissipation relation is known as an Einstein–Hefland moment. This follows from time reversal symmetry, and results in a traditional Green–Kubo expression for the response of a current in terms of an integrated time correlation function. The Gaussian form of $I_0(j)$ is valid only for small j, as is the subsequent linear response relationship that follows. Their utility derives from their thermodynamic origin, in which the entropy production uniquely determines the response. This endows linear response relationships with great generality. They are equally valid independent of the specific dynamics of the system, provided the large deviation form holds. In practice, this requires that correlation times for the current are finite. Nonlinear relationships can be extracted from this framework as well, however their generality is more limited.

Beyond linear response, kinetic properties—not just thermodynamic ones—become important.

Exercise 9.13: Use Lebowitz–Spohn symmetry, $\psi_E(\lambda) = \psi_E(\beta E - \lambda)$, to prove that the mean current to second order in the field strength

$$\frac{\langle J \rangle_E}{\tau} = \frac{\langle \delta J^2 \rangle_0}{2\tau}\beta E + \frac{1}{2\tau}\frac{d\langle \delta J^2 \rangle_0}{d\beta E}(\beta E)^2$$

depends on how the current fluctuations change under an applied field.

It is straightforward to generalize these results to cases where multiple driving fields act in conjunction with multiple resultant fluxes to produce entropy. If we have m different fields, the entropy can be written as

$$S/k_{\mathrm{B}} = \beta \sum_{i=1}^{m} J_i E_i$$

where E_i is the conjugate field for current J_i. These could be macroscopic, for example applied gradients in pressure, chemical potential, and temperature, or microscopic reflective of individual cycles of transitions that a discrete state Markov model could be decomposed into. A large deviation function encoding the statistics of each current could be defined as

$$\psi_E(\boldsymbol{\lambda}) = -\lim_{\tau \to \infty}\frac{1}{\tau}\ln\left\langle \exp\left[-\sum_i \lambda_i J_i\right]\right\rangle_E$$

where $\boldsymbol{\lambda} = \{\lambda_1, \lambda_2 \ldots\}$ are the counting variables for each distinct current. The large deviation function obeys a Lebowitz–Spohn symmetry,

$$\psi_E(\boldsymbol{\lambda}) = \psi_E(\beta\mathbf{E} - \boldsymbol{\lambda})$$

where $\mathbf{E} = \{E_1, E_2 \ldots E_m\}$. The cumulant generating function can be expressed through its Taylor series

$$\tau\psi_E(\boldsymbol{\lambda}) = \sum_{i=1}^{m} \langle J_i \rangle_{\mathbf{E}}\, \lambda_i - \frac{1}{2}\sum_{i,j=1}^{m} \langle \delta J_i \delta J_j \rangle_{\mathbf{E}}\, \lambda_i \lambda_j + \ldots$$

$$= \sum_{i=1}^{m} \langle J_i \rangle_{\mathbf{E}}\,(\beta E_i - \lambda_i) - \frac{1}{2}\sum_{i,j=1}^{m} \langle \delta J_i \delta J_j \rangle_{\mathbf{E}}\,(\beta E_i - \lambda_i)(\beta E_j - \lambda_j) + \ldots$$

where we have truncated the expansion at second order and in the second line used Lebowitz–Spohn symmetry. Taking the λ_i'th derivative of the second expression, we find

$$\tau \frac{d\psi_E(\boldsymbol{\lambda})}{d\lambda_i} = \langle J_i \rangle_{\mathbf{E}}$$

$$= -\langle J_i \rangle_{\mathbf{E}} + \langle \delta J_i \delta J_i \rangle_{\mathbf{E}} \left(\beta E_i - \lambda_i \right) + \frac{1}{2} \sum_{j \neq i} \langle \delta J_i \delta J_j \rangle_{\mathbf{E}} \left(\beta E_j - \lambda_j \right) + \dots$$

where in the first line we have used the definition of the cumulant generating function. Equating the two lines and taking the derivative with respect to E_j, we find

$$\chi_{ij} = \frac{1}{\tau} \frac{d \langle J_i \rangle_{\mathbf{E}}}{dE_j} \bigg|_{\mathbf{E}=0} = \frac{\beta}{4\tau} \left[\langle \delta J_i \delta J_j \rangle_0 + \langle \delta J_j \delta J_i \rangle_0 \right]$$

that the response function for generating current i due to applied field E_j in equilibrium is given by a current–current correlation function. This function is symmetric under exchange of $i \to j$, or $\chi_{ij} = \chi_{ji}$ which is known as Onsager reciprocity. The derivation here is due to Gaspard. These relations show that if the two currents are correlated then applying a force conjugate to one can generate a flux in another. Further, the size of the flux generated by a non-conjugate force is the same in both directions. Concretely, if one drove a mass flux by applying a density gradient, then a heat flux will also be generated. Similarly if one drove a heat flux by applying a temperature gradient, then a mass flux could follow. The reciprocal relations state that the size of the heat flux generated by the density gradient is the same as the mass flux generated by the temperature gradient, a dynamical version of a Maxwell relation.

9.12 Macroscopic fluctuation theory

Evaluating current large deviation functions for interacting systems is difficult. However, there are limiting cases where this can be done by coarse-graining the dynamics of the system. Rather than working with the microscopic variables, if we work with fluctuating fields, there are some significant simplicities that arise. This is akin to the developments in Chapter 4 where evaluating the free energies for interacting models is intractable, but approximate techniques are available for field theories. The relevant field theoretic perspective that one must adopt here is a dynamic one. If we are interested in currents of macroscopic variables, like the fluctuations of the total mass current of a system, fluctuating hydrodynamics provides an amenable description.

Consider for example the equation of motion derived in Chapter 6 for the fluctuating density field. As a conserved quantity, it obeys a continuity equation

$$\frac{\partial}{\partial t} \rho(\mathbf{r}, t) = -\nabla \cdot j(\mathbf{r}, t)$$

where the mass current j is given by

$$j(\mathbf{r}, t) = -D\rho(\mathbf{r}, t)\nabla \beta \mu(\mathbf{r}) - \sqrt{2D\rho(\mathbf{r}, t)}\boldsymbol{\eta}(\mathbf{r}, t)$$

which consists of two parts. The first is a deterministic gradient flow dependent on another field $\beta\mu(\mathbf{r})$, the local chemical potential. This field arose from the forces present in the initial system. The second drift is a random noise, with statistics $\langle \boldsymbol{\eta}(\mathbf{r}, t) \rangle = 0$

and $\langle \boldsymbol{\eta}(\mathbf{r}, t) \otimes \boldsymbol{\eta}(\mathbf{r}', t') \rangle = 1\delta(t - t')\delta(\mathbf{r} - \mathbf{r}')$ that are delta correlated in space and time. The corresponding stochastic path integral for this equation of motion can be constructed using the methods in the previous section. Explicitly this follows as

$$P[\rho, j] = \int d\eta\, p(\eta)\delta\left[j + D\rho\nabla\beta\mu + \sqrt{2D\rho}\eta\right]$$

$$= \int \mathcal{D}[\hat{\rho}] \int d\eta\, p[\eta] \exp\left\{-i \int_{\mathbf{r}} \int_t \hat{\rho}\left[j + D\rho\nabla\beta\mu + \sqrt{2D\rho}\eta\right]\right\}$$

$$\propto \int \mathcal{D}[\hat{\rho}] \exp\left\{-i \int_{\mathbf{r}} \int_t \hat{\rho}\left[j + D\rho\nabla\beta\mu\right] - 2D \int_{\mathbf{r}} \int_t \hat{\rho}^2\rho\right\}$$

$$= \exp\left\{-\int_{\mathbf{r}} \int_t \left[j + D\rho\nabla\beta\mu\right]^2 / 4D\rho\right\}$$

where in the first line we have identified the relevant path probability in the density and current as an integral over the noise, and in the second line we have introduced an auxiliary field, $\hat{\rho}$, in order to represent the delta function. The expression in the third line, which follows by doing the Gaussian integral over the noise, is known as the Martin–Siggia–Rose path integral. In some cases, working directly with the two fields, ρ and $\hat{\rho}$ is a convenient way to proceed. Here we have integrated $\hat{\rho}$ out in the fourth line, resulting in a quadratic action in space and time. This functional, typically prescribed with a phenomenological density-dependent chemical potential field, is known as *macroscopic fluctuation theory.*

We can use a saddle point approximation to work out a corresponding rate function for current fluctuations within macroscopic fluctuation theory. In the steady state limit, the time dependence of the current and density fields vanishes, and the resulting path action is equal to the observation time times an integral over space. In the long time, large deviation limit, we can minimize the action with respect to the fluctuating density field in what is known as the contraction principle of large deviation theory

$$I(j) = \min_{\rho(\mathbf{r})} \int d\mathbf{r}\, \frac{[j + D\rho(\mathbf{r})\nabla\beta\mu(\mathbf{r})]^2}{4D\rho(\mathbf{r})}$$

in order to compute the current rate function $I(j)$. In the absence of gradients in $\beta\mu(\mathbf{r})$ the rate function is simply quadratic in the current, $I(j) = j^2 V/4D\rho$ where V is the volume of the system, reflective of the Gaussian nature of the theory.

Exercise 9.14: Show that for a one-dimensional system in steady-state in contact with boundaries imposing a chemical potential μ_l at $x = 0$ and μ_l at $x = L$ Gallavotti–Cohen symmetry holds

$$I(j) - I(-j) = -\beta j \left(\mu_h - \mu_l\right)$$

where $-\beta j \left(\mu_h - \mu_l\right)$ is identified as the entropy production rate.

More generally, if a system is not translationally invariant because it is in contact with spatially separated reservoirs, the current fluctuations will not be Gaussian.

However, solving for the rate function is not straightforward as the form of the path integral above is not closed because the density and current depend on the local chemical potential field. As we did with Landau theory, we can ignore correlations and make a local approximation to $\beta\mu(\mathbf{r})$. If we assume that the chemical potential follows an ideal density dependence, $\beta\mu = \ln\rho(\mathbf{r})/\rho_0$ where ρ_0 is a constant, then the resulting path action is

$$P[\rho, j] = \exp\left\{-\int_{\mathbf{r}}\int_{t}[j + D\nabla\rho]^2/4D\rho\right\}$$

which is amenable to explicit analysis. In one-dimension, the optimal density profile, $\bar{\rho}(r)$, satisfies a Euler–Lagrange equation

$$\frac{\delta}{\delta\rho(r)}\int_0^L dr\left[j + D\frac{d\rho(r)}{dr}\right]^2\frac{1}{4D\rho(r)} = -\frac{j^2 - D^2[\bar{\rho}'(r)]^2 + 2D^2\bar{\rho}(r)\bar{\rho}''(r)}{4D\bar{\rho}^2(r)}$$

where the primes denote spatial derivative. Setting this expression equal to 0 and rearranging

$$j^2 = -2D^2\bar{\rho}(r)\frac{d^2\bar{\rho}(r)}{dr^2} + \left(D\frac{d\bar{\rho}(r)}{dr}\right)^2$$

we have an explicit expression for the argument of the integral entering into the path action. Making this substitution,

$$I(j) = \frac{1}{2}\int dr\left[j\frac{d\ln\bar{\rho}(r)}{dr} - D\frac{d^2\bar{\rho}(r)}{dr^2} + \frac{D}{\rho(r)}\left(\frac{d\bar{\rho}(r)}{dr}\right)^2\right]$$

where we have used the chain rule to introduce the derivative of $\ln\bar{\rho}$. To proceed any further we need to define some boundary conditions.

Let's imagine that the system is contact with two reservoirs, one located at $r = 0$ and one located at $r = L$. These reservoirs impose a densities $\bar{\rho}(0) = \rho_0$ and $\bar{\rho}(L) = \rho_L$. The first integral can be done with the fundamental theorem of calculus without knowing the explicit form of the density profile,

$$I(j) = \frac{1}{2}j\ln\frac{\rho_L}{\rho_0} + \frac{D}{2}\int_0^L dr\left[\frac{1}{\bar{\rho}(r)}\left(\frac{d\bar{\rho}(r)}{dr}\right)^2 - \frac{d^2\bar{\rho}(r)}{dr^2}\right]$$

where in the limit $\rho_L = \rho_0$ we find that fluctuations in the current result from spontaneously generated curvature in the density profile. Solving the Euler–Lagrange equation using Laplace transforms, we find an optimal profile that is quadratic

$$\bar{\rho}(r) = \rho_0(1 - r/L)^2 + \rho_L(r/L)^2 + (r/L)(1 - r/L)\sqrt{4\rho_0\rho_L + j^2L^2/D^2}$$

which is shown for a few example currents in Figure 9.12. Notice that the profile is invariant under $j \to -j$, reflecting time-reversal symmetry.

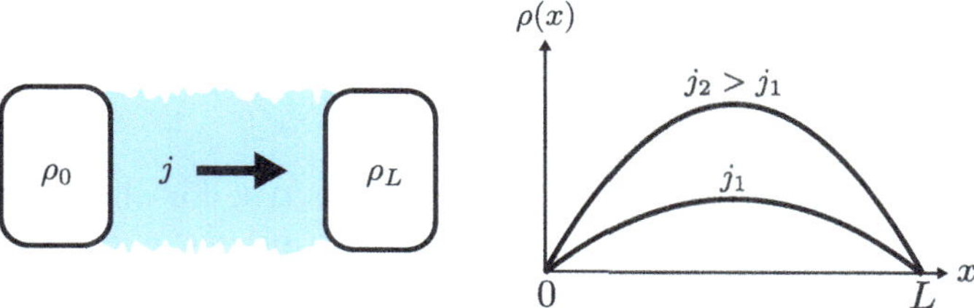

Fig. 9.12 For a $1d$ system connected to two reservoirs at fixed density, the optimal density profiles for given current fluctuations are shown, for two current values and $\rho_0 = \rho_L$.

Inserting the optimal density profile into the expression for the current rate function and performing the integral we arrive at

$$I(j) = \frac{j}{2}\ln\left(\frac{\rho_L}{\rho_0}\right) + \frac{j}{2}\ln\left(\frac{\sqrt{j^2 + K^2} + j}{\sqrt{j^2 + K^2} - j}\right) - \sqrt{j^2 + K^2} + \frac{D(\rho_L + \rho_0)}{L}$$

where $K = D\sqrt{\rho_L \rho_0}/L$, and is manifestly non-Gaussian but obeys Gallavotti–Cohen symmetry. The mean current $\langle j \rangle = D(\rho_L - \rho_0)/L$ and the variance in the limit $\rho_L = \rho_0 = \rho$, $\langle \delta j^2 \rangle = 2D\rho/L$, are easily evaluated from the extrema of $I(j)$ and variations around it. Notice that at large values of j, the tails of the distribution are linear in j. Note that while we have worked with the density field and associated mass currents, the theory is general and can be applied with suitable modifications for the conservation laws and response functions to other currents as well.

9.13   Nonequilibrium response theory

Macroscopic fluctuation theory provides a phenomenological means of understanding current fluctuations arbitrarily away from equilibrium. In order to understand them from a microscopic perspective we need a response theory that works within nonequilibrium steady states. This is gnerally complicated by the fact that we do not have a closed form expression for the steady-state distribution function, or how it deforms under additional applied fields. To alleviate this complication, we will employ ensembles of trajectories and consider how their stochastic actions change upon perturbations. This perspective was first articulated by Maes and closely related to the idea of Mallivin weights introduced by Warren and Allen.

Consider a system undergoing stochastic dynamics, which in the absence of a constant applied force samples a thermal equilibrium. We would like to know how an average changes upon turning on an applied force, E. We can explicitly write down the average of an arbitrary observable, O, in the presence of E, as

$$\langle O \rangle_E = \int \mathcal{D}[\mathbf{X}(\tau)] O[\mathbf{X}(\tau)] P_E[\mathbf{X}(\tau)]$$

$$= \int \mathcal{D}[\mathbf{X}(\tau)] O[\mathbf{X}(\tau)] e^{\beta \Delta \Gamma_E[\mathbf{X}(\tau)]} P_0[\mathbf{X}(\tau)] = \left\langle O[\mathbf{X}(\tau)] e^{\beta \Delta \Gamma_E[\mathbf{X}(\tau)]} \right\rangle_0$$

and relate that to an average in the absence of E where

$$e^{\beta \Delta \Gamma_E[\mathbf{X}(\tau)]} = \frac{P_E[\mathbf{X}(\tau)]}{P_0[\mathbf{X}(\tau)]}$$

is the relative ratio of path probabilities. This relation is true for all values of E, and relates averages within two distinct dynamical ensembles, or ensembles of trajectories. It implies that fluctuations within the equilibrium system contain information about fluctuations arbitrarily far from it. The relation is true for any observable, for example, setting $O = 1$ results in

$$\left\langle e^{\beta \Delta \Gamma_E[\mathbf{X}(\tau)]} \right\rangle_0 = 1$$

a sum rule for the change of measure similar to the fluctuation theorem. Taking $O = \exp[-\lambda \Delta \Gamma_E]$

$$\left\langle e^{-(\lambda \Delta \Gamma_E)} \right\rangle_E = \left\langle e^{(\beta - \lambda) \Delta \Gamma_E} \right\rangle_0$$

we find a relation $\psi_E(\lambda) = \psi_0(\lambda - \beta)$ between the large deviation function for relative action in the driven system and that in the equilibrium system.

The relative ratio of path probabilities in general contains terms that are symmetric and asymmetric in time. Let us separate those two pieces by defining a time-reversed path, $\mathbf{X}^*(\tau)$. Following from the fluctuation theorem, the entropy production is the asymmetric part

$$S_E = (\Delta \Gamma_E[\mathbf{X}(\tau)] - \Delta \Gamma_E[\mathbf{X}^*(\tau)])/2$$

where we expect that $S_E/k_\mathrm{B} = \beta E J[\mathbf{X}(\tau)]/2$ where J is a current associated with the force E. In equilibrium, thermal response theories worked out in Chapter 3, such thermodynamic terms completely determined the response of observables. However, more generally there is additionally a time reversal-symmetric part

$$K_E = (\Delta \Gamma_E[\mathbf{X}(\tau)] + \Delta \Gamma_E[\mathbf{X}^*(\tau)])/2 \,,$$

that is often called the frenesy or traffic. Both S_E and K_E are excesses over the system for $E = 0$. This decomposition is particularly useful when we consider expanding the average of O in a series of higher orders of E. Assuming that the exponential of $\Delta \Gamma_E$ is smooth,

$$\langle O \rangle_E = \langle O \rangle_0 + \beta E \left\langle O \frac{\partial \Delta \Gamma_E}{\partial E} \right\rangle_0 + \frac{\beta E^2}{2} \left\langle O \frac{\partial^2 \Delta \Gamma_E}{\partial E^2} \right\rangle_0$$
$$+ \frac{\beta^2 E^2}{2} \left\langle O \left(\frac{\partial \Delta \Gamma_E}{\partial E} \right)^2 \right\rangle_0 + \mathcal{O}(E^3)$$

we find that the deviation of O from its equilibrium value can be written as a linear response piece, proportional to E, as well as additional higher order terms. If we insert the decomposition of $\Delta \Gamma_E$ into this expression we find,

$$\Delta\langle O\rangle_E = \frac{\beta E}{2}\langle OJ\rangle_0 + \beta E\langle OK'_E\rangle_0 + \frac{\beta E^2}{2}\langle OK''_E\rangle_0 + \frac{\beta^2 E^2}{8}\langle OJ^2\rangle_0$$
$$+ \frac{\beta^2 E^2}{2}\left\langle O(K'_E)^2\right\rangle_0 + \frac{\beta^2 E^2}{2}\langle OJK'_E\rangle_0 + \mathcal{O}(E^3)$$

where the prime on the K_E denotes differentiation with respect to E. At this point, the decomposition seems only to have made the expression more complicated. However, consider averaging a specific observable, like the current J generated by the applied force. In this case, owing to the fact that equilibrium averages must be overall time-reversal symmetric, the expression simplifies to,

$$\langle J\rangle_E = \beta E\langle J^2\rangle_0/2 + \beta^2 E^2\langle J^2 K'_E\rangle_0/2 + \mathcal{O}(E^3)$$

where we find the response of the current to an applied force is proportional to its variance at first order, which is nothing more than a Green–Kubo relationship, and at second order, depends on both the current as well as the frenesy. This expansion looks just like what we found in equilibrium. Higher-order response depends on higher order cumulants. The subtlety here is that the applied force, E, is not conjugate to the current J within the path ensemble. Therefore, generically correlations between the current as well as the frenesy both enter into the response formula.

Second order dynamical response relative to equilibrium is equivalent to linear response about a nonequilibrium steady state. Let's consider an explicit example of the such a nonequilibrium steady-state linear response relation, by considering a tracer particle a convective flow. Such a system can exhibit anomalous phenomena like negative differential mobility, or a decrease in the average velocity in the presence of an applied force. This behavior is forbidden at equilibrium, where the second law requires that currents are aligned with their driving forces. The system is pictured in Figure 9.13, as is a cartoon of its response.

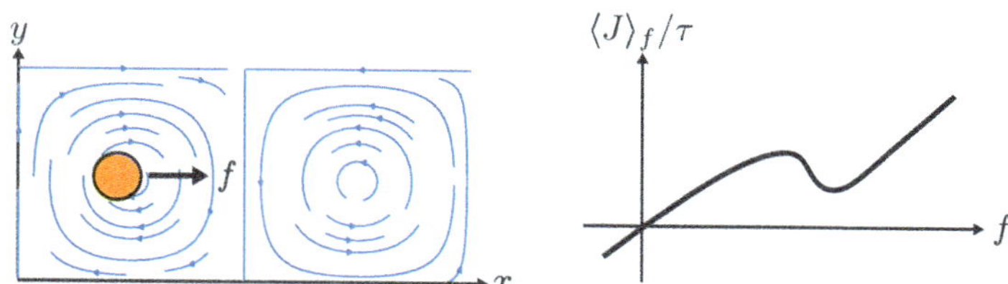

Fig. 9.13 Illustration of a tracer particle moving in a convective flow, and the resultant current-force relationship.

The equation of motion for such a particle is taken to be

$$m\dot{v}_x(t) = -\gamma(v_x - \partial_y\Phi(\mathbf{r})) + f + \eta \qquad \langle\eta\rangle = 0 \qquad \langle\eta(t)\eta(t')\rangle = 2k_BT\gamma\delta(t-t')$$

where $\Phi(\mathbf{r})$ is the stream function, essentially a vector potential generated by the steady state flow profile of the fluid, and f is the constant applied force to the tracer particle. The path action associated with this equation of motion for just x is

$$P_f[\mathbf{X}(\tau)] = \exp\left[-\beta \int_0^\tau dt \, (m\dot{v}_x(t) + \gamma(v_x - \partial_y\Phi(\mathbf{r})) - f)^2 / 4\gamma\right]$$

and the associated ratio of path actions with and without the force is

$$\beta\Delta\Gamma = \beta f \int_0^\tau dt \, (m\dot{v}_x/2\gamma + v_x/2 - \partial_y\Phi(\mathbf{r})/2 - f^2/4\gamma)$$

where we can identify the entropy production,

$$S_f = \int_0^\tau d' \, fv_x(t)/2$$

and frenetic term, dropping the boundary term from $\dot{v}_x$,

$$K_f = -\int_0^\tau dt \, f\partial_y\Phi(\mathbf{r})/2 - f^2/4\gamma$$

where the second term is independent of the specific trajectory.

If we are interested in the integrated current, or in this case the mean displacement,

$$\langle J(\tau)\rangle_f = \int_0^\tau dt \, \langle v_x\rangle = \langle x(\tau) - x(0)\rangle_f$$

then from our expression above, we find to first-order in f

$$\langle J(\tau)\rangle_f = \frac{\beta f}{2} \int_0^\tau dt \int_0^\tau dt' \, \langle v_x(t)(v_x(t') - \partial_y\Phi(\mathbf{r}(t)))\rangle_0 \, .$$

The first term correlates the velocities at two times, while the second term correlates the velocity with the external drift, evaluated in the steady-state with $f = 0$. In the long time limit, we can define the mobility as

$$\lim_{\tau\to\infty} \frac{1}{\tau} \frac{d\langle J(\tau)\rangle_0}{df} \equiv \mu$$

and using the stationarity of equilibrium averages, our expression becomes

$$\mu = \beta \int_0^\infty dt \, \langle v_x(0)v_x(t)\rangle_0 - \beta \int_0^\infty dt \, \langle v_x(0)\partial_y\Phi[\mathbf{r}(t)]\rangle_0$$

$$= \beta D - \beta \int_0^\infty dt \, \langle v_x(0)\partial_y\Phi[\mathbf{r}(t)]\rangle_0$$

which clarifies the linear response around a nonequilibrium is distinct from that found in equilibrium. In equilibrium the Einstein relation $\beta D = \mu$ relates the mobility to the diffusivity by $k_{\mathrm{B}}T$. This relation guarantees that $\mu > 0$. However, out of equilibrium the Einstein relation break down, and the mobility includes a correlation between the velocity and the drift given by the streaming function. In cases where the flow pattern supports convective cells, these correlations can be large enough to drive $\mu < 0$. This relationship for the mobility is closely related to the Speck–Seifert formulation of nonequilibrium response developed from a Fokker–Planck equation formalism.

Further reading

A discussion of master equations can be found in Robert Zwanzig's *Nonequilibrium statistical mechanics* and Nico van Kampen's *Stochastic processes in physics and chemistry*. Large deviation theory has been carefully reviewed by Hugo Touchette in *The large deviation approach to statistical mechanics*. The underpinning of stochastic thermodynamics and are discussed in Gavin Crooks' thesis, *Excursions in statistical dynamics*. Other material on stochastic thermodynamics can be found in Udo Seifert's early review *Stochastic thermodynamics: principles and perspectives* in the European Physical Journal B 2008. Luca Peliti and Simone Pigolotti have written *Stochastic thermodynamics: an introduction book* that contains related material. Macroscopic fluctuation theory is reviewed in the literature by Bertini, De Sole, Gabrielli, Jona-Lasinio, and Landim, *Macroscopic fluctuation theory*, Reviews of Modern Physics 2015.

Additional exercises

Exercise 9.15: Here, we will consider a discrete two-level system coupled to two different thermal reservoirs with which it can exchange mass and energy. Specifically, consider a quantum dot in contact with two electrodes from which electrons can jump off of or onto. Imagine the dot has only one accessible energy level, ϵ, and thus admits only two internal states, 1 and 0, corresponding to an electron occupying the energy level or not. The two electrodes are at fixed temperatures $T^{(1)}$ and $T^{(2)}$ and fixed electronic chemical potentials $\mu^{(1)}$ and $\mu^{(2)}$. The rates to hop on or off of the dot from the $\nu = 1, 2$ electrode are

$$k_{01}^{(\nu)} = \gamma F_\nu(\epsilon) \qquad k_{10}^{(\nu)} = \gamma \left[1 - F_\nu(\epsilon)\right]$$

where γ is the coupling constant to the electrode and $F_\nu(\epsilon)$

$$F_\nu(\epsilon) = \frac{1}{e^{(\epsilon - \mu^{(\nu)})/k_{\mathrm{B}} T^{(\nu)}} + 1}$$

is the Fermi distribution that describes the energy levels accessible within the ν electrode. The corresponding master equation is

$$\frac{dp_1}{dt} = (k_{01}^{(1)} + k_{01}^{(2)})(1 - p_1) - (k_{10}^{(1)} + k_{10}^{(2)})p_1$$

where p_1 is the probability of an electron occupying dot. From normalization, $p_1 + p_0 = 1$, allowing us to eliminate p_0.

1. Determine the stationary distributions, $\bar{p}_0$ and $\bar{p}_1$.
2. From the stationary distribution and the equations in the main text, show that the rate of entropy production is equal to

$$\dot{S}/k_{\mathrm{B}} = \frac{\gamma}{2} \left[F_1(\epsilon) - F_2(\epsilon)\right] \left[(\epsilon/T^{(2)} - \mu^{(2)}/T^{(2)}) - (\epsilon/T^{(1)} - \mu^{(1)}/T^{(1)})\right]$$

and identify the relevant current J and conjugate force F.

3. In this system, the electron is responsible for transferring both energy and mass. This is known as the so-called strong-coupling limit between two fluxes. While one condition of thermal equilibrium would correspond to $T^{(2)} = T^{(1)}$ and $\mu^{(2)} = \mu^{(1)}$, identify an alternative condition that would also satisfy detailed balance, or admit an equilibrium steady state.

4. The rate of chemical work done by the system is $\dot{\mathcal{W}}_N = \left(\mu^{(2)} - \mu^{(1)} \right) J_N$ where J_N is the average mass flux from electrode 1 into electrode 2. If we take electrode 1 to be the hot reservoir, with heat flux out of it $\dot{\mathcal{Q}}_1$, the efficiency of this process can be defined as

$$\zeta = -\dot{\mathcal{W}}_N / \dot{\mathcal{Q}}_1 \, .$$

Show that in the steady state this efficiency is

$$\zeta = 1 - (1 - \zeta_c) \frac{\left(\epsilon/T^{(2)} - \mu^{(2)}/T^{(2)} \right)}{\left(\epsilon/T^{(1)} - \mu^{(1)}/T^{(1)} \right)}$$

where ζ_c is the Carnot efficiency.

5. Under what conditions is the Carnot efficiency recovered?

Exercise 9.16: To illustrate the fluctuation theorems we have developed throughout the chapter, we will perform an exact calculation whose symmetry we can interrogate. Specifically, we will consider an overdamped particle moving in a harmonic potential. In order to drive this system out of equilibrium, we will imagine that we can change the minimum of the harmonic potential with some arbitrary protocol we will denote as $\Lambda(t)$. The potential that the particle feels is,

$$U(t) = \frac{k}{2} \left[x - \Lambda(t) \right]^2$$

where x denotes the position of the particle, and k is the spring constant. For simplicity, we will consider an equation of motion for an overdamped, or Brownian particle,

$$\gamma \dot{x} = -\partial_x U + \eta$$

where γ is friction from the solvent, and η is random noise that depends on temperature T, and γ, in such a way as to obey the fluctuation–dissipation relation.

1. Using the definition of the work, $\mathcal{W}$, from earlier in the chapter, determine an expression for $\mathcal{W}$ in terms of an integral over $x(t)$. Identify analogous expressions for the total change in energy ΔU and the heat $\mathcal{Q}$.

2. To solve for the work as a function of observation time, we need to solve the equations of motion of the particle under the protocol. Here, there is a reference frame that simplifies the calculation significantly. Specifically, we can write the particles' position as a sum of two terms,

$$x(t) = y(t) + \Delta x(t)$$

where $y(t)$ moves with the minimum of the potential, and $\Delta x(t)$ is the difference between that co-moving frame and the actual position. Determine equations of motion for $y(t)$ and $\Delta x(t)$ and show that their integrated form satisfies

$$y(t) = \Lambda(t) - \int_0^t dt' e^{-(t-t')/\tau} \dot{\Lambda}(t') \qquad \Delta x(t) = 1/\gamma \int_0^t dt' e^{-(t-t')/\tau} \eta(t')$$

provided $y(0) = \Lambda(0) = x(0) = 0$, and $\tau = \gamma/k$.

3. Since the mean noise is zero and uncorrelated with the position, $\langle \Delta x(t) \rangle$ can be evaluated. Use this to show that the average work is given by

$$\langle \mathcal{W}(t) \rangle = k \int_0^t dt' \int_0^{t'} dt'' \dot{\Lambda}(t') \dot{\Lambda}(t'') e^{-(t'-t'')/\tau}$$

where the mean work can be integrated provided a form of $\Lambda(t)$.

4. Compute the variance of the work and show explicitly

$$\langle \delta \mathcal{W}(t)^2 \rangle = 2k_{\mathrm{B}}T \langle \mathcal{W}(t) \rangle$$

namely that it is given by the average work up to a factor of $2k_{\mathrm{B}}T$.

5. Acknowledging the Gaussian statistics of the process, and the relationship between the mean and variance, show that

$$\frac{P(\mathcal{W})}{P(-\mathcal{W})} = e^{\beta \mathcal{W}}$$

or the Crooks work fluctuation theorem holds.

Exercise 9.17: For dynamics in a continuous space there are only a few exactly solvable problems for which the generalized eigenvalue equation for the large deviation function can be determined. In this exercise, we will consider two of these and in so doing discuss in more detail the strategies for solving them.

1. Consider the displacement fluctuations for a particle undergoing free diffusion in the overdamped limit. The tilted operator satisfies

$$\mathcal{L}_\lambda = D \left(\frac{d}{dx} + \lambda \right)^2$$

where D is the diffusion constant, x the particle's position and λ the counting parameter. Diagonalize this operator using a Fourier basis. Show that the dominate real eigenvalue is $\psi = -D\lambda^2$ using the Perron–Frobenius theorem that guarantees the dominate eigenvalue is a real number.

2. Next take an Ornstein–Uhlenbeck process, where a particle moves in a harmonic potential. Let's consider the large deviations associated with the time averaged particle position. The associated tilted operator is

$$\mathcal{L}_\lambda = -\beta Dkx \frac{d}{dx} + D \frac{d^2}{dx^2} - \lambda x$$

where k is the spring constant and $\beta = 1/k_{\mathrm{B}}T$. Verify that the dominate right eigenvector is

$$\phi_\lambda(x) = e^{-\lambda x/\beta Dk - 3\lambda^2/2\beta^3 D^2 k^3}$$

and determine its associated eigenvalue.

Exercise 9.18: Here we will work out a simple model for a motor protein walking on a microtubial, which we will model as an asymmetric random walk. In particular, you will work out the large deviation rate function for entropy production fluctuations in the steady-state. Consider a one-dimensional lattice, on which a particle can hop to the right with rate k_r and to the left with rate k_l. Each site is identical and there are an infinite number of sites. The hopping process obeys a Markovian master equation, so that each hop is uncorrelated with the previous one.

1. Denoting the number of hops to the right as n_r and the number of hops to the left as n_l, show that joint probability $P(n_r, n_l, \tau)$ for observing the system over an observation time τ is

$$P(n_l, n_r, \tau) = \frac{M!}{(M - n_l)!\, n_l!} k_l^{n_l} k_r^{n_r} \frac{t^M}{M!} e^{-(k_l + k_r)\tau}$$

   where $M = n_l + n_r$

2. Provided the ratio of rates satisfies a local detailed balance, $\ln k_r/k_l = \beta f$, where f is related to the energy expended to direct the motor forward, show that the average entropy production can be written as $\langle \Delta S \rangle / \tau = k_{\mathrm{B}} k_r (1 - e^{-\beta f})\beta f$.

3. Assuming that $n_r, n_l \gg 1$, show that the rate function $I(n_l, n_r) = -\ln P(n_l, n_r, \tau)/\tau$ can be written as

$$I(n_l, n_r) = -k_l \left[-j_l \ln j_l + j_l - j_r e^{\beta f} \ln j_r + j_r e^{\beta f} - 1 - e^{\beta f} \right]$$

   where $j_\alpha = n_\alpha/k_\alpha \tau$ is the $\alpha = \{r, l\}$ component to the current.

4. For a fixed rate of entropy production, currents j_r and j_l are not independent. Determine a relationship between j_r, j_l and $\Delta S/\tau$ in terms of β, f, k_r and fundamental constants.

5. We can use a saddle point approximation to determine the rate function. To proceed, find the j_l that makes a given rate of entropy production most likely.

6. Eliminating j_l with its optimal value, show that the rate function for entropy can be written as

$$I(\bar{s}) = -k_l e^{\beta f} \left[2\cosh(\beta f/2) - \bar{s}\sinh(\beta f/2)\ln h(\bar{s}) + g(\bar{s}) \right]$$

   where

$$h(\bar{s}) = e^{-\beta f} \left[1 + \bar{s}^2 \cosh(\beta f) - \bar{s}^2 + \bar{s}g(\bar{s})\sinh(\beta f/2) \right]$$

   and

$$g(\bar{s}) = \sqrt{4 - 2\bar{s}^2 + 2\bar{s}^2 \cosh(\beta f)}$$

   and $\bar{s} = \Delta S/\langle \Delta S \rangle$.

7. Show that $I(\bar{s})$ satisfies Gallavotti–Cohen symmetry.

8. Make a plot of $I(\bar{s})$ for various values of βf, using $k_l = 1$ to set the fundamental unit of time. Comment on its form for negative values of $\bar{s}$.

10

Molecular dynamics simulations

Molecular dynamics simulations are a computational tool used to study the behavior of complex systems at the atomic and molecular levels. Unlike the Monte Carlo algorithms presented previously in Chapter 5, molecular dynamics involve the solution of equations of motion for a set of interacting particles, and thus provides direct access to kinetic and nonequilibrium phenomena. For an autonomously evolving system complex enough for the ergodic hypothesis to be valid, simulating the dynamics over time can simultaneously provide insights into the structure and thermodynamics of a wide range of systems. The basic idea of molecular dynamics is to approximate an equation of motion by discretizing it, so that time is updated in discrete steps. When the discretization is chosen appropriately, the emergent behavior of many interacting particles can be viewed within a statistical mechanics framework. In the same spirit as the chapter on Monte Carlo, in this chapter the standard algorithms employed in the study of molecular dynamics are reviewed and their limitations explored. Particular emphasis is placed on the properties of specific approximations to the equations of motion. Practical aspects associated with efficient simulation, as well as a few more advanced sampling algorithms that can be employed with molecular dynamics simulations are discussed. We will only consider classical dynamics, as algorithms to effectively simulate quantum dynamics employ similar principles, but are more complicated to implement and more limited in scope.

10.1 Discretizing Hamiltonian mechanics

Our discussion of approximating classical equations of motion will begin by reconsidering how we initially formulated the dynamics of a complex system. We have previously introduced the Liouvillian within the context of classical mechanics, as a function that acts on phase space, the $6N$ dimensional vector of positions and momentum, $\mathbf{x} = \{\mathbf{r}^N, \mathbf{p}^N\}$. We saw that the Liouville operator underpinned statistical mechanics, as its existence was a sufficient condition for the stationarity of the Boltzmann distribution and all of the results that follow from it. As a reminder, the Liouville operator $\mathcal{L}$ generates displacements in phase space, for example,

$$\mathbf{x}(t) = e^{\mathcal{L}t}\mathbf{x}(0) \quad \text{or} \quad A[\mathbf{x}(t)] = e^{\mathcal{L}t}A[\mathbf{x}(0)]$$

for an arbitrary observable A. For a classical system, the Liouvillian is composed of two parts,

$$\mathcal{L} = \mathcal{L}_r + \mathcal{L}_p$$

a piece that acts on the positions,

$$\mathcal{L}_r = \sum_i \dot{\mathbf{r}}_i \cdot \frac{\partial}{\partial \mathbf{r}_i}$$

and a piece that acts on the momentum,

$$\mathcal{L}_p = \sum_i \dot{\mathbf{p}}_i \cdot \frac{\partial}{\partial \mathbf{p}_i}.$$

where both $\dot{\mathbf{r}}_i$ and $\dot{\mathbf{p}}_i$ are evolved with a Hamiltonian. The positional part acts on a function, $g(r,p)$, to advance the positions by a time t,

$$e^{t\dot{r}\frac{d}{dr}} g(r,p) = g(r + \dot{r}t, p)$$

and similarly,

$$e^{t\dot{p}\frac{d}{dp}} g(r,p) = g(r, p + \dot{p}t)$$

the momentum part advances the momentum to a time t. However, the amount the positions are advanced depends on $\dot{r}$ which is $\mathbf{p}/m$, and similarly, the amount the momentum is advanced depends on $\dot{p}$ which is the force $\mathbf{F}(\mathbf{r})$. Therefore, these operators do not commute, and subsequently,

$$e^{\mathcal{L}} \neq e^{\mathcal{L}_r} e^{\mathcal{L}_p} .$$

This is the common situation for linear differential operators. In general this means we need to simultaneously solve the equations of motion for positions and momenta.

For a very small time displacement, the Liouvillian can be factorized

$$e^{\mathcal{L}t} = \lim_{M \to \infty} \left[e^{\mathcal{L}_p \Delta t/2} e^{\mathcal{L}_r \Delta t} e^{\mathcal{L}_p \Delta t/2} \right]^M$$

which is known as a *Strang splitting* with $\Delta t = t/M$. We have seen this in the quantum case, in our development of path integral theories in Chapter 3. This factorization is not unique, as different splittings are possible. This discretization over each Δt,

$$e^{\mathcal{L}\Delta t} \approx e^{\mathcal{L}_p \Delta t/2} e^{\mathcal{L}_r \Delta t} e^{\mathcal{L}_p \Delta t/2}$$

can be interpreted as a sequential update rule,

$$\mathbf{p}(t + \Delta t/2) = \mathbf{p}(t) + \dot{\mathbf{p}}(t)\Delta t/2$$
$$\mathbf{r}(t + \Delta t) = \mathbf{r}(t) + \dot{\mathbf{r}}(t + \Delta t/2)\Delta t$$
$$\mathbf{p}(t + \Delta t) = \mathbf{p}(t + \Delta t/2) + \dot{\mathbf{p}}(t + \Delta t)\Delta t/2$$

which is amenable to implementation on a computer where representations of functions are discrete. Thus by considering a splitting of the Liouvillian over small time increments, which can subsequently be iterated, we have in effect derived a means of evaluating solutions of Hamilton's equations of motion numerically. We will see that this strategy of approximating Liouville's equation endows the resultant numerical algorithm with a number of desirable properties, inherited from the relationships found back in Chapter 6. This particular update rule is known as the *velocity Verlet algorithm*.

Exercise 10.1: Show that the velocity Verlet algorithm is time-reversal symmetric. Specifically, write down the update rule associated with a forward timestep, Δt, and it for a backward timestep, $-\Delta t$, and show that these are equivalent. What does this imply about the conservation of energy?

A representation of Verlet's method for solving the equations of motion of many interacting degrees of freedom is shown in Algorithm 7. In it the positions, velocities, and forces are stored at each timestep. Functions that initialize the positions and velocities are indicated, as is a function that evaluates the force, *Force*. Note that only one force evaluation is needed per iteration, because we have chosen to evolve r over the full time step.

Algorithm 7 Velocity Verlet algorithm

1: $\mathbf{r} = \text{Initialize Lattice}()$
2: $\mathbf{v} = \text{Initialize Velocities}()$
3: $\mathbf{F} = \text{Force}(\mathbf{r})$ ▷ Compute forces
4: **while** $t \leq \tau$ **do**
5:      $\mathbf{v} = \mathbf{v} + \mathbf{F}\Delta t/2m$ ▷ Update velocities
6:      $\mathbf{r} = \mathbf{r} + \mathbf{v}\Delta t$ ▷ Update positions
7:      $\mathbf{F} = \text{Force}(\mathbf{r})$ ▷ Compute forces
8:      $\mathbf{v} = \mathbf{v} + \mathbf{F}\Delta t/2m$ ▷ Update velocities
9:      $t = t + \Delta t$ ▷ Update time
10: **end while**

In order to propagate the set of N particle positions $\mathbf{r}^N$ and velocities $\mathbf{v}^N$, we need an initial condition. As before in the Monte Carlo sampling of hard disks, a convenient initial position for structureless particles is to place them on a regular lattice. The function *Initialize Lattice* is envisioned as doing just that. But how ought we initialize the velocities? One way to do so is to appeal to our notion of temperature. Within a thermal equilibrium, we expect that on average the kinetic energy is related to the temperature through $T = m \langle v^2 \rangle /k_{\mathrm{B}}$ where m is the mass of the particle. It seems sensible to define an instantaneous measure of the temperature as

$$T(t) = \frac{m}{Ndk_{\mathrm{B}}} \sum_{i=1} |\mathbf{v}_i(t)|^2$$

where d is the dimensionality. Provided a set of velocities, and a target temperature, we can rescale the velocities to ensure that they coincide. However, in thermal equilibrium the velocities are Gaussian distributed. In order to transform uniform random numbers that are easy to generate into Gaussian random numbers, we need the so called Box–Muller transform. If U_1 and U_2 are two independent uniform random numbers on the interval $[0, 1]$, then

$$R_1 = \sqrt{-2\ln U_1}\cos(2\pi U_2) \qquad R_2 = \sqrt{-2\ln U_1}\sin(2\pi U_2)$$

provide two independent Gaussian random numbers R_1 and R_2 with mean 0 and unit variance. While an infinite sampling of these Gaussian random numbers would ensure that there is no net center of mass velocity in the system, for a finite N a net momentum is possible. Therefore, a full description for initializing the velocities requires the generation of Gaussian random numbers, the subtraction of any net velocity, and finally a rescaling to a desired temperature. These steps are illustrated in pseudocode in Algorithm 8.

Algorithm 8 Initialize velocities at temperature T_{target}

1:   $\mathbf{v}_{\text{com}} = 0$ ▷ Initialize center of mass velocity
2:   $T = 0$ ▷ Initialize temperature
3:   **for** $n = 1,\ n \le N/2$ **do**
4:      $\mathbf{U}_1 = \text{Rand}()$ ▷ Draw uniform random variable
5:      $\mathbf{U}_2 = \text{Rand}()$
6:      $\mathbf{R}_1 = \sqrt{-2\ln\mathbf{U}_1}\cos(2\pi\mathbf{U}_2)$ ▷ Box Muller transform
7:      $\mathbf{R}_2 = \sqrt{-2\ln\mathbf{U}_1}\sin(2\pi\mathbf{U}_2)$
8:      $\mathbf{v}_{2n-1} = \mathbf{R}_1$ ▷ Initialize particle $2n-1$'s velocity
9:      $\mathbf{v}_{2n} = \mathbf{R}_2$ ▷ Initialize particle $2n$'s velocity
10:     $\mathbf{v}_{\text{com}} = \mathbf{v}_{\text{com}} + \mathbf{v}_{2n-1}/N + \mathbf{v}_{2n}/N$
11:     $T = T + m\mathbf{v}_{2n-1}^2/2dk_{\text{B}}N + m\mathbf{v}_{2n}^2/2dk_{\text{B}}N$
12:   **end for**
13:   $\alpha = \sqrt{T_{\text{target}}/T}$ ▷ Determine scale factor for target temperature
14:   **for** $n = 1,\ n \le N$ **do**
15:     $\mathbf{v}_n = (\mathbf{v}_n - \mathbf{v}_{\text{com}}) \times \alpha$ ▷ Shift and rescale velocity
16:   **end for**

10.2   Verlet's algorithm

How does the Verlet algorithm compare to something more straightforward? Beginning with Newton's equations of motion,

$$\dot{\mathbf{r}} = \mathbf{v} \qquad m\dot{\mathbf{v}} = \mathbf{F}(\mathbf{r})$$

we could Taylor expand the change in position and velocity for small Δt. Expanding up to second order in the timestep, for the position we get,

$$\mathbf{r}(t + \Delta t) \approx \mathbf{r}(t) + \dot{\mathbf{r}}(t)\Delta t + \ddot{\mathbf{r}}(t)\Delta t^2/2 + \mathcal{O}(\Delta t^3)$$
$$= \mathbf{r}(t) + \mathbf{v}(t)\Delta t + \mathbf{F}(t)\Delta t^2/2m + \mathcal{O}(\Delta t^3)$$

where we have identified the time derivatives as the velocity and force. Expanding the velocity, also to second order,

$$\mathbf{v}(t + \Delta t) \approx \mathbf{v}(t) + \dot{\mathbf{v}}(t)\Delta t + \ddot{\mathbf{v}}(t)\Delta t^2/2 + \mathcal{O}(\Delta t^3)$$
$$= \mathbf{v}(t) + \mathbf{F}(t)\Delta t/m + \dot{\mathbf{F}}(t)\Delta t^2/2m + \mathcal{O}(\Delta t^3)$$

where for this truncation we need a way of computing the time derivative of the force. Taylor expanding it,

$$\mathbf{F}(t + \Delta t) \approx \mathbf{F}(t) + \dot{\mathbf{F}}(t)\Delta t$$
$$\dot{\mathbf{F}}(t) = [\mathbf{F}(t + \Delta t) - \mathbf{F}(t)]/\Delta t$$

where in the second line we have rearranged the expansion and found a finite difference formula. Substituting this approximation to the force back into the velocity expansion,

$$\mathbf{r}(t + \Delta t) = \mathbf{r}(t) + \mathbf{v}(t)\Delta t + \mathbf{F}(t)\Delta t^2/2m$$
$$\mathbf{v}(t + \Delta t) = \mathbf{v}(t) + [\mathbf{F}(t + \Delta t) + \mathbf{F}(t)]\Delta t/2m$$

we have an equation that is correct to second order in Δt. These equations are known as the *Leap Frog algorithm*. Rearranging them

$$\mathbf{v}(t + \Delta t/2) = \mathbf{v}(t) + \mathbf{F}(t)\Delta t/2m$$
$$\mathbf{r}(t + \Delta t) = \mathbf{r}(t) + \mathbf{v}(t + \Delta t/2)\Delta t$$
$$\mathbf{v}(t + \Delta t) = \mathbf{v}(t + \Delta t/2) + \mathbf{F}(t + \Delta t)\Delta t/2m$$

we arrive at the velocity Verlet algorithm, precisely the update rule we derived by the Strang splitting. When Verlet wrote down this algorithm in 1967, it was admittedly by a method of Taylor expansions. However, noting that it also follows from an approximation to the Liouvillian, we can understand a number of desirable properties it has that we will expand upon below.

The Verlet algorithm is a second-order expansion, but the local error in the positions is $\mathcal{O}(\Delta t^4)$. This follows from rewriting the Leap Frog version for two adjacent steps,

$$\mathbf{r}(t + \Delta t) = \mathbf{r}(t) + \mathbf{v}(t)\Delta t + \mathbf{F}(t)\Delta t^2/2m + \dot{\mathbf{F}}(t)\Delta t^3/6m$$
$$\mathbf{r}(t - \Delta t) = \mathbf{r}(t) - \mathbf{v}(t)\Delta t + \mathbf{F}(t)\Delta t^2/2m - \dot{\mathbf{F}}(t)\Delta t^3/6m$$

Summing them together,

$$\mathbf{r}(t + \Delta t) = 2\mathbf{r}(t) - \mathbf{r}(t - \Delta t) + \mathbf{F}(t)\Delta t^2/m + \mathcal{O}(\Delta t^4)$$

we see that local time reversibility cancels the third-order term. Globally however, the algorithm is accurate to second order, which follows from

$$\text{Error}[\mathbf{r}(t + n\Delta t)] = \frac{n(n + 1)}{2}\mathcal{O}(\Delta t^4)$$
$$\text{Error}[\mathbf{r}(t + \tau)] = \left(\frac{\tau^2}{2\Delta t^2} + \frac{\tau}{2\Delta t}\right)\mathcal{O}(\Delta t^4) \sim \mathcal{O}(\Delta t^2)$$

where the second line can be proved by induction, carrying the local error.

The time-reversal symmetry of the algorithm results in a conserved quantity. The conserved quantity associated with the full Liouvilian, $\mathcal{L}$, is the total energy (i.e., $\mathcal{L}\mathcal{H} =$

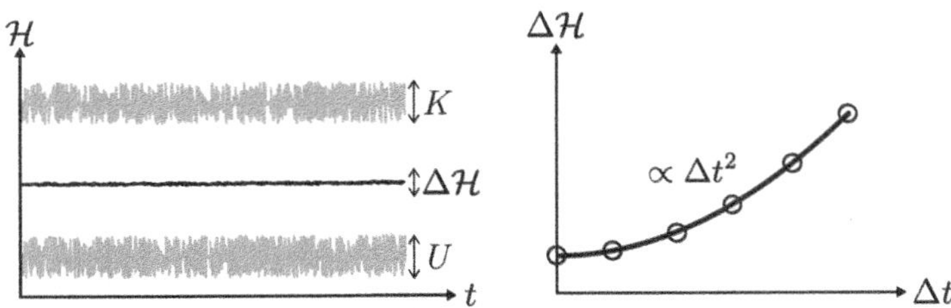

Fig. 10.1 Illustration of the conservation of the total energy as the sum of the kinetic (top) and potential (bottom) energies, and its timestep dependence for the Verlet algorithm.

0). Similarly, if $\mathcal{L}_s$ is the discrete Liouvillian, there is a so-called *shadow Hamiltonian,* $\mathcal{H}_s$ that is conserved under its action (i.e., $\mathcal{L}_s\mathcal{H}_s = 0$). For finite Δt, the energy is not conserved, but it is close to the shadow Hamiltonian,

$$|\mathcal{H}_s - \mathcal{H}| \sim \mathcal{O}(\Delta t^2)$$

where the closeness is of order Δt^2. As illustrated in Figure 10.1, the individual kinetic and potential energy contributions will fluctuate, but the fluctuations of their total go to 0 as $\Delta t^2 \to 0$. Indeed the establishment of this scaling of the fluctuations of the energy about its conserved shadow quantity is a reasonable way to choose Δt. While generally the explicit evaluation of the shadow Hamiltonian is difficult, for a linear system of the form

$$\dot{r} = v \qquad m\dot{v} = -kr$$

the conserved quantity of the velocity Verlet algorithm is

$$\mathcal{H}_s = \frac{m}{2}\left(1 - k\Delta t^2/4m\right)v^2 + \frac{k}{2}r^2$$

which converges to $\mathcal{H}$ in the small timestep limit.

Exercise 10.2: Confirm that that the shadow Hamiltonian $\mathcal{H}_s$ for the harmonic oscillator is a constant of motion within the Verlet algorithm.

While the exact energy is not conserved, the total momentum,

$$P(t) = \sum_{i=1}^{N} m_i v_i(t)$$

is. This can be shown simply by writing down the momentum $\Delta t/2$ from t,

$$P(t + \Delta t/2) = \sum_{i=1}^{N} m_i\left[v_i(t) + \frac{\Delta t}{2m_i}F_i(t)\right]$$

$$= P(t) + \frac{\Delta t}{2}\sum_{i=1}^{N}F_i(t) = P(t)$$

where the third line follows if $\sum_i F_i(t) = 0$. There is a more sophisticated view of conservation of momentum, that results from the existence of translational symmetry. If there is translational symmetry, then,

$$U(\mathbf{r}^N) = U(\mathbf{r}^N + \Delta r)$$

where U is the potential energy and the equality is true for any Δr. Taking a limit of small Δr,

$$U(\mathbf{r}^N) = U(\mathbf{r}^N) + \Delta \mathbf{r} \sum_i \nabla_i U(\mathbf{r}^N)$$

where the above equality only holds if $\sum_i \nabla_i U(\mathbf{r}^N) = -\sum_i \mathbf{F}_i(t) = 0$. Again a symmetry produces a conserved quantity. While the Verlet algorithm also preserves angular momentum, the existence of a cubic simulation domain breaks rotational symmetry, causing the angular momentum to be not conserved.

Perhaps most importantly, the Verlet algorithm preserves the norm of phase space. In the continuum time limit, this is a property that for Hamiltonian dynamics results in Liouville's theorem. One way to state this is the conservation of probability. Namely for an initial region in phase space, R_0, and its time-evolved region R_t, conservation of probability can be written

$$\int_{R_0} d\mathbf{x}(0) f[\mathbf{x}(0)] = \int_{R_t} d\mathbf{x}(t) f[\mathbf{x}(t)]$$

which has to be true for all t. For small dt,

$$\mathbf{x}(dt) = \mathbf{x}(0) + \dot{\mathbf{x}}(0)dt$$

defining a change of variables

$$\mathbf{x}' \equiv \mathbf{x}(0) + \dot{\mathbf{x}}(0)dt$$

and inserting into the time evolved side of the equality,

$$\int_{R_{dt}} d\mathbf{x}' f[\mathbf{x}'] = \int_{R_0} d\mathbf{x} \left| \frac{d\mathbf{x}'}{d\mathbf{x}} \right| f[\mathbf{x}]$$

where in doing this change of variables we have to insert a Jacobian for the transformation. In order for the above to be true, that factor must be identically 1, or $\det(J) = 1$ where

$$J = \begin{pmatrix} \frac{dx_i'}{dx_i} & \frac{dx_i'}{dx_j} & \cdots \\ \vdots & \frac{dx_j'}{dx_j} & \cdots \\ \vdots & \vdots & \ddots \end{pmatrix} = \begin{pmatrix} \frac{dr_i(t+\Delta t)}{dr_i(t)} & \frac{dr_i(t+\Delta t)}{dv_i(t)} & \cdots \\ \vdots & \frac{dv_i(t+\Delta t)}{dv_i(t)} & \cdots \\ \vdots & \vdots & \ddots \end{pmatrix}$$

which for the Verlet algorithm is indeed unity. Time reversibility and the fact that it preserves norm, or is *symplectic*, are necessary conditions for the Verlet algorithm to accurately evolve a thermal distribution over long times.

Exercise 10.3: Demonstrate that the velocity Verlet algorithm is norm preserving by explicitly evaluating the Jacobian of the transformation and demonstrating that it is 1.

10.3 Atomic and molecular models

The Verlet algorithm provides a means to evolve an interacting system of particles. However, to evaluate its update rules, we need a description of the forces between atoms and molecules. A very common assumption is to posit that the potential between two atoms is pairwise additive,

$$U(\mathbf{r}^N) = \frac{1}{2} \sum_{i \neq j} u_2(|\mathbf{r}_i - \mathbf{r}_j|)$$

where u_2 is a pair potential that depends only on the distance between two particle centers. While fundamental particles (neutrons, electrons, protons, etc.) only interact in a pairwise manner, composite bodies such as atoms and molecules do not. In general, the total interaction potential can be written as a sum of many body interaction potentials (two-body, three-body, etc.). Often the sum converges rapidly, and only two-body interactions contribute. Provided a pair potential, the force acting on particle i is given by,

$$\mathbf{F}_i = - \sum_{j \neq i} u_2'(|\mathbf{r}_i - \mathbf{r}_j|) \cdot \hat{\mathbf{r}}_{ij}$$

where $\hat{\mathbf{r}}_{ij}$ is the unit vector, $(\mathbf{r}_i - \mathbf{r}_j)/|\mathbf{r}_{ij}|$.

A very commonly adopted pair potential is the Lennard–Jones potential shown in Figure 10.2,, has the form

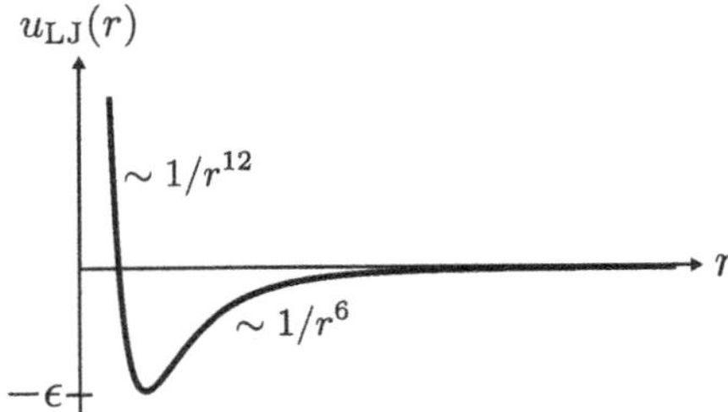

Fig. 10.2 Illustration of the Lennard–Jones potential for pairwise additive forces, where ϵ sets the strength of the attraction and the intercept $r = \sigma$ sets the effective diameter of a particle.

$$u_{\mathrm{LJ}}(r) = 4\epsilon \left[\left(\frac{\sigma}{r}\right)^{12} - \left(\frac{\sigma}{r}\right)^{6} \right]$$

which has a basic energy scale ϵ, and lengthscale σ. The form can be motivated by noting that for point particles without a fixed charge, or dipole, the first order Coulomb term is the induced dipole dispersion interaction that is attractive and decays as $\sim r^{-6}$. The repulsive part at short ranges, that scales like $\sim r^{-12}$, mimics the Pauli exclusion principle and enforces volume exclusion, albeit not very accurately. It is easy to solve for the minimum of the potential at $\ell = 2^{1/6}\sigma \sim 1.12\sigma$, with depth, $u_{\mathrm{LJ}}(\ell) = -\epsilon$.

Exercise 10.3: Show that an alternative form for the Lennard–Jones potential that separates the attractive and repulsive branches is $u_{\mathrm{LJ}}(r) = u_{\mathrm{r}}(r) + \lambda u_{\mathrm{a}}(r)$ where

$$u_{\mathrm{r}}(r) = \begin{cases} 4\epsilon \left[\left(\frac{\sigma}{r}\right)^{12} - \left(\frac{\sigma}{r}\right)^{6} \right] + \epsilon & r \le 2^{1/6}\sigma \\ 0 & r > 2^{1/6}\sigma \end{cases}$$

$$u_{\mathrm{a}}(r) = \begin{cases} -\epsilon & r \le 2^{1/6}\sigma \\ 4\epsilon \left[\left(\frac{\sigma}{r}\right)^{12} - \left(\frac{\sigma}{r}\right)^{6} \right] & r > 2^{1/6}\sigma \end{cases}$$

for $\lambda = 1$ and demonstrate that it is a smooth function for $r > 0$.

Provided a Lennard–Jones potential, there is a natural set of reduced units that can be adopted,

$$
\begin{array}{ll}
\text{length} & \bar{r} = r/\sigma \\
\text{energy} & \bar{T} = k_{\mathrm{B}}T/\epsilon \\
\text{force} & \bar{F} = F\sigma/\epsilon \\
\text{time} & \bar{t} = t/\sqrt{m\sigma^2/\epsilon} \\
\text{velocity} & \bar{v} = v\sqrt{m/\epsilon}
\end{array}
$$

where the unit of time is derivable from non-dimensionalizing the equation of motion,

$$\sigma\bar{r}(\bar{t} + \Delta\bar{t}) = \sigma\bar{r}(\bar{t}) + \sqrt{m\sigma^2/\epsilon}\,\Delta\bar{v}(\bar{t})\frac{\sigma}{\sqrt{m\sigma^2/\epsilon}} + \frac{\Delta\bar{t}^2 m\sigma^2}{2m\epsilon}\bar{F}(\bar{t})\frac{\epsilon}{\sigma}$$

which reduces to

$$\bar{r}(\bar{t} + \Delta\bar{t}) = \bar{r}(\bar{t}) + \Delta\bar{t}\bar{v}(t) + \frac{\Delta\bar{t}^2}{2}\bar{F}(\bar{t})\,.$$

These equations can then be implemented straightforwardly to simulate a simple Lennard–Jones fluid, which has served as a principal model for molecular dynamics simulations.

In addition to the forces, we need to define a domain or volume to simulate the system in. If we employ periodic boundary conditions as we did for the hard disk Monte Carlo, we find a difficulty associated with the fact that the Lennard–Jones potential is finite for all $r < \infty$. A natural solution to deal with a finite box length, L, would be to truncate the potential at some value $r < L/2$ and use the nearest image convention as we did with the hard disks. However, the Verlet algorithm requires forces in order to update particle velocities or positions and for the algorithm to make sense these forces need to be finite. Simply truncating the potential at some cutoff, r_c, would result in a step discontinuity in the potential and a corresponding divergence in the force. A tractable solution is to truncate and then shift the potential to smoothly go to zero at the cutoff distance. If we define $\tilde{u}_{\text{LJ}}(r)$ to be the truncated and shifted potential, then its relation to the original potential is given by

$$
\tilde{u}_{\text{LJ}}(r) = \begin{cases} 4\epsilon \left[\left(\frac{\sigma}{r}\right)^{12} - \left(\frac{\sigma}{r}\right)^{6} \right] - u_{\text{LJ}}(r_c) & r \le r_c \\ 0 & r > r_c \end{cases}
$$

where r_c is typically taken to be around 3-5σ for typical fluid conditions.

The algorithm for calculating the forces for a truncated and shifted Lennard–Jones potential is shown in Algorithm 9. While specialized for this particular pair potential, the structure is general. After initializing the energy and forces, two nested for loops iterate over all unique pairs. The Verlet list we implemented back in Chapter 5 could be used as well. After checking whether a given pair is within the cutoff using the nearest image convention for a system with periodic boundary conditions, the force is summed. Noting the fact that the force on particle i from particle j is equal and opposite to the force on particle j from particle i, Newton's third law, we can save time by only computing each pair force once. Note that in Algorithm 9 we have made explicit ϵ and σ, which if using Lennard–Jones units would result in each being 1.

Algorithm 9 Calculation of Lennard–Jones forces

1: $U = 0$ ▷ Initialize potential energy
2: **for** $n = 1, n \le N$ **do**
3:      $\mathbf{F}_n = 0$ ▷ Initialize forces
4: **end for**
5: **for** $n = 1, n \le N - 1$ **do**   ▷ Loop over unique pairs
6:      **for** $m = n + 1, m \le N$ **do**
7:          $\mathbf{r} = \mathbf{r}_n - \mathbf{r}_m$ ▷ Compute displacement
8:          $\mathbf{r} = \mathbf{r} - L \operatorname{int}(\mathbf{r}/L)$ ▷ Employ periodic boundary conditions
9:          $r = |\mathbf{r}|$
10:          **if** $r < r_c$ **then**   ▷ Check if distance is within cutoff
11:              $F = 48\epsilon[(\sigma/r)^{12} - (\sigma/r)^{6}]r^{-2}$ ▷ Compute magnitude of force
12:              $\mathbf{F}_n = \mathbf{F}_n + F\mathbf{r}$ ▷ Use Newton's third law
13:              $\mathbf{F}_m = \mathbf{F}_m - F\mathbf{r}$
14:              $U = U + 4\epsilon[(\sigma/r)^{12} - (\sigma/r)^{6}] - U(r_c)$ ▷ Sum potential energy
15:          **end if**
16:      **end for**
17: **end for**

The truncation of the Lennard–Jones potential means that we are underestimating the cohesive energy of the system, as we are neglecting parts of the attractions that occur at long distances. If we assume that beyond the cutoff distance, positional correlations have decayed, which would be reasonable for a fluid away from its critical temperature provided r_c is more than a couple particle diameters, then we can estimate the correction to the potential energy. If U_{tail} is the remaining potential energy, not included due to the truncation of interactions, then for density ρ

$$U_{\text{tail}} = 4\pi\rho N \int_{r_c}^{\infty} dr\, r^2 u_{\text{LJ}}(r)$$

provides this correction up to quadrature. This expression results from the general relationship between structure and thermodynamics explored in Chapter 3. Also explored in that chapter is a molecular representation of the pressure for a system interacting through pairwise additive forces. The virial expression for the pressure p is

$$\begin{aligned}
p &= \frac{m}{dL^3} \sum_i |\mathbf{v}_i|^2 + \frac{1}{dL^3} \sum_i \mathbf{r}_i \cdot \mathbf{F}_i \\
&= \frac{m}{dL^3} \sum_i |\mathbf{v}_i|^2 + \frac{1}{2dL^3} \sum_{i \neq j} (\mathbf{r}_i - \mathbf{r}_j) \cdot \mathbf{F}_{ij}
\end{aligned}$$

where we have assumed the simulation domain has volume L^3, and $\mathbf{F}_{ij}$ is the force between particles i and j. The second line follows from Newton's third law, and sums the contributions from the potential and kinetic energies of the particles to produce a pressure that will fluctuate as the particles interact.

Exercise 10.5: Derive this expression for the tail correction for the potential energy using the relationship between the average energy and the radial distribution function, and determine an equivalent relationship for the pressure by taking the appropriate derivative of U_{tail}.

The Lennard–Jones fluid is a reasonable model for atomic particles that either lack internal structure, or that are sufficiently rigid and compact that their internal structure can be ignored. There are many cases, however, where we would wish to model particles that have an internal structure, which can fluctuate. As a canonical example of such a system, we can consider a model of a polymer known as a worm-like chain, for which monomer units are bound together with flexible bonds. An image of such a polymer is shown in Figure 10.3. Worm-like chains are reasonable models of polymers that are semi-flexible, where segments point in roughly the same direction. Models of worm-like chains are used routinely to simulate bio-molecular polymers like DNA and RNA.

For a single worm-like chain, the potential energy includes intramolecular terms that describe bond stretching, U_{b}, and bending, U_{a}, as well as dispersion and volume exclusion described by Lennard–Jones interactions. For a configuration of N monomers, $\mathbf{r}^N$, the potential can be written

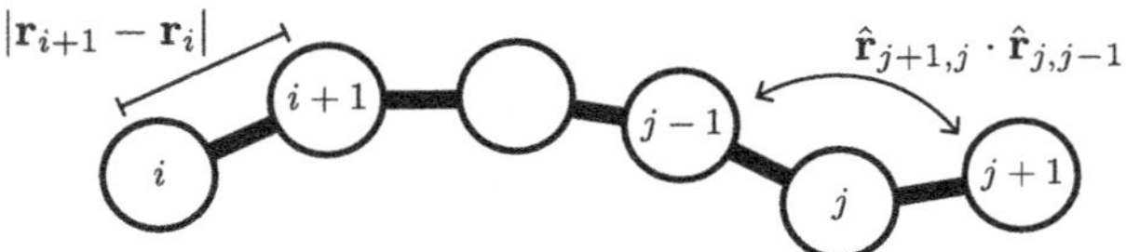

Fig. 10.3 Schematic of a worm-like chain polymer with stretching forces that depend on adjacent monomer displacements and bending forces that depend on angles between triplets of monomers.

$$U\left(\mathbf{r}^N\right) = U_{\mathrm{b}}\left(\mathbf{r}^N\right) + U_{\mathrm{a}}\left(\mathbf{r}^N\right) + U_{\mathrm{LJ}}\left(\mathbf{r}^N\right)$$

where the bond stretching potential is harmonic

$$U_{\mathrm{b}}\left(\mathbf{r}^N\right) = \frac{\kappa_{\mathrm{b}}}{2} \sum_{i=1}^{N-1} \left(|\mathbf{r}_{i+1} - \mathbf{r}_i| - \ell\right)^2$$

as is the bending potential

$$U_{\mathrm{a}}\left(\mathbf{r}^N\right) = \frac{\kappa_{\mathrm{a}}}{2} \sum_{i=2}^{N-1} \left[\hat{\mathbf{r}}_{i+1,i} \cdot \hat{\mathbf{r}}_{i,i-1} - \cos(\theta)\right]^2$$

and κ_{b} encodes the stiffness of the bond stretching potential and κ_{a} the bending rigidity. The parameters ℓ and θ encode the typical monomer–monomer distance and curvature. For a worm-like chain typically $\theta = 0$ and $\ell = \sigma$ the diameter in the Lennard–Jones potential. If $\epsilon = 0$ and $\kappa_{\mathrm{a}} = 0$ then the model would reduce to a freely jointed chain, or an ideal Gaussian polymer. The extent to which either are finite implies there are correlations between monomers.

Exercise 10.4: Write down the explicit equation of motion for the ith monomer of the polymer by determining the forces due to the bond, angle, and Lennard–Jones potentials.

For either the Lennard–Jones fluid or the worm-like chain, a common goal of molecular simulation is to evaluate time correlation functions. Such functions encode the response of systems, how they relax, and how they transport mass, energy, or charge. Computationally the evaluation of time correlation functions can be cumbersome. Consider a fluctuating dynamic variable $A(t)$ whose mean is 0. We can define an autocorrelation function for its fluctuations as

$$C_{AA}(t) = \langle A(0)A(t)\rangle$$

where the averaging is in thermal equilibrium. For a trajectory integrated over $n\Delta t$ timesteps, we could approximate the average of $C_{AA}(t)$ using the time-translational invariance of the equilibrium average

$$C_{AA}(i\Delta t) \approx \frac{1}{n-i} \sum_{j=0}^{n-i} A(j\Delta t) \cdot A((j+i)\Delta t)$$

for each i. If we want an estimate of the correlations over a significant fraction of n, the number of operators is on the order of n^2. This is computationally inconvenient, as the initial propagation of the trajectory only requires n calculations. Figure 10.4 shows how this double sum results from time-translational invariance.

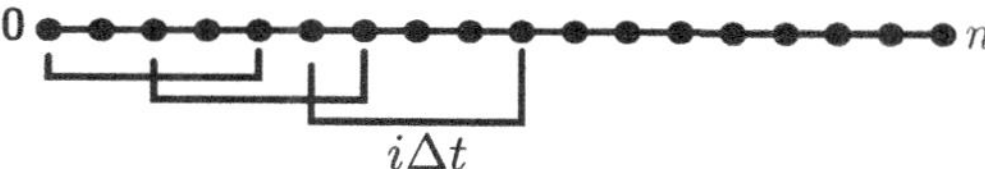

Fig. 10.4 Illustration of the use of time-translational invariance in computing time correlation function.

We can circumvent this time complexity by using the *Wiener–Khinchin theorem*. This theorem relates the autocorrelation function to the Fourier transform of the modulus of the timeseries. Defining the Fourier transform of $A(t)$ as

$$\hat{A}(\omega) = \int_{-\infty}^{\infty} dt\, A(t) e^{i\omega t}$$

then its autocorrelation function is

$$C_{AA}(t) = \int_{-\infty}^{\infty} d\omega\, |\hat{A}(\omega)|^2 e^{-i\omega t}$$

which is the inverse Fourier transform of $|\hat{A}(\omega)|^2$. Discretizing this requires having to do only two n length sums, reducing the number of operations.

10.4 Simulating at constant temperature

The algorithms we have thus far introduced evolve Hamilton's equations of motion. They conserve, approximately, the total energy. Thus, they sample a microcanonical ensemble. If we were to make repeated measurements of the total energy, or even the total potential energy, these global quantities would not be Boltzmann distributed. We know however from our initial developments in Chapter 1 that the canonical ensemble emerges from a microcanonical one, provided that we can divide our system up into a local region of interest and a surrounding bath. Indeed, it is empirically known that measurements of local quantities, like a single particle's velocity, will appear canonically distributed provided even modestly sized systems—large enough that the surrounding particles can act as an effective bath. We also know from the developments in Chapter 6 that a Boltzmann distribution is stable under the evolution of Hamiltonian mechanics. If we could prepare an initial equilibrium distribution of states at a fixed temperature the subsequent dynamics would continue sampling the canonical ensemble.

In principle, if we desired to extract properties from a canonical distribution we could employ a Verlet algorithm directly. However, there are two practical difficulties associated with doing so. First, we need a system that is *large enough*. What this means is difficult to anticipate, and clearly computationally costly to scale up arbitrarily. Additionally, simulating a system with fixed energies will complicate the analysis of extensive quantities that will depend on global measurements of the system and conserved quantities. The second difficulty is associated with preparing an initial equilibrium configuration. While for a classical system we know how to sample from the velocity portion of the Boltzmann distribution, it is not typically easy to sample from the configuration portion. Indeed, the whole development of Monte Carlo approaches in Chapter 5 was devoted to this task. For these reasons it is advantageous to develop a methodology to sample from the canonical ensemble directly.

One naive means of doing so would be to use the procedure we developed to initialize the system, namely after drawing an initial set of velocities we could rescale them periodically to a target temperature. Repeated rescaling of the temperature may result in the correct mean properties of a system, however the fluctuations would be incorrect. A simple way of understanding this is to consider the expected fluctuations in our measurement of the temperature in a canonical ensemble,

$$\langle \delta T^2 \rangle = \frac{m^2(\langle v_i^4 \rangle - \langle v_i^2 \rangle^2)}{k_{\mathrm{B}}^2 dN} = \frac{2\langle T \rangle^2}{dN}$$

which is finite in general, but would be quenched if we insisted that $T = T_{\text{target}}$ at every time.

Rather than rescaling the velocities, the velocities could be redrawn. If this is done randomly, the ensemble and its fluctuations would be correct. Generally, the colloquial way of referring to algorithms that impose an ensemble at fixed temperature is a *thermostat*. The so-called Andersen thermostat proposed to randomly sample velocities according to a Poisson process with some fixed rate, $\hat{\gamma}$. This procedure, though it generates correct static properties, is not easily connected to a microscopic dynamical process, leaving time-dependent properties difficult to interpret. If the rate of resampling the velocities is low, the dynamical behavior may be only slightly modified. This is illustrated in Figure 10.5 where the expected behavior of the diffusion coefficient D for a dense Lennard–Jones system is shown for a range of resampling rates, $\hat{\gamma}$. At very high rates, the velocities will be quenched immediately resulting in a vanishing diffusion constant, while if $\hat{\gamma} \ll 1/\tau_c$ where τ_c is the intrinsic correlation time of the velocities, the impact is diminished.

An alternative approach for thermostating molecular simulations that has an understood dynamic origin is to employ the Langevin equation deduced in Chapter 7 and used throughout our discussions of dynamics in a constant temperature ensemble. For a system evolving with an underdamped Langevin equation, the Fokker–Planck operator can be decomposed into three parts,

$$\mathcal{L} = \mathcal{L}_r + \mathcal{L}_v + \mathcal{L}_b$$

where the first two are those that appear in the deterministic Liouvillian,

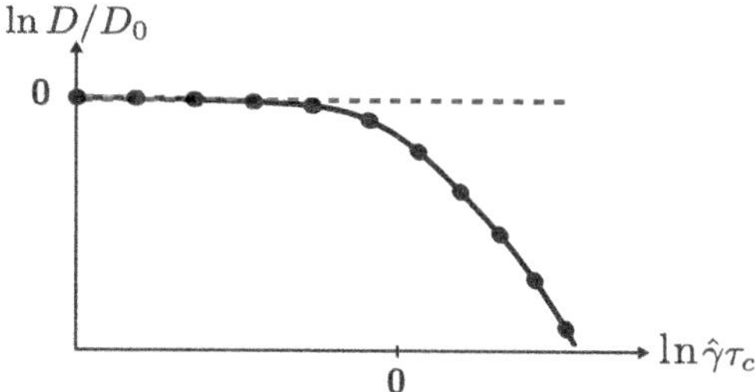

Fig. 10.5 Dependence on the diffusion constant at fixed temperature, D, relative to its value at constant energy, D_0, as a function of the velocity rescaling rate $\hat{\gamma}$ in units of the intrinsic momentum decorrelation time τ_c.

$$\mathcal{L}_r = \sum_i \mathbf{v}_i \frac{\partial}{\partial \mathbf{r}_i} \qquad \mathcal{L}_v = \sum_i \frac{\mathbf{F}_i}{m_i} \frac{\partial}{\partial \mathbf{v}_i}$$

and the last operator encodes the action of the bath

$$\mathcal{L}_b = \sum_i \frac{\gamma_i}{m_i} \cdot \frac{\partial}{\partial \mathbf{v}_i} \mathbf{v}_i + \frac{\gamma_i k_\mathrm{B} T}{m_i^2} \frac{\partial^2}{\partial \mathbf{v}_i^2}$$

where the first term dissipates energy through the friction γ and the second injects it with temperature T. As we did for the Verlet algorithm, we can construct a discrete update rule by approximating the action of the Fokker–Planck operator over some small time, Δt. An analogous Trotterization yields,

$$e^{(\mathcal{L}_v + \mathcal{L}_r + \mathcal{L}_b)\Delta t} \approx e^{\mathcal{L}_b \Delta t/2} e^{\mathcal{L}_v \Delta t/2} e^{\mathcal{L}_r \Delta t} e^{\mathcal{L}_v \Delta t/2} e^{\mathcal{L}_b \Delta t/2}$$

which is one symmetric or Strang splitting of many possible. The inner three operators in this approximation are identical to the velocity Verlet algorithm, while the outer two are new. They correspond to updating the velocity of each particle independently in a manner consistent with an Ornstein–Uhlenbeck process. These two steps require some special attention.

The Langevin equation equivalent to the Ornstein–Uhlenbeck Fokker–Planck operator, $\mathcal{L}_b$, is

$$m_i \dot{\mathbf{v}}_i = -\gamma_i \mathbf{v}_i + \boldsymbol{\eta}_i \qquad \langle \boldsymbol{\eta}_i \rangle = 0 \qquad \langle \boldsymbol{\eta}_i(t) \otimes \boldsymbol{\eta}_j(t') \rangle = 2k_\mathrm{B} T \gamma_i \mathbf{1}\delta(t - t')\delta_{ij}$$

for each particle i. This equation of motion can be integrated yielding for each component,

$$v_i(t) = e^{-\gamma_i t/m_i} v_i(0) + \frac{1}{m_i} \int_0^t dt' \, e^{-\gamma_i(t-t')/m_i} \eta_i(t')$$

which is true for any t. The first term is a deterministic relaxation of the velocity with a characteristic relaxation time m_i/γ_i. The second term is an integral over the Gaussian random variable η_i. The integral of a Gaussian random variable is also a

Gaussian random variable. In this case, if we look at the integral over a time $\Delta t/2$, and define a new variable $R_i(t)$ as

$$R_i(t) = \frac{1}{m_i} \int_t^{t+\Delta t/2} dt'\, e^{-\gamma_i(t+\Delta t/2-t')/m_i}\eta_i(t')$$

then using the properties of η_i,

$$\langle R_i(t)\rangle = 0 \qquad \langle R_i(t)R_j(t')\rangle = \frac{k_{\mathrm B}T}{m_i}\left(1 - e^{-\gamma_i\Delta t/m_i}\right)\delta_{ij}\delta(t-t')$$

we arrive at a new Gaussian random variable with timestep-dependent variance. For compactness, let's define $a_i = \exp[-\gamma_i\Delta t/2m_i]$.

Putting this result together with the Trotter splitting, we arrive at a generalization of the velocity Verlet algorithm that samples a canonical ensemble by imposing a fixed temperature. The resultant rule is broken up into five steps

$$\mathbf{v}'(t + \Delta t/2) = \mathbf{a}\cdot\mathbf{v}(t) + \mathbf{R}(t)$$
$$\mathbf{v}(t + \Delta t/2) = \mathbf{v}'(t + \Delta t/2) + \mathbf{F}(t)\Delta t/2m$$
$$\mathbf{r}(t + \Delta t) = \mathbf{r}(t) + \mathbf{v}(t + \Delta t/2)\Delta t$$
$$\mathbf{v}''(t + \Delta t) = \mathbf{v}(t + \Delta t/2) + \mathbf{F}(t + \Delta t)\Delta t/2m$$
$$\mathbf{v}(t + \Delta t) = \mathbf{a}\cdot\mathbf{v}''(t + \Delta t) + \mathbf{R}(t + \Delta t)$$

where in order to keep track of the different intermediate steps over the Δt integration, we have introduced additional intermediate velocities $\mathbf{v}'(t+\Delta t/2)$ and $\mathbf{v}''(t+\Delta t)$ that can help with book keeping. We have also introduced $(\mathbf{a})_{ij} = a_i\delta_{ij}$ as a diagonal matrix.

Exercise 10.6: Show that the mean-squared displacement for a free particle integrated with the Langevin thermostat with timestep Δt is equal to

$$\langle|r(n\Delta t) - r(0)|^2\rangle = n\frac{(\Delta t)^2 k_{\mathrm B}T}{m}\frac{1 + a^2}{1 - a^2}$$

in the long time limit $t = n\Delta t \gg 1$, and determine its limiting behavior for $\Delta t \to 0$.

For each time step we have one force evaluation and draw two random numbers that are Gaussian distributed with a timestep-dependent variance. The Gaussian distribution can be sampled using the Box–Muller scheme and multiplying the resultant random number by $\sqrt{k_{\mathrm B}T(1 - a_i^2)/m_i}$ to create the correct variance. Algorithm 10 presents the resulting algorithm. Note that just as with the Anderson thermostat, if the friction γ is large, then time-dependent quantities will be dependent on the parameters of the thermostat. In general, if a particular time correlation function has a characteristic decay time of τ_c then m_i/γ_i needs to be much larger than this time in order to mitigate the effect of the thermostat on the intrinsic dynamics generated in its absence.

Algorithm 10 Langevin thermostated molecular dynamics

1: $\mathbf{r} = \text{Initialize Lattice}()$
2: $\mathbf{v} = \text{Initialize Velocities}()$
3: $\mathbf{F} = \text{Force}(\mathbf{r})$ ▷ Compute forces
4: **while** $t \leq \tau$ **do**
5:      $\mathbf{U}_1 = \text{Rand}()$ ▷ Draw uniform random variable
6:      $\mathbf{U}_2 = \text{Rand}()$
7:      $\mathbf{R}_1 = \sqrt{k_{\mathrm{B}}T(1-\mathbf{a}^2)/m}\sqrt{-2\ln \mathbf{U}_1}\cos(2\pi\mathbf{U}_2)$ ▷ Box Muller transform
8:      $\mathbf{R}_2 = \sqrt{k_{\mathrm{B}}T(1-\mathbf{a}^2)/m}\sqrt{-2\ln \mathbf{U}_1}\sin(2\pi\mathbf{U}_2)$
9:      $\mathbf{v} = \mathbf{v}\,\mathbf{a} + \mathbf{R}_1$ ▷ Add random force
10:     $\mathbf{v} = \mathbf{v} + \mathbf{F}\Delta t/2m$ ▷ Update velocities
11:     $\mathbf{r} = \mathbf{r} + \mathbf{v}\Delta t$ ▷ Update positions
12:     $\mathbf{F} = \text{Force}(\mathbf{r})$ ▷ Compute forces
13:     $\mathbf{v} = \mathbf{v} + \mathbf{F}\Delta t/2m$ ▷ Update velocities
14:     $\mathbf{v} = \mathbf{v}\,\mathbf{a} + \mathbf{R}_2$ ▷ Add random force
15:     $t = t + \Delta t$ ▷ Update time
16: **end while**

With a thermostat, energy is not conserved, so understanding the energetics is slightly more complicated. Specifically, we have mechanisms for the energy to change due to the Hamiltonian not being exactly conserved for finite timestep, and additionally the thermostat acts as a bath which means heat can be exchanged between it and the system. A particularly illuminating way to think about the energetics of the thermostated molecular dynamics is to view changes in energy due to the discretization error as work, while energy exchange with the bath is of course heat. The resultant first law can be viewed as

$$\Delta \mathcal{H} = \mathcal{W}_{\mathrm{s}} + \mathcal{Q}$$

where the heat is given by energy changes over the steps where the bath acts

$$\mathcal{Q} = \frac{m}{2}\left\{[\mathbf{v}'(t+\Delta t/2)]^2 - [\mathbf{v}(t)]^2 + [\mathbf{v}(t+\Delta t)]^2 - [\mathbf{v}''(t+\Delta t)]^2\right\}$$

and the *shadow work* is associated with the changes in energy during the deterministic steps

$$\mathcal{W}_{\mathrm{s}} = \frac{m}{2}\left\{[\mathbf{v}''(t+\Delta t)]^2 - [\mathbf{v}'(t+\Delta t/2)]^2\right\} + U[\mathbf{r}(t+\Delta t)] - U[\mathbf{r}(t)]$$

which is expected to scale like Δt^2, from the results of the original Verlet algorithm. Note because the interaction with the bath is integrated exactly, we do not need to consider shadow work in those steps. If in addition, an explicitly time-dependent potential is used in the simulation, an updating step for the potential can be added symmetrically by breaking up the position update into two pieces,

$$\mathbf{r}(t+\Delta t/2) = \mathbf{r}(t) + \mathbf{v}(t+\Delta t/2)\Delta t/2$$
$$U[\mathbf{r}(t+\Delta t/2),t] \rightarrow U[\mathbf{r}(t+\Delta t/2),t+\Delta t]$$
$$\mathbf{r}(t+\Delta t) = \mathbf{r}(t+\Delta t/2) + \mathbf{v}(t+\Delta t/2)\Delta t/2$$

and computing the physical work done by the time-dependent potential as

$$\mathcal{W} = U[\mathbf{r}(t + \Delta t/2), t + \Delta t] - U[\mathbf{r}(t + \Delta t/2), t]$$

where consistent with our notion of work as the change of the energy due only to changes in the Hamiltonian, it is evaluated at fixed particle positions. These discrete means of following energy flows into and out of the system, allows for the simulation of nonequilibrium processes described in Chapter 9.

There are other means of sampling from a canonical distribution, and even methods to fix the pressure, using molecular dynamics. These algorithms typically employ auxiliary variables and Gauss' principle of least constraint to write coupled deterministic equations of motion for the system and the variables that control the energy or the domain size. These deterministic equations however are not generally derivable from an underlying microscopic model and are not guaranteed to be ergodic, unlike the Langevin equation.

10.5 Free energy calculations

An important use of molecular simulations is the evaluation of free energies of interacting models, as they underpin phase behavior, solubility, equilibrium constants, and a host of other phenomena. The evaluation of an absolute free energy is not possible from any technique that samples from a Boltzmann distribution, since it cannot be written as an expectation value. However, free energy differences are readily computable.

To explore common means of computing free energy differences, let's consider the free energetics associated with manipulating the end-to-end distance of the worm-like chain model, $R(\mathbf{r}^N) = |\mathbf{r}_N - \mathbf{r}_1|$. In particular, let us envision manipulating R by applying an external force Λ. Access to the free energy as a function of Λ would provide the equation of state of the polymer, relating the force to the extension. In the presence of Λ the potential energy is

$$U_\lambda = U_0(\mathbf{r}^N) - \Lambda R(\mathbf{r}^N)$$

where $U_0(\mathbf{r}^N)$ is the interaction potential in the absence of the extra force, which might include bond, angle, and Lennard–Jones terms. The change in the free energy of the system with and without the added force is

$$A(\Lambda) - A(0) = -k_\mathrm{B}T \ln \frac{Q(\Lambda)}{Q(0)}$$
$$= -k_\mathrm{B}T \ln \frac{\int d\mathbf{r}^N \, e^{-\beta U_\Lambda}}{\int d\mathbf{r}^N \, e^{-\beta U_0}}$$

which is just a ratio of configurational partition functions, given the classical system where momentum fluctuations are irrelevant. This ratio of partition functions can be recast into a number of computable expectation values. One such expectation value is

$$\frac{Q(\Lambda)}{Q(0)} = \left\langle e^{\beta \Lambda R} \right\rangle_0$$

where we note that the common contribution of U_0 to the Boltzmann factors in both partition functions can be viewed as the averaging operation, where the exponential average is in an ensemble generated under just U_0.

While in principle this exponential average can be computed, doing so directly is typically untenable because exponential averages converge very poorly. One useful means of evaluating the average is to use the molecular dynamics version of the umbrella sampling algorithm introduced in Chapter 5. In particular, we can add a series of biasing potentials of the form

$$U_B = \frac{k}{2} \left[R_b - R(\mathbf{r}^N) \right]^2$$

which are harmonic with a spring constant k and center R_b. Introducing an added potential introduces an extra force into the equations of motion, which for a thermostated system would be

$$\dot{\mathbf{r}}_i = \mathbf{v}_i \qquad m_i \dot{\mathbf{v}}_i = -\nabla_i U_0(\mathbf{r}^N) - \nabla_i U_B(\mathbf{r}^N) - \gamma \mathbf{v}_i + \boldsymbol{\eta}_i .$$

Molecular dynamics can then be used to generate histograms of R under these biased potentials and construct the probability distribution $P_0(R)$ in an unbiased system following a histogram reweighting procedure. With access to the full distribution function, the exponential average can be obtained directly

$$\left\langle e^{\beta \Lambda R} \right\rangle_0 = \int dR \, P_0(R) e^{\beta \Lambda R}$$

where in practice the integral would be done numerically.

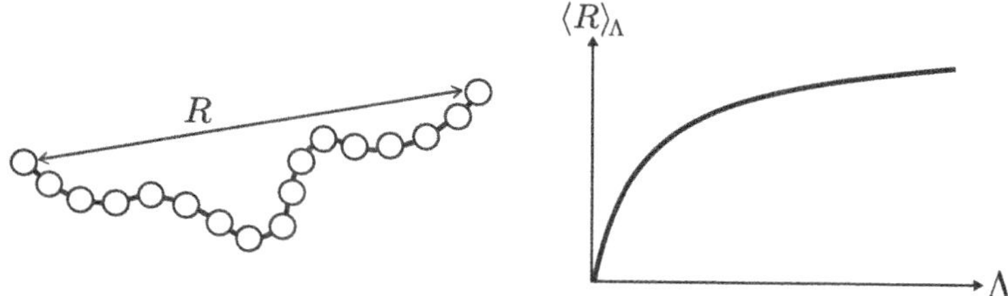

Fig. 10.6 Force extension curve for a worm-like chain that when integrated yields the free energy change $A(\Lambda) - A(0)$.

While umbrella sampling provides a robust means of evaluating free energy differences even when deeply metastable states exist, when the distribution is better behaved, simpler alternatives exist. For example, taking the derivative of the free energy $A(\Lambda)$ with respect to Λ

$$\frac{\partial A(\Lambda)}{\partial \Lambda} = \left\langle \frac{\partial U_\Lambda}{\partial \Lambda} \right\rangle_\Lambda$$

we find it is given by another simple average, this time in the ensemble generated at finite Λ. Integrating both sides

$$A(\Lambda) - A(0) = -\int_0^\Lambda d\Lambda' \left\langle \hat{R}[\mathbf{r}^N] \right\rangle_{\Lambda'}$$

we use the fact that A is a state function to yield the exact difference on the left-hand side. The right-hand side of the expression then only requires that we evaluate the average end-to-end distance on a grid of Λ' between 0 and Λ and perform the numerical integral. A typical force-extension curve is shown in Figure 10.6. This method is known as thermodynamic integration. If $P_0(R)$ is unimodal, thermodynamic integration will be accurate and efficient. Otherwise, the lack of explicit forces to help overcome barriers can cause slow convergence. A way to ensure that the calculation has converged is to perform the transformation $0 \to \Lambda$ and also its reverse $\Lambda \to 0$ and confirm that these are equal and opposite.

> **Exercise 10.7:** Derive a generalization of the above expression for a potential of the form
>
> $$U_\Lambda = (1 - \Lambda)U_0 + \Lambda U_1$$
>
> and relate the resultant free energy change to an integral of a gradient of the potential.

Both of the previous free energy estimators rely on the relationship between free energies and partition functions. An alternative perspective is available mechanically, as a free energy is also of course the reversible work to transform a system. We could for example, determine a protocol for which to vary Λ between 0 and some finite value. If that protocol was endowed with a time dependence, $\Lambda(t)$, then the work associated with varying it is related to the free energy,

$$\mathcal{W} = \int_0^\tau dt' \, \dot{\Lambda}(t')\frac{\partial U}{\partial \Lambda} \qquad \beta\Delta A = -\ln\left\langle e^{-\beta\mathcal{W}} \right\rangle$$

where the second relation is the Jarzynski equality, and the average is an average over trajectories whose initial conditions are Boltzmann distributed with $\Lambda = 0$. Just as the static exponential average is difficult to evaluate, this dynamic average requires $\mathcal{O}(\exp(\beta|\langle\mathcal{W}\rangle - \Delta A|)$ number of realizations to converge. If the mean work is large relative to its reversible limit, this will preclude its use. If $\Lambda(t)$ changes slowly enough that the system remains in equilibrium,

$$\Delta A \approx \langle\mathcal{W}\rangle$$

and all realizations will yield the same value of the work which is equal to the free energy. This reversible limit can be checked by evaluating higher-order cumulants of the work. If $\beta\left\langle \delta\mathcal{W}^2 \right\rangle$ is negligible compared to its mean, then the reversible work limit is satisfied.

10.6 Importance sampling trajectories

The algorithms presented here require a number of computations that scale linearly with the observation time of the simulation. Provided a desire to study a system for twice as long, twice as many floating point operations will need to be evaluated. While for studying the typical behavior of a system, this is sufficient, there are many cases where one is interested in rare dynamical processes. As discussed in Chapter 8, the characteristic timescale τ for observing a fluctuation associated with a change in free energy $\mathcal{O}(\beta\Delta A)$ is exponentially large in that change, $\tau \sim \exp(\beta\Delta A)$. So even with a linear scaling algorithm, the type of spontaneous events we can probe is necessarily limited to those that do not have significant activation energies. The free energy techniques described previously allow us to circumvent this problem for configurational observables in equilibrium by adding additional forces to render atypical fluctuations likely. However, using such techniques to inform dynamical observables is difficult, since in general the underlying equations of motion are altered, and thus the resultant dynamics can be difficult to interpret.

A canonical example of a dynamical rare event problem is the study of reactions. We previously developed the notion of a reaction path ensemble, or the collection of all trajectories conditioned on starting in some abstract state A and ending in some state B. For Markovian dynamics, the probability of a specific trajectory $\mathbf{X}_n(\tau)$ within that ensemble is

$$P_{AB}[\mathbf{X}_n(\tau)] = p(\mathbf{x}_0^n) \prod_{i=0}^{\tau/\Delta t - 1} P(\mathbf{x}_{(i+1)\Delta t}^n | \mathbf{x}_{i\Delta t}^n) h_A(\mathbf{x}_0^n) h_B(\mathbf{x}_\tau^n)$$

where h_X are the indicator functions that filter trajectories for only those that satisfy the conditioning of starting in a reactant state and ending in a product state. The initial distribution $p(\mathbf{x}_0^n)$ and transition probabilities $P(\mathbf{x}_{(i+1)\Delta t}^n | \mathbf{x}_{i\Delta t}^n)$ can in general be arbitrary, chosen from the discrete or continuous equations of motion we have thus far considered. If over the observation time τ the reaction is unlikely, then sampling reactive trajectories from an unconditioned ensemble is going to be difficult.

A method called *transition path sampling* was invented by Bolhius, Dellago, Geissler, and Chandler to rectify this problem. The idea is relatively simple. Provided a measure like the trajectory probability $P_{AB}[\mathbf{X}_n(\tau)]$, one can use a Markov chain Monte Carlo procedure to sample from that distribution. Using the notion of detailed balance, one could construct a dynamics in *trajectory space* by attempting changes from some existing trajectory $\mathbf{X}_n(\tau)$ to generate a new trajectory $\mathbf{X}_m(\tau)$ and enforce an acceptance probability determined to ensure that the resulting steady-state distribution of trajectories was consistent with $P_{AB}[\mathbf{X}_n(\tau)]$. The specific detailed balance condition is then

$$P_{AB}[\mathbf{X}_n(\tau)]P_{\mathrm{att}}(\mathbf{X}_n \to \mathbf{X}_m)P_{\mathrm{acc}}(\mathbf{X}_n \to \mathbf{X}_m) =$$
$$P_{AB}[\mathbf{X}_m(\tau)]P_{\mathrm{att}}(\mathbf{X}_m \to \mathbf{X}_n)P_{\mathrm{acc}}(\mathbf{X}_m \to \mathbf{X}_n)$$

which appears exactly as its configurational analog used to sampled form an equilibrium distribution.

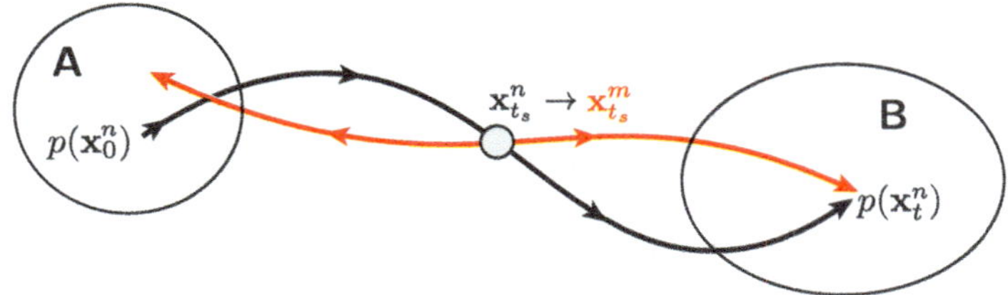

Fig. 10.7 Schematic illustration of the shooting move in transition path sampling.

As with any Monte Carlo, procedure however, the utility of such an algorithm depends strongly on the details of how the sampling is done, or what moves are proposed. For generating new trajectories, the primary move is a so-called shooting move. This is illustrated in Figure 10.7. A shooting move in transition path sampling consists of selecting a point along the trajectory at random, altering that point in some way, and then integrating the trajectory forward and backward in time away from that point. The resultant attempt probability consists of a product of the probability to select the shooting point, denoted t_s, often taken as uniform over $\{0, \tau/\Delta t\}$, and the probabilities of generating the new segments of the trajectories away from t_s. The probability of generating the new segment of the trajectory in the forward direction is a product of transition probabilities, and the probability to generate the new segment of the trajectory in the backward direction from the shooting point is a product of time-reversed transition probabilities. Putting this all together,

$$P_{\mathrm{att}}(\mathbf{X}_n \to \mathbf{X}_m) = \frac{\Delta t}{\tau} \prod_{i=0}^{t_\mathrm{s}-1} P((\mathbf{x}^*)_{i\Delta t}^m | (\mathbf{x}^*)_{(i+1)\Delta t}^m) \prod_{i=t_\mathrm{s}}^{\tau/\Delta t-1} P(\mathbf{x}_{(i+1)\Delta t}^m | \mathbf{x}_{i\Delta t}^m)$$

where $P((\mathbf{x}^*)_{i\Delta t}^m | (\mathbf{x}_{(i+1)\Delta t}^*)^m)$ is the time-reversed transition probability for generating the new trajectory segment.

From the developments in Chapter 9, the ratio of a transition probability to its time reverse is given by

$$P((\mathbf{x}^*)_{i\Delta t}^m | (\mathbf{x}^*)_{(i+1)\Delta t}^m) = P(\mathbf{x}_{(i+1)\Delta t}^m | \mathbf{x}_{i\Delta t}^m) e^{\beta \mathcal{Q}_{i+1,i}^m}$$

where $\mathcal{Q}_{i+1,i}^m$ is the heat dissipated into an environment at fixed temperature in the new trajectory over the transition indexed by i and $i + 1$. If a deterministic, detailed balanced dynamics is used then there is no heat generated, and the forward and backward transition probabilities are equal and functionally given by the inversion of the velocities $\mathbf{v}^N \to -\mathbf{v}^N$. Solving for the ratio of acceptance probabilities,

$$\frac{P_{\mathrm{acc}}(\mathbf{X}_n \to \mathbf{X}_m)}{P_{\mathrm{acc}}(\mathbf{X}_m \to \mathbf{X}_n)} = \frac{P_{AB}[\mathbf{X}_m(t)] P_{\mathrm{att}}(\mathbf{X}_m \to \mathbf{X}_n)}{P_{AB}[\mathbf{X}_n(t)] P_{\mathrm{att}}(\mathbf{X}_n \to \mathbf{X}_m)}$$

$$= \frac{h_A(\mathbf{x}_0^m) h_B(\mathbf{x}_\tau^m)}{h_A(\mathbf{x}_0^n) h_B(\mathbf{x}_\tau^n)} \frac{p(\mathbf{x}_0^m)}{p(\mathbf{x}_0^n)} e^{\beta \sum_{i=0}^{t_s-1} \mathcal{Q}_{i+1,i}^n - \mathcal{Q}_{i+1,i}^m}$$

we find it generally depends on the conditions imposed by the indicator functions, the ratio of initial condition probabilities and the difference in heat generated in the backward segment. For the special case of a canonical distribution of initial conditions, we can evaluate the second ratio

$$\frac{p(\mathbf{x}_0^m)}{p(\mathbf{x}_0^n)} = e^{-\beta[E(\mathbf{x}_0^m) - E_n(\mathbf{x}_0^n)]}$$

which is just given by the difference in the initial energies. For a system in which no work is being done

$$\sum_{i=0}^{t_s-1} \mathcal{Q}_{i+1,i}^n - \mathcal{Q}_{i+1,i}^m = E(\mathbf{x}_{t_s}^n) - E(\mathbf{x}_0^n) - E(\mathbf{x}_{t_s}^m) + E(\mathbf{x}_0^m)$$

the sum of the heat exchanged with the bath is equal to just the change in energy, from the first law. This change in energy has contributions from the initial conditions that cancel with those from $p(\mathbf{x}_0^m)$ and $p(\mathbf{x}_0^n)$. Energetically, all that would be left is the change in energy from the shooting point. If $E(\mathbf{x}_{t_s}^n) = E(\mathbf{x}_{t_s}^m)$, then the acceptance probability in a Metropolis form reduces to

$$P_{\mathrm{acc}}(\mathbf{X}_n \to \mathbf{X}_m) = \min\left[1, h_A(\mathbf{x}_0^m) h_B(\mathbf{x}_\tau^m)\right]$$

namely, just the conditioning applied to the trajectory ensemble. Thus in a shooting move for an equilibrium system in which a change to the system at the shooting point does not change the energy of the system, the acceptance criterion just requires one to check whether or not the new trajectory is reactive. Example moves might be to swap velocities between particles in deterministic dynamics, or to do nothing and simply evolve new trajectories with a new noise history in stochastic dynamics.

The resultant shooting algorithm is summarized in Algorithm 11. First, as with any Monte Carlo procedure, an initial condition is required. In trajectory sampling, this requires the generation of an initial trajectory that satisfies the conditioning. The initial trajectory can be prepared artificially by interpolating a system along some predefined coordinate that spans A and B, or by preparing the system near a putative transition state by means of umbrella sampling, and integrating it forward and backward in time. From that initial trajectory, over a number of Monte Carlo trials, a random shooting point is selected uniformly over the interval $(0, \tau)$, an energy-conserving change to the shooting point is made, and a new trajectory is generated by integrating forward and backward in time. For a detailed balance-preserving dynamics, the acceptance criterion is just given by checking that the reactive conditioning is satisfied.

Many other transition path sampling moves are available and can be effective depending on the system's equation of motion. So-called shifting moves leave most of the old trajectory in place by effectively advancing or retracting time. These are accomplished by drawing a trajectory segment near $t = 0$ or $t = \tau$, defining it as the new initial or final time and integrating the opposite end point forward or backward in time to restore the fixed length of the trajectory. For detailed balance-preserving dynamics, this can be done with an acceptance criterion that depends only on the fulfillment of the conditioning. Generally performing transition path sampling for a system driven out-of-equilibrium is difficult due to the accumulation of the heat in

---

Algorithm 11 Transition path sampling shooting algorithm

1: $\mathbf{X}_n$ = Initialize trajectory
2: **for** $i = 1$, $i \leq$ steps **do**
3:   t_s=Int$[\tau/\Delta t \times$ Rand$() + 1]$ ▷ Pick random shooting position
4:   $\mathbf{x}_m(t_s)$ = Change$[\mathbf{x}_n(t_s)]$ ▷ Randomly change shooting point
5:   $\mathbf{X}_m[t_s : \tau/\Delta t]$ = Integrate Forward$[\mathbf{x}_m(t_s)]$
6:   $\mathbf{X}_m[0 : t_s]$ = Integrate Backward$[\mathbf{x}_m^*(t_s)]$
7:   h_A = Check A$[\mathbf{x}_m(0)]$
8:   h_B = Check B$[\mathbf{x}_m(\tau/\Delta t)]$
9:   **if** $h_A \times h_B = 1$ **then**   ▷ Accept based on reaction
10:    $\mathbf{X}_n \rightarrow \mathbf{X}_m$ ▷ Replace old trajectory with new one
11:   **end if**
12: **end for**

---

the backward shooting direction. Typically in these cases, one employs methods developed around only integrating trajectories forward in time. For example, independent initial conditions can be generated by evolving unconditioned dynamics and then shot forward with the noise history of a previous reactive trajectory.

At its essence, transition path sampling is a means of importance sampling trajectory spaces, and so can be used for a variety of other rare dynamic event problems. For example, it can also be used to sample the tilted ensembles that appear in the study of dynamical large deviation functions. For stochastic dynamics, the trajectory reweighting principles elaborated in past chapters can be unified with path sampling to aid cases out of equilibrium where traditional methods are difficult to apply. As path ensembles and stochastic dynamics inform so much of the modern view of nonequilibrium statistical mechanics, as presented in the last few chapters, path sampling tools are no doubt poised to play a growing role in the computational study of molecular systems. For now, it and everything else in this text should provide a sufficient basis for pursuing knowledge in ever more advanced areas in the rich and ever-changing field of statistical mechanics, the closest we have to a theory of everything.

Exercise 10.8: Work out the acceptance probability for a shooting move, for an ensemble whose trajectory distribution is

$$P_\lambda[\mathbf{X}_n(\tau)] = p(\mathbf{x}_0^n) \prod_{i=0}^{\tau/\Delta t-1} P(\mathbf{x}_{(i+1)\Delta t}^n | \mathbf{x}_{i\Delta t}^n) e^{-\lambda j_{i+1,i} + \psi(\lambda)\Delta t}$$

where $j_{i+1,i}$ is the increment of a time extensive current and $\psi(\lambda)$ is the large deviation function that normalizes the tilted trajectory distribution, with a trajectory generated with equilibrium transition probabilities.

Further reading

Molecular dynamics is widely applied in physical and biological sciences. Many textbooks exist that provide a solid foundation, and additional practical techniques. Of these Daan Frenkel and Berend Smit's *Understanding molecular simulation: From algorithms to applications*, and Michael Allen and Dominic Tildesley's *Computer simulations of liquids* are both very useful. The perspective of deriving integrators from Trotterization of the Liouvillian is due to Mark Tuckerman and his *Statistical mechanics: Theory and molecular simulation* provides additional exposition. The analysis of the Langevin thermostat in terms of heat and work can be read in the primary literature in work by Sivak, Chodera, and Crooks, Journal of Chemical Physics, B, 2014.

Additional exercises

Exercise 10.9: In this exercise we will explore the relationship between the classical action associated with Newton's equation of motion and Verlet's algorithm. Consider the path of the classical coordinate $r(t)$, where the time t extends from 0 to τ. The action $S[r(t)]$ of that path is

$$S[r(t)] = \int_0^\tau dt\, L[r(t), \dot{r}(t)]$$

where $L[r(t), \dot{r}(t)]$ is the so-called Lagrangian

$$L[r(t), \dot{r}(t)] = \frac{m}{2}\dot{r}^2(t) - U[r(t)]$$

with $U(r)$ denoting the potential energy. Hamilton's principle is that the classical path between two fixed points is the path that minimizes the action. If $\Delta r(t)$ denotes a small deviation from the classical path at time t, this means that

$$0 = \delta S = \int_0^\tau dt\, \left[\Delta r(t)\frac{\partial L}{\partial r(t)} + \Delta \dot{r}(t)\frac{\partial L}{\partial \dot{r}(t)} \right].$$

1. Show that for the variational equation to hold for all small $\Delta r(t)$, positive or negative, then
$$0 = -m\ddot{r}(t) - \frac{\partial U}{\partial r(t)}, \qquad 0 \leq t \leq \tau.$$

2. Suppose that you want to apply a low-order quadrature procedure to evaluate the integral in the equation for the action. The path is then described by a sequence of P discrete points, r_j, where $r_j = r(t)$ at time slice $t = j\Delta t$, $1 \leq j \leq P$. Show that for small enough Δt, the action can be expressed as
$$S = \sum_{i=1}^{P} \left[\frac{1}{2}\left(\frac{m}{\Delta t}\right)(r_i - r_{i-1})^2 - \Delta t\, U(r_i) \right].$$

3. Using Hamilton's principle with the discrete action, find an expression for r_{i+1} in terms of r_i and r_{i-1}. How does your result compare with the Verlet algorithm?

Exercise 10.10: Next, we will consider "Molecular" dynamics of a harmonic oscillator. No introduction to dynamics would be complete without a study of harmonic oscillators. In this problem you will apply Verlet's algorithm to a one dimensional, classical harmonic oscillator, approximately evolving its position r and velocity v in time. Specifically, consider a particle of mass m moving in a one-dimensional potential

$$U = \frac{1}{2}kr^2$$

Its equation of motion should be very familiar:

$$\dot{r} = v \quad \dot{v} = F/m$$

where $F = -dU/dr$ is the force.

1. Adopt ℓ and $\tau = \sqrt{m/k}$ as units of length and time, respectively. Beginning with the Verlet equations for advancing r and v from time t to time $t + \Delta t$, write out corresponding equations for the dimensionless quantities

$$\bar{r} = r/\ell \quad \text{and} \quad \bar{v} = v\tau/\ell$$

   Your answer should involve only $\bar{r}(\bar{t} + \Delta\bar{t}), \bar{r}(\bar{t}), \bar{v}(\bar{t} + \Delta\bar{t})$, and $\bar{v}(\bar{t})$, and dimensionless $\bar{t} = t/\tau$, and $\Delta\bar{t} = \Delta t/\tau$.

2. From now on, we will work with these dimensionless quantities exclusively, i.e., the overbars on r, v, t, and Δt will be implied. Implement your update equations from part 1 on a computer. As an initial condition, take

$$r(0) = 1 \quad v(0) = 0$$

   and use a time step of $\Delta t = 0.1$, to propagate the system for enough time steps to complete 10 periods of oscillation. Make a plot of r and v as functions of time. Does the period of oscillation match your expectations? Explain. Also make a plot of the kinetic, potential, and total energies as functions of time.

3. Make a parametric plot of $\{r(t), v(t)\}$. That is, plot all of the ordered pairs $\{r, v\}$ on a graph with r on the horizontal axis, v on the vertical axis. This type of plot, known as a phase portrait, can often reveal important structure about dynamical systems. If the energy of our harmonic oscillator were conserved, what shape would this plot trace out and how does it compare to your observations?

4. Compute a histogram of the oscillator's position r, accumulated at regular time intervals over a long trajectory. Compute the probability distribution $p(r)$, by appropriately normalizing this histogram from your simulated trajectory and plot your result.

5. The correct form of $p(r)$, in the limit $\Delta t \to 0$, can be obtained from a simple argument: Consider a small interval in the oscillator's position, between r and $r + dr$. The amount of time the oscillator spends in this interval, $p(r) \propto 1/|v(r)|$, is inversely proportional to its speed as it passes through. Explain why this is so. Using this, together with the conservation of energy, write $p(r)$ as a function of r.

Exercise 10.11: We have performed several exact calculations based on Langevin dynamics for very simple systems. Here you will derive and implement basic methods for advancing overdamped Langevin dynamics numerically, as required for systems with complicated potentials. We will focus on the Langevin equation in one-dimension

$$\gamma\dot{q} = -\frac{dw}{dq} + \eta \qquad \langle\eta(t)\eta(t')\rangle = 2k_{\mathrm{B}}T\gamma\delta(t - t')$$

where all quantities are dimensionless, e.g., the coordinate q is implicitly expressed relative to some length scale ℓ. The simplest approach to integrating this equation of motion, known as the Euler method, assumes that the force $F(q) = -dw/dq$ does not vary during the time step Δt, and draws time-step dependent random forces.

1. Let q_t be the value of q at the beginning of an integration time step, $q_{t+\Delta t}$ be the value of q at the end of the step, and $F_t = F(q_t)$. Taking the force to be constant (i.e., $F = F_t$) for all times between t and $t+\Delta t$, derive an equation relating $q_{t+\Delta t}$ to q_t, F_t, Δt, γ, and an integrated random force R defined as

$$R = \gamma^{-1}\int_t^{t+\Delta t} dt'\,\eta(t')$$

2. Show that $\langle R^2 \rangle = 2D\Delta t$, where $D = k_{\mathrm{B}}T/\gamma$. This defines the variance of a time-step dependent random force, suitable for use in numerical integration.

3. Consider a Brownian particle, whose position q evolves through the overdamped Langevin equation, subject to a constant force $\bar{F}$. Its average drift velocity over a period t,

$$\bar{v} = \langle(q_t - q_0)\rangle / t$$

is proportional to the driving force, $\bar{v} = \mu\bar{F}$. Using your result from parts (1) and (2), show that the mobility μ is related to the particle's diffusion constant, specifically, $\mu = \beta D$.

4. Implement the discrete equation of motion derived above into a molecular dynamics code. Derive the force from the potential,

$$w(q) = q^2(1 - q)^2 + 5q$$

   Using a time step of $\Delta t = 0.001$, a friction of $\gamma = 1.0$ and a temperature of $k_{\mathrm{B}}T = 1$, evolve a short trajectory (10000 Δt) and make a plot of $q(t)$. Using a longer trajectory, make a histogram of q and compute $\langle q \rangle$, $\langle(\delta q)^2\rangle$, and the third order cumulant, $\langle(\delta q)^3\rangle$, where the averages are the equilibrium averages.

5. Using $H_0 = w(q)$ as a reference Hamiltonian, derive the static linear response relation for the change in $\Delta\langle q\rangle_F = \langle q\rangle_F - \langle q\rangle$ to an additional constant force, $H = H_0(q) - Fq$. Add a constant force to your molecular dynamics simulation, and recompute $\langle q\rangle_F$ with $F = \{0.1, 0.5, 1, 1.5, 2, 3\}$. Make a plot of $\langle q\rangle_F$ as a function of F, and compare it to the linear response and second order response approximation using your results from part 4.

6. Derive the first order dynamical response of $\Delta q(t) = \bar{q}(t) - \langle q\rangle$, to the instantaneous application of $-Fq$, in terms of a time correlation function, $C_{qq}(t) = \langle\delta q(0)\delta q(t)\rangle$.

7. Compute and plot the normalized correlation function $C_{qq}(t)/C_{qq}(0)$ in the absence of the force. Good enough statistics should require a trajectory of $10^7 \Delta t$.

8. Using an initial condition of $q(0) = 0.0$, which is far from its equilibrium value, average over 1000 realizations of it relaxing (using trajectories around $5\times 10^3\,\Delta t$). Plot the time dependent average divided by its equilibrium average $\Delta q(t)/|\langle q \rangle|$. Compare this to the normalized correlation function computed in part 7. Comment on the agreement or disagreement.

Exercise 10.12: In Exercise 10.11, you derived a discrete equation of motion that could be implemented on a computed to integrate an overdamped Langevin equation. Using the code you wrote for that previous problem, implement a bistable potential of the form

$$w(q) = (q + 1)^2 (q - 1)^2 ,$$

which features minima at $q_A = -1$ and $q_B = 1$. With this system you will study splitting probabilities, $\phi_B(q_0)$, numerically. For the following set $\gamma = 1$ and $\Delta t = 0.001$.

1. For temperature $k_{\mathrm{B}}T = 0.1$, use the exact form in Chapter 8 to calculate $\phi_B(q_0)$ for ending in q_B numerically for a range of coordinate values between q_A and q_B. Plot your result for the splitting probability as a function of q_0. Repeat the calculation for temperatures $k_{\mathrm{B}}T = 0.2$ and $k_{\mathrm{B}}T = 0.4$.

2. For temperature $k_{\mathrm{B}}T = 0.1, 0.2$ and 0.4, use simulations to determine $\phi_B(q_0)$ for several values of q_0 between q_A and q_B. For each value of q_0, generate a large number of trajectories initiated with $q = q_0$, advancing time until q reaches either q_A or q_B. Plot your simulation result for the splitting probability as a function of q_0. Are your simulation results consistent with the exact expression?

3. How accurate is the harmonic barrier approximation for this potential? To address this question, perform a second-order Taylor expansion of $w(q)$ about its maximum at $q = 0$ and determine an appropriate value of $m\omega^2$. This approximation was used previously in Exercise 8.13 part 5. Within the harmonic approximation estimate $\phi_B(q_0)$ for values of q_0 between q_A and q_B. Carry out this calculation for the temperatures $k_{\mathrm{B}}T = 0.1$, $k_{\mathrm{B}}T = 0.2$, and $k_{\mathrm{B}}T = 0.4$. In each case, plot the harmonic estimate together with your numerical evaluation of the exact expression. Why is the approximation better in some cases than others?

Exercise 10.13: In this problem you will compare the structure and dynamics of Lennard–Jones systems with and without attractions. In particular, write a molecular dynamics code using the velocity Verlet algorithm with the Langevin thermostat to simulate a Lennard–Jones fluid in two spatial dimensions. With this code, using a system of at least $N = 100$ particles, address the following questions.

1. Prepare and equilibrate a Lennard–Jones fluid with a cutoff, $r_c = 3\sigma$ using the method of truncating and shifting the potential. What is the average potential energy and pressure of a system whose density is $\rho\sigma^2 = 0.75$ and temperature $k_{\mathrm{B}}T/\epsilon = 0.43$?

2. Compute the radial distribution function, and plot it over a range of r.

3. Given the radial distribution function, is the tail correction to the potential energy derived in the text valid?

4. Keeping the temperature and density the same, simulate the Lennard–Jones system with a cutoff, $r_c = 2^{1/6}\sigma$. Such a system contains only repulsive interactions. Reevaluate the average potential energy and pressure of a system.

5. Recompute the radial distribution function for this purely repulsive fluid and plot it alongside the system whose cutoff was $r_c = 3\sigma$. You should find that the two are remarkably similar. Explain the origin of their similarities.

6. The use of a purely repulsive reference potential is the basis of the Weeks–Chandler–Andersen theory of the liquid state. Using a decomposition of the Lennard–Jones potential into a purely attractive and purely repulsive part, apply first order perturbation theory to evaluate the correction to the potential energy from the attractive part of the potential. You will likely find it useful to refer to the result of Chapter 3. The formal result is given as an integral you will need to evaluate numerically.

7. Compare the perturbative result in part 6 to the computed value of the average potential energy in part 1 and comment on their agreement.

Exercise 10.14: In this problem, you will model a single molecule pulling experiment using a simple Gaussian chain in a thermal bath to explore the Jarzynski equality. In particular, we will employ a purely harmonic potential energy function for beads along the chain

$$ U\left(\mathbf{r}^N\right) = \frac{\kappa_\mathrm{b}}{2} \sum_{i=1}^{N-1} \left(\mathbf{r}_{i+1} - \mathbf{r}_i\right)^2 $$

where κ_b is the stiffness of the bond between the monomers, whose rest length has been set to 0. To mimic the pulling experiment, we will constrain the first monomer to stay at the origin $\mathbf{r}_1 = 0$ by not evolving its equation of motion. The final monomer, $\mathbf{r}_N$, will be manipulated directly $\mathbf{r}_N(t) = vt\hat{\mathbf{z}}$ by pulling at a constant speed v in the z direction in order to perform work on the system. We will evolve the system with

$$ m\dot{\mathbf{v}}_i = -\nabla_i U - \gamma \mathbf{v}_i + \boldsymbol{\eta}_i \qquad \langle \boldsymbol{\eta}_i \rangle = 0 \qquad \langle \boldsymbol{\eta}_i(t) \otimes \boldsymbol{\eta}_j(t') \rangle = 2k_\mathrm{B}T\gamma_i \mathbf{1}\delta(t - t')\delta_{ij} $$

and adopt a unit system where γ/κ_b defines a unit of time, $\sqrt{k_\mathrm{B}T/\kappa_\mathrm{b}}$ defines a unit of length, and mass $\gamma^2/\kappa_\mathrm{b}$, all of which are set to 1.

1. For a constant pulling velocity, derive an expression for the work, $\mathcal{W}$.

2. Using $N = 20$, and $m = 0.02$, generate an equilibrium configuration of the polymer with an end-to-end distance $|\mathbf{r}_N - \mathbf{r}_0| = 1$. After that initial relaxation, start pulling the polymer with a range of constant velocities until the end-to-end distance is 10. Individually compute the work, change in energy and heat released through the thermostat, and plot each. Show that for a timestep small enough, the first law is satisfied for all time along the trajectory.

3. Choose a small velocity and timestep, and create a histogram of work values using the procedure in part 2, using multiple initial conditions and trajectories.

4. Fit your distribution of work values to a Gaussian, and numerically evaluate $-\ln \langle \exp[-\beta\mathcal{W}] \rangle$. Does this agree with your expectations for $\beta\Delta A$ from the Jarzynski equality? Why might a Gaussian be correct functional form for this system?

Index

The manufacturer's authorised representative in the EU for product safety is Oxford University Press España S.A. of El Parque Empresarial San Fernando de Henares, Avenida de Castilla, 2 - 28830 Madrid (www.oup.es/en or product.safety@oup.com). OUP España S.A. also acts as importer into Spain of products made by the manufacturer.
Printed and bound by CPI Group (UK) Ltd, Croydon, CR0 4YY
18/05/2026
02112605-0001